Pharmaceutical Discovery, Development and Manufacturing Forum 2017

Core Programming Area at the 2017 AIChE Annual Meeting

Minneapolis, Minnesota, USA
29 October – 3 November 2017

ISBN: 978-1-5108-5812-1

Printed from e-media with permission by:

Curran Associates, Inc.
57 Morehouse Lane
Red Hook, NY 12571

CURRAN ASSOCIATES INC. proceedings.com

Printed by Curran Associates, Inc. (2018)

AIChE
120 Wall Street, FL 23
New York, NY 10005-4020

Phone: (800) 242-4363
Fax: (203) 775-5177

www.aiche.org

Additional copies of this publication are available from:

Curran Associates, Inc.
57 Morehouse Lane
Red Hook, NY 12571 USA
Phone: 845-758-0400
Fax: 845-758-2633
Email: curran@proceedings.com
Web: www.proceedings.com

TABLE OF CONTENTS

Some abstracts in this book refer to images that were not made available by the publisher

497316 Sustained Supersaturation of Erlotinib Sdd Ternary Amorphous Systems

Abstract:

As more low-solubility drug candidates enter the pharmaceutical development pipeline, the need for solubilization-enhancing technologies is ever greater. In an amorphous dispersion, the solubility of a drug is enhanced by kinetically trapping the molecules into the higher-energy amorphous state, often realizing a 5-50x enhancement in solubility relative to the crystalline state. Spray dried dispersions (SDDs), containing amorphous drug and a concentration-enhancing matrix polymer are a scalable and mature technology which takes advantage of the amorphous drug solubility enhancement. Typically, an SDD must meet numerous performance metrics including rapid dissolution rate, sustainment of supersaturation, physical stability, chemical stability and manufacturability. We have successfully formulated over 1,000 low-solubility compounds as SDDs over the past two decades.

Despite this success, for some compounds it can be challenging to meet all of the performance requirements using only one excipient in the SDD system. In these cases, a multi-component amorphous system approach may warranted, in which two excipients are utilized, each tailored to separately achieve specific performance metrics. In this presentation, we discuss the case study of erlotinib, a rapidly-crystallizing compound which requires both fast dissolution and sustained supersaturation. In this case, erlotinib SDD is spray dried with a fast-dissolving matrix polymer (Eudragit L100) and formulated with a crystallization-inhibiting polymer (HPMC) to maximize both dissolution rate and sustainment. The combination of Eudragit L100 and HPMC succeeds by increasing the in vitro test AUC over ten-fold compared to binary SDD formulations.

488954 Connecting the Product-Process-Performance Interplay for Improved Understanding in Development of a Spray Dried Dispersion Drug Product

Abstract:

The pharmaceutical industry has faced a shift in the physicochemical properties of pre-clinical drug candidates rendering them less soluble and bioavailable. Spray drying has become an important unit operation in the pharmaceutical industry to produce spray dried amorphous dispersions (SDD) which can improve apparent solubility versus the crystalline drug as well as dissolution rate, thus improvingthe bioavailability of poorly soluble active pharmaceutical ingredients (API). The development of drug products consisting of spray dried solid dispersions is complex as it involves the need for strengthened understanding of the solid dispersion and drug product formulation attributes as they relate to enhancement of drug dissolution rate and maintenance thereof in the drug product. In contrast to standard drug particle erosion as main dissolution mechanism in conventional crystalline API formulation, solid dispersions can release drug in various types of species ranging from free drug, nano-particles, bile salt micelles, precipitated drug, and so on[1]. The characteristic dissolution mechanisms are impacted by API physiochemical properties and formulation as well as the process conditions utilized to make them.

This work will cover: 1. different mechanisms of dissolution and species formed for various APIs, 2. how the spray drying process influences dissolution and amorphous product stability, 3. utility of novel spectroscopic and other characterization techniques to study dispersion quality and performance, 4. influence of dispersion and drug product formulation properties on dissolution and tablet properties, 5. end to end considerations for ensuring how key quality and performance attributes are designed in and maintained in the drug product.

Spray dried dispersion particles and powders were manufactured on various scales ranging from single particle levitation, ProCepT 4M8, GEA mobile minor and GEA PSD-2 spray dryers. Multiple process conditions resulting in a range of drying rates were studied and will be presented. The resulting powders will be characterized by standard techniques such as modulated differential scanning calorimetry (mDSC) and x-ray powder diffraction (XRPD) to evaluate bulk amorphous dispersion uniformity. In several cases, additional techniques such as focused ion beam scanning electron calorimetry (FIB-SEM), x-ray photoelectron spectroscopy (XPS), and energy dispersive x-ray spectroscopy (EDX) were utilized to look a particle level homogeneity. Drug product formulations were made by standard tumble blenders, dry granulated, lubricated and compressed into tablets. Discriminating dissolution methodologies were used to study drug release from the dispersions and tablet formulations. Environmental stressing was conducted to look at stability and impact on drug release from tablets.

The choice of polymers and surfactants that make up the SDD composition was shown to have a significant impact on the amorphous quality, particle level uniformity, resulting phase stability and dissolution behavior. The choice of solvent composition and spray drying conditions was shown to have a modest to profound impact on these quality attributes. This was shown through dissolution studies coupled with human PK data as well as phase uniformity issues discovered immediately following manufacture or after post-process stressing. The API has inherent physio-chemical properties that tend to give rise to characteristic speciation behavior during dissolution of SDD powder. The rate of release of overall drug into these entities was negatively impacted as the dispersion load in the tablet increases, tablet void fraction exceeded critical levels, and tablet disintegrant system was not properly designed.

In conclusion, spray dried dispersion based drug products require end to end study and understanding of the API behavior, formulation and process components, characterization methodology for amorphous quality and dissolution methodology and environmental stresses to be successful. Discriminating dissolution methodologies can help understand the specific mechanisms of dissolution enhancement and novel characterization methods can be useful in characterizing and ensuring maintenance of amorphous quality in the SDD and final drug product.

References:

1: D.T Friesen et al. Hydroxypropyl Methylcellulose Acetate Succinate-Based

Spray-Dried Dispersions: An Overview. Molecular Pharmaceutics. 2008. VOL. 5, NO. 6, pg. 1003â€"1019

495763 Strategically-Designed Polymer Synthons As Pharmaceutical Excipients in Oral Drug Delivery

Abstract:

Synthetic polymer excipients, equipped with tunable structures and properties, can enable more bioavailable, inexpensive, and safe oral drug formulations when rationally designed to solubilize active pharmaceutical ingredients in amorphous solid dispersions. However, the underlying polymer-drug interactions in these materials to control drug supersaturation generation and maintenance currently remain poorly understood. We present a new approach to discover specialized excipients for a highly-crystalline drugs, driven by molecular evolution-inspired ideas to rapidly construct polymer synthons (macromolecules with structural subunits correlating to desired noncovalent interactions, analogous to the retrosynthetic analysis taught in organic chemistry) using high-throughput synthesis and screening protocols. In this manner, the molar mass and chemical composition state spaces were sampled to identify an optimal excipient synthon among 70 copolymers to solubilize phenytoin. The leading system contained N-isopropylacrylamide (NIPAm, responsible for inhibiting phenytoin crystallization) and N,N-dimethylacrylamide (DMA, providing polymer hydrophilicity), and polymer-drug interactions were revealed using NOESY and DOSY NMR spectroscopy. Complementary all-atom molecular dynamic simulations corroborated experimental evidence of NIPAm synthons strongly binding to phenytoin. When administered to rats, we observed a record 23-fold increase in area-under-the-curve bioavailability over neat drug with excellent in vitro-in vivo correlations. Altogether, these discoveries demonstrate how existing synthesis tools in the polymer chemistry landscape can accelerate the translation of new drug candidates (or even failed compounds) into more effective and accessible therapeutic treatments.

497295 Pill Burden Reduction through Engineering the Mechanical Properties of Spray-Dried Amorphous Solid Dispersions

Abstract:

Drug candidates with poor oral bioavailability due to low aqueous solubility and dissolution rate are common in the pharmaceutical industry. One of the most mature and well understood techniques to improve oral absorption of drug candidates is to formulate the drug as a spray-dried amorphous solid dispersion (SDD). Given SDDs are diluted with excipients to, primarily, stabilize the amorphous form of the drug, a common challenge in SDD oral delivery of tablets is pill burden. One promising solution to reduce pill burden is engineering the SDD particles to replace the use of tablet fillers, which primarily attenuate undesirable mechanical and powder flow properties of crystalline drugs. Given that the mechanical and powder flow properties of SDDs can be significantly tuned during the spray drying process, the use of fillers can be substantially reduced. However, tuning the particle properties of SDDs is a hierarchical multidimensional optimization problem, and for that reason, a clear relationship between particle properties and physical and chemical state stability of the drug and in vivo performance is generally established prior to optimization of mechanical and powder flow properties. There have been several recent reviews on this subject, but there lacks the direct connection between spray-drying process conditions and their impact on final tablet formulation [1-2].

The aim of the current study is to establish relationships between spray drying process parameters and the resulting physical and mechanical properties of the SDD particles, and secondly, to use these relationships to formulate high SDD loaded (greater than 70%) immediate release tablets. Felodipine spray-dried with PVP VA64 was used as the model SDD system. The powder mechanical properties measured include compression profiles, SEM imaging, modified in-die compression analysis [3], BET surface analysis, and Hiestand indices.

Results showed that, for a given SDD formulation, specific surface area of the powder was controllable by tuning the particle size, particle wall thickness, and to a lesser extent, particle morphology through drying kinetics and atomization during spray drying. This resulted in SDDs that spanned a 10-fold difference in specific surface area. The resulting mechanical properties of these high surface area SDDs included significant increases in tensile strength and bonding index (BI) when compressed, due to larger bonding area between particles [4].

The material properties were strongly dependent on the drug loading, with higher drug loading SDDs having diminished compactibility compared to their placebo counterparts. These results suggest that, counter to selecting the highest drug loading SDD with acceptable physical stability and in vivo performance, a lower drug loading SDD formulation could allow for less filler in the final drug product, resulting in an overall higher loading within the tablet. Consequently, this also improves the physical stability robustness.

In addition, it was observed that, when compressed as the neat material, the compacts had varying degrees of anisotropy, which resulted in greater axial elasticity. High elastic recovery decreases the bonding area between particles, reducing the strength of the tablets and increasing the risk for capping and lamination. Therefore, decreasing the degree of anisotropy could reduce the risk of these tablet defects and should be considered when engineering SDD particles for integration into tablet formulations [5].

Overall, the powder properties resulted in various immediate release tablet formulations with as high as 80% concentration of SDD. It was demonstrated that the SDD could span the tabletability range of typical fillers depending on how it was spray dried, allowing it to â€œdo the jobâ€ of different types of fillers. The study ultimately concludes that rationally selecting spray-drying conditions to target specific particle properties, such as high surface area, high compactibility, and isotropic geometries, allows tablet formulations that can successfully decrease pill burden.

[1] Patel, B.B. et al. Revealing facts behind spray dried solid dispersion technology used for solubility enhancement. Saudi Pharm. J. 2015, 23 (4), 352-365.

[2] Singh, A. and Van den Mooter, G. Spray drying formulation of amorphous solid dispersions. Adv. Drug Deliv. Rev. 2015.

[3] Katz, J. M., Roopwani, R., Buckner, I. S. A Material-Sparing Method for Assessment of Powder Deformation Characteristics Using Data Collected During a Single Compression-Decompression Cycle. J. Pharm. Sci. 2013, 102 (10), 3687-3693.

[4] Osei-Yeboah, F., Chang, S., and Sun, C.C. A critical Examination of the Phenomenon of Bonding Area - Bonding Strength Interplay in Powder Tableting. Pharm Res. 2016, 33, 1126-1132.

[5] Sun, C.C. Microstructure of Tablet â€“ Pharmaceutical Significance, Assessment, and Engineering. Pharm. Res. 2017, 34, 918-928.

500057 Reprocessing of Spray Dried Dispersions: Minimizing Resources in Process Development and Dealing with the Possible Failure

Abstract:

Reprocessing of spray dried dispersions: minimizing resources in process development and dealing with the possible failure

T. Porfirio[1,2*], I. Duarte[1], R. Ferreira[1], B. Henriques[1], J. Vicente[1], V. Semião[2]

[1] Hovione Farmaciência SA, Sete Casas, 2674-506 Loures, Portugal; *tporfirio@hovione.com

[2]LAETA, IDMEC, Mechanical Engineering Department, Instituto Superior Técnico, Universidade de Lisboa, Av. Rovisco Pais 1, 1049-001 Lisbon, Portugal

Spray drying is a well-established manufacturing process to produce amorphous solid dispersions to improve the bioavailability of poorly soluble drugs. However, the scale-up of the process is often considered a resource-intensive challenge requiring large quantities of expensive drugs [1]. Therefore, the possibility of reprocessing spray dried dispersions (SDD) is very appealing since it minimizes the material's requirements during the process development. Furthermore if the feasibility of the reprocessing steps is demonstrated during development it can be applied during commercial routine as risk mitigation strategy.

However, the reprocessing can have significant potential to adversely affect the identity, strength, quality, purity or other attribute of the drug and excipients [2]. The reprocessing should be performed on the cases where there is an improvement in one or more quality attributes without adverse effect on the others. The drive of the present work is to understand the effect of successive reprocessing steps on chemical and physical properties of spray dried solid dispersions using different stabilizing polymers commonly used in the production of amorphous solid dispersions.

Three different systems were produced and reprocessed (as shown in Figure 1) using itraconazole (ITZ) as model drug and HPMC E3, HPMCAS-MG and Copovidone as different stabilizing polymers. After each spray drying step a sample was collected for analytical characterization to assess the physicochemical attributes of spray dried material viz. particle size, bulk density and content of residual solvents, assay, and dissolution profile. The remaining spray dried product was reprocessed to prepare the feed solution of the subsequent spray drying step. This study carried out two reprocessing in order to understand when the material begins to present modifications in its properties.

Figure 1 – Scheme of the adopted method

All the tests were performed in a lab-spray dryer equipped with a two-fluid nozzle and using the same drying conditions (such as outlet temperature, drying gas flowrate, feed flowrate and atomization flowrate). A solvent mixture of dichloromethane and methanol was used to dissolve the material in a solid concentration of 5 wt.%.

Placebo tests with polymer-only were also performed to assess the physiochemical properties, such as molecular weight, monomer content and heavy metal levels of the stabilizing polymers. The same conditions were used as in the tests with active substance.

The use of different formulations and placebos tests will clarify the role of the polymer type on the SDD reprocessing.

Reference:

[1] Hallow, D., Mudryk, B., Braem, A., Tabora, J., Lyngberg, O. & Tummala, S., 2010. An Example of Utilizing Mechanistic and Empirical Modeling in Quality by Design. Journal of Pharmaceutical Innovation, 5(4), pp.193–203.
[2] Guidance for Industry - Drug Substance Chemistry, Manufacturing, and Controls Information, U.S. Department of Health and Human Services – Food and Drug Administration

494745 *In silico* Development of Amorphous Solid Dispersions for Optimal Performance and Stability

Abstract:

The focus of the present work lies on screening formulations for amorphous solid dispersions (ASDs) with a computational tool which allows a comprehensive set of virtual experiments to be performed without API expenditure and with a fast turnover and consequently to explore the design space in full prior to manufacturing prototypes in the laboratory.

ASDs of poorly water soluble active pharmaceutical ingredients (APIs) in hydrophilic polymeric matrices are a commonly used strategy to increase the solubility and dissolution rate of oral-dosage forms [1]. However, successful development of amorphous solid dispersions for optimal performance and stability depends on both the formulation (e.g. drug load, polymer system and surfactants) and on the manufacturing technology/process conditions. For example, adequate solvent mixtures used in the process offer the possibility of influencing particle morphology and drying rate [2] and have been reported to offer advantages in the solid state drug-polymer miscibility and on the stability of the ASDs [3,4]. On the other hand, excessively complex formulations can lead to undesirable phase separation which can compromise physical stability [5]. However, optimizing the formulation (e.g. drug load, polymer system and solubilizers) and the process (e.g. drying kinetics and droplet/particle size) requires extensive laboratory work, which is often compromised by time and cost constraints. Therefore the ability to perform virtual experiments is highly advantageous and cost-effective.

In this presentation we cover the mathematical modeling of the various concurrent physical phenomena that a solution of API and polymer (and sometimes other excipients) undergo prior to forming an amorphous solid dispersion, such as atomization, co-solvent evaporative dynamics, propensity for phase separation and shell formation [6,7]. We show a set of case studies of *in silico* screening benchmarked against data from solvent casting and spray drying experiments and also show a case-study where an excessively complex ASD formulation led to phase separation and compromised the stability of the amorphous intermediate.

Figure 1: Sketch of various stages throughout the droplet drying history and the concomitant propensity for phase separation and crystalization if the formulation and/or process are not adequately designed.

References

[1] Perrie Y., Rades T. "Pharmaceutics - Drug Delivery and Targeting", Pharmaceutical Press, 2010.

[2] Wulsten, E., Kiekens, F, van Dycke, F., Voorspoels, J. & Lee, G., "Levitated single-droplet drying: Case study with itraconazole dried in binary organic solvent mixtures", Int. J. Pharmaceutics, 2009, 378(1-2) pp. 116-121

[3] Janssens, S, Anné, M., Rombaut, P. & Van den Mooter, G., "Spray drying from complex solvent systems broadens the applicability of Kollicoat IR as a carrier in the formulation of solid dispersions", Eur. J. Pharm. Sci. 2009, 237, pp. 241-248.

[4] Paudel, A. & Van den Mooter, G., "Influence of Solvent Composition on the Miscibility and Physical

Stability of Naproxen/PVP K 25 Solid Dispersions Prepared by Cosolvent Spray-Drying", Pharm. Res. 2012, 29, pp. 251-270.

[5] Janssens, S., Nagels, S., de Armas, H.N., D'Autry, W., Van Schepdael, A. & Van den Mooter, G., "Formulation and characterization of ternary solid dispersions made up of Itraconazole and two excipients, TPGS 1000 and PVPVA 64, that were selected based on a supersaturation screening study", Eur. J. Pharm. Biopharm, 69:1, 2008

[6] Duarte, Í., Santos, J. L, Pinto, J. G & Temtem, M., "Screening methodologies for the development of spray-dried amorphous solid dispersions", Pharm. Res., 2015, 32 pp. 222 – 237, 2015.

[7] Prudic, A., Ji, Y. & Sadowski, G., "Thermodynamic Phase Behavior of API/Polymer Solid Dispersions", Mol. Pharmaceutics, 2014, 11(7), pp. 2294-2304

502687 Developing Improved Enzymes for Industrial Biocatalysis

Abstract:

Pharmaceutical agents typically contain multiple functional groups with a few stereocenters that must be synthesized in high chemical purity. The field of Biocatalysis uses enzymes to perform highly selective chemical transformations. These enzymes are mainly evolved around a specific metabolic or chemical transformation and in many cases screening panels of these enzymes results in just enough activity for small-scale delivery and proof-of-concept experiment, not large-scale manufacturing. To overcome this challenge, we utilize the power of high-throughput rapid screening technologies and machine-learning algorithms during reaction and protein engineering to develop an optimized, highly selective biochemical transformation that enables large-scale manufacturing. Recent success from our laboratories will be discussed.

495941 Technoeconomic Optimisation, Antisolvent Selection and Comparative Environmental Evaluation for Continuous Paracetamol Crystallisation

Abstract:

Pharmaceutical production has long relied solely on batch processing, which offers many benefits including equipment flexibility and extensive know-how of a mature technology. Nevertheless, in an environment of growing R&D expenditure and an ever-increase drive towards sustainability, strong research interest is evident towards Continuous Pharmaceutical Manufacturing (CPM) (Gutmann et al., 2015). Continuous production can achieve higher yields at the required high purities purities, better heat and mass transfer, decreased processing times, and improved efficiency and reliability. The multifacetted research effort in developing CPM technologies includes demonstrations of novel continuous synthesis routes (Baumann and Baxendale, 2015), the development of cutting-edge microscale flow devices (Jensen, 2017), advances in downstream product formulation (Ierapetritou et al., 2016), and demonstrations of full end-to-end continuous production lines for approved pharmaceutical formulations (Mascia et al., 2013).

The need for cost-effective R&D methodologies brings process modelling and simulation to the forefront of detailed, knowledge-based process option evaluation. Furthermore, they are also pivotal towards evaluating alternative design parameters for existing or newly developed processes (Benyahia et al., 2012) and for developing control strategies, including model-predictive control (MPC) (Lakerveld et al., 2013; Mesbah et al., 2015).

Our recent publications have illustrated the use of systematic process modelling, plantwide simulation and detailed costing for technoeconomic evaluation of CPM processes (Jolliffe and Gerogiorgis, 2016). Moreover, we have employed nonlinear optimisation on the basis of rigorous process models to determine optimal design and operating parameters for key product separation operations (Jolliffe and Gerogiorgis, 2017a, 2017b). The present paper focuses on systematically evaluating a range of solvents, antisolvents and operating conditions for the crystallisation of paracetamol.

The solvents and antisolvents studied are those that are relatively benign and are most preferable to use, and for which thorough experimental solubility data have been compiled (Granberg and Rasmuson, 1999, 2000; Hojjati and Rohani, 2006). These substances are namely water, toluene, methanol, ethanol, 1-propanol, 2-propanol, 1-butanol, acetone, acetonitrile and ethyl acetate. Where data are not available, rigorous thermodynamic estimation tools (such as the UNIFAC and NRTL activity coefficient prediction methods) have been employed. The impact of crystallisation temperature and antisolvent addition ratio are also considered. The cases are formulated into a unified nonlinear optimisation (NLP) framework to determine optimal design variables and process conditions. In addition to technical metrics for evaluation such as product recovery, the sustainability of the different cases has been evaluated by monitoring quantitative green chemistry metrics, such as the E-factor. The use of the solvent and antisolvent in plug flow and Continuous Oscillatory Baffled Crystallisers (COBC) has relied on published models thereof, because these novel unit operations have been shown to offer promising scalability and excellent heat and mass transfer control (McGlone et al., 2015; Su et al., 2015). The solved optimisation cases illustrate the relative benefits and drawbacks of different combinations of solvent and antisolvent, in both technical and sustainability terms. Accordingly, the considered crystalliser designs and unit volume/size variations underscore the benefits of COBC technology, and its potential for production-scale implementation.

LITERATURE REFERENCES

Baumann, M., Baxendale, I.R., 2015. The synthesis of active pharmaceutical ingredients (APIs) using continuous flow chemistry. Beilstein J. Org. Chem. 11, 1194–1219.

Benyahia, B., Lakerveld, R., Barton, P.I., 2012. A plant-wide dynamic model of a continuous pharmaceutical process. Ind. Eng. Chem. Res. 51, 15393–15412.

Granberg, R.A., Rasmuson, Å.C., 2000. Solubility of paracetamol in binary and ternary mixtures of water + acetone + toluene. J. Chem. Eng. Data 45, 478–483.

Granberg, R.A., Rasmuson, Å.C., 1999. Solubility of paracetamol in pure solvents. J. Chem. Eng. Data 44, 1391–1395.

Gutmann, B., Cantillo, D., Kappe, C.O., 2015. Continuous-flow technology—A tool for the safe manufacturing of active pharmaceutical ingredients. Angew. Chem. Int. Ed. 54, 6688–6728.

Hojjati, H., Rohani, S., 2006. Measurement and prediction of solubility of paracetamol in water–isopropanol solution. Part 1. Measurement and data analysis. Org. Process Res. Dev. 10, 1101–1109.

Ierapetritou, M., Muzzio, F., Reklaitis, G., 2016. Perspectives on the continuous manufacturing of powder-based pharmaceutical processes. AIChE J. 62, 1846–1862.

Jensen, K.F., 2017. Flow chemistry—Microreaction technology comes of age. AIChE J. 63, 858–869. doi:10.1002/aic.15642

Jolliffe, H.G., Gerogiorgis, D.I., 2017a. Technoeconomic optimisation of a conceptual flowsheet for continuous separation of an analgaesic Active Pharmaceutical Ingredient (API). Ind. Eng. Chem. Res., in press.

Jolliffe, H.G., Gerogiorgis, D.I., 2017b. Technoeconomic optimisation and comparative environmental evaluation of continuous crystallisation and antisolvent selection for artemisinin recovery. Comput. Chem. Eng., in press.

Jolliffe, H.G., Gerogiorgis, D.I., 2016. Plantwide design and economic evaluation of two Continuous Pharmaceutical Manufacturing (CPM) Cases: ibuprofen and artemisinin. Comput. Chem. Eng. 91, 269–288.

Lakerveld, R., Benyahia, B., Braatz, R.D., Barton, P.I., 2013. Model-based design of a plant-wide control strategy for a continuous pharmaceutical plant. AIChE J. 59, 3671–3685.

Mascia, S., Heider, P.L., Zhang, H., Lakerveld, R., Benyahia, B., Barton, P.I., Braatz, R.D., Cooney, C.L., Evans, J.M.B., Jamison, T.F., Jensen, K.F., Myerson, A.S., Trout, B.L., 2013. End-to-End continuous manufacturing of pharmaceuticals: integrated synthesis, purification, and final dosage formation. Angew. Chem. Int. Ed. 52(47): 12359–12363.

McGlone, T., Briggs, N.E.B., Clark, C.A., Brown, C.J., Sefcik, J., Florence, A.J., 2015. Oscillatory Flow Reactors (OFRs) for continuous manufacturing and crystallization. Org. Process Res. Dev. 19, 1186–1202.

Mesbah, A., Paulson, J.A., Lakerveld, R., Braatz, R.D., 2015. Plant-wide model predictive control for a continuous pharmaceutical process, in: 2015 American Control Conference (ACC), pp. 4301–4307.

Su, Q., Nagy, Z.K., Rielly, C.D., 2015. Pharmaceutical crystallisation processes from batch to continuous operation using MSMPR stages: Modelling, design, and control. Chem. Eng. Process. Process Intensif. 89, 41–53.

498830 Metric-Driven Process Development of a Late Stage Drug Substance

Abstract:

Batch chemical manufacture of a pharmaceutical intermediate typically includes three common elements: reaction, work-up, and isolation. At the start of development, resources are often divided among these elements to enable work in parallel. While such practice allows investigations to proceed quickly, there is a risk that choices made for each element are not necessarily optimal for processing nor environmentally sound when examining the process in its entirety. This occurs because the criterion for a successful reaction differs for work-up and isolation. For example, consider solvent selection. It is often desirable to choose a solvent system that affords homogenous reaction conditions, becomes immiscible during work-up to remove water-soluble waste, and ultimately proves insoluble for the product yet highly soluble for impurities. This presentation will contextualize these challenges and share experimental practices through a case study of a process to synthesize a late-stage active pharmaceutical ingredient (API). The process described was constrained to deliver a specific crystal form and of course meet the general criteria of high impurity purge and high yield. Material properties of the API â€“ a BCS Class I material â€“ as well as of the reagents further challenged development by requiring a thorough understanding of the impact of both pH and the presence of water. A combination of automated and manual experiments streamlined work and serves as a platform for future development. Several metrics such as greenness, cost, overall cycle time, and process mass intensity were used to evaluate the options and guide development decisions.

501507 Scalable Continuous Synthesis of Enantiomerically Pure Amine APIs in an Enzyme Membrane Reactor

Abstract:

Chiral amine groups were found in 11% of the 100 highest selling drugs in 2015. Notable examples include oseltamivir phosphate (sold as Tamiflu®), used for treating influenza symptoms, dexmethylphenidate (sold as Focalin®), an ADHD medication, and sitagliptin (sold as Januvia®), an anti-type II diabetes medicine. Amine dehydrogenases (AmDHs), which were first developed in 2012[1], catalyze the synthesis of a chiral amine through reductive amination of a prochiral ketone with ammonia as the nitrogen source and NADH as cofactor, to produce an *(R)*-amine, with NAD^+ and water as by-products. To date, the largest scale at which AmDHs have been used was in a 100 mL batch process with an isolated yield of 213 mg of amine product. Because of the high cost of NADH, formate dehydrogenase (FDH) as a cofactor regeneration enzyme to convert NAD^+ to NADH. Atom economy for this process is extremely high.

We present the first example of a continuous process for chiral amine synthesis with AmDHs, employing an enzyme membrane reactor (EMR). The EMR consists of a continuously stirred reaction volume with a semipermeable UF membrane at the outlet which traps macromolecules within the reactor, but allows the small molecule substrates and products to pass[2-3]. Conversion of substrates is measured by following the change in the optical rotation of the outlet stream with a polarimeter. This rapid, online measurement allows for constant control of conversion, which is crucial for process optimization.
In this presentation, we outline the initial process development for this new reactor system for the synthesis of a model chiral amine compound. This begins with the characterization of the kinetic parameters of both enzymes which will determine rates and inhibition patterns. After a stable operating point is determined, the space-time yield (STY) can be determined. As we examine the enzyme deactivation behavior over a period of multiple days, the total turnover number (TTN) for both enzymes is obtained and will be reported. Data from these experiments will be used to optimize STY and TTN. Lastly, we will analyze continuous amine production via EMR with Green Chemistry metrics.

1. Abrahamson, M. J.; Vazquez-Figueroa, E.; Woodall, N. B.; Moore, J. C.; Bommarius, A. S., Development of an Amine Dehydrogenase for Synthesis of Chiral Amines. *Angewandte Chemie-International Edition* **2012,** *51* (16), 3969-3972.

2. Bommarius, A. S.; Drauz, K.; Gunther, K.; Knaup, G.; Schwarm, M., L-methionine related L-amino acids by acylase cleavage of their corresponding N-acetyl-DL-derivatives. *Tetrahedron-Asymmetry* **1997,** *8* (19), 3197-3200.

3. Kragl, U.; Kruse, W.; Hummel, W.; Wandrey, C., Enzyme engineering aspects of biocatalysis: Cofactor regeneration as example. *Biotechnology and Bioengineering* **1996,** *52* (2), 309-319.

Development of a Biocatalytic Cascade for the Deracimization of a Racemic Alcohol

Shane Grosser[1], Jeff Moore[2], Amude Kassim[2], Rob Leighty[3], and Wenyong Chen[1]

[1]Process Research and Development, Merck & Co., Inc., Rahway, NJ USA

[2]Chemical Biotechnology and Structure, Merck & Co., Inc., Rahway, NJ USA

[3]Global Vaccines and Biologics Commercialization, Merck & Co., Inc., Kenilworth, NJ USA

This submission is focused on the development and lab-, kilo- and pilot- scale demonstration of a biocatalytic deracimization process for the generation of an optically pure chiral alcohol, the key precursor to a novel dimeric catalyst used in the endgame of an active API synthesis. In particular, the team was tasked with developing a biocatalytic cascade capable of performing the requisite oxidation and reduction chemistry, followed by establishing a process to purify the chiral alcohol from the resulting complex reaction media.

What is particularly novel about this chemistry is that it uses two different ketoreductase enzymes, one which is S-selective, and another which is R-selective, to stereoselectively oxidize the S-enantiomer of the racemic alcohol substrate to the corresponding ketone, and then to stereoselectively reduce that ketone back to the desired R-enantiomer. In addition to the two ketoreductase enzymes, this process also uses two different enzymatic cofactor recycling systems for maintenance of the NAD^+ and $NADP^+$ co-factors in the appropriate oxidation states. The orthogonal nature of these recycling systems is afforded through an aerobic oxidation of NADH back to NAD^+ by NADH oxidase (NOX) for the first reaction, and the reduction of $NADP^+$ to NADPH by phosphite dehydrogenase (PDH) in the second reaction. In the absence of fully optimized enzyme variants with perfect selectivity for NAD^+ vs. $NADP^+$, this system is executed as a sequential catalytic cascade. As both ketoreductases are capable of running in either the oxidation or the reduction directions, successful conversion to >98%ee is controlled by both a staged charging approach, as well as control of the cofactor recycling systems. For example, following complete conversion of the undesired enantiomer to the ketone during the oxidation step, the oxidation's zinc dependent ketoreductase enzyme is deactivated by the addition of EDTA, a powerful zinc chelator. Since the second ketoreductase enzyme is not zinc dependent, it is unaffected by the presence of residual EDTA and proceeds without issue once the reduction enzyme, and it's recycling enzyme PDH, have been charged.

Additional obstacles overcome in the development of this process center around the initial oxidation reaction and its catalyst recycling system. The primary oxidant in the NAD^+ recycle happens to be molecular oxygen, delivered through compressed air to the aqueous reaction. As the moderately slow rate of O_2 delivery tends to impose a mass transfer limitation for this step, precise control at the lab scale, as well as thoughtful considerations of equipment geometry and process execution were critical in ensuring a robust process across scales. Towards that end, significant effort was placed on the refinement and characterization of a reactor setup with line of sight to a commercial synthetic chemistry reactor, including online monitoring and automated control of reaction performance via air flow monitoring and control, dissolved oxygen monitoring and control, as well as reaction component monitoring (via in-situ fluorescence spectroscopy and online LC sampling.) Employing these reactor designs, and further leveraging data-rich experimentation practices, the team developed a reaction fingerprint for both the oxidation and reduction reactions, successfully demonstrating each of them across easymax, optimax, 20L and 200 gal reactors. This includes refinement of kLa regression methodologies across scales for the O_2 dependent oxidation reaction, as well as reaction profiling using online HPLC. This body of information was critical in ensuring appropriate reactor configuration and process execution and has even enabled real-time modifications to the pilot scale equipment configuration and operational parameters based on at-line profiling data.

Enzyme removal is frequently a challenge in free solution biocatalysis, particularly in instances where the product possesses significant aqueous phase solubility. This was certainly true for this chemistry, which presented a number of such challenges regarding protein/product purification. Ultimately, the team developed and implemented a robust protein/product separation, including a novel kosmotropic enzyme precipitation and extraction regimen. Specifically, acetonitrile was added to the aqueous end of reaction mixture, followed by the kosmotropic salt potassium phosphate. This induced a number of phase separation phenomena, which were exploited to enact a robust purification. First, the addition of the potassium phosphate effectively salted-out the protein, rendering a precipitated protein phase which could be readily filtered for enzyme removal. Secondly, the salt also helped to induce a phase split between the otherwise miscible acetonitrile/water mixture, and lastly, the salt also helped to shuttle the highly water soluble product into the organic layer, thereby permitting extractive workup away from the rest of the reaction byproducts. From this point, moderately standard techniques, such as carbon treatment for color removal, and a distillative crystallization could be employed resulting in a robust, scalable, and highly portable process for the deracimization of the racemic alcohol.

497589 Design of New Reactors to Produce Levan Biopolymer through Enzymatic Catalysis: Kinetics and Mass Transfer

Abstract:

Biopolymers are rising in importance during last years because of its potential application in biomedicine and therapeutics. Most of them are usually produced by microbial fermentation and that implies some drawbacks such as sterilization control, high-cost downstream, problems with uniformity of different parameters, etc. One of the most promising biopolymers is levan due to its amphiphilic properties. However, its production has been only focused on microorganisms and its value on market is too high so as to continue researching on its potential applications.

In this work; levan enzymatic way was explored by using the enzyme involved in the process, levansucrase from *Bacillus subtilis*. Levan synthesis by biocatalysis was studied and adjusted to a kinetic model which explains the behavior. Moreover, a complete analysis of reaction conditions was carried out by studying temperature, agitation speed, and pH. Furthermore, the differences between hydrolysis reaction and transfructosylation reaction were analyzed so as to a better understanding of catalytic mechanism. Results show that hydrolysis is clearly the limiting step and for that reason, the use of fructose instead of sucrose is presented as a successful alternative in order to improve yield.

Moreover, levansucrase was immobilized in two supports (a packed bead reactor and a monolithic reactor) so as to a better use and recycling of the enzyme. The effect of immobilization in production rate was also studied and a model of mass transfer through both reactors was also proposed. Mass transfer model shows not limiting transfer phenomena in external transport for both reactors, and, concerning internal transport, monolith reactor shows successful results with an effectiveness factor greater than 0.9. Additionally, the enzyme activity when it was recovered from monolith reactor was studied and it shows a low reduction in catalysis rate compared with the native one.

These results can allow reducing its high value in market and a greater implementation in biomedicine and therapeutics.

509826 Automated Systems for Screening, Kinetics, and Optimization of Chemical Synthesis and On Demand Production

Abstract:

Over the two past decades, small-scale continuous synthesis has matured from simple demonstration examples to applications in pharmaceuticals and fine chemicals. The field has moved beyond single transformations to continuous multistep synthesis of active pharmaceutical ingredients (APIs) by incorporating in-line workup techniques. Integration of on-line measurements of reactant flows, reactor temperature, and outlet concentrations with feedback control systems has enabled automated optimization of reaction yields as well as determining kinetic information. The present contribution starts with automated screening and optimization of chemical reaction in microsystems and continues with integration of small-scale reactors and separators in on-demand continuous synthesis of APIs. Automated continuous/oscillating droplet flow systems are demonstrated for simultaneous optimization of discrete process variables (e.g., catalyst, ligands, and solvents) and continuous conditions (temperature and reagent concentrations). Coupling the droplet reactor with a light source converts the system into an automated microfluidic platform for exploring and optimizing visible-light photoredox catalysis.

As an example of process integration for on demand manufacturing, we present a plug-and-play, reconfigurable, refrigerator-sized manufacturing platform for on-demand synthesis of common pharmaceuticals, e.g. ciprofloxacin. This flexible system is capable of complex multi-step synthesis, in-line purification, post-synthesis work-up and formulation. Multistep synthesis occurs at elevated temperatures and pressures to enhance reaction rates, and the resulting residence times are a few minutes, in contrast to the multiple hour-long synthesis typically needed for batch. Typical production rates are grams/hour sufficient to produce thousands of doses per day of common pharmaceuticals. Finally, we discuss opportunities for integration of machine learning for synthesis planning with fully automated chemical synthesis systems.

509827 Can Laboratory Automation Finally Fulfill the Promise of Revolutionizing Pharmaceutical Discovery?

Abstract:

From Bruce Merrifield's pioneering work in the automation of peptide synthesis in the 60's to the boom and bust of combinatorial chemistry in the 90's, laboratory automation technology for chemical synthesis in early stage drug discovery research has played a role of both hero and villain. This lecture will unearth insights into this rich history and will detail Lilly's research experience with discovery-scale laboratory automation. A specific emphasis will be on our more than a decade of experience with our one of a kind remote access fully integrated Automated Synthesis Laboratory (ASL). Discussion of the key learnings, successes, and challenges in our ASL experience will serve as the backdrop for a bold new vision for the future of remote access, fully integrated laboratory science. The audience will learn about Lilly's ongoing investment in the construction of the Lilly Life Sciences Studio (L2S2), which we believe will significantly change pharmaceutical research in both industry and academia.

509828 Emerging Technology: A Key Enabler for Modernizing Pharmaceutical Manufacturing and Advancing Product Quality

Abstract:

Over the past decade, there has been substantial progress towards modernizing pharmaceutical manufacturing. However, pharmaceutical manufacturing continues to confront a number of challenges, which result in unacceptably high occurrence of product quality issues. Nearly two thirds of all drug shortages relate to quality issues surrounding product manufacturing or facilities. These quality issues expose patients to unnecessary risk and negatively impact public health. Working with industry to encourage the modernization of pharmaceutical manufacturing technology is a mutually beneficial approach to increase manufacturing efficiency, improve overall drug quality, avoid shortages, and prepare for future regulatory challenges.

To encourage the adoption of innovative approaches in pharmaceutical manufacturing and to prepare for reviews and inspections involving technology for which there is little experience, the Agency established the Emerging Technology Program. This program features a cross-functional Emerging Technology Team (ETT) which works directly with industry to identify and resolve potential scientific and policy issues that may impact technologies new to the pharmaceutical industry. In a relatively short period of time, the Emerging Technology Program has facilitated the approvals of the first 3D-printed drug product, the first drug product made via continuous manufacturing, and the first switch to a continuous manufacturing process for an approved drug product.

Advancements in regulatory science are essential to FDA's mission and enable the ETT to issue clear recommendations to sponsors about the risks and benefits of products that involve new technologies. FDA's research program aims to address any knowledge and experience gaps to support these needed improvements. The knowledge gained from the internal and sponsored research will inform policy, review, and inspection activities, ensuring that FDA regulatory policies reflect state-of-the-art manufacturing science.

In this talk, FDA's emerging technology program is described including a summary of industry interactions and observed trends. FDA's manufacturing science research programs are highlighted e.g. continuous manufacturing, 3D printing, and process modeling to demonstrate how advances in regulatory science support the emerging technology program.

490628 Accurate and Efficient Representation of Intramolecular Energy in *Ab Initio* Generation of Crystal Structures

Abstract:

The CrystalPredictor I[1,2] and II[3] codes have been used to explore the space of crystal structures successfully in several crystal structure prediction (CSP) investigations in recent years[4], including in the series of blind tests organised by the Cambridge Crystallographic Data Centre[5,6] and in the prediction of the crystal structures of pharmaceutically-relevant molecules[7,8].

We present a summary of CrystalPredictor, focussing on improvements to the lattice energy evaluation that the most recent blind test[9], as well as our own investigations into a variety of flexible polymorphic molecules, has prompted. These improvements aim to achieve greater accuracy in the initial ranking of potential crystal structures, while managing computational cost so that a thorough exploration of the search space is possible. We discuss the smoothing of the intramolecular potential in a Non-Uniform context[10]; an innovation in CrystalPredictor II that allows the most efficient use of computational effort to cover a flexible moleculeâ€™s conformational space. We use blind test molecule 26 as an example, as well as short case studies on the polymorphic landscapes of Bromhexine HCl and Aripiprazole.

(1) Karamertzanis, P. G.; Pantelides, C. C. *J Comput Chem* **2005**, *26*, 304.

(2) Karamertzanis, P. G.; Pantelides, C. C. *Mol Phys* **2007**, *105*, 273.

(3) Habgood, M.; Sugden, I. J.; Kazantsev, A. V.; Adjiman, C. S.; Pantelides, C. C. *J Chem Theory Comput* **2015**, *11*, 1957.

(4) Pantelides, C. C.; Adjiman, C. S.; Kazantsev, A. V. *Top Curr Chem* **2014**, *345*, 25.

(5) Day, G. M.; Cooper, T. G.; Cruz-Cabeza, A. J.; Hejczyk, K. E.; Ammon, H. L.; Boerrigter, S. X. M.; Tan, J. S.; Della Valle, R. G.; Venuti, E.; Jose, J.; Gadre, S. R.; Desiraju, G. R.; Thakur, T. S.; van Eijck, B. P.; Facelli, J. C.; Bazterra, V. E.; Ferraro, M. B.; Hofmann, D. W. M.; Neumann, M. A.; Leusen, F. J. J.; Kendrick, J.; Price, S. L.; Misquitta, A. J.; Karamertzanis, P. G.; Welch, G. W. A.; Scheraga, H. A.; Arnautova, Y. A.; Schmidt, M. U.; van de Streek, J.; Wolf, A. K.; Schweizer, B. *Acta Crystallogr B* **2009**, *65*, 107.

(6) Bardwell, D. A.; Adjiman, C. S.; Arnautova, Y. A.; Bartashevich, E.; Boerrigter, S. X. M.; Braun, D. E.; Cruz-Cabeza, A. J.; Day, G. M.; Della Valle, R. G.; Desiraju, G. R.; van Eijck, B. P.; Facelli, J. C.; Ferraro, M. B.; Grillo, D.; Habgood, M.; Hofmann, D. W. M.; Hofmann, F.; Jose, K. V. J.; Karamertzanis, P. G.; Kazantsev, A. V.; Kendrick, J.; Kuleshova, L. N.; Leusen, F. J. J.; Maleev, A. V.; Misquitta, A. J.; Mohamed, S.; Needs, R. J.; Neumann, M. A.; Nikylov, D.; Orendt, A. M.; Pal, R.; Pantelides, C. C.; Pickard, C. J.; Price, L. S.; Price, S. L.; Scheraga, H. A.; van de Streek, J.; Thakur, T. S.; Tiwari, S.; Venuti, E.; Zhitkov, I. K. *Acta Crystallogr B* **2011**, *67*, 535.

(7) Vasileiadis, M.; Pantelides, C. C.; Adjiman, C. S. *Chem Eng Sci* **2015**, *121*, 60.

(8) Vasileiadis, M.; Kazantsev, A. V.; Karamertzanis, P. G.; Adjiman, C. S.; Pantelides, C. C. *Acta Crystallographica Section B-Structural Science* **2012**, *68*, 677.

(9) Reilly, A. M. et. al. *Acta Crystallographica Section B: Structural Science, Crystal Engineering and Materials* **2016**.

(10) Sugden I, Adjiman CS, Pantelides CC *Acta Crystallographica Section B-Structural Science* **2016**, *72*, 864.

A New Approach in Applying CSP for the Pharmaceutical Industry

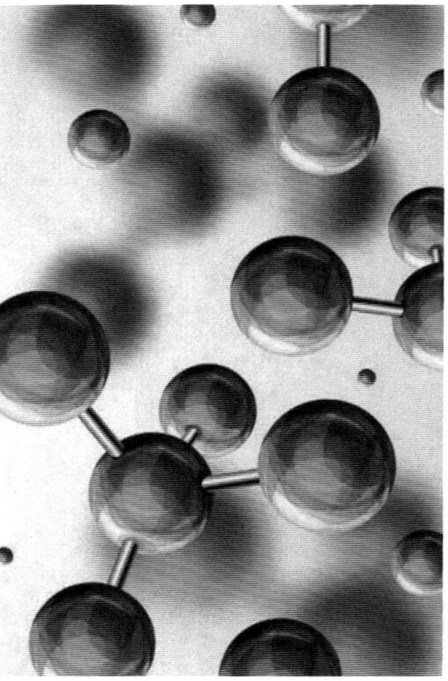

Alan Jiang, MD, PhD & Shuhao Wen, PhD
XtalPi Inc.

AlChe Conference
October 30, 2017

Outline

XtalPi

- Industry challenges
- Technology platform and solution
- Output and case studies
- Summary

Industry Outlook

XtalPi

The impact of successful solid form development

Drug solid form study is crucial to pre-formulation, formulation, and clinical trial success, and is pertinent throughout the lengthy and trying drug R&D process.

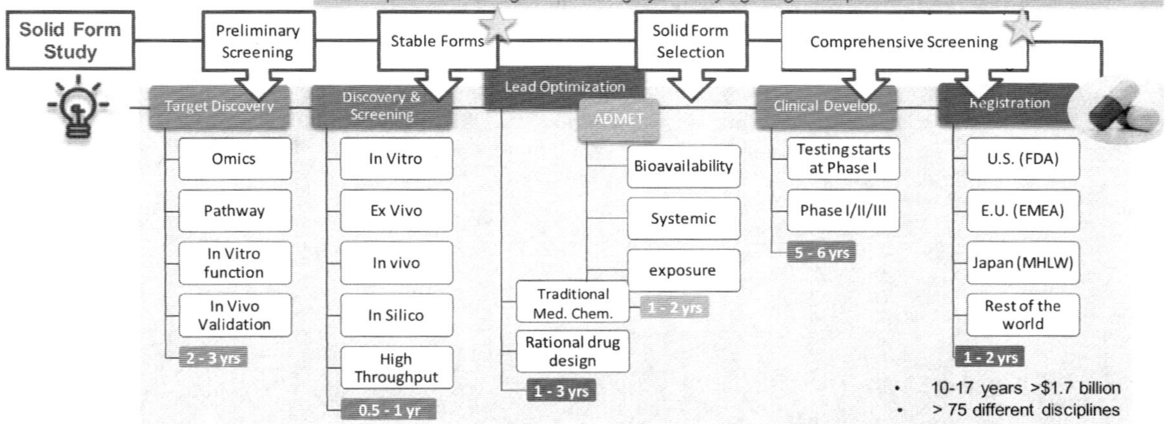

- 10-17 years >$1.7 billion
- > 75 different disciplines
- < 10% overall probability of success in clinical trials

XtalPi

The Cost of Poor Solid Form Research

Case study: Axitinib

The current polymorph screening method is neither effective nor adequate in finding the most stable solid form for complex molecules. Experiments may take years, yet it is impossible to claim certainty.

- During early development, Form IV was considered as the most stable form.
- Later Form XXV was discovered to be more stable than Form IV.
- Eventually form XLI was discovered to be the most stable and was chosen for further development.
- There are in total 5 anhydrous polymorphs and 66 solvates polymorphs being reported for Axitinib.

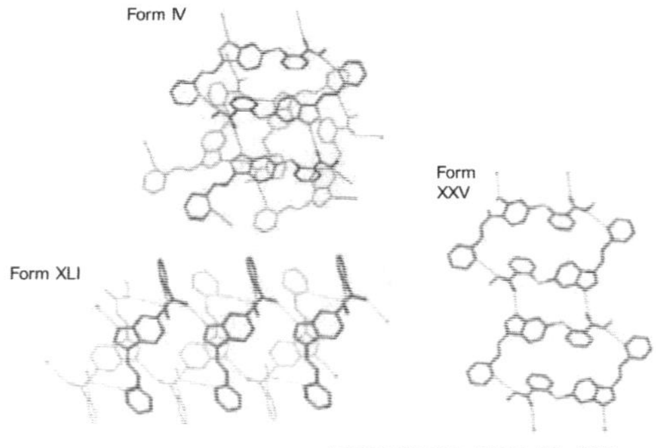

Anthony Campeta, Pharm. Sci., 2010
Manolis Vasileiadis, Chem. Engr. Sci., 2015

The Cost of Poor Solid Form Research

Case study: Ritonavir

- In 1996, Abbott releases Ritonavir using its only polymorph (form I)
- In 1998, Ritonavir's form I started to convert to a more stable form II with different solubility and bioavailability, rendering the drug ineffective.
- Ritonavir was withdrew from the market, causing huge losses in sales. Abbott could not give an explanation why the transform started.
- Upon further crystallization experiments, 5 different solid forms were identified.

Form	Melting point, °C	ΔH_{fus}, J/g	Solid-state structure
I*	122	78.2	Monoclinic
II*	122	87.8	Orthorhombic
III	78–82	60.3	Monoclinic†
IV	116	59.8	Not assigned
V	97	32.0	Monoclinic†

Three more polymorphs were discovered years later (2002).

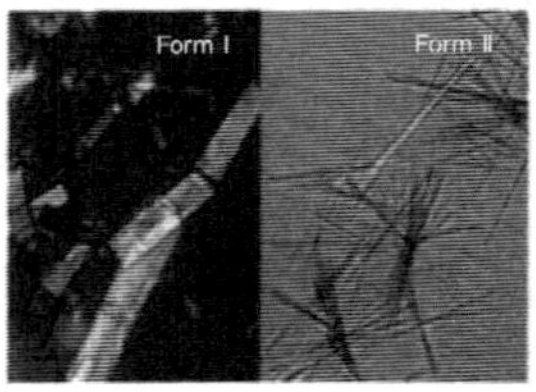

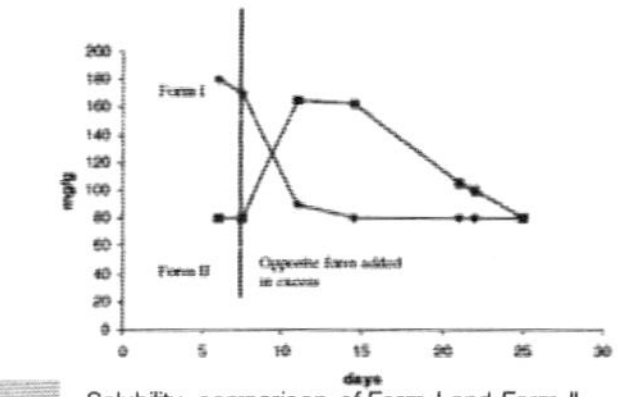

Solubility comparison of Form I and Form II

Even the most competent innovative company may fail in its drug solid form research. Sometimes with post-market drugs, causing devastating loss.

Sherry L. Morissette, PNAS, 2003
John Bauer, Pharm. Res.,2001

Challenges & Opportunities

Providing the long-awaited game-changing solutions

The bigger picture:

- A shortage of new drugs
- Dire drug development success rate
- Decreasing pipelines
- Imminent patent expirations
- Pharmaceutical companies facing economic pressure

XtalPi's Goal:

Expedite solid form development with lower cost and risk
Empower robust pipelines and enhance life-cycle management

With exhaustive calculation and accurate prediction

XtalPi Technology - Crystal Structure Prediction

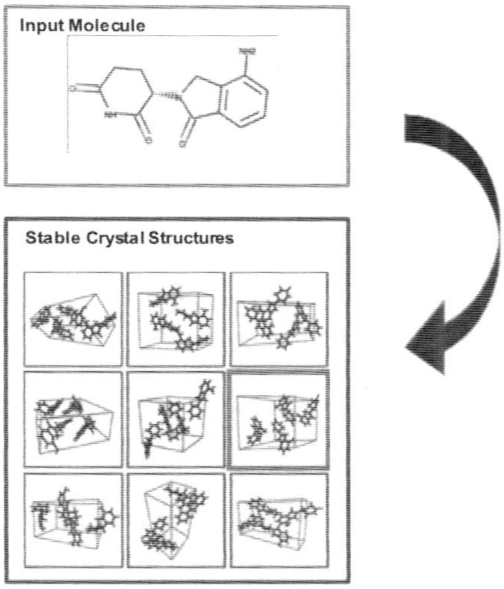

- Automated and powerful *cloud computing* platform
- Intelligent *conformation analysis*
- Accurate *force field* training
- Global/local *conformational search*
- Efficient *clustering* algorithm
- ***Multi-stage ranking*** screen
- ***Completeness*** test
- ***Solid informatics***

XtalPi

XtalPi Technology - The information relevant to experiment

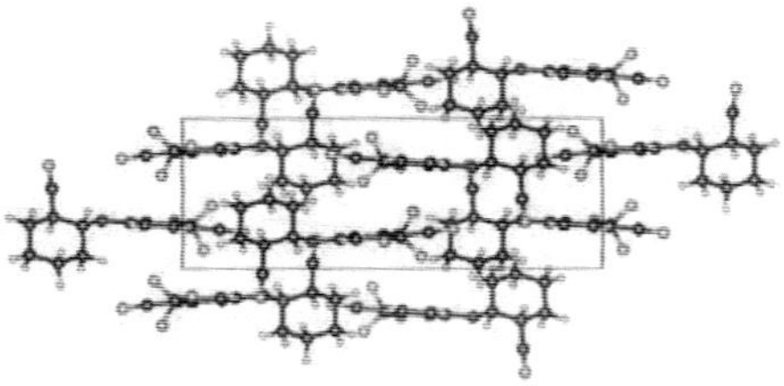

Structure

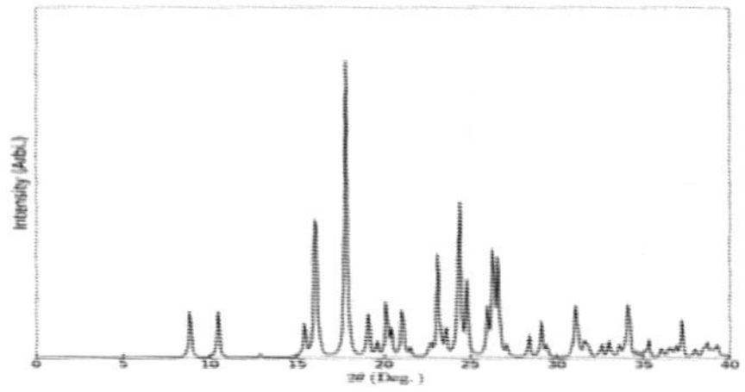

XRD

The following properties can also be provided to link the gap between theory and experiment:

1. Relative stability and solubility of different crystal forms;
2. Selection of solvents for crystallization and coformers for co-crystal;
3. Spectroscopies including IR and Raman as reference for online/in-situ/time-dependent measurements;
4. Nucleation process with long-time brute force or enhanced sampling molecular dynamics simulations;
5. Enthalpy changes between different predicted forms;

XtalPi

XtalPi Technology - High Performance Computing on AWS

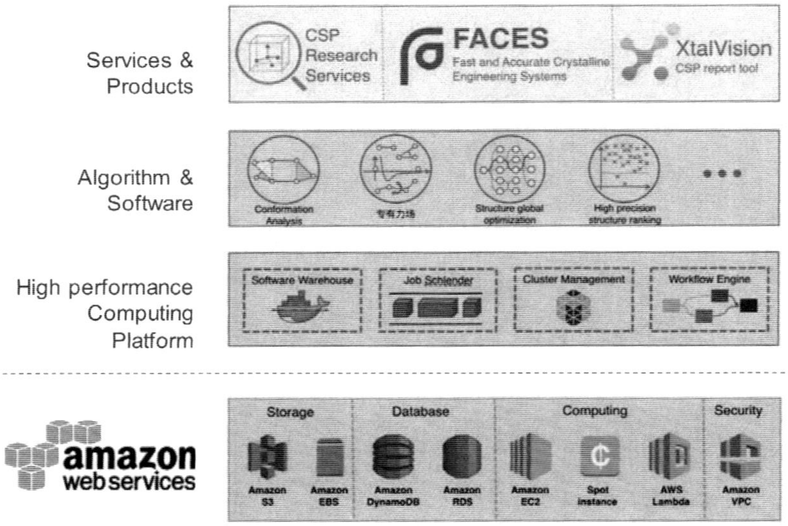

XtalPi

HPC on AWS - Extremely large computing resources

16 Regions in Global

to find the best nodes

Millions Cores

to build a cluster for computing

In seconds

to scale the cores of cluster

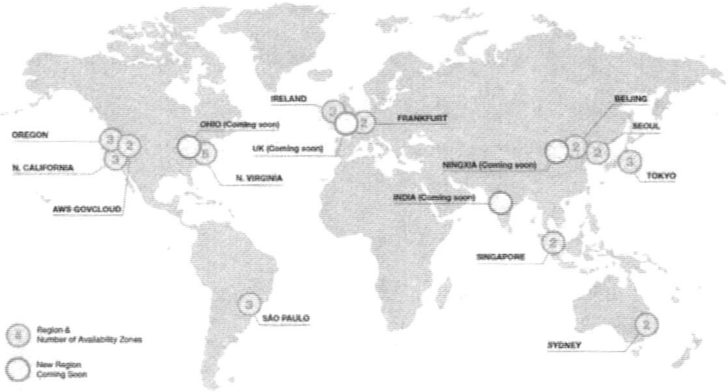

XtalPi

HPC on AWS - Security

- Based on AWS Shared Security Responsibility Model
- Network is isolated with Virtual Private Cloud (VPC)
- All traffic outside the VPC is encrypted

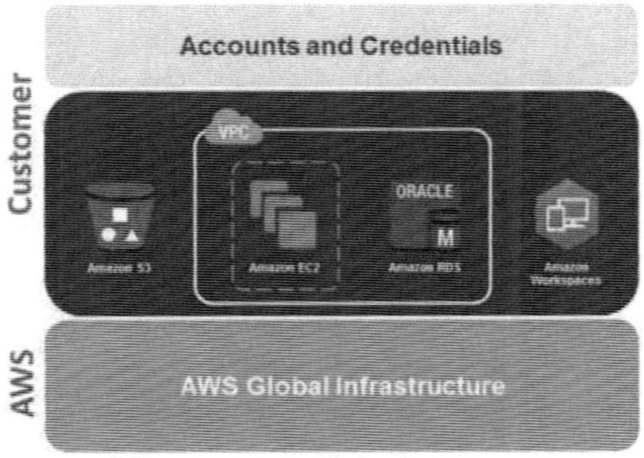

Figure 1: AWS Shared Security Responsibility Model

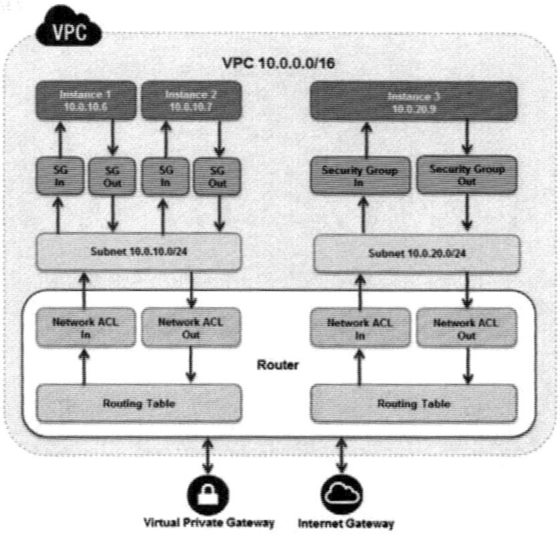

Figure 6: Flexible Network Topologies

XtalPi

HPC on AWS - Generate Report Data for XtalVision

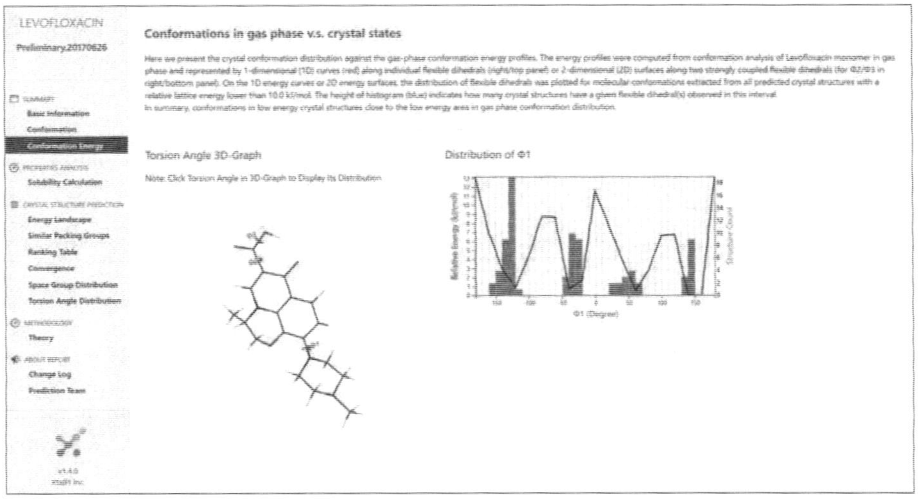

XtalPi

HPC on AWS - Generate Report Data for XtalVision

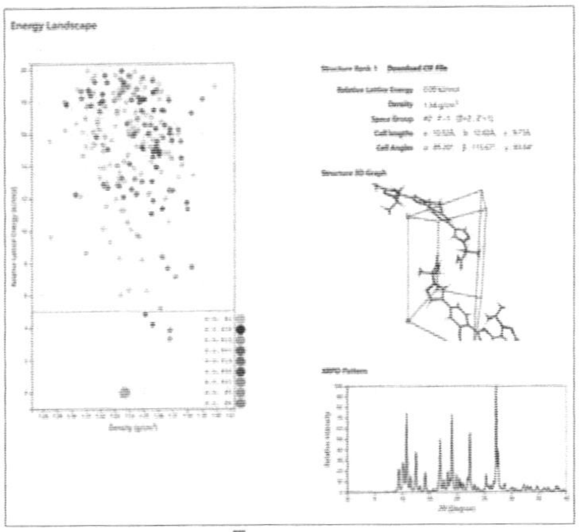

Energy Landscape

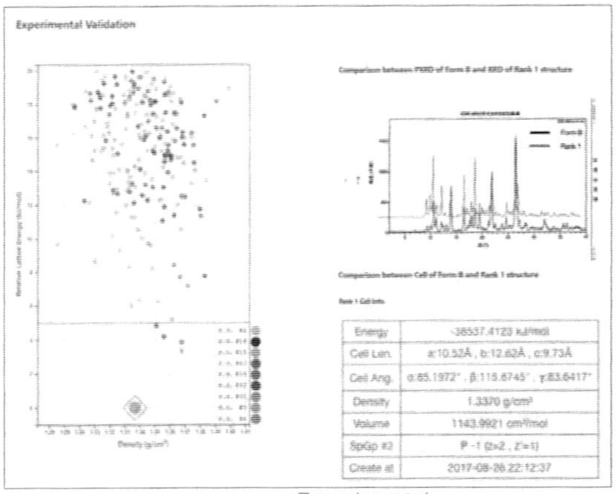

Experimental Validation

XtalPi

Brigatinib

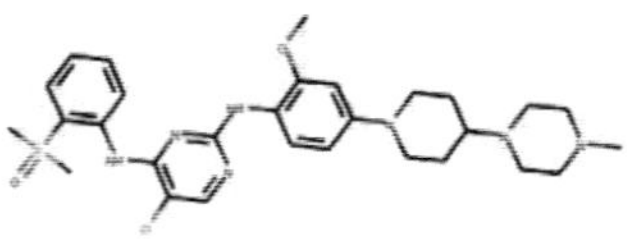

$C_{29}H_{39}ClN_7O_2P$	
MW	584.105 g/mol
HD	2
HA	9
FRB	8

Brigatinib has been filed by Ariad in US for the treatment of patients with anaplastic lymphoma kinase (ALK)-positive locally advanced or metastatic non-small cell lung cancer (NSCLC) who has been resistant to crizotinib.

Brigatinib received Breakthrough Therapy designation from the FDA for the treatment of patients with ALK+ NSCLC whose tumors are resistant to crizotinib, and was granted orphan drug designation by the FDA for the treatment of ALK+ NSCLC. The Company is seeking accelerated approval for brigatinib from the FDA and plans to request a priority review of the applic

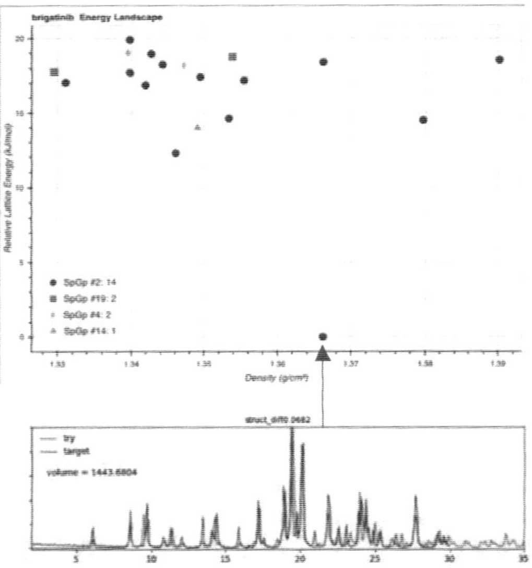

Comparison between PXRD of Form A and XRD of refined Rank 1 structure

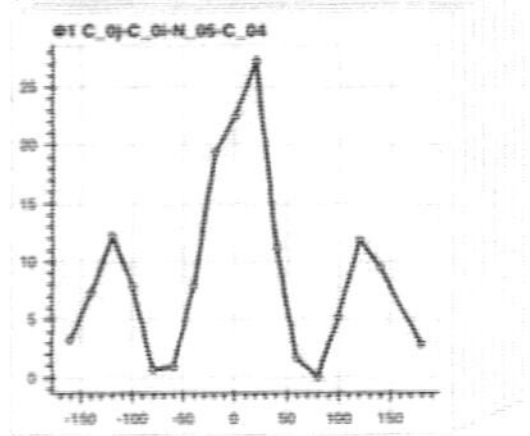

Conformation Energy Analysis

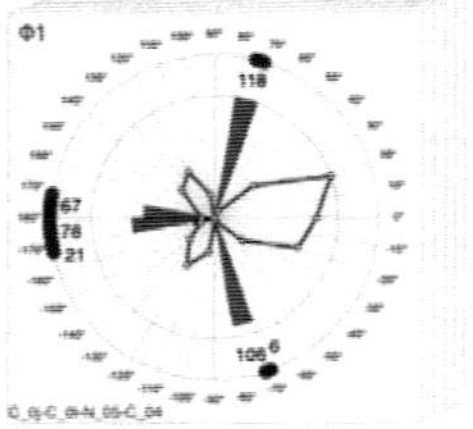

Torsion Angle Distribution

XtalPi

Dapagliflozin Propylene Glycol Solvate Monohydrate

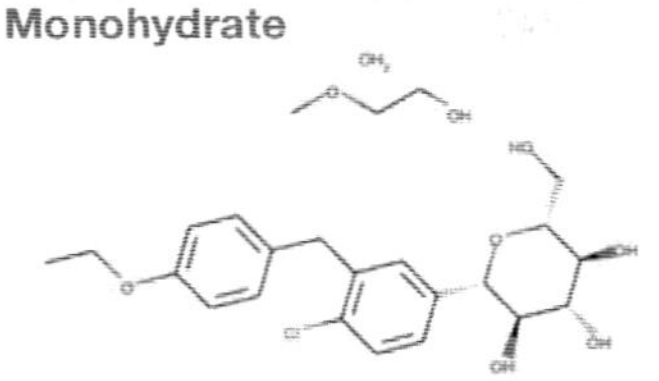

$C_{21}H_{25}ClO_6{\bullet}C_3H_8O_2{\bullet}H_2O$	
MW	502.988 g/mol
HD	4 + 1
HA	6 + 2
FRB	10

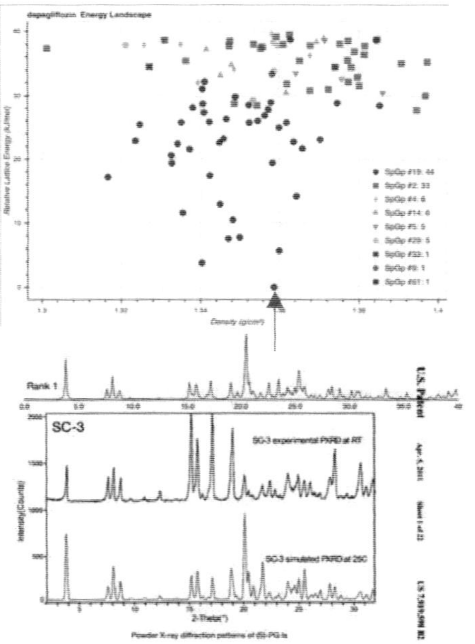

Comparison between PXRD of SC-3 and XRD of refined Rank 1

Dapagliflozin is indicated for the management of diabetes mellitus type 2, and functions to improve glycemic control in adults when combined with diet and exercise. Dapagliflozin is a sodium-glucose cotransporter 2 inhibitor, which prevents glucose reabsorption in the kidney. Using dapagliflozin leads to heavy glycosuria (glucose excretion in the urine), which can lead to weight loss and tiredness. Dapagliflozin was approved by the FDA on Jan 08, 2014. Dapagliflozin is not recommended for patients with type 1 diabetes mellitus or for the treatment of diabetic ketoacidosis.

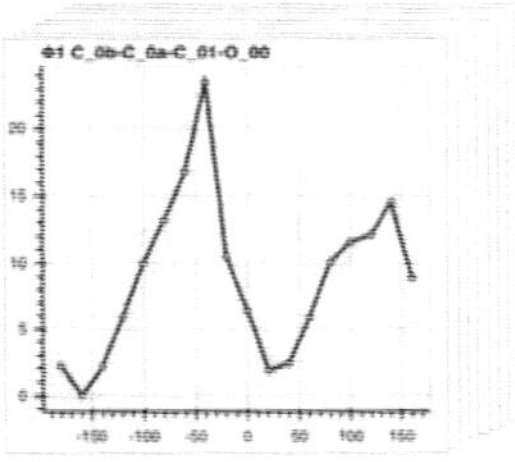

Conformation Energy Analysis

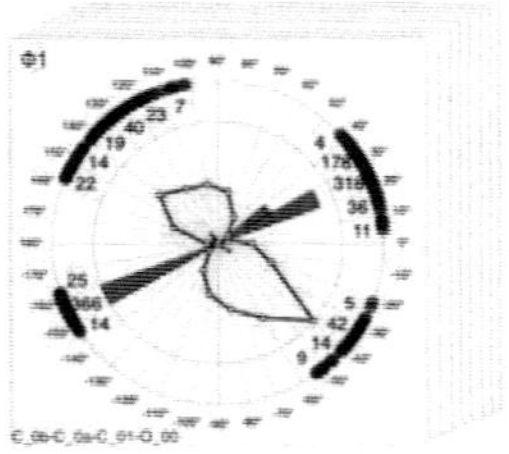

Torsion Angle Distribution

Veliparib (ABT-888)

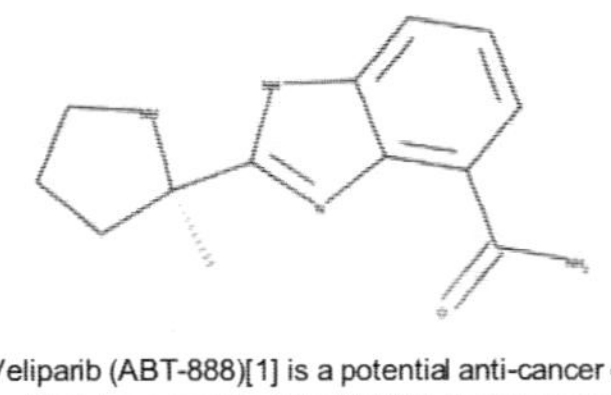

$C_{13}H_{18}Cl_2N_4O$	
MW	317.214 g/mol
HD	5
HA	3
FRB	2

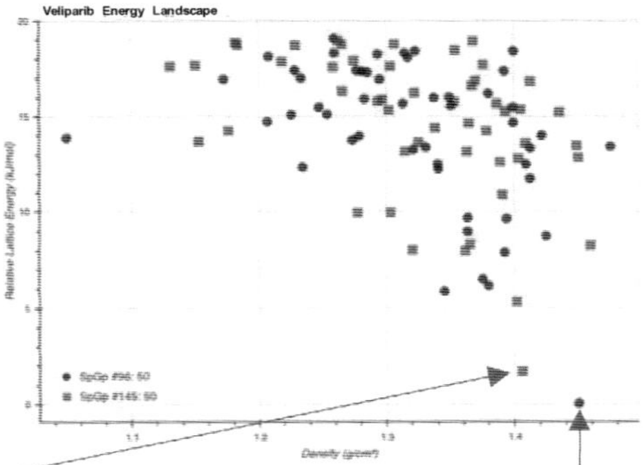

Veliparib (ABT-888)[1] is a potential anti-cancer drug acting as a PARP inhibitor. It kills cancer cells by blocking a protein called PARP, thereby preventing the repair of DNA or genetic damage in cancer cells and possibly making them more susceptible to anticancer treatments. Veliparib may make whole brain radiation treatment work more effectively against brain metastases from NSCLC. It has been shown to potentiate the effects of many chemotherapeutics, and as such has been part of many combination clinical trials.

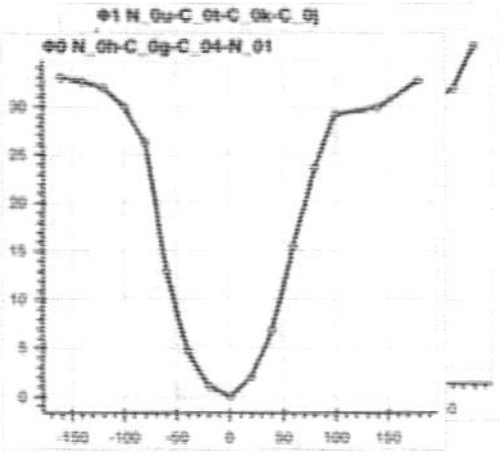

Conformation Energy Analysis

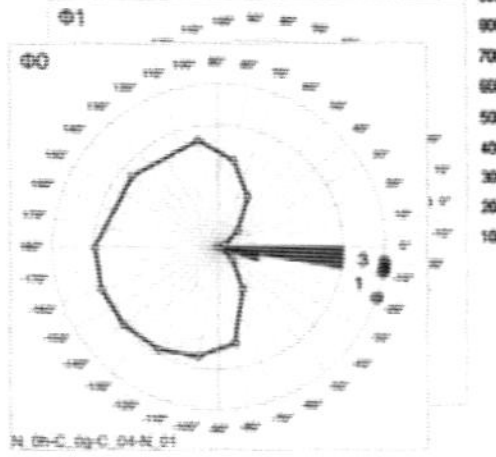

Torsion Angle Distribution

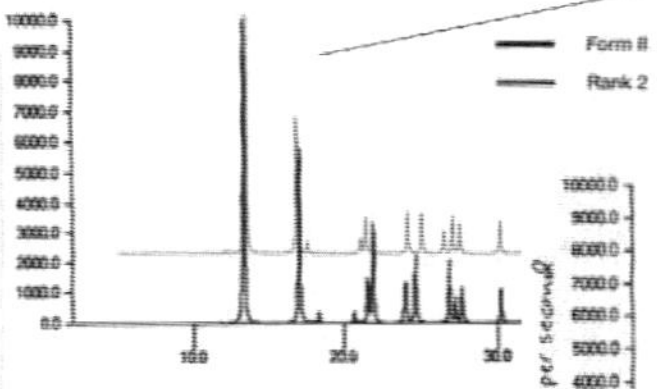

Comparison between PXRD of Form I & II and XRD of refined Rank 1 & 2 structure

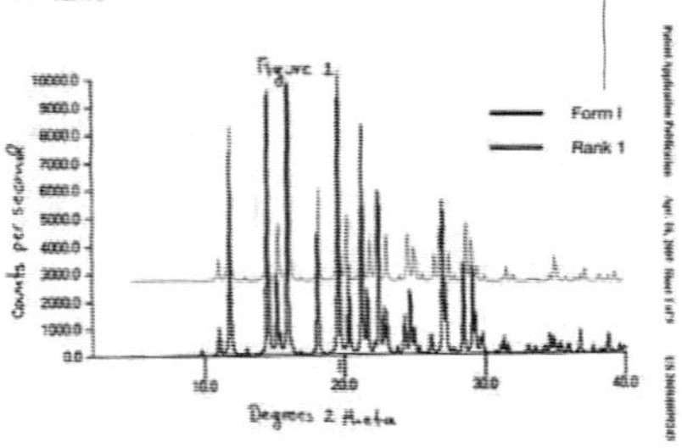

Taselisib (GDC-0032)

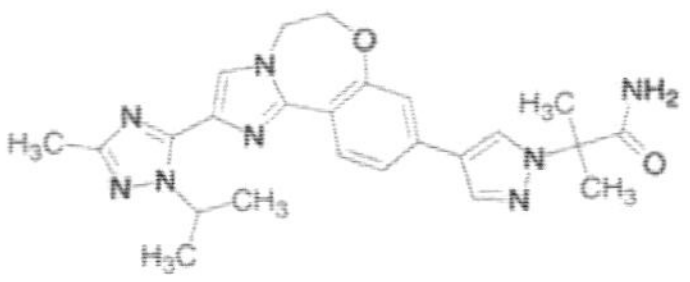

$C_{24}H_{28}N_8O_2$	
MW	460.542 g/mol
HD	1
HA	9
FRB	5

Taselisib (GDC-0032) is an experimental cancer drug in development by Roche. It is a small molecule inhibitor targeting phosphoinositide 3-kinase subtype PIK3CA. Taselisib is in phase II clinical trials for treatment of metastatic breast cancer and non-small cell lung cancer.

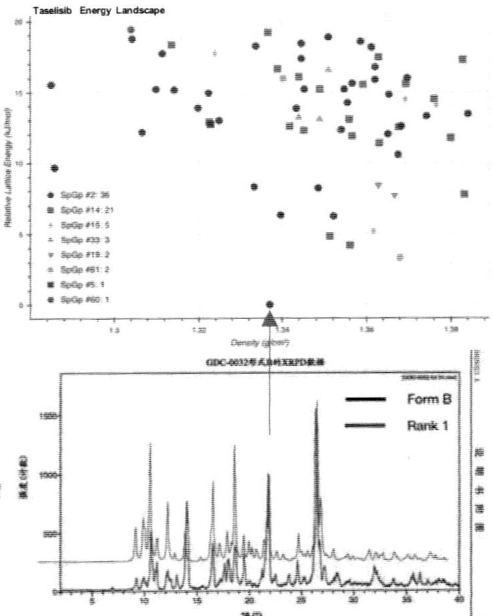

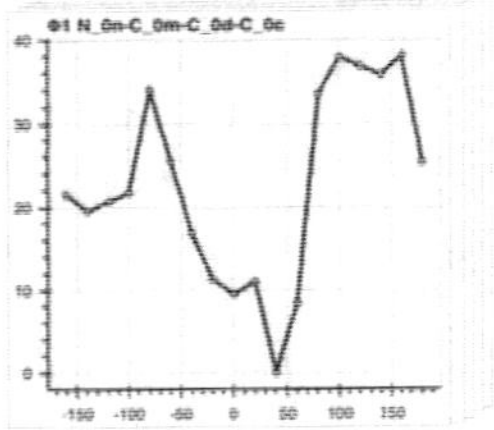

Conformation Energy Analysis

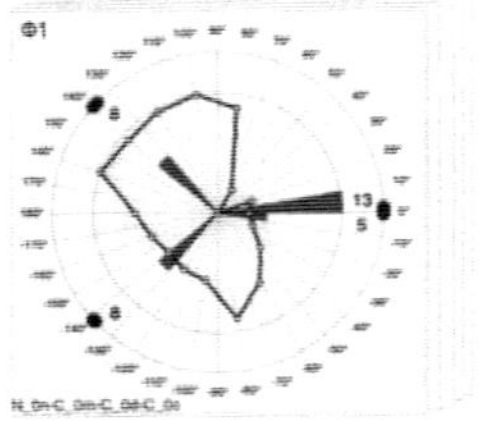

Torsion Angle Distribution

Comparison between PXRD of Form B and XRD of refined Rank 1 structure

Other Case We Researched

Case	Formula	Type	MW	FRB
Tadalafil	$C_{22}H_{19}N_3O_4$	Z'=1	389.411 g/mol	1
Brivaracetam	$C_{11}H_{20}N_2O_2$	Z'=1	212.293 g/mol	5
Tavaborole	$C_7H_6BFO_2$	Z'=1 (With B)	151.931 g/mol	-
Mycophenolate Mofetil	$C_{23}H_{31}NO_7$	Z'=1	433.501 g/mol	10
ZZZMUC01	$C_7H_5N_3O_6$	Z'=2	227.132 g/mol	3
Lenalidomide	$C_{13}H_{13}N_3O_3$	Z'=1 / Hemihydrate / Co-Crystal	259.265 g/mol	1
Gabapentin	$C_9H_{17}NO_2$	Solvate / Solvate Monohydrate	171.240 g/mol	3
ASAXUO	$C_{14}H_{14}N_4O_5$	Hemihydrate	318.29 g/mol	-
EVORIR	$C_{15}H_{16}N_4O_2$	Dihydrate	284.32 g/mol	2
KUXQEB	$C_{10}H_{11}ClN_2O_3$	Chlorhydrate	242.66 g/mol	3
XP-101	-	Z'=1 / Z'=2 / Monohydrate	415.433 g/mol	7
XP-102	-	Salt trihydrate	677.75 g/mol	6 + 6
XP-103	-	Z'=1 / Z'=2	467.349 g/mol	6
......	......	......	......	......

Realization of CSP in industry

XtalPi

The current CSP solution directly impacts in the drug development process

Development Process	Clinical Trial Success	IP Security and Shelf Life
• Identify the best solid form candidate in the early stage will support decision-making on solid form development in the later stages, such as formulation and manufacturing, and contributes to increased probability of success	• Better understanding of the molecule packing in cocrystals enable rational design of free forms, hydrates, salts and cocrystals with better solubility and stability, crucial for successful clinical trials.	• Understanding the complete landscape of stable and metastable forms minimizes the risk of patent challenges, maximizing the patent life and prolonging the drug's market exclusivity period.

501548 Capturing the Role of Temperature and the Sensitivity to Energy Function Complexity in Crystal Polymorph Stability Using Molecular Modeling

Abstract:

Solid materials with more than one polymorph can restructure in response to a change in external conditions, fundamentally altering material properties like solubility and hardness. Such restructuring events have hindered pharmaceutical solid form development in numerous therapeutic candidates and have led to costly market recalls. Computer modeling has the potential to identify all observable solid forms and predict relevant restructuring events in the early stages of development. However, standard crystal structure prediction methods neglect the effect of temperature on relative crystal stabilities, which is vital for predicting temperature-mediated polymorphic transformations. In addition, discussions on the level of complexity to use in the energy function have focused on the effects on crystal lattice energies, rather than crystal free energies. In this study, we use molecular dynamics simulations to capture the effects of temperature and entropy on the stabilities of numerous polymorphic materials at different levels of energy function complexity.

Twelve small molecule organic systems with known temperature-mediated transformations were modeled with the point-charge OPLS-AA potential. Our simulations correctly identify the high temperature solid form as having a larger entropy than the low temperature form in all systems examined. The estimated entropy differences in the classical point-charge potential are significantly closer to experimental measurements than estimated enthalpy differences. This result suggests that entropy difference estimates are less sensitive to the complexity of the simulation potential than the corresponding enthalpy estimates.

We further probe the energy function sensitivity by directly computing the added effects of a more accurate polarizable Hamiltonian on polymorph free energies. We show that the change in free energy to the more complex potential comes predominately from enthalpic rather than entropic contributions in the system examined.

501105 Can Lattice Dynamics with Anisotropic and Isotropic Thermal Expansion Accurately Estimate Thermodynamic Properties of Crystals Pharmaceutics Compared to Molecular Dynamics?

Abstract:

We compare three lattice dynamics methods to determine to what extent they can capture the free energy differences of organic polymorphs in comparison to full molecular dynamics (MD). Efficient methods to determine favorable crystalline structures of small organic molecules is pharmaceutically relevant in the drug development process. For different polymorphs the physical and chemical properties can vary significantly, which can lead to legal and production issues. Experimental approaches to capture all polymorphs are costly and time consuming. Alternatively, current computational methods are able to exhaustively search for the lattice minimum structures, but fail to emphasize the importance of entropic effects. While MD provides accuracy limited by the potential to capture thermal motion, the computational cost of the (quasi-)harmonic approximation (lattice dynamics) is far more feasible to use in a full scale crystal structure prediction.

We have implemented three lattice dynamic methods to determine what level of thermal expansion is required to compute the thermodynamic properties of small organic polymorphs in varying size and torsional degrees of freedom. The fastest method is the harmonic approximation (HA), which neglects thermal expansion but includes the entropy of harmonic vibrations, giving at least a crude estimate for polymorphic free energy differences. We also examine the quasiharmonic approximations, including both isotropic and anisotropic expansion, using new methodology to make the anisotropic expansion computationally feasible. Finally, we compare the relative free energies versus temperature curves generated with MD to compare all lattice dynamic models to.

For small and rigid molecules (1,4-Diiodobenzene and resorcinol), where the entropic difference between polymorphs is constant, we found that isotropic QHA was sufficient in determining polymorph free energy differences within error of MD out to 300K. However, for more flexible molecules (chlorpropamide and tolbutamide) anisotropic-QHA is required to correct for failures with an isotropic model. We hypothesize that failures between our anisotropic thermal expansion model and MD is due to anharmonic motions, especially in the case of disordered crystals.

487235 Prediction of Phase Transition Thermodynamics for Crystals of Pharmaceutical Compounds

Abstract:

Phase transition properties of pharmaceutical crystals are affecting their thermodynamic aqueous solubility, which is considered to be one of the major factors in determining the ultimate success or failure of drug compounds. Solubility is important because it restricts the maximal drug concentration in the human body upon taking the drug. In addition, solubility in organic solvents is a crucial property for both process and medicinal chemistry as well as for formulation design. Many computational models of aqueous solubility have been developed but the challenge of accurately predicting aqueous solubility for diverse molecules as a critical component of drug discovery and development remains in place.

Solubility prediction requires the knowledge of free energy of fusion ($\boldsymbol{\Delta G_{fus}}$) and free energy of mixing ($\boldsymbol{\Delta G_{mix}}$). Using equilibrium thermodynamics and chemical potentials, it is possible to predict $\boldsymbol{\Delta G_{mix}}$ accurately. Unfortunately, there is no rigorous way to predict $\boldsymbol{\Delta G_{fus}}$, which translates into inaccuracies in predictive models of solubility. Free energy of fusion is often cannot be measured experimentally too.

To enable accurate $\boldsymbol{\Delta G_{fus}}$ prediction we have combined accurate *ab initio* quantum-chemical calculation of crystals with conventional machine learning (ML) and Quantitative Structure Property Relationships (QSPR) approaches. Based on quantum-chemical calculations of all intermolecular interactions in the unit cell we construct the energy vector diagrams (EVD) of the crystal structure. The EVDs are used to encode intermolecular interactions, basic structural motif of the crystal, and the interactions between such motifs in the form of crystallographic fingerprints that are used as system descriptors in developing QSPR models.

Models have been developed for an aqueous solubility dataset consisting of approximately 40,000 experimental appoints for about 6500 organic molecules after thorough manual curation and standardization; this dataset was constructed from numerous publically available sources. We show that our QSPR models afford accurate prediction of $\boldsymbol{\Delta G_{fus}}$ with less than 1 kcal/mol unsigned error of prediction. In addition to free energy of fusion, we have modeled additional important endpoints such as enthalpy of sublimation, enthalpy of vaporization, heat capacity in liquid and solid phases.

491226 Crystal Engineering Applications of COSMO-RS

Abstract:

Since its first introduction in 1995 COSMO-RS theory has become a standard tool for the prediction of molecular thermodynamic properties in pure liquids and mixtures, mainly due to its efficient and accurate treatment of intermolecular interactions in solution.[1]

Although COSMO-RS originally is a liquid phase thermodynamic theory, there are meanwhile many applications which are of relevance for crystal engineering and drug development in general. In addition to solubility prediction and the screening for suitable solvents, a rather recent application is the computational screening of cocrystal forming compounds (coformers).

Here, interactions within the cocrystal are approximated by assuming a hypothetical subcooled stoichiometric mixture of API and coformer. The resulting mixing enthalpy is a highly useful quantity giving the propensity of that mixture to form a cocrystal.[2]

Hence it is now possible to efficiently pre-select from thousands of potential coformers in order to focus experimentally only on the most promising candidates for co-crystallization.

A concise overview of this approach will be given including its strengths and weaknesses with a subsequent illustration of the latest COSMO-RS applications in the field of crystal engineering such as the prediction of solvates. Another potential area of interest is the discovery of co-amorphous materials by means of COSMO-RS which has been reported recently.[3] A further straight-forward application of COSMO-RS is the computational exploration of ternary phase diagrams for API, coformer and solvent or solvent mixtures.

1. Klamt A. The COSMO and COSMO-RS solvation models. WIREs: Comput. Mol. Sci. 2011;1:699â€“709.

2. Abramov YA, Loschen C, Klamt A. Rational coformer or solvent selection for pharmaceutical cocrystallization or desolvation. J. Pharm. Sci. 2012;101:3687.

3. Corner PA, Harburn JJ, Steed JW, McCabe JF, Berry DJ. Stabilisation of an amorphous form of ROY through a predicted co-former interaction. Chem. Commun. 2016;52:6537â€“40.

488462 Enabling Rational Morphology Design and Crystal Engineering with Addict Software for Mechanistic Crystal Growth Models

Abstract:

Enabling rational morphology design and crystal engineering with ADDICT software for mechanistic crystal growth models

P. Zhu, J. Li, C. Tilbury, Y. Sun, K. Girard, Y. A. Abramov, M. F. Doherty

Mechanistic crystal growth models offer the potential for improved engineering of crystalline products. These methods adopt an accurate description of the underlying surface growth physics& solid-state chemistry, resulting in the ability to predict effects of the growth environment on crystal shapes and enabling a systematic sweep of the design space to discover optimal growth conditions. However, this increased fidelity comes at a price of, in general, requiring significant time invested in developing expertise and then routines that implement these methods. We have developed proof-of-concept software that delivers an automated implementation of the mechanistic spiral growth model for non-centrosymmetric molecules. This software, called ADDICT (Advanced Design and Development of Industrial Crystallization Technology), can make predictions for any neutral organic molecule; developments are underway to broaden its applicability to organic salts.

ADDICT provides two significant benefits. First, it requires little expertise to use, yet yields the power of mechanistic predictions and conveys increased ability for rational design of crystal products. Only the crystallography (predictions are for a specific polymorph) and a partial charge calculation are required, then ADDICT can predict the crystal shape at specified growth conditions; over 30 solvents are included within the program. ADDICT also contains a visualization tool to indicate hypotheticalcrystal shapes for the crystallography, which can guide selection of design conditions to modify relative growth rates appropriately and realize an optimal product. Second, ADDICTenables increased testing of mechanistic methods for a variety of crystal systems. This can guide model development; one of the key advantages of adopting a mechanistic approach is the ability to systematically improve the underlying engine driving shape predictions, which has guided concurrent research.

In this presentation, we provide an overview of our automation strategy and software capabilities. We demonstrate shape predictions for three systems: two complex organic aromatics[1,2] and a Pfizer pharmaceutical molecule, and then compare to experimental morphologies.

1: 1,2,3,4,6,7-Hexahydro-5H-benzo(f)cyclopenta(c)isoquinolin-5-one

CSD: ABACAI

Hongbin Li, Hua Yang, J.L. Petersen, K.K. Wang, J. Org. Chem. (2004), 69, 4500, doi:10.1021/jo049716t

2: 3-(Acetoxymethyl)-1,4-diphenylazetidin-2-one

CSD: ABAFAL

A. Basak, S.S. Bag, P.A. Mazumdar, V. Bertolasi, A.K. Das, J. Chem. Res. (2004), 318

503670 Digital Alchemy for Assembly Engineeringc

Abstract:

Back in the day, Stuart Churchill was a pioneer in the use of computers for modeling chemical and thermal processes. Since that time, the role of computing in chemical engineering has become indispensable. In this talk, we present new computational methods we call "digital alchemy" that allow for predictive modeling of thermodynamic assembly, and permit the design of building blocks optimized for assembly into target structures with desired properties. In this approach, standard thermodynamic ensembles are generalized to include alchemical variables and their thermodynamic conjugates, allowing for rapid, real-time computational investigation of large, complex design spaces.

503671 Improved Algebraic, Numerical, and Graphical Representations in Fluid Mechanics

Abstract:

The concept of equivalents is applied to fully developed flows through straight channels, and two applications are introduced: the equivalents that produce equal rates of flow through channels of different cross-section, and the diminishment that occurs between the regimes of laminar and turbulent flow. The validity of the eddy diffusivity and the invalidity of the mixing-length as measures of the shear stress due to turbulence have perhaps never before been demonstrated so clearly. The possible choices of primary variables and the consequences are also examined. The objective of these explorations is an improved understanding of the basic fluid mechanics by teachers, students, and industrial practitioners.

503672 The Scaling of Turbulence Near the Wall and the Churchill Turbulent Flux Correlation: Insights with Lagrangian Simulations

Abstract:

The evaluation of mean velocity and mean temperature in wall-bounded turbulent flows has been based on the "classical" theory of turbulence and the law of the wall. The viscous wall parameters, namely the friction velocity, the friction temperature, and the kinematic viscosity of the fluid have been used to scale not only the mean values of velocity, but also other mean turbulent quantities close to the wall, including the Reynolds stresses and the turbulent velocity and temperature fluctuations. Stuart Churchill, as he questioned the validity of the Reynolds analogy between momentum and heat or mass transfer in his works [1], he realized that extending the scaling of momentum to the scaling of scalar quantities in turbulence was not a sound approach. He proposed a scaling of turbulence that depends on the turbulent momentum flux and the turbulent heat flux. In his AIChE Institute Lecture (in Miami, 1998), Churchill presented a summary of his work on this topic, and closed form solutions for the mean velocity and temperature [2]. The turbulent Prandtl number, the most significant parameter for the modeling of turbulent transport not only in theory but also in modern CFD commercial software packages, was also interpreted physically based on these ideas [3]. We have used direct numerical simulation results and results from Lagrangian simulations [4] for the transport of passive scalars in turbulent channel flows to explore Churchill's ideas on turbulent flow scaling and to develop an updated theory for the scaling of heat transport in turbulent flows [5-7]. His passion to remove empiricisms and approximations from the development of predictive correlations in turbulent transport has been an inspiration. In this talk, we will revisit the Churchill scaling in view of our simulation results, and will discuss the counterintuitive finding that wall turbulence can be used to separate rather than mix substances under special circumstances [8].

1. Churchill, S. W., 1996, "A Critique of Predictive and Correlative Models for Turbulent Flow and Convection," Ind. Eng. Chem. Res., 35, pp. 3122–3140.
2. Churchill, S.W., 2000, "Progress in the Thermal Sciences: AICHE Institute Lecture," AIChE Journal, 46(9), pp. 1704-1722.
3. Churchill, S. W., 2002, "A reinterpretation of the Turbulent Prandtl Number," Ind. Eng. Chem. Res., 41, pp. 6393–6401.
4. Papavassiliou, D.V, 2002, "Turbulent transport from continuous sources at the wall of a channel," Int. J. Heat Mass Transfer, 45(17), pp. 3571-3583.
5. Le, P.M., and Papavassiliou, D.V., 2008, "On the scaling of heat transfer using thermal flux gradients for fully developed turbulent channel and Couette flows," Int. Commun. Heat Mass Transfer, 35(4), pp. 404-412.
6. Srinivasan, C., and Papavassiliou, D. V., 2011, "Prediction of Turbulent Prandtl Number in Wall Flows With Lagrangian Simulations," Ind. Eng. Chem. Res., 50(15), pp. 8881–8891.
7. Srinivasan, C., and Papavassiliou, D.V., 2013, "Heat transfer scaling for wall bounded turbulent flows," Applied Mechanics Review 65(3), Art. 031002.

8. Nguyen, Q., Srinivasan, C., and Papavassiliou, D.V., 2015, "Flow induced separation in wall turbulence", Phys Rev E, 91, Art. 033019.

503674 Flow Boiling Using a Piranha Pin Fin Heat Sink

Abstract:

Flow Boiling using a Piranha Pin Fin Heat Sink

Corey Woodcock, Xiangfei Yu, Yoav Peles[1] and Joel Plawsky

Rensselaer Polytechnic Institute; [1]University of Central Florida

The *Piranha Pin Fin* (*PPF*) microdevice is a heat exchanger engineered to dissipate extreme thermal loads using either dielectric fluids such as HFE7000 or water. Fabrication of the device is CMOS compatible and the first generations of the device have been fashioned from silicon substrates. The PPF operates on the principle that heat transfer would be more efficient if heated fluids could be removed nearly as fast as they are created. This process would keep the driving force as high as possible. Thus the heat exchanger is built as two layers; a primary substrate that dissipates the bulk of the heat and shunts the heated fluids to a booster section that continues to heat the fluid and extends the total surface area of the device. Three generations of devices were fabricated. The first two generations were single-layer systems designed to help understand the operation of the primary layer and optimize the shape and location of the pin fins. The third generation was the first, two-layer system to be fabricated and tested.

Figure 1 Representative PPF array

Correlations for the heat transfer coefficient, h_{tp}, based on the Boiling Number and the Weber Number in the PPF design was developed and validated with Gen II. The discrepancy of heat transfer coefficient between predicted and measured values was within 20%. The ratio of two-phase to single-phase pressure drop was correlated with the Boiling Number (Bo) and the Jakob Number (Ja). The discrepancy of the two-phase multiplier between prediction and experimental measurements were within 12%.

The third generation *PPF* heat sinks were the first series to provide both a primary substrate populated with *PPF*s, as well as a booster substrate populated with solid micro pin fins. The substrates are micro-aligned and fusion-bonded to provide a full three dimensional (3D) architecture with bulk silicon properties. Experimental results from the Gen III *PPF* microdevices provide a proof of concept of the true *PPF* heat sink concept with fluid venting into multiple substrates. At low flowrates (G = 670 [kg/s/m^2]) the Gen III *PPF* heat sink has been shown to dissipate heat loads approaching 1.1 [kW/cm^2] at an exit quality of 95%, while at higher flowrates (G = 3,300 [kg/s/m^2]) the thermal limit of the heat sink remains unknown due to exceeding the available power supply at >1.4 [kW/cm^2].

503673 Transport Problems in the Spirit of Stuart Churchill for Teaching and Research at the University of Michigan

Abstract:

Stuart Churchillâ€™s solutions to important heat transport problems have greatly influence both pedagogy and research. The correlation of Churchill and Chu for the natural convection heat transfer to or from a vertical plate, over an astonishing range of Rayleigh numbers is based partly on a similarity solution of a type common in heat, mass, and momentum transfer for problems in chemical engineering. One such problem is asphaltene or wax transport and deposition onto the wall of an oil pipeline, which can be solved using similar methods, as will be discussed. In addition, such problems, and the wealth of their applications make for excellent motivations and examples in the teaching of graduate or undergraduate transport, as will be discussed in the talk.

500209 A Compartmentalized, Numerical Model of Continuous Blending

Abstract:

Here we present an empirically-derived, numerical model of a continuous blender capable of describing both the mass flow rate and residence time distribution of a given powder as a function of processing conditions. The approach discretizes the blending process in both time and space, describing a stepwise progression of powder through segmented portions of the equipment, which are defined primarily by the geometry of the mixing blades. The model defines three distinct blending regimes within the process: A conveying mode, where powder is rapidly transported from the blender inlet to outlet; a mixing mode, where intermixing of powder between nearby blender segments is more completely described; and a transition regime between the two. Further, we show that coupling the model with a given materials system on a specific blender provides an approach to identifying appropriate process ranges in the context of desired blending regimes. Moreover, the model is shown to serve as a useful analytical tool for defining design spaces in the context of QbD, affording greater mechanistic insights into blending behavior than a standard linear regression analysis of an orthogonal design of experiments.

491186 Impact of Raw Material and Blend Properties on the Screw Feeding, Continuous Blending and Tableting Unit Operation of an Integrated Continuous Direct Compression Platform

Abstract:

Continuous manufacturing will likely change the way how drug products are developed and commercialized. Although continuous processes simplify the development process by removing the need of multiple unit operations and scale up, they also bring in new challenges. Only limited resources are available early in the development stage and hence the formulation and process needs to be developed in a short time frame with only a limited quantity of active ingredient. This is particularly challenging for continuous processes because most of the equipment is designed at production scale. Drug product development for continuous processes requires therefore innovative strategies and simulation to maximize the information with a minimal amount of active. Understanding and modeling the impact of material and blend properties at each unit operation is therefore the key step to develop drug products of high quality with minimal resources.

Direct compression is by far the simplest and most cost efficient manufacturing route for oral solid dosage forms. Although direct compression is an inherently continuous technique, simple unit operations preceding tableting (i.e. weighing and blending) are historically performed batch wise. To enable continuous direct compression, the integration of continuous powder feeding units, continuous dry powder mixers and a tablet press is required.

This work aims to leverage raw material and blend properties with required settings for optimal feeding, blending and tableting in an integrated continuous direct compression platform. First, the raw material properties of a wide range of fillers for direct compression and active pharmaceutical ingredients (API) were determined. The characterization included determination of particle size and shape (laser diffraction, static image analysis, scanning electron microscopy), specific surface area, tapped, bulk and true density, flowability (drained angle of repose, dynamic angle of repose, flow function coefficient, basic flowability energy, specific energy, aeration index, permeability), compressibility, charge density, moisture content (loss-on-drying) and dynamic vapor sorption. Subsequently, principal component analysis was applied on the API and filler dataset. This allowed to select 6 APIs (acetaminophen micronized, powder and dense power, caffeine powder, metoprolol micronized and etravirine spray dried) and 5 DC fillers (Avicel PH 101 and 200, Emcompress, Pearlitol 100SD, Tablettose 80) with unique properties. These were combined together with magnesium stearate in 30 tertiary mixtures that were also characterized for their properties.

Continuous direct compression trials were performed with these formulations to evaluate the impact of blending (impeller speed and configuration), die filling (paddle wheel speed) and compression (main compression force) parameters on responses of the process (fill level, mean residence time and number of blade passes) and product (blend and content uniformity, tablet weight variability and tensile strength).

The hopper refill performance could be related to the flowability of the powder. Materials with a flow function coefficient below 2 (i.e. no powder flow) gradually reduced the apparent volume of the refill valve over time. At the feeding stage, the amount of material dispensed per screw revolution was related to the bulk and tapped density whereas the variability on the feeder flow rate was correlated with the flow function coefficient of the material. At the first or intensive mixing stage, the fluidization state of the powder bed depended on the process settings. The residence mass in the first continuous mixer was positively correlated with the bulk and tapped density of the blend when the mixer operated in the densely stirred powder bed flow regime. Additionally, more permeable blends resulted in lower residence masses as these blends fluidized more easily. At the second blending or lubrication stage, all blends were completely fluidized and as a result this response was not correlated with material properties. At the die filling stage of the tableting unit operation, the residence mass in the feed frame of the tablet press was a strong function of the bulk density of the blend (R^2=0.97) as the powder bed was in a densely stirred regime during processing. In contrast, the fill depth required for achieving target tablet weight (175 mg) was negatively correlated with the bulk density (R^2=-0.90). In contrast, tablet weight variability depended on the compressibility (R^2=0.68), unconfined yield strength (R^2=0.72) and flow function coefficient (R^2=0.56) of the blends. Overall, blends characterized by a low density and poor flowability yielded the highest tablet weight variability at low paddle speed.

The tabletability was strongly dependent on the selected filler. The cellulose based fillers resulted overall in tablets with the highest tensile strength. Additionally, Avicel PH101 showed better tabletability compared to PH200 driven by the higher bonding area of the smaller particles. However, the cellulose based fillers exhibited a high lubricant sensitivity. As the cellulose particles were subjected to a higher number of paddle wheel revolutions, the tensile strength decreased with 45% for some blends. Pearlitol 100SD also resulted in tablets of good strength and showed only a limited lubrication sensitivity. However, high ejection forces were observed with pearlitol 100 SD and higher levels of magnesium stearate (1.25 % w/w) were required to limit the ejection force during tableting. Formulations consisting of Tablettose 80 or Emcompress fillers with the elastically deforming acetaminophens resulted in relatively weak tablets.

The database will be further completed with blend and content uniformity data. Subsequently, advanced multivariate models such as T-PLS will be applied to link the API and Filler properties to the process parameters and all responses of the continuous direct compression process. These models are valuable tools to select excipient properties as a function of API properties during drug product development for continuous direct compression processes.

493141 Feeder Characterization and Model Development Accounting for Incoming Material Properties

Abstract:

In previous years, research focus has been placed in characterizing feeder performance and tooling selection using a formulation based approach [1-3]. This approach consists on studying feeder performance for a specific ingredient (e.g., excipient, lubricant, or API) based on its percentage in a formulation and a desired process flow rate [4, 5]. Although efficient for the early stages of continuous manufacturing, this approach does not provide process developers with sufficient information to ensure process performance on instances where process throughput is changed and/or material properties change from the ones measured during characterization. The formulation based approach also does not cut down significantly the amount of work needed to design new processes for new formulations. To address these issues a novel approach based on the characterization of feeders for a broad range of materials and flow rates (i.e., independent of formulation requirements) was performed by our group. The goal of this work is to develop models and tools that predict flow rate variability and feeding capacity over a broad range of ingredient material properties, which serve as inputs to the model.

In this presentation, we describe a characterization procedure applied to a GEA compact feeder using a series of materials relevant to the pharmaceutical industry. Our experimental plan involved flow rate characterization and Residence Time Distribution (RTD) characterization. Feeder flow rate characterization is performed by measuring the accuracy and variability of material dispensing as a function of volumetric throughput and feeder hopper fill level. Twenty (20) different materials were run in a GEA compact feeder. The materials span a broad range of material properties, as shown by study of our internal material property database using Principal Component Analysis (PCA). Eight (8) feeder configurations based on the screw size and gear box configuration were studied. At each feeder configuration, flow rate experiments were carried out in both volumetric and gravimetric flow. The objective was to understand the effect of gear box, screw size, and material properties on the outgoing flow of the feeder (i.e., feed factor deviation, mean flow rate, and standard deviation) to develop a model capable of predicting the behavior. The flow rate model, the flow rate data analysis, the procedure for calibration, and the overall flow rate model will be introduced in this presentation.

The RTD of the feeder was characterized for a variety of materials. Feeder RTD is important for two main reasons: (1) traceability of raw materials and (2) characterization of the dynamic response time to changes in incoming material properties. In both cases, the goal is to understand how long it takes for a new material (e.g., lot added due to refill or lots with new material properties) introduced into the feeder to "wash out" the material that was already in the feeder. The RTD for the GEA feeder was characterized using four (4) different materials (i.e., tracer-base combinations) and two (2) different flow rates. The aim was to understand if material properties had an impact on the RTD profile (i.e., mixing behavior) inside of the feeder. The results from the RTD profile were used to model the mixing behavior inside of the feeder using a tanks-in-series model with side compartments. Regressed coefficients were used to compare the mixing behavior between the materials and then correlated to those materials' properties.

References

1. Boukouvala, F., et al., *An integrated approach for dynamic flowsheet modeling and sensitivity analysis of a continuous tablet manufacturing process.* Computers & Chemical Engineering, 2012. **42**: p. 30-47.

2. Toebermann, J., .-C., Rosenkranz, J., Werther, J., Gruhn, G., *Block-oriented process simulation of solids processes.* Computers & Chemical Engineering, 2000. **23**: p. 1773-1782.

3. Hartage, E.U., Pogodda, M., Reimers, C., Schwier, D., Gruhn, G., Werther J., *Flowsheet simulation of solids processes.* KONA, 2006. **24**: p. 146-156.

4. Engisch, W.E. and F.J. Muzzio, *Method for characterization of loss-in-weight feeder equipment.* Powder Technology, 2012. **228**: p. 395-403.

5. Yang, S. and J.R.G. Evans, *Metering and dispensing of powder; the quest for new solid freeforming techniques.* Powder Technology, 2007. **178**: p. 56-72.

494769 Disintegration and Dissolution Modeling for Accelerated Drug Product Process Development

Abstract:

We present a methodology to streamline tableting process development complemented by a mathematical model of disintegration and dissolution. In a state-of-the-art tableting process development the main goal is to achieve an optimal tradeoff between mechanical strength, disintegration time, friability, dissolution profile, tablet weight uniformity and any other performance variable of interest. Excluding dissolution testing, the assessment of tensile strength, disintegration time, friability and tablet weight uniformity is based on simple and fast tests which can be performed during the tablet manufacturing and the results can be promptly used for the process optimization. In contrast, the dissolution assessment is laborious, time consuming and can only be reliably performed in analytical laboratories. Therefore, taking into account the dissolution profile in the optimization loop largely increases the development cycle-time which is often not feasible at early process development stages. However, this is particularly important when formulating amorphous drug product intermediates to optimize the release rate and supersaturation maintenance which are paramount for drug performance [1].

In the proposed methodology the early assessment of dissolution profile is performed with a first principles mathematical model which takes into account intrinsic material properties such as drug diffusivity in the polymer matrix and in the dissolution media [2], the polymer swelling/erosion rate [3] and whether the API is crystalline or is in an amorphous solid dispersion. The disintegration dynamics are handled with a Population Balance Equation [4] and are coupled to the dissolution model. Whenever direct assessment of intrinsic material properties is not possible, these parameters are obtained via curve-fitting the model to experimental dissolution data. The curve-fitting is performed using a particle swarm optimization algorithm, which is suitable for multivariate and multi-objective problems.

We present a case-study on the application of the methodology to the downstream process development of an amorphous solid dispersion which illustrates the advantages of complementing laboratory development with a computational tool in terms of process and drug performance without hindering the development cycle-time.

Figure 1: Dissolution profile of a tablet formulated from an amorphous solid dispersion intermediate obtained both experimentally and from the presented mathematical model.

References

[1] Henriques, J., Valente, P. & Winters, C. "Formulating Amorphous Solid Dispersions: Bridging Particle Engineering and Formulation", American Pharmaceutical Review, Nov/Dec. 2016.

[2] Wang, Y., Abrahamsson, B., Lindfors, L. & Brasseur, J.G. "Comparison and Analysis of Theoretical Models for Diffusion-Controlled Dissolution", Mol. Pharmaceutics, 9, 1052-1066, 2012.

[3] Siepmann, J., Kranz, H., Peppas, N. A., & Bodmeier, R. "Calculation of the required size and shape of hydroxypropyl methylcellulose matrices to achieve desired drug release profiles", Int. J. of Pharm., 201(2), 151–164, 2000.

[4] Wilson, D., Wren, S. & Reynolds G., "Linking Dissolution to Disintegration in Immediate Release

Tablets Using Image Analysis and a Population Balance Modelling Approach", Pharm. Res., 29:198-208, 2012.

490283 Modeling of Flow and Drying of Aqueous Polymer Coatings on Porous Pharmaceutical Tablets

Abstract:

The application of polymer coating films on the surface of tablet cores is crucial in the manufacture of solid dosage drugs. Understanding the coating film behaviour at the point of contact between the liquid solution and the porous tablet core can reduce the number of defective tablets being discarded and allow a better prediction of the optimal operating conditions for the coating process. In our work, to estimate the coating film attributes and uniformity, we developed a new mathematical model that defines and describes the spreading, absorption and evaporation of polymer solutions onto and from a tablet.

Several researchers (Weidner et al., 1995; Schwarz et al., 2001) have considered polymer coating film drying but have not accounted for changes in coating density owing to solvent evaporation. In our work, we considered that the film density increases as the solvent (water) evaporates from the surface of the tablet. The influence of solvent evaporation on important coating spreading parameters, such as the viscosity of the aqueous solution and the diffusion coefficient of the polymer in the solvent, was also investigated.

To better describe the behaviour of the film on the surface of the tablet we divided the overall process into three stages. For the first stage, we considered the coating solution impacting on the tablet surface and then spreading on the surface and absorbing into the tablet. We modelled the coating solution as a liquid of changing density and viscosity due to solvent (water) evaporation. The absorption during the first stage is affected by polymer particle retention in the tablet pores, the volume fraction of the latter not being necessarily uniform in the tablet. We assumed that the viscosity of the coating solution becomes infinite when the mass fraction of the polymer reaches a certain value (Weidner et al., 1995). At this point, the film spreading and absorption are hindered and the second stage commences. During the latter, we assumed that only solvent evaporation takes place from the tablet surface. This stage lasts until the polymer concentration reaches a new critical value, at which point a polymer particle crust develops at the surface of the coating film (Kadja and Bergeles, 2003). In the final stage, we considered the evaporation of the solvent (water) from beneath the coating crust (i.e., tablet porous core).

We derived an expression for the movement of the absorbed coating fluid inside the porous tablet based on a kinematic boundary condition of the free wetting front surface (Alleborn and Raszillier, 2004). The pressure and the vertical and radial velocities inside the porous substrate were calculated from the Laplace and Darcy equations, respectively. We also took into account the effect of the tablet porosity and permeability on the polymer particle retention.

We implemented our novel model in gPROMS, employing the Modelbuilder modelling platform (Process Systems Enterprise Ltd., 2015). The model estimated the water and polymer content inside the porous tablet core and the dry coating film position and height. The numerical results were in good agreement with experimental data taken from the literature (Schwartz et al. 2001; Möltgen et al. 2012). In future work, our approach could be extended to also predict the spreading of non-Newtonian coating liquids employed by the pharmaceutical industry as there are many different models that propose relationships between viscosity, stress and strain rate.

References:

Alleborn, N. and Raszillier, H. (2004) "Spreading and sorption of a droplet on a porous substrate", Chemical Engineering Science, 59(10), pp. 2071-2088.

Kadja, M., Bergeles, G., (2003) "Modelling of slurry droplet drying", Applied Thermal Engineering, 23(7) pp. 829-844.

Möltgen, C., Puchert, T., Menezes, J.C., Lochmann, D. and Reich, G. (2012) 'A novel in-line NIR spectroscopy application for the monitoring of tablet film coating in an industrial scale process', Talanta, 92, pp. 26–37.

Process Systems Enterprise Ltd. gPROMS User Guide. London, UK, December 2015.

Schwartz, L.W., Roy, R.V., Eley, R.R. and Petrash, S. (2001) "Dewetting patterns in a drying liquid film", Journal of Colloid and Interface Science, 234(2), pp. 363-374.

Weidner, D.E., Schwartz, L.W. and Eley, R.R. (1996) "Role of surface tension gradients in correcting coating defects in corners", Journal of Colloid and Interface Science, 179(1), pp. 66-75.

17AIChE Dr.Reddy's

Modeling and simulation efforts in scale up and characterisation of semi-solid dosage forms

AICHE Annual Meeting 2017

Saurav S Rath, Birendra K David and Ravi C Palaparthi
Process Engineering, **Dr. Reddy's Laboratories, Hyderabad**

17AIChE Dr.Reddy's

Executive summary **Generic drug development**

Challenges faced by generic pharmaceutical industry

- ✓ Tight timelines of product development
- ✓ Market constraints
- ✓ Manufacturing capabilities
- ✓ Regulatory restrictions

Getting it **right the first time** demands detailed product & process understanding and implementing appropriate mechanistic modeling and Quality-by-design (QbD) approaches in product life cycle, to minimize downstream risks.

This work provides case studies in **topical dosage products**.

Introduction

Typical process flow and composition

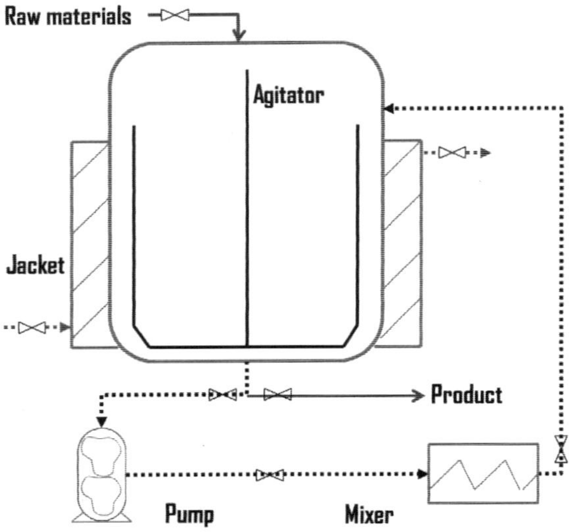

Raw materials	
Aqueous phase	**Oil phase**
Water, surfactant, suspending agent	Wax, fatty alcohol, anti-oxidant, **drug**

Unit operations

Heating, mixing, homogenization, cooling

Formulation

Emulsion, solution, suspension

WHY mechanistic modeling?

Mapping of CPPs, CMAs and CQAs

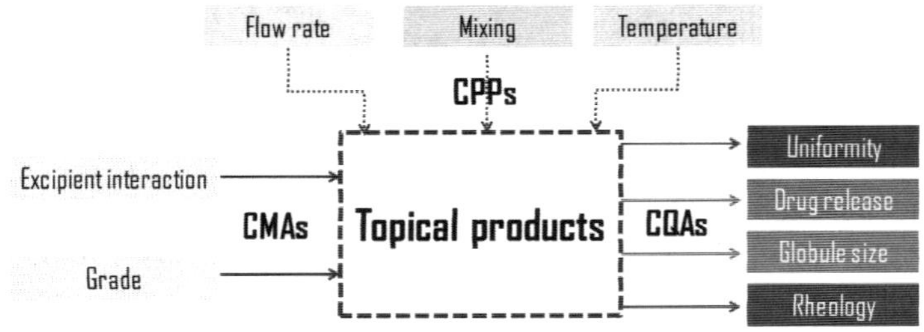

In practical sense, these QbD elements are inter-linked. For example in emulsion (O/W) systems, the non-Newtonian rheology varies not only with CPPs and CMAs but also is an implicit function of globule size.

Complex relation

Excipients + Process + Rheology → Globule size

This calls for various mechanistic models, to help predict the product behaviour.

Scale-up heuristics — Modeling and simulation capability

17AIChE Dr.Reddy's

Prediction of globule size distribution – Population balance modeling (PBM)

Drawbacks of general trend

X Can't predict the distribution (d_{10}, d_{50} and d_{90})

X Unsuitable for both internally and externally circulated mixers

Highlights of PBM

✓ Valid for both Newtonian and non-Newtonian fluid based systems

✓ Working for both laminar or turbulent flow regimes

✓ Can be used for all types of available high shear homogenizers

✓ What-if studies

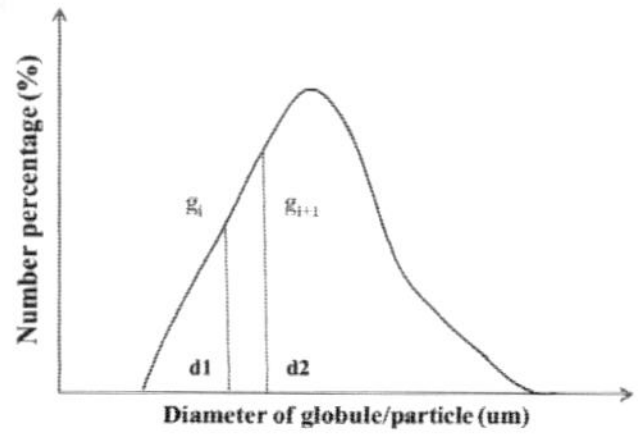

Number percentage distribution

Number of drops in [v,v+dv] per unit volume — *Breakage rate* — *Probability of forming daughter droplet (v) from mother (v')*

$$\frac{\partial n(v,t)}{\partial t} = -\,g(v)n(v,t) + \int_0^\infty \beta(v,v')g(v')n(v',t)dv'$$

$$^{1}g(d_p) = c_{1,b}\frac{\epsilon^{1/3}}{d_p^{2/3}}\exp\left(-\frac{c_{2,b}\gamma}{\rho_d\,\epsilon^{2/3}d_p^{5/3}}\right), where\ \ \varepsilon = [\mu\dot{\gamma}^2 v + Po\,(QN^2D^2\rho_c)]/V$$

[1] *Determination of breakage rates using single drop experiments, Matthias Kraume*

Case study – I — Predicting globule size distribution

17AIChE Dr.Reddy's

System – O/W emulsion

Inputs of PBM

✓ Shear and velocity profile in tank (CFD)

✓ Premix concentration (Microscopy or any relevant equipment)

✓ Geometry and process parameters

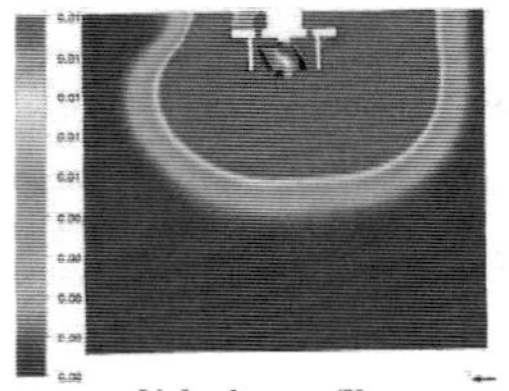

Velocity profile

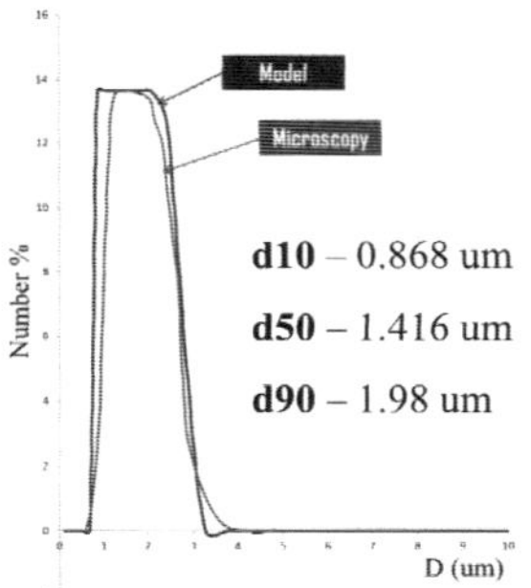

Model vs Microscopy data (Number %)

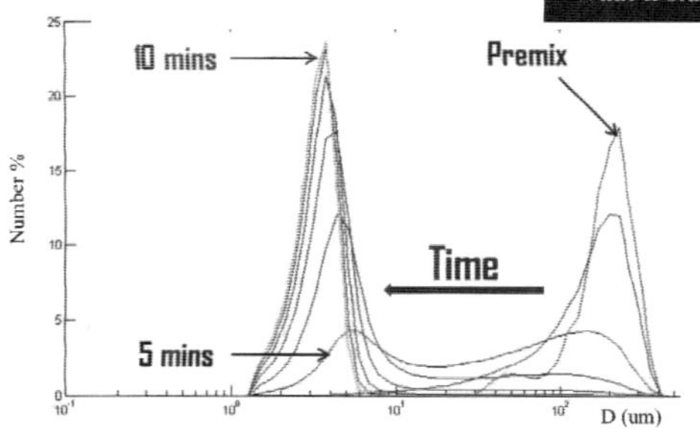

Shift of number % distribution with time

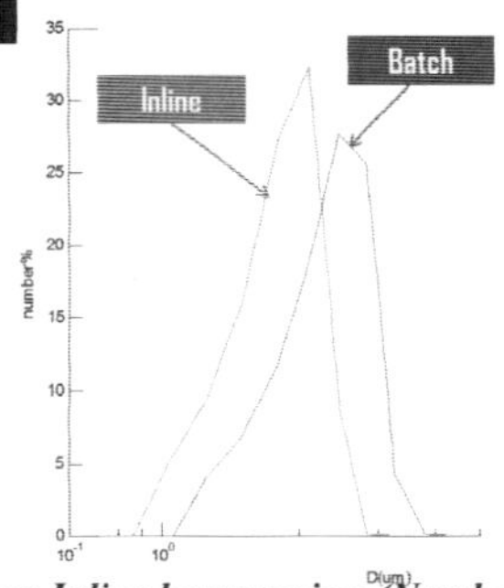

Batch vs Inline homogenizer (Number %)

Scale-up heuristics

Modeling and simulation capability

17AIChE Dr.Reddy's

Shear Strain – Scale-up correlation for homogenization

Drawback of general trend

X Doesn't take into account the effect of viscosity on other CQAs (GSD)

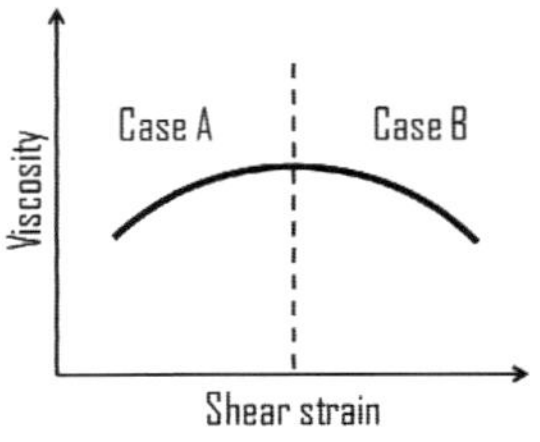

Overview of shear strain correlation

Case A: Shear < Shear $_{critical}$

With increase in shear strain, the apparent viscosity increases because of decrease in globule size. Order of (increase in viscosity due to increasing shear) >> Order of (decrease in viscosity due to thixotropic behaviour)

Case B: Shear > Shear $_{critical}$

With increase in shear strain, the globule size seizes to reduce and it reaches a plateau[2]. Increase RPM beyond max point → More shear imparted to the system →Viscosity reduces

Shear rate

$$Shear\ strain = \beta.\left(\frac{N.D}{\delta}\right).t, \text{where}$$

D	m	Rotor diameter
δ	m	Rotor-stator gap
N	s^{-1}	Homogenizer speed
t	s	Homogenizer time

[2]Deformation and breakup of drops in simple shear flows. Bruijn, de, R.A. and High shear mixers: A review of typical applications Jinli Zhang

Case study – II

Estimating shear strain

17AIChE Dr.Reddy's

CQA [Viscosity ~ 20,000 cP (measured at 2/s) and average globule size ~ 3.9 – 4 um]

Given the CQA, the required shear ~ β x 110 (**Point A**). Using the **algorithm**, homogenizer speed and time are calculated.

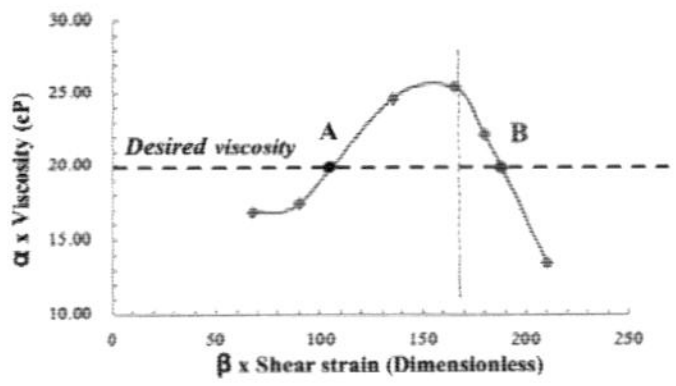

Variation of viscosity with shear

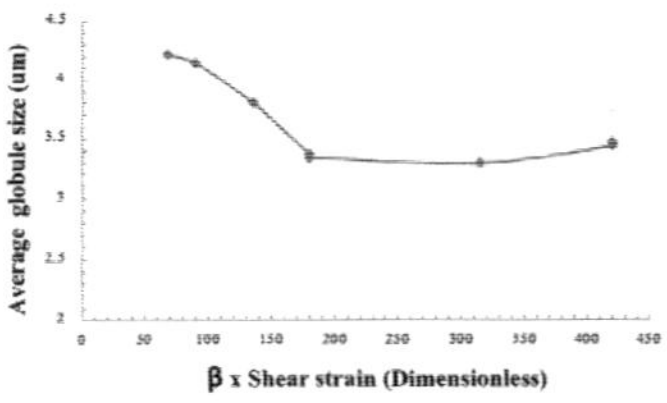

Variation of globule size with shear

Case Study – III

Mapping of shear and equipment

For shear sensitive products, it is imperative to know the possibilities by which shear can be applied to the system.

- ✓ Homogenizer (Size reduction and homogeneity)
- ✓ Lobe pump (Recirculation) ~ **5-10 % of homogenizer**
- ✓ Impeller (Uniformity of bulk) ~ 0.05-3 % of homogenizer
- ✓ Others – 0.01-0.05 % of homogenizer

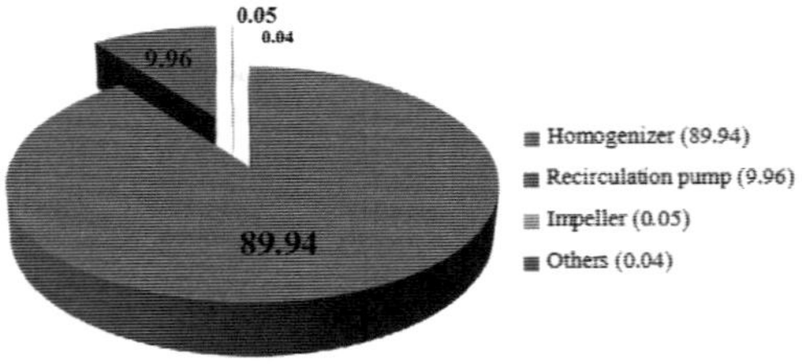

Shear imparted by various equipment for one such emulsion

Scale-up heuristics — Modeling and simulation capability

17AIChE Dr.Reddy's

Algorithm for calculation of homogenizer parameters

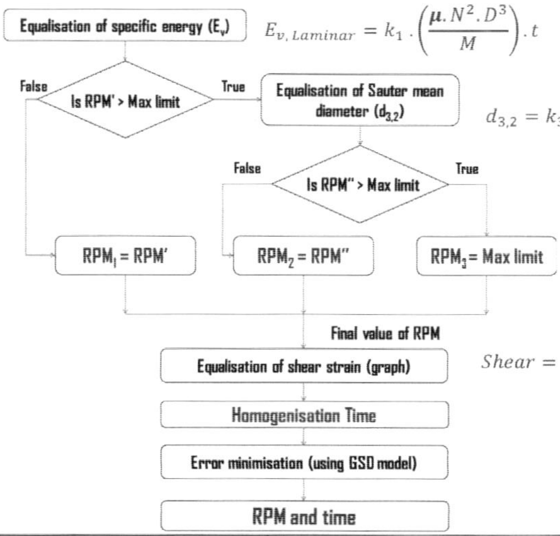

$$E_{v,Laminar} = k_1 \cdot \left(\frac{\mu . N^2 . D^3}{M}\right) . t \qquad E_{v,Turbulent} = k_2 \cdot \left(\frac{N^3 . D^5}{M}\right) . t$$

$$d_{3,2} = k_3 \cdot D . \left(\frac{\rho . N^2 . D^3}{\sigma}\right)^{-3/5}$$

Weber number

$$Shear = \beta . \left(\frac{N.D}{\delta}\right) . t$$

Glossary

D	(m)	Rotor diameter
δ	(m)	Rotor-stator gap
N	(s^{-1})	Homogenizer speed
t	(s)	Homogenizer time
ρ	(kg/m^3)	Density of phase
μ	(Pa.s)	Viscosity of phase
σ	(N/m)	Surface tension
M	(Kg)	Mass of phase
Q	(Kg/s)	Flow rate of phase
β	-	Correction factor

Case study- IV — Effect of cooling rate on product CQA

17AIChE Dr.Reddy's

System – O/W emulsion

Cooling of bulk can be controlled by modifying the following:

- ✓ Flow rate of coolant fluid
- ✓ Temperature of coolant fluid
- ✓ Mixing parameters

IV (b) Cooling time may not directly affect product viscosity.

IV (a) Fast cooling → Higher viscosity

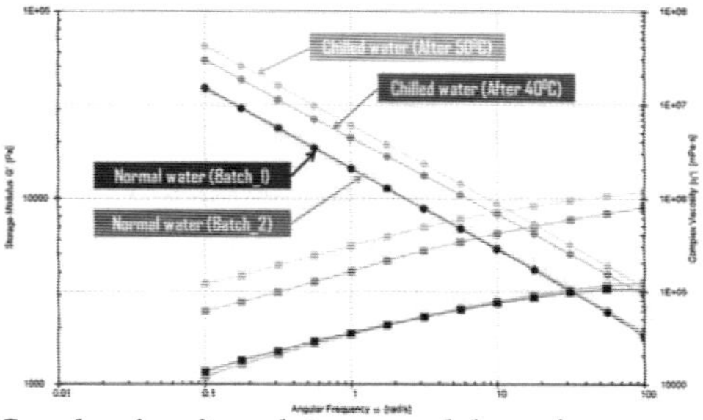

Complex viscosity and storage modulus vs frequency

Cooling rate	Average viscosity (cP) at $2s^{-1}$	
	Lab	Plant
Fast	82,312	100,000
Slow	81,800	69,000

The lab-scale data shows that cooling rate doesn't impact viscosity. In plant scale, the shear due to extra recirculation (>40 mins) resulted in drop of viscosity (shear-viscosity relation).

It is critical to optimise shear and cooling rate to attain desired CQA.

Scale-up heuristics — Modeling and simulation capability

17AIChE Dr.Reddy's

Algorithm for calculation of cooling time

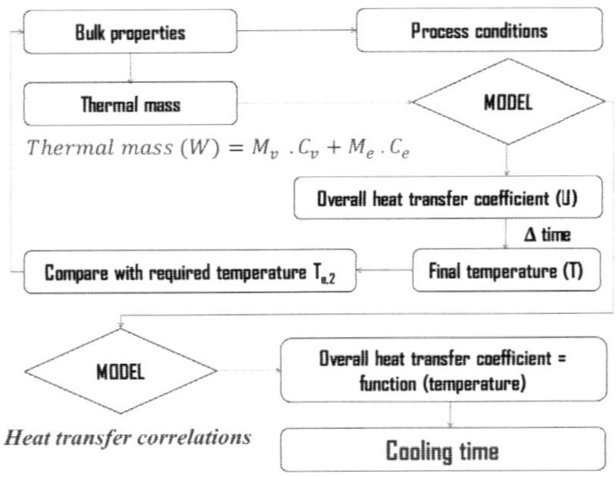

$$Cooling\ time\ (t) = W\ . \frac{exp\left(\frac{u.a}{m.C}\right)}{m.C.\left(exp\left(\frac{u.a}{m.C}\right)-1\right)}\ .\ ln\left(\frac{T_{e,1}-T_{w,1}}{T_{e,2}-T_{w,2}}\right)$$

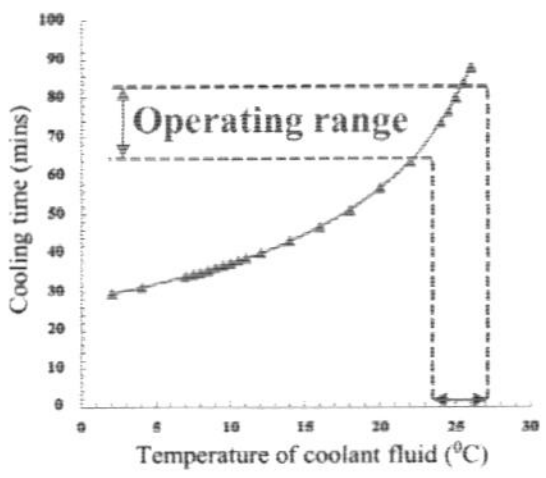

Time vs inlet fluid temperature

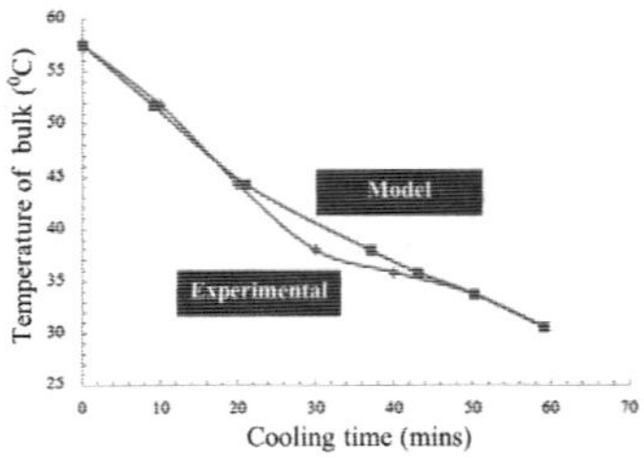

Model vs experiment (cooling time)

Glossary

u	W/(m^2.K)	Overall heat transfer coefficients
a	m^2	Surface area
m	Kg/s	Flow rate of coolant
$T_{e,1}/T_{e,2}$	K	Initial/Final temperature of bulk
$T_{w1}/T_{w,2}$	K	Inlet/Outlet temperature of coolant
M_e/M_v	Kg	Mass of bulk/Vessel
C_e/C_v	J/Kg.K	Specific heat capacity of bulk/Vessel

Scale-up heuristics — Modeling and simulation capability

17AIChE Dr.Reddy's

Correlations for mixing parameters – Effect on bulk uniformity

The mapping of right equipment (impeller, baffles and geometry) and process parameters is crucial for attaining homogeneity (*reducing dead zones*) in bulk.

- ✓ Minimum suspension velocity [N_{js}] – Suspensions
- ✓ Circulation velocity and solubility/miscibility time – Solutions

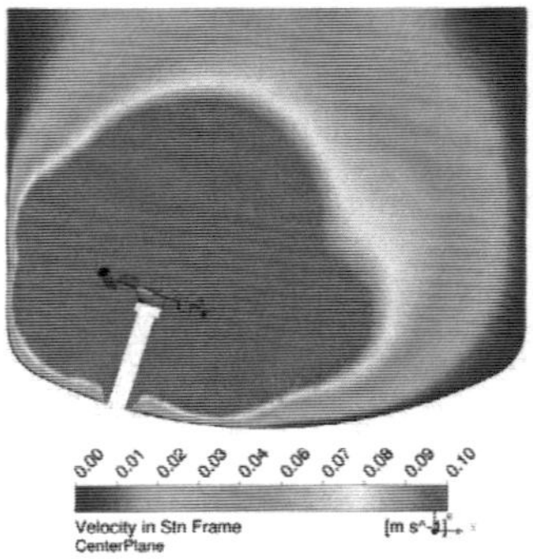

CFD analysis of mixing tank

Case study- V — Defining rate limiting step in mixing

$Circulation\ velocity\ (V_s) = N_q.N.D^3/A$

$Circulation\ time\ (t) \sim 5L/V_s$

Characteristic length

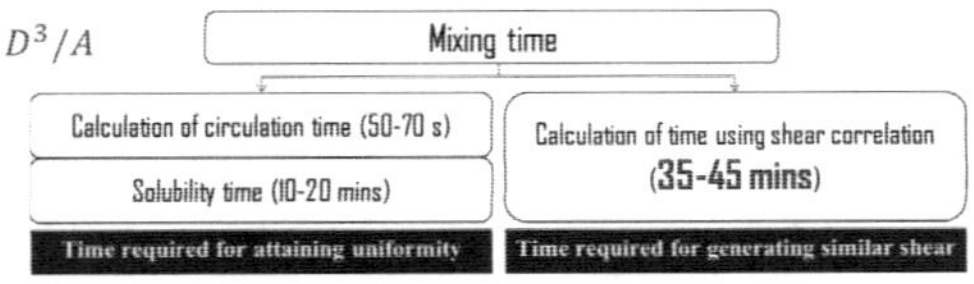

Glossary

N	s^{-1}	Impeller speed
N_q	-	Flow number
D	m	Impeller diameter
A	m^2	Surface area
L	m	Characteristic length

Mixing parameters	200 kg	1000 kg	Correlations
Speed (RPM)	500	700-800	Power/volume
Time (mins)	15	40	Shear
Viscosity	**55,000**	57,000	Viscometer (at 2/s)

System properties

Ointment (Wax base with soluble API)

Density = 990 Kg/m^3

k – 191.35 and n – 0.3 (Shear thinning)

Conclusion

Generic drug development

17AIChE Dr.Reddy's

This work focusses on models obtained from computational fluid dynamics (CFD) coupled with population balance modeling (PBM) and constitutive models (like shear, energy density).

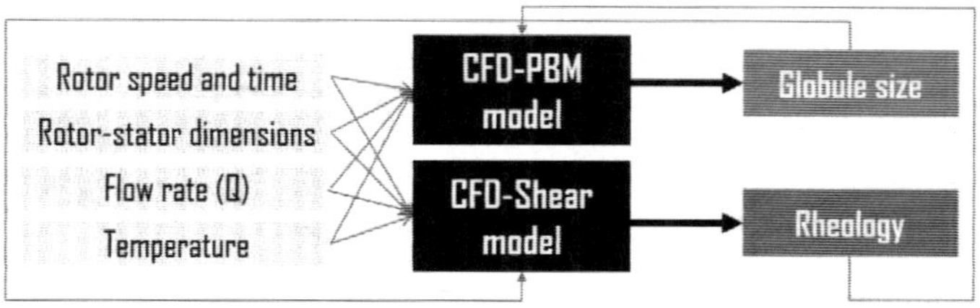

Defining, and scaling up the right manufacturing process with a given set of ingredients, to achieve the right product characteristics presents as a challenge to the process engineer.

Thank You

Back up slides

Batch vs Inline homogenizer

In case of inline homogenizer, the energy given into the system is more guided. Given the similar mixing efficiency for both batch and inline process, the amount of dead zones are more in case of batch process (CFD).

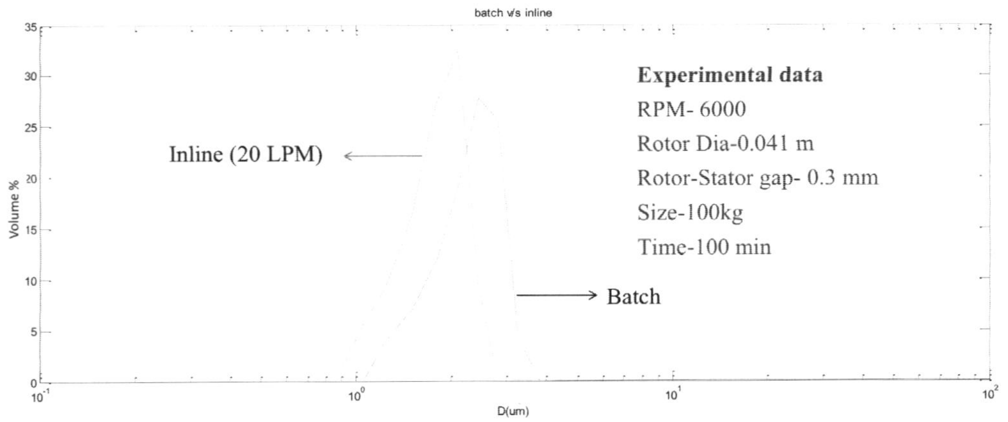

Scale-up heuristics — Modeling and simulation capability

17AIChE Dr.Reddy's

Prediction of globule size distribution – **Population balance modeling (PBM)**

Drawbacks of general trend

- X Can't predict the distribution (d_{10}, d_{50} and d_{90})
- X Unsuitable for both internally and externally circulated mixers

Highlights of PBM

- ✓ Valid for both Newtonian and non-Newtonian fluid based systems
- ✓ Working for both laminar or turbulent flow regimes
- ✓ Can be used for all types of available high shear homogenizers
- ✓ What-if studies

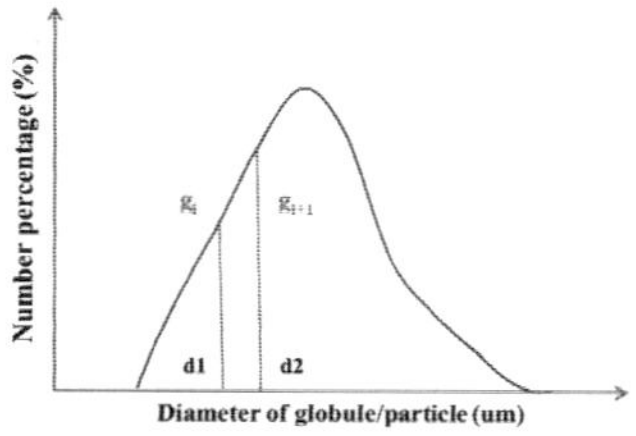

Number percentage distribution

Number of drops in [v,v+dv] per unit volume — *Breakage rate* — *Probability of forming daughter droplet (v) from mother (v')*

$$\frac{\partial n(v,t)}{\partial t} = -\,g(v)n(v,t) + \int_0^{\infty} \beta(v,v')g(v')n(v',t)dv'$$

$$^1 g(d_p) = c_{1,b}\frac{\epsilon^{1/3}}{d_p^{2/3}}\exp\left(-\frac{c_{2\,b}\gamma}{\rho_d\,\varepsilon^{2/3}d_p^{5/3}}\right), where\ \ \varepsilon = [\mu\dot{\gamma}^2 v + Po\,(QN^2D^2\rho_c)]/V$$

[1] *Determination of breakage rates using single drop experiments, Matthias Kraume*

Population balance modeling (Framework)

Method Approximations

i. Coalescence is neglected (Concentration of surfactant >> CMC)

ii. Material properties (ρ,σ) are assumed based on experiments and literature.

iii. GSD for premix was assumed to be that of the simply shaken mixture.

PBM Solution Approximations

i. Use of Trapezoidal rule for solving integral part of PBM equation.

ii. C1,C2 - constants in PBM equation are set using result of a particular product and need to be refined using parameter estimation.

iii. The prediction of GSD is done on a discrete basis where input is given in form of data points (nodes), accuracy of the system can be increased by including more nodes.

Population balance modeling (Framework)

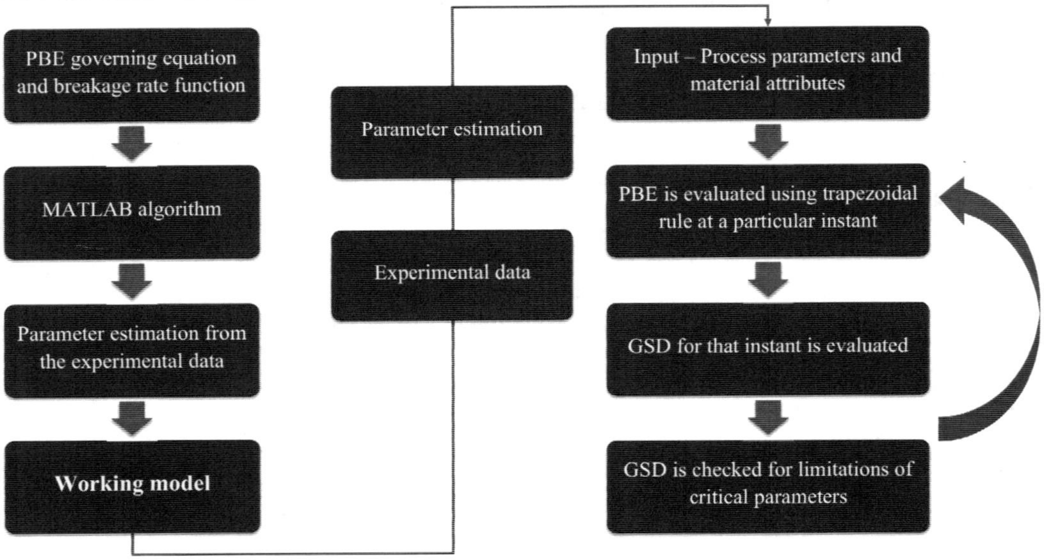

Effect of cooling rate (Literature)

Mixing speeds and cooling rates can impact of scale-up and influence physical properties of semisolids, such as its viscosity [1]

High cooling rate → High viscosity [2]

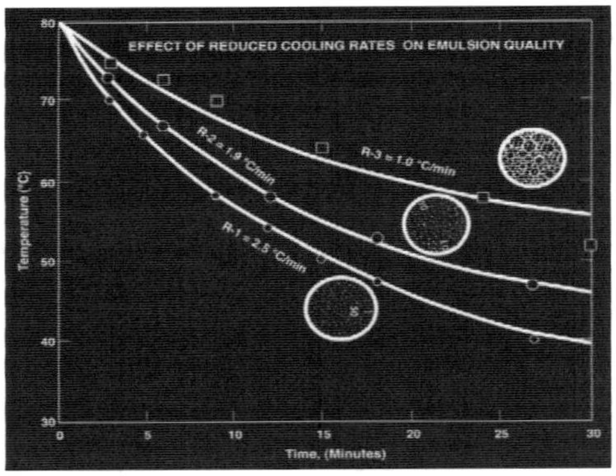

(1) Wessels P, Holz M, Erni F, Krummen K and Ogorka J. 1997. Statistical Evaluation of Stability Data of Pharmaceutical Products for Specification Setting. Drug Development and Industrial Pharmacy, 23(5): 427-439

(2) T. J. Lin, "Rheology Fundamentals and Applications in Cosmetic Formulation, Manufacturing and Quality Control" in The Chemistry and Manufacture of Cosmetics, Vol. 1 Basic Science (2000)

489812 Dissolution Kinetics Modelling of a BCS Class II Active Pharmaceutical Ingredient

Abstract:

Mechanisms of dissolution kinetics of active pharmaceutical ingredient (API) crystals have been extensively studied in the pharmaceutical domain, as these have a significant impact upon the bioavailability of drugs within the body. In such studies, the pharmaceutical dosage form of the API is typically suspended in a stirred vessel (in which the hydrodynamic conditions are well controlled), allowing the rate of dissolution to be measured using either UV or conductivity methods, providing an insight into the drug's behavior within the body. Dissolution kinetics can also be used to improve industrial crystallization processes, as the selective dissolution of fines could lead to narrower crystal size distributions (termed kinetic ripening). However, while the dissolution of APIs is of critical importance in the pharmaceutical industry, very few studies have been conducted to date which combine experimental methods with numerical simulations in order to achieve a comprehensive understanding of the principle influencing factors. This study attempts to address this shortcoming by examining the dissolution performance of Ibuprofen, a BCS Class II API, with PAT techniques experimentally and mathematical modelling.

The dissolution rate of Ibuprofen with different particle size distribution (PSD) and crystal habit was measured experimentally at different agitation rates. In order to simulate the environment encountered by the drug in the human body, all experiments were carried out in water at 37ËšC using a phosphate buffer to maintain a pH of 7.20 (in accordance with U.S. Pharmacopoeia requirements). Different crystal habits were obtained by recrystallizing Ibuprofen in methanol and hexane respectively. Crystals generated using methanol as the solvent produce polyhedral crystals while those generated using hexane are needlelike. Different agitation conditions have been investigated while the effect of particle size upon the dissolution rate was examined by using four different size fractions (sub 150Âµm, 150-300Âµm, 300-500µm and 500-850µm).

Experiments were carried out in a 100mL, temperature controlled Easymax vessel. Agitation in the vessel was achieved using a stainless-steel pitched blade impeller. Solvent and crystals were added into the glass vessel and the rate of dissolution was *in-situ* measured using on-line UV probe to track the liquid phase Ibuprofen concentration, while focused beam reflectance measurement (FBRM) was applied to track the solid phase particle size by measuring chord length distribution (CLD) and particle vision measurement (PVM) was used to monitor the shape and structure of the crystals. The initial and final PSD of Ibuprofen were also measured *off-line* using a Malvern Mastersizer 2000.

As dissolution is the reverse process of crystal growth, it involves two steps: (1) surface reaction and disintegration of the surface species; (2) mass transfer of these species into the bulk solution across the diffusion layer. It is typically assumed that the overall process is limited by the slower step. Based on this theory, a kinetic model has been devised to describe the dissolution behavior of a polydisperse powder under non-sink conditions. Two rate limiting regimes, surface detachment controlling and mass transfer controlling, were derived respectively. This model was solved using numerical methods and combined with a fraction of the collected experimental data in order to obtain a shape

modulated dissolution rate and an overall correlation for the dissolution rate. The remaining experimental data was used to verify the accuracy of the model, demonstrating a very good agreement between theory and experiment. This dissolution kinetic model, which was derived from fundamental principles, is of great use to identify favorable crystal characteristics enabling higher rates of dissolution. It can be used to inform parameter selection during Quality by Design (QbD), for instance, the prediction of PSD of API crystals could be applied to guide the crystallization process for API manufacturing while the estimation of the API concentration changing with time provides useful information for formulation development.

Keywords: Dissolution kinetics; Mathematical modelling; API manufacturing and formulation

509054 Solve this! Fundamental Approach to Problem Solving in Industrial Processes I (Invited Talks)

Abstract:

Solving complex plant problems is an everyday exercise that affects all projects and budgets, and it is very rarely introduced to students at any level. In fact, by the time young engineers are few years into working in industry, most have lost the engineering skills that they developed in school. We hope to encourage future engineers to retain those skills and learn the importance of plant problem-solving.

504263 Large-Scale Synthesis of Dihydrostreptomycin via Hydrogenation of Streptomycin in a Membrane Dispersion Microreactor

Abstract:

In this work, a membrane dispersion microreactor was applied for the hydrogenation of streptomycin (STR) to produce dihydrostreptomycin (DHS). The reductant was dispersed into the continuous phase containing STR, leading to a significant increase in the mass-transfer specific surface area, and the conversion rate of STR reached 99.81 wt% within 3 min. In contrast, in stirred tanks, the conversion rate of STR only reached 99.5% in 30 min. Moreover, the minimun amount of KBH_4 used in the membrane dispersion microreactor was about 35.71% lower than that in traditional tanks, when getting the same conversion rate at about 99.5%.

487921 Direct Comparison Between Batch Bulk Mixing and Continuous Millifluidics in the Synthesis of Amorphous Drug Nanoparticles

Abstract:

Amorphous drug nanoparticles in the form of drug-polysaccharide nanoparticle complex (or drug nanoplex) had emerged as an ideal supersaturating delivery system of poorly-soluble drugs attributed to its many attractive characteristics. Herein we presented for the first time direct comparison between two nanoplex synthesis platforms, i.e. millifluidics and bulk mixing, representing continuous and batch production modes, respectively. They were compared by the resultant nanoplexâ€™s (1) physical characteristics (size, zeta potential, payload), (2) preparation efficiency, (3) storage stability, (4) dissolution rate/supersaturation generation, and (5) production consistency. For this purpose, the effects of key variables in drug-polysaccharide complexation (pH, charge ratio) were investigated in both platforms. Perphenazine and dextran sulfate were used as the drug and polysaccharide models, respectively.The results showed that both platforms shared similar dependences on pH and charge ratio with similar optimal preparation conditions. The pH was found as the governing variable through its influence on size and zeta potential, while the other variables (e.g. charge ratio, millifluidicsâ€™ residence time, tube diameter) were insignificant. The other aspects of nanoplex were minimally affected by the process variables. Nanoplexes having mostly similar characteristics were produced by both platforms, except for payload and supersaturation generation. Nevertheless, millifluidics was favored owed to its superior production consistency and well-demonstrated scalability.

Novel Method of Evaluating Liquid Absorption with Intra-Particle Pore of Pharmaceutical Porous Materials using a Capillary Rise Method

Takuma Oba [a),c)], Yasutomi Kato [b)], Ryoichi Sonoda [a)], Kohei Tahara [c)], Yoshiaki Kawashima [d)], Hirofumi Takeuchi [c)]

[a)]Formulation Department, CMC Center, Kaken Pharmaceutical Co. Ltd., [b)]R&D Planning & Project Management Department, Kaken Pharmaceutical Co. Ltd., [c)]Gifu Pharmaceutical University, [d)]Aichi Gakuin University

17AIChE Annual Meeting, Minneapolis, MN

Summary

Introduction

Porous materials are attractive stuff for pharmaceutical particulate designing, such as for powdering oily or semi-solid drugs and improving the solubility of poorly soluble drugs. Understanding the mechanism and behavior of liquid absorption into intra pores and interstices of porous carrier particles should be important to effectively load active pharmaceutical ingredients (APIs) with the carriers. We established a simple and practical method to measure the amount of liquid or API solution absorbed intra-particle pores and trapped in interstices of particles separately.

Purposes

- To understand the absorption mechanism and behavior of liquid with the porous material
- To establish a simple and practical method to evaluate the liquid absorption of porous carriers
- To predict the amount of model drug loaded with particles by using the established method.

Experimental

Materials

- Porous Carrier: Porous calcium silicate ($CaSiO_3$) with petal-like pores (Florite®-RE, FR, from Tomita Pharmaceutical Co., Ltd.)
- Model API as a model of poorly water soluble drug: Itraconazole (ITCZ from Hanmi Pharm. Co., Ltd.)
- Solvent: Water, Ethanol, Dichloromethane and Glycerin

Physicochemical Properties Evaluated

Measurement of Contact angle (Sessile drop method)
— Contact angle meter (DSA30, KRÜSS GmbH)

- Absorption experiment (Capillary Rise Method)
  — Force tensiometer (K12, KRÜSS GmbH)

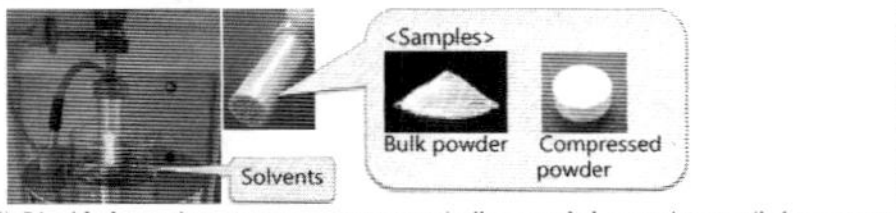

1) Liquid absorption amount was automatically recorded over time until the amount became constant
2) K was obtained from the slope of absorption kinetics
3) Penetration rate (cm²/sec) was calculated from modified Washburn equation below

$$\frac{h^2}{t} = K \cdot \left(\frac{V_{void}\,\rho}{l_{powder}}\right)^2$$

- h: height reached by the liquid front at time t
- l_{powder}: the apparent height of packed powder sample
- V_{void}: the volume of intra-particle pores and interstices of particles
- ρ: the density of liquid

- Mercury intrusion porosimetry
  — Mercury Porosimeter (9305 pore sizer, Shimadzu Co.)
- Preparation of ITCZ loaded FR (ITCZ-FR)

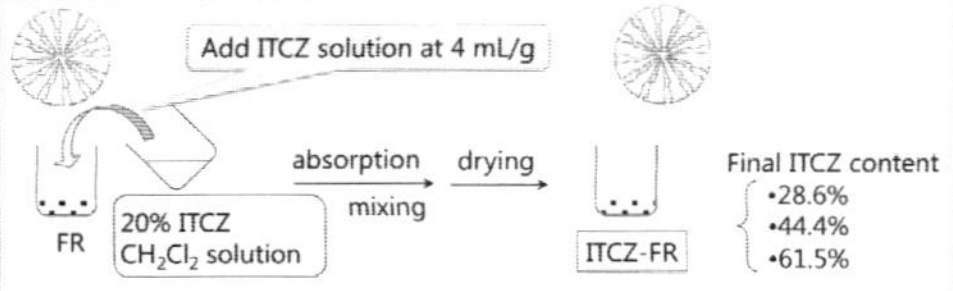

- Evaluation of ITCZ loaded FR (ITCZ-FR)
  — Morphology by Scanning Electron Microscopy (JSM-6390LA, JEOL Ltd.,)
  — Specific surface area by gas absorption method (Gemini2375, Shimadzu Co.)
  — Crystal states by X-ray diffraction (RINT2200UltimaII/PC, Rigaku Co.,Ltd.,)

Conclusions

- ✓ **A liquid absorption experiment using a compressed sample of porous carriers was possible to predict the available intra-particle pore volume of the porous carriers.**
- ✓ **Prepared API loaded particle based on the findings from this experiment was found to adsorb API efficiently in the intra-particle pores in amorphous state.**
- ✓ **This novel method was found to be useful for understanding the liquid absorption mechanism to the porous carriers and for designing the API loaded particles for medicinal drug products.**

Results & Disccussion

Absorption experiment using Bulk FR powder

Table 1 Physicochemical properties of solvents employed

Sample	Water	Ethanol	Dichloro-methane	Glycerin
Viscosity (mPa·s)	0.89	1.10	0.44	924.7
Surface tension (mN/m)	72.3	22.5	28.6	65.2
Density(g/cm³)	0.99	0.79	1.33	1.26
Contact Angle(°)*	24.4 (cosθ = 0.91)	18.0 (cosθ=0.95)	17.7 (cosθ=0.95)	69.6 (cosθ=0.35)

* Contact angle to Calcium Silicate surface

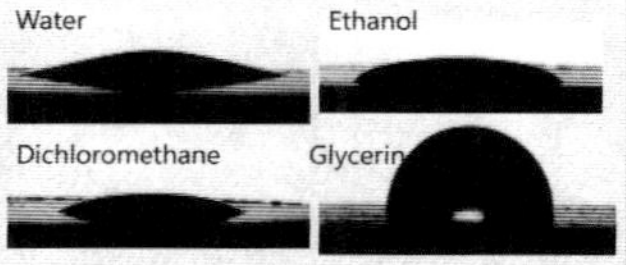

Fig. 1 Images of solvents drop placed on the compressed calcium silicate surface

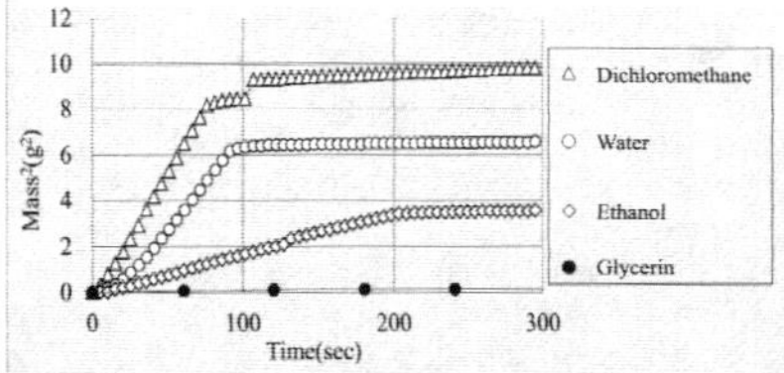

Fig. 2 Kinetics of capillary rise of solvents in FR powder bed

Table 2 Results of Absorption experiment of FR with various solvents

Sample	Water	Ethanol	Dichloro-methane	Glycerin
Absorption amount	7.96 mL	7.47 mL	7.19 mL	7.39 mL
Absorption time	2.1 min	4.7 min	1.8 min	430.6 min
K from slope part (g²/sec)	0.0857	0.0160	0.1149	0.0004
Penetration rate (normalize value*)	0.094 (1.00)	0.011 (0.12)	0.223 (2.37)	0.001 (0.01)

*normalized by the result of Water

- The wettability of FR with various solvents was clarified by the Penetration rate.
- Penetration rate was strongly affected by the viscosity of the solvent.
- The absorption amount of liquid from this method was apparent amount including both penetrated intra particle pores and trapped in interstices voids of particles.

⇒ Liquid absorption test using a compressed porous carriers was established for precise evaluation

Absorption experiment using compressed FR

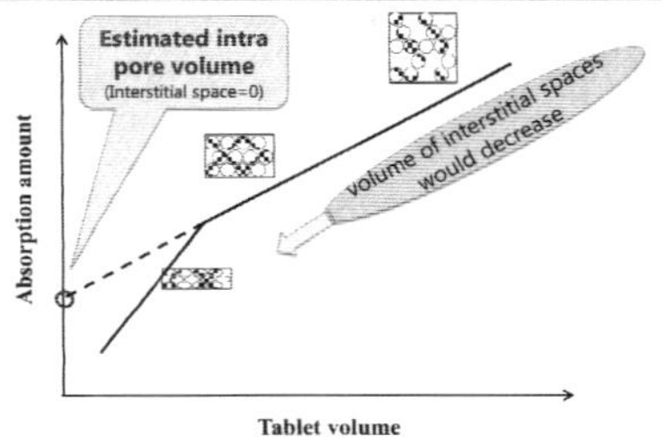

Fig. 3 Expected diagram of the relationship among tablet volume, absorption amount and inner states of tablet

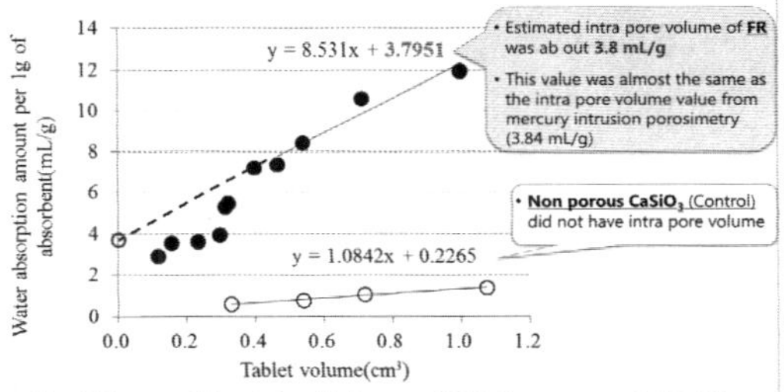

Fig. 4 Diagram of the relationship between FR(●) & non porous $CaSiO_3$(○) tablet volume and water absorption amount.

- This method can evaluate the absorption of intra particle pores and interstices voids of particles separately.
- FR was found to be able to absorb about 3.8 mL liquid into its intra-particle pore per 1 g FR.

⇒ It was possible to predict available intra-particle pore volume of the porous carriers by using this method

Preparation and evaluation of ITCZ loaded FR

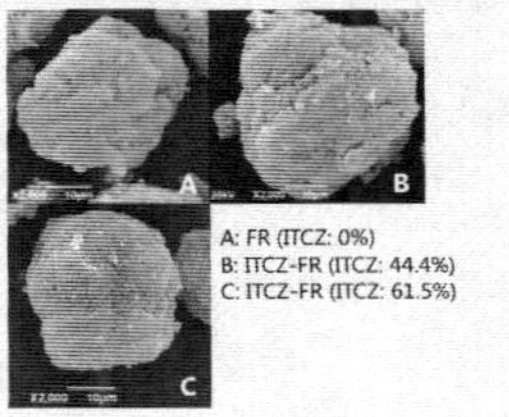

Fig. 5 SEM images of ITCZ-FR

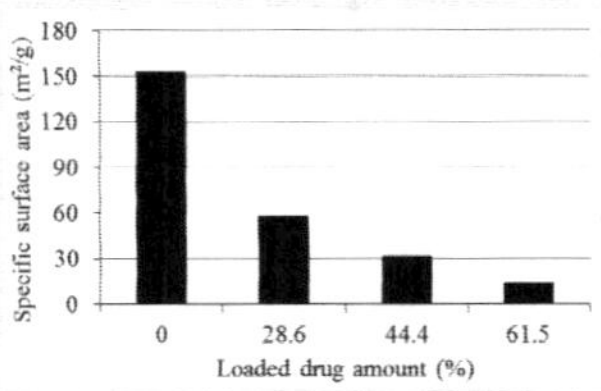

Fig. 6 Effect of loaded ITCZ amount on specific surface area of FR

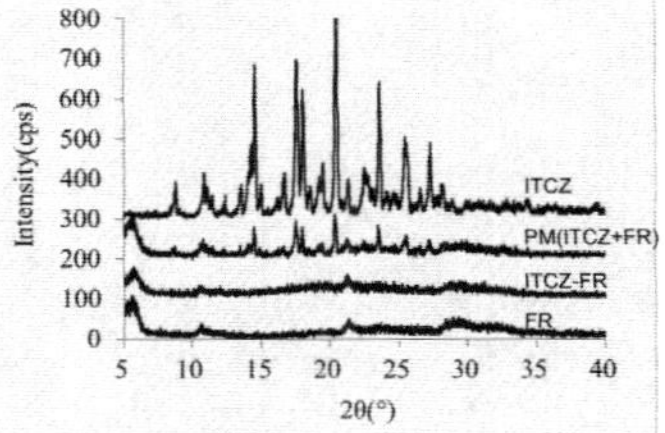

Fig. 7 Powder XRD Patterns of ITCZ-FR

- Absorption test using compressed FR can predict the amount of ITCZ solution to be absorbed inside the pore of FR particles.
- The ITCZ absorbed in the pore of FR was found to be amorphous state.

⇒ Our novel method to measure the intra-pore volume was found to be practically useful for formulation study

489919 Evaluation of Experimental Methods for Solubility Determination of Carbamazepine in Ethanol for Cooling Crystallization Process DesignÂ

Abstract:

Solubility data is important for pharmaceutical formulation and process design. For example, solubility data is essential for estimating the supersaturation and understanding the relative stability of crystal forms. The determination of an accurate solubility curve is critical for efficient design of the crystallization process. Inaccurate determination can lead to use of a higher or lower supersaturation level that would result in undesired extra nucleation or crystal growth. However, there are various solubility measurement techniques. Here we will evaluate several popular ones and discuss how to obtain accurate solubility data.

In this study, carbamazepine (CBZ) is chosen as the model drug due to its various polymorphic forms. The goal was to evaluate the effect of (a) dissolving methods, (b) sampling separation, and (c) quantification technique on the accuracy of solubility data. Also, the solubility curve was developed for form III and II at various temperatures for future crystallization applications. The two dissolving methods examined for Form III CBZ are: (1) the shake-flask, (2) mechanical stirring. The shake-flask method was performed as the base scenario to examine solubility differences by two off-line sampling separation techniques: (1) centrifuging or (2) filtering. The concentration was determined by three quantification techniques: (1) off-line gravimetrically, or (2) off-line high liquid chromatography method (HPLC) or (3) on-line Raman method. In addition, mechanical stirring, filtering and gravimetrical options were chosen to measure the solubility data of form II for a comprehensive understanding of solubility curves of CBZ form II and III.

The mechanic stirring method demonstrated similar mean values but with less variance compared to the shake-flask method. Through analysis of means, there were no significant differences between using centrifugation or filtration as the sampling separation method. But it is noted that a minimum of 30 minutes is required for settling time to avoid significant effect on the equilibrium concentration. Between the two off-line concentration quantification methods (gravimetric and HPLC), the HPLC method always showed lower mean values, but their difference from the gravimetric values are < 4.6%; the standard deviation for gravimetric methods is ~ 1.5% while the standard deviation for HPLC is ~ 3%. The reason for the standard deviation difference is probably due to the multiple-step sample dilution procedure for HPLC. On-line measurements in mechanic stirring with Raman spectroscopy was done to determine the solubility curve of form III as a function temperature (5 - 35 °C). The solubility curve was compared to off-line concentration quantification methods, and the differences for the mean values are normally < 5%. This on-line Raman method allows for quick determination of solubility curve from a single experimental study, and could be used as a quick solubility method if a low accuracy is acceptable. In addition, the dissolution profile of the metastable form (form II) was analyzed between 5 °C and 35 °C in order to develop determine whether the solubility curve for CBZ form II and form III are mono-tropic or enantio-tropic. The results suggest that the system shows a mono-tropic system between 5 °C and 35 °C, while the solubility ratio decreases from 1.20 at 35 °C to 1.12 at 5 °C. This work presents a comprehensive study on the solubility data of CBZ and the solubility determination methods that are important during the process development and design of the crystallization process.

490293 Continuous Solvent Exchange Washing of Pharmaceutical Intermediates

Abstract:

Continuous Solvent Exchange Washing of Pharmaceutical Intermediates

M. Kreimer[1], I. Aigner[1], S. Sacher[1], M. Krumme[2], T. Mannschott[2], P. van der Wel[3], A. Kaptein[3], J.G. Khinast[1,4]

[1]Research Center Pharmaceutical Engineering (RCPE) GmbH, Inffeldgasse 13, 8010 Graz, Austria

[2]Novartis Pharma AG, Novartis Campus, 4056 Basel, Switzerland

[3]Hosokawa Micron B.V., Gildenstraat 26, 7005 BL Doetinchem, Netherlands

[4]Graz University of Technology, Institute for Process and Particle Engineering, Inffeldgasse 13, 8010 Graz, Austria

In recent years, a lot of effort has been invested by the pharmaceutical industry to move away from batch processing to more agile and flexible continuous manufacturing processes. Going continuous has been discussed and described intensively by academia, industry and regulators, as cost-efficient manufacturing methods and easy up-scaling is strongly demanded. An entire continuous manufacturing plant includes several process steps from API production to the final oral dosage form such as synthesis, crystallization, washing, filtration, drying, milling, blending, granulation and tableting/capsule filling.

After API synthesis the input material for solid dosage form production can be prepared by crystallization through different methods like cooling, evaporation or sublimation. The produced suspension is usually a mixture of several liquids and solids, including the crystallized API. Before drying of the suspension mixture, various washing and filtration steps are involved to purify the drug product. The amount of impurities has to be decreased below a certain threshold to ensure patient safety and drug efficacy. Furthermore, tailored particle properties such as particle size distribution must not be influenced by washing to ensure consistent product quality and process ability within downstream process steps.

A typical washing and filtration process is arranged in cascades to guarantee a highly uniform and pure crystalline material. Therefore, anti-solvent is added in a precisely defined mixing ratio to the mother liquor. An optimal process design is necessary to decrease the number of multiple washing/filtration steps to a minimum to avoid changes of the solid product by mechanical stresses. In addition, considering environmental aspects, energy consumption and production of hazardous waste can be diminished as well.

In this work a newly developed, continuous solvent exchange washing process was established suitable for low

throughputs in the range of 2-12 l/h. A robust process design can be shown by means of model substances with a particle size below 100 µm. Variations in the composition of the mother liquid suspension and the washing medium revealed the influence of the solvent exchange washing process on the product particle size distribution. Within the given design space, suitable configurations were achieved for consistent product quality. In the specific case, changes of the particle size distribution were diminished below 15%.

493431 Design and Characterization of Fast-Release Clofazimine Nanoparticles to Improve Bioavailability

Abstract:

Clofazimine is a lipophilic riminophenazine antibiotic recommended by World Health Organization (WHO) as multidrug therapy of leprosy. To improve the drug dissolution rate, clofazimine was encapsulated in nanoparticles prepared with three affordable surface stabilizers, hypromellose acetate succinate (HPMCAS), lecithin, and zein, through the Flash Nanoprecipitation (FNP) process. Lyophilization and spray drying were applied and optimized to solidify nanoparticles. Dissolution tests were carried out to evaluate release profile of clofazimine nanoparticles prepared by different surface coatings in simulated gastric and intestinal fluids, comparing to raw clofazimine powder and commercial product Lamprene®. Remarkably improved dissolution rate and supersaturation levels were observed with all clofazimine nanoparticles despite different release kinetics. In addition, differential scanning calorimetry (DSC) was explored to study the physical state of clofazimine in nanoparticles. The results strongly suggest that the nanoparticles prepared with affordable coatings can be used as viable cryptosporidiosis therapeutics.

494873 Design of Efficient Metal Nanocatalysts for Continuous Synthesis of Drug Substances Via Cross Coupling Reactions

Abstract:

Design of Efficient Metal Nanocatalysts for Continuous Synthesis of Drug Substances via Cross Coupling Reactions

Andishaeh Dadgar, Farshid Mohammadparast, Marimuthu Andiappan, School of Chemical Engineering, Oklahoma State University, Stillwater, OK, USA.

Cross coupling reactions are widely used reactions in pharmaceutical industry. Heterogeneous metal nanocatalysts can be used as a platform to switch cross-couplings from traditionally-used batch mode to continuous flow mode. As the use of metal nanocatalysts in cross coupling reactions continues to grow, there is becoming a greater need for fast and efficient characterization methods. There is a group of metal nanocatalyst that exhibit a phenomena called surface plasmon resonance (SPR). SPR is a unique trait to each catalyst based on composition, size, and shape of the particles. SPR then shows up in the UV-Vis region of light easily measureable using UV-Vis spectroscopy. This gives a powerful tool to use in characterizing plasmonic metal nanocatalyst.

In this contribution, we show that surface plasmon resonance (SPR) spectroscopy can be used as platform to characterize nanometal catalyst quickly and efficiently. Using finite-difference time-domain optical simulations, we show the use of UV-Vis extinction spectrums to monitor nanoparticle geometry to predict the spectroscopic pattern we will see in experiments. We demonstrate that for a variety of nanocatalysts there is a measurable SPR peak shift and/or change in SPR intensity in the UV-Vis extinction spectrum that can be observed and predicted when the catalyst change size or leach. This novel and simple UV-Vis surface plasmon spectroscopic technique will help to (a) fundamentally understand the catalytic pathway and (b) develop efficient catalysts for a variety of reaction sets.

496128 Downstream Processing of Extrudates: Polymer Platform Development for Hot-Melt Extrusion/Tableting Via in-Line Monitoring of Compaction Properties

Abstract:

INTRODUCTION

As the number of applications for polymers in pharmaceutical development is increasing, there is need for fundamental understanding on how such compounds behave during processing. This research is focussed on the tableting behaviour of polymers used as carriers for solid dispersions manufactured via hot-melt extrusion (HME). Despite the fact that HME has been widely used for solid dispersion technologies there is still limited knowledge on how this processing technique might influence the downstream processing such as the tableting behaviour of pharmaceutical polymers and their formulations (Iyer et al., 2013). Therefore, a novel approach was developed to monitor and analyse 'in-die' compaction properties directly on a rotary tablet press during downstream processing of milled extrudates. This is advantageous due to the possibility of performing in-line measurements during the tableting process. During the first part of this research, a polymer platform was developed from which researchers can select the adequate polymer for both HME and tableting purpose by using formulation properties and in-line measured compaction properties (Grymonpré et al., 2017). Afterwards, a promising polymer was selected from this platform and subjected to a quality-by-design (QbD) approach with different active pharmaceutical ingredients (API's) to understand the influence of process and formulation parameters on the critical quality attributes (CQA) of extrudates and tablets of a hot-melt extruded solid dispersion formulation.

MATERIALS AND METHODS Soluplus (SOL), Kollidon VA 64 (VA 64) and Eudragit EPO (EPO) were selected as amorphous polymers, PVA 4-88 as semi-crystalline polymer, while Celecoxib (CEL) was chosen as BCS class II model drug for development of the polymer platform. Neat polymers and physical mixtures (35% drug load) were processed via HME (Prism Eurolab 16, Thermo Fisher, Germany) using co-rotating twin-screws containing 3 mixing zones.

The extrusion temperature was determined based on the extrudable range of each material as defined via a rheological screening, i.e. the temperature range corresponding to a complex viscosity between 1000 and 10000 Pa.s (Gupta et al., 2015). After cooling the extrudates were milled and sieved to obtain powders with comparable particle sizes as the neat polymer. For each polymer, 3 formulations (neat polymer, neat polymer extrudates and polymer-CEL extrudates) were compressed to tablets (270 ± 10 mg) on a rotary tablet press (MODUL^{TM} P, GEA Pharma Systems) equipped with cylindrical flat-faced Euro B punches of 10 mm diameter, using 6 main compaction pressures without a pre-compression step. All tablets were analysed for 'out-of-die' properties (tabletability, compressibility, compactibility) immediately after ejection. Punch deformation at each compaction pressure was calculated and corrected for during this study. An experimental approach was developed for in-line analysis of the compaction properties (plasticity factor (PF), in-die elastic recovery (IER) and Heckel analysis) on a rotary tablet press using linear variable displacement transducers (LVDT) incorporated inside the turret and clamped onto one pair of punches and a complementary data acquisition analysis system (CDAAS, GEA Pharma Systems). This approach enabled the monitoring of punch stroke movements during a compression cycle directly on a rotary tablet press. Principal component analysis (PCA) was executed on the relevant characterization and compaction data in order to classify the different materials according to their flowability and compaction behaviour using the multivariate data analysing software SIMCA 13.0.3 (Umetrics, Umeå, Sweden).

Furthermore, a statistical Design of Experiments (DoE) was used to evaluate the influence of three HME process parameters: barrel temperature (160-200 °C), screw speed (50-200 rpm), throughput (0.2-0.5 kg/h) and one formulations parameter: drug load (0-20 %) on the CQA's of extrudates and tablets of the selected SOL-CEL solid dispersion formulation.

RESULTS AND DISCUSSION

Solid state characterization showed that HME yielded stable glassy solutions for all amorphous polymer formulations containing 35 % CEL as these were stable for at least 6 months under stress conditions (40 °C and 75% relative humidity). In general, higher tablet tensile strengths were observed for glassy solutions formulated with all polymers due to higher interparticulate bonding strengths per unit bonding area, while no changes were detected in the interparticulate bonding areas between formulations. There was need for analysing the 'in-die' tablet compaction properties of these formulations to understand the mechanisms which modified the mechanical properties such as tablet tensile strength. At higher compaction pressures, glassy solutions of SOL and EPO underwent more plastic deformation which was correlated to a higher compactibility and tabletability of these formulations. Additionally, the formation of stable glassy solutions altered the formulations towards more fragmentary behaviour under compression which was beneficial for the tabletability. No differences in plasticity factor were detected when tableting the neat polymers compared to their extrudates, indicating that HME did not alter the volume reduction mechanism of the amorphous polymers without API. However, this was not the case when using the semi-crystalline polymer (PVA 4-88) as a significant decrease in tabletability was detected after HME of the neat polymer which was linked with a change in crystalline fraction (32 to 21.4%). With the inclusion of amorphous CEL, the tablet tensile strength (TS) was slightly recovered due to similar transition towards a more fragmentary behavior. Principal component analysis (PCA) was applied to summarize the behaviour during HME, milling, feeding and compaction of all formulations, enabling the selection of polymers which have the highest potential as carrier during HME for solid dispersions with CEL in respect of further downstreaming by tableting. By application of PCA on this polymer platform, SOL and VA 64 were selected as more favourable polymers compared to EPO and PVA 4-88 for HME/tableting purpose. The QbD-study on SOL-CEL enabled to evaluate the HME-process as well as the impact of the tableting behaviour on this formulation. Torque was mainly influenced by barrel temperature and drug load since both factors impacted the melt viscosity. In general, drug load had the most significant impact on the extrudate properties compared to the process parameters indicating that for this specific formulation the HME process was very robust. Tabletability was clearly affected by the drug load, while changing the process parameters had no impact on this property. Tablets manufactured from the glassy solutions yielded a significantly higher TS. Drug load also altered the 'shape' of the curves (i.e. inflection point), indicating changes in the mechanical properties of the formulations. Conclusion

Monitoring the punch movement using the described instrumentation and CDAAS-software was an effective tool for in-line measurement of compaction properties on a rotary tablet press. By combining 'in-die' and 'out-of-die' techniques, it was possible to investigate in a comprehensive way the impact of HME on the tableting behaviour of pharmaceutical polymers and their formulations. While HME had only a limited influence on the compaction properties of the amorphous polymers when no drug was included, HME changed the compaction properties of glassy solutions towards a more fragmentary behaviour, independent of the polymer type. The semi-crystalline PVA 4-88 encountered more impact of HME on tabletability of the neat polymer due to a change in crystalline fraction. In respect to the milling efficiency of the extrudates, the used amorphous polymers were significantly beneficial over the semi-crystalline polymer since they were more brittle. A polymer platform was successfully established using PCA on the large dataset, enabling selection of the most suitable polymers for a specific formulation in respect to HME, milling and further downstream processing towards tablets. Afterwards, a promising polymer was selected from this platform and successfully subjected to a quality-by-design (QbD) approach which emphasized that HME was a robust process for the combination SOL-CEL. However, the quality of extrudates and tablets can be optimized by adjusting specific formulation parameters (e.g. drug load).

REFERENCES

1. **1.** Iyer, R., et al. The impact of hot melt extrusion and spray drying on mechanical properties and tableting indices of materials used in pharmaceutical development. J. Pharm. Sci. 102, 3604–3613 (2013).
2. **2.** Grymonpré, W., et al. In-line monitoring of compaction properties on a rotary tablet press during tablet manufacturing of hot-melt extruded amorphous solid dispersions. Int. J. Pharm. 517, 348–358 (2017).
3. **3.** Gupta, S.S.., et al. Effect of carbamazepine on viscoelastic properties and hot melt extrudability of Soluplus. Int. J. Pharm. 478, 232-239 (2015).

498661 Virtual High-Throughput Screening Pipeline: Size and Classification Distribution Effects on Experimentally Validated Hit-Rates

Abstract:

Drug leads are a small fraction of small organic compounds, which itself is a small fraction of conceivable organic molecules. High-Throughput screening (HTS) is used to screen multitudes of compounds in parallel (test more compounds at the same time) and in experimental working volumes (reduce the amount of reagents and biological substances used). However, HTS has a nominal hit-rate of a few percent at best (Dobson, 2004) and thus most resources are used to test inactive compounds. In efforts to increase efficiency, computational methods are used to pre-screen compound libraries: likely active molecules are focused on and likely inactive or negative molecules are removed.

Previously, we developed a virtual HTS (vHTS) pipeline: using available experimental data, models predicting class and activity are trained and applied to compound databases to identify compounds likely to be active and to focus experimental efforts on (Chen and Visco, 2016), with the option of retraining the models to improve performance (hit-rate) or to identify more leads. The pipeline was applied several NCBIâ€™s PubChem Bioassay datasets: AID 825 (target: Cathepsin L, 1st iteration hit-rate: 19%, 2nd iteration hit-rate: 75%) (Chen and Visco, 2016), AID 728 (target: Factor XIIa, 1st iteration hit-rate: 43%, 2nd iteration hit-rate: 100%) (Chen and Visco), and AID 846 (target: Factor Xia, 1st hit-rate: 27%, 2nditeration hit-rate: 62%) (Chen and Visco).

To determine the hit-rate enrichment ability of the pipeline, we have applied our pipeline on more datasets, specifically targeted to examine how the pipeline responds to datasets of different sizes and classification distributions. Determining the effect of these two dataset parameters will indicate what datasets the pipeline has the most enrichment value on and/or what the expected enrichment ability is for a given dataset.

In this poster, we present work characterizing the effects of size and classification distribution. While controlling for one parameter (size or classification distribution), the other is varied. Experimental validation is conducted based on the protocol specified by the original dataset to determine vHTS hit-rates. The hit-rates determined using the same experimental protocol will allow for direct comparison of hit-rates and to identify the hit-rate enrichment ability of the pipeline for a given dataset. Based on the results, there will be a clearer idea what datasets will have the most impact on and what an expected hit-rate for a given dataset would be. We also aim to show, indirectly, the pipelineâ€™s hit-rate enrichment ability is repeatable and the pipeline is robust enough to handle many different kinds of datasets.

Chen J.J.F., Visco D.P. Jr., Developing an in silico pipeline for faster drug candidate discovery: Virtual high throughput screening with the Signature molecular descriptor using support vector machine models, *Chemical Engineering Science*, 2 March 2016. http://dx.doi.org/10.1016/j.ces.2016.02.037.

Dobson, C.M., 2004. Chemical space and biology. Nature 432, 824-828.

Chen, J.J.F, Visco, D.P. Jr. Identifying Novel Factor XIIa Inhibitors With PCA-GA-SVM Developed vHTS Models. *Manuscript in preparation.*

Chen, J.J.F, Visco, D.P. Jr. Identifying Novel Factor XIa Inhibitors With PCA-GA-SVM Developed vHTS Models. *Manuscript in preparation.*

499892 PAT on Oscillatory Systems: Monitor and Control Continuous Crystallization with Fourier Transform Infrared (FTIR) Spectrometer

Abstract:

Most pharmaceutical manufacturing processes include a crystallization step to purify active pharmaceutical ingredients (APIs). Despite its long history, difficulties associated with crystallization control prevail due to its complex kinetics.[1] However, process analytical technologies (PAT) have improved swiftly allowing for crystallization monitoring and control.[2] Traditionally, crystallization is carried out in batch which often results in batch-to-batch variation.[3] Continuous crystallization, on the other hand, is promising for its enhanced reproducibility and narrower property distributions amongst other improvements.[4] Continuous crystallization has become a key step in the current trend to reform the pharmaceutical industry from batch to continuous manufacturing. A well-studied system for continuous crystallization is the Mixed-Suspension-Mixed-Product-Removal (MSMPR) crystallizer in which the hydrodynamics are controlled with an agitator. There are many challenges associated with stirred tank systems such as poor local mixing, low heat and mass transfer, and varying shear rate.[5] To address these challenges, a novel commercial oscillatory baffled reactor (OBR) has been studied. It consists of a reactor tank and oscillating â€˜donutâ€™ shaped baffles to provide more uniform mixing.[6] Such systems have been well studied for API synthesis and are gaining interest for crystallization given their enhanced mass and heat transfer characteristics while imposing less shear rate, ideal for crystallization.[7] However, control strategies for continuous crystallization in such oscillatory systems have yet to be studied in the literature. The difficulty controlling continuous crystallization in an OBR stems from inadequate measurement of key factors. Vibrations caused by the oscillating motion often interfere with the signals of the PAT tools which consequently result in inconsistent readings.

A model-free control strategy for continuous cooling crystallization of paracetamol in the OBR is implemented in this study. Supersaturation control (SSC) is a model-free technique where concentration is monitored and supersaturation level is controlled by adjusting operating conditions such as temperature for the case of cooling crystallization. SSC has been well studied in MSMPR systems.[8,9] An SSC implementation was carried out for continuous crystallization in the OBR via CryMOCO, a commercially available integrated control software, a Huber thermo-regulator with a temperature sensor and a FTIR spectrometer. The FTIR is equipped with the Interferogram correcting the signal interference from oscillations. It provides online measurements of solute concentration in solution used by CryMOCO to control supersaturation. This study serves as a proof-of-concept for continuous crystallization SSC control in oscillatory systems.

1. Braatz RD. Advanced control of crystallization processes. *Annu Rev Control*. 2002;26 I:87-99. doi:10.1016/S1367-5788(02)80016-5.

2. FDA. Guidance for Industry PAT: A Framework for Innovative Pharmaceutical Development, Manufacuring, and Quality Assurance. *FDA Off Doc*. 2004;(September):16. doi:http://www.fda.gov/CDER/guidance/6419fnl.pdf.

3. Quon JL, Zhang H, Alvarez A, Evans J, Myerson AS, Trout BL. Continuous crystallization of Aliskiren hemifumarate. *Cryst Growth Des*. 2012;12(6):3036-3044. doi:10.1021/cg300253a.

4. Alvarez AJ, Singh A, Myerson AS. Crystallization of Cyclosporine in a Multistage Continuous MSMPR Crystallizer. *Cryst Growth Des*. 2011;11(10):4392-4400. doi:10.1021/cg200546g.

5. Kacker R, Regensburg SI, Kramer HJM. Residence Time Distribution of Dispersed Liquid and Solid Phase in a Continuous Oscillatory Flow Baffled Crystallizer. *Chem Eng J*. 2017;317:413-423. doi:10.1016/j.cej.2017.02.007.

6. Hewgill MR, Mackley MR, Pandit AB, Pannu SS. Enhancement of gas-liquid mass transfer using oscillatory flow in a baffled tube. *Chem Eng Sci*. 1993;48(4):799-809. doi:10.1016/0009-2509(93)80145-G.

7. Ni X, Mackley MR, Harvey AP, Stonestreet P, Baird MHI, Rama Rao NV. Mixing Through Oscillations and Pulsationsâ€"A Guide to Achieving Process Enhancements in the Chemical and Process Industries. *Chem Eng Res Des*. 2003;81(3):373-383. doi:10.1205/02638760360596928.

8. Yang Y, Nagy ZK. Combined Cooling and Antisolvent Crystallization in Continuous Mixed Suspension, Mixed Product Removal Cascade Crystallizers: Steady-State and Startup Optimization. *Ind Eng Chem Res*. 2015;54(21):5673-5682. doi:10.1021/ie5034254.

9. Yang Y, Nagy ZK. Advanced control approaches for combined cooling/antisolvent crystallization in continuous mixed suspension mixed product removal cascade crystallizers. *Chem Eng Sci*. 2015;127:362-373. doi:10.1016/j.ces.2015.01.060.

500265 Continuous Manufacturing of Oral Disintegrating Films: A Quality By Design Approach

Abstract:

The ICH guidelines and FDA guidances have encouraged pharmaceutical industries to implement risk based (ICH Q9), systematic and science–based approaches (ICH Q8 (R2) and ICH Q11), a robust pharmaceutical quality system (ICH Q10) and introduce Quality by Design (QbD) concepts, emphasizing science and risk-based approaches to assure product quality. Innovative continuous manufacturing (CM) technologies have a great potential to improve control over quality, reduce costs, enhance process safety, and significantly reduce the timelines currently involved across the medicines' supply chain. When compared to traditional batch manufacturing, continuous manufacturing also has enhanced ability to control the process and product quality through knowledge of residence times and the implementation of process analytical technology (PAT) tools. The objective of this work is to understand the relationships between product and process variables and critical quality attributes (CQAs) of ondansetron (a model drug compound) to support development and review of control strategies for continuous manufacturing of ODFs. This can aid in the understanding of common failure modes that provide the foundation for systematic risk assessment of continuous film processing.

The ODFs of ondansetron were prepared in-house using the solvent casting method in a continuous tape caster. These films were characterized in-line, at-line or off-line for drug assay, content uniformity, disintegration, dissolution, morphology, surface roughness, wettability, thickness, % moisture uptake, % moisture content, mechanical properties.

Principles of QbD along with appropriate design of experiments (DoE) were used to obtain a comprehensive understanding about the effect of critical material attributes (CMA's) such as % w/w of plasticizer, surfactant and critical process parameters (CPP's) such as mixing speed, coat thickness, air flowrate and drying temperature on the quality and performance of ondansetron ODF. Suitable Design of Experiments (DoE) was used to screen and optimize critical parameters that impact the product quality of ODFs. Data analysis using ANOVA and multifactor analysis was performed in elucidating interactions between different variables, rank order the various CMA's and CPP's, and help provide a predictive model for the process. A design space was established and verified by performing a failure mode analysis.

500607 Controlled Delivery of Fluorophores from 3D Two-Photon Photolithographic Printed Poly(Ethylene Glycol) Methacrylate Scaffolds

Abstract:

Controlled drug delivery systems have been utilized to enhance the therapeutic effects of many current drugs by effectively delivering drugs in a time-dependent and repeatable manner. With the aid of 3D printing technology, a novel drug delivery device was fabricated and tested to appraise functionality. Poly(ethylene glycol) methacrylate(PEGMA) scaffolds were fabricated using a Photonic Professional GT 3D printer. These scaffolds were printed with varying degrees of hatching, slicing, and pore sizes in order to observe the effects of these parameters on drug released. To test for controlled drug release, rhodamine B was used as a model drug and was homogenously mixed into the PEGMA to be trapped inside the polymer matrix after photopolymerization. Scaffolds were analyzed using a fluorescence microscope and quantified through an imaging software. Scaffolds were also imaged under the scanning electron microscope(SEM) to visualize the ultra-morphology and fine resolution of the micron size printed scaffolds. PEGMA scaffolds were then investigated for toxicity, where MTS assays were performed on induced pluripotent stem cells (IPSCs). The variability of the hatching, slicing, and pore size of the fabricated PEGMA scaffolds provided controlled release of rhodamine. When incubated with the iPSC, the scaffolds were observed to have no cytotoxic effects. The SEM images reaffirms the ability to fabricate scaffolds at the micron and nano- scale with high precision. Using the two-photon photolithographic printer, complex and miniature scaffolds can be printed with fine details, while providing control drug release.

500678 Controlled and Sequential Delivery of Fluorophores from 3D Printed Alginate-PLGA Tubes

Abstract:

Controlled drug delivery systems, that include sequential and/or sustained drug delivery, have been utilized to enhance the therapeutic effects of many current drugs by effectively delivering drugs in a time-dependent and repeatable manner. In this study, with the aid of 3D printing technology, a novel drug delivery device was fabricated and tested to evaluate sequential delivery functionality. With an alginate shell and a poly(lactic-co-glycolic acid) (PLGA) core, the fabricated tubes displayed sequential release of distinct fluorescent dyes and showed no cytotoxicity when incubated with the human embryonic kidney (HEK293) cell line or bone marrow stromal stem cells (BMSC). The controlled differential release of drugs or proteins through such a delivery system has the potential to be used in a wide variety of biomedical applications from treating cancer to regenerative medicine.

Journal of chemical & engineering data

Article

pubs.acs.org/jced

Pressurized-Synthetic Methodology for Solubility Determination at Elevated Temperatures with Application to Paracetamol in Pure Solvents

Brian de Souza,* Leila Keshavarz, Giuseppe Cogoni, and Patrick J. Frawley

Solid State Pharmaceutical Cluster (SSPC), Materials & Surface Science Institute (MSSI), Lonsdale Building, Department of Mechanical, Aeronautical and Biomedical Engineering, University of Limerick, Castletroy, Co. Limerick, Ireland

ABSTRACT: This paper describes a new nonintrusive method for the determination of high-temperature solubility data. Accurate high-temperature solubility data is vital to many industrial manufacturing processes such as cooling crystallization with direct implications for yield, throughput, and solvent usage. However, the provision of such data is notably absent from published literature for many active pharmaceutical ingredients. Pressurized-synthetic methodology is presented as a new technique for determining high-temperature solubility data. Paracetamol (acetaminophen) is used as a reference active pharmaceutical ingredient to validate the methodology. Solubility data determined using the pressurized-synthetic approach is reported for several pure solvents across a significantly extended temperature range. In the case of methanol, solubility data is obtained up to 354.15 K, above the atmospheric boiling point of the solvent, 337.65 K, and far in excess of the temperature range for which data exists in the literature, 268.15–303.15 K. The data obtained using the pressurized-synthetic method is validated against an extended gravimetric data set at temperatures up to the atmospheric boiling point for each solvent. Sensitivity studies were conducted to determine the influence of factors such as temperature gradient on the ultimate solubility determination. A temperature-based standard deviation of 0.1 K was established for paracetamol in 2-propanol at 303.15 K, comparing favorably with the temperature-based equivalent standard deviation of 0.2 K for the gravimetric approach. Binary interaction parameters for the pressurized-synthetic solubility data are derived and estimated for four different activity coefficient models, namely Margules, Van-Laar, Wilson, and non-random two-liquid (NRTL), along with the empirical solubility equation of Apelblat. For each solvent, the quality of fit of each of the activity coefficient models is analyzed. The NRTL model was found to best fit the experimental data for methanol, ethanol, 2-propanol, and acetone with mean square errors of 5.73×10^{-5}, 3.00×10^{-4}, 1.70×10^{-4}, and 7.35×10^{-5}, respectively. The pressurized-synthetic approach provides a nonintrusive, validated, and readily automated approach for the provision of valuable high-temperature solubility data that can be readily extended to binary and ternary systems.

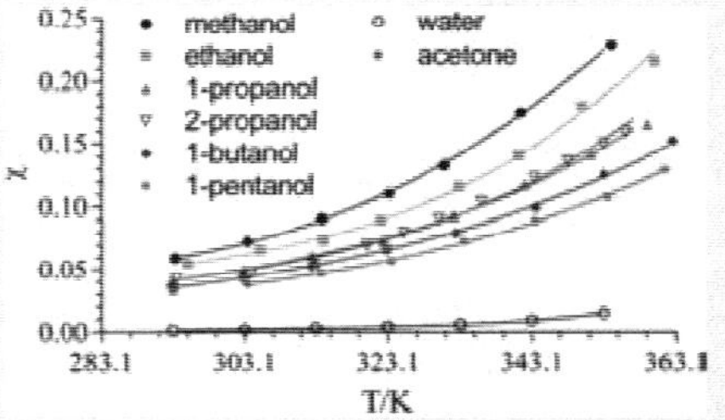

1. INTRODUCTION

The solubility of bioactive compounds is of critical importance to the pharmaceutical industry. Not only is accurate solubility data vital in terms of quantifying drug absorption and bioavailability, it is also essential for the development of manufacturing operations such as separation, liquid extraction, and crystallization.

In the case of the pharmaceutical cooling crystallization process, the difference in solubility as measured between upper and lower temperature bounds is by definition implicitly linked with the critical crystallization design metrics of yield, throughput, and solvent usage. For maximizing yield and throughput of cooling crystallization while minimizing solvent usage, a wide temperature span is normally desired. The provision of high- and low-temperature solubility data is therefore of upmost importance for industrial crystallization design. In general, it can be seen that API solubility curves show the most pronounced gradients at elevated temperatures. Thus, small increases in the upper temperature can potentially lead to significant process improvements. Accurate high-temperature solubility data is therefore imperative for designing crystallization processes; however, remarkably, the provision of high-temperature solubility data is notably absent from published literature to date.

There are several widely used methods for thermodynamic solubility determination, the most common of which include the gravimetric and shake flask methods.[1] In the gravimetric approach, the temperature of equilibration is fixed, and a saturated solution is achieved by contacting excess solute with solvent. Equilibrium is approached asymptotically, whereupon a sample of the equilibrated solution is filtered and weighed. The solvent is then gently evaporated, thus allowing for the evaluation of residual crystalline material mass and solubility. In the case of the shake flask method, again, an excess of drug solute is added to the solubility medium and the phases are again separated using filtration or centrifugation. Quantification

Received: January 31, 2017
Accepted: April 11, 2017
Published: April 19, 2017

of drug concentration in the saturated solution is measured by techniques such as ultraviolet (UV) spectrophotometric analysis or high-pressure liquid chromatography (HPLC).

Multiple drawbacks are associated with both gravimetric and shake flask methods. At elevated temperatures, and with volatile solvents, solubility measurements using either of these methods can be compromised by virtue of evaporation of the solvent. Unavoidable cooling occurs despite efforts to minimize the time between sample collection, filtering, and gravimetric weighing. Filtration of fine particles may not be entirely successful with some particles passing to the sample of equilibrated solution. Neither approach lends favorably to high-throughput screening studies. The maximum temperature for which solubility data can be obtained using both approaches is bound by the atmospheric boiling point of the solvent. In the case of UV or HPLC, samples may need to be diluted further for analysis, introducing the potential for measurement error. Regulation and maintaince of equilibrium temperature during analysis is a further potential concern. For viscous solvents in particular, settling of particles for analysis of clear solution may not occur.

This paper introduces a newly developed methodology for solubility determination, namely the pressurized-synthetic method. This nonintrusive method is based on the dissolution and consequent disappearance of the solute as monitored using laser instrumentation. The use of reactor pressurization extends the capability of the synthetic method, providing an innovative way to determine solubility even beyond the atmospheric boiling point of the solvent. The approach overcomes all the drawbacks previously listed for the gravimetric and shake flask methods. It does not require equilibrated sample collection or filtration. The approach can be readily automated, and the methodology does not occupy resources of specialized instrumentation.

For evaluating the pressurized-synthetic methodology, a reference active pharmaceutical ingredient (API) with well-established solubility data was required. Paracetamol (also known as acetaminophen) was chosen for this purpose. Paracetamol is an important analgesic and antipyretic drug. As a benchmark compound, it has, and continues to be, widely applied in crystallization research.[2−4] A range of pure solvents are employed in the manufacturing of paracetamol; in this paper, solubilities of paracetamol in eight of the most commonly used pure solvents are considered (methanol, ethanol, 1-pentanol, 1-propanol, 2-propanol, 1-butanol, water, and acetone).

There are several examples of solubility studies regarding paracetamol at low temperatures. However, an extensive literature search revealed no experimental data at elevated temperatures. Perhaps the most relevant of these studies is that of Granberg and Rasmuson,[5] who determined the solubility of paracetamol in a range of pure solvents using the gravimetric approach. The range of temperatures in that study was limited to 268.15−303.15 K. Likewise, Fernandez[6] gravimetrically determined the effect of solubility of paracetamol in ethanol for solution temperatures of 283.15−308.15 K, returning a slightly lower solubility when compared to the Granberg and Rasmuson data. A further study by Granberg and Rasmuson[7] examined the solubility of paracetamol of binary and ternary mixtures of water + acetone + toluene, again with a temperature range of 268.15−303.15 K.

Shakeel et al.[8] did present data covering a higher temperature range of 298.15−333.15 K for paracetamol using the shake flask method. However, the elevated temperature in that study was facilitated by a transcutol-water mixture, where transcutol has a high boiling point under atmospheric conditions of 475.15 K.

No previous literature was sourced in which the solubility of paracetamol was experimentally established at a temperature in excess of the atmospheric boiling point of the solvent.

The aims set out for the present work were to (1) establish solubility data for paracetamol up to and beyond the boiling point in pure solvents using the new pressurized-synthetic methodology; (2) validate the approach with comparisons with earlier and new gravimetric data, if necessary; (3) provide thermodynamic modeling of the extended solubility data; and finally (4) perform a comparative evaluation of activity coefficient models in terms of quality of fit.

2. MATERIALS AND METHODS

2.1. Materials. Paracetamol (BioXtra ≥99% NMR) was obtained directly from Sigma-Aldrich [CAS 103-90-2]. Paracetamol is known to exist in three polymorphic forms: monoclinic, commercially available, and room temperature relatively thermodynamically stable form I, orthorhombic (form II, metastable), and the highly unstable form III, which is rarely observed outside of fusion reactions. The crystal structures for both monoclinic and orthorhombic polymorphs have previously been reported in the literature.[9] Powder X-ray diffraction was utilized as per Figure 1 to confirm the crystal

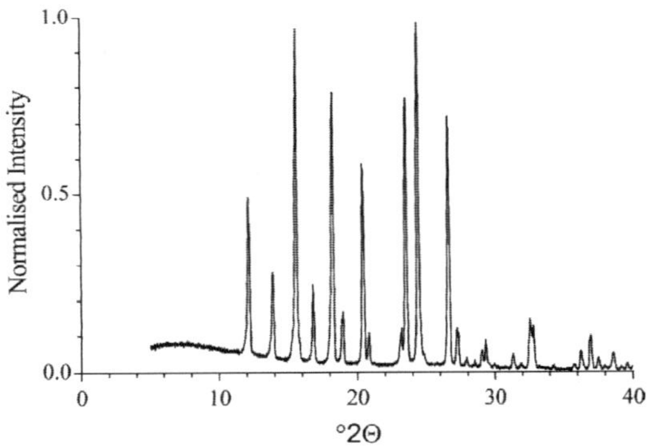

Figure 1. Experimental powder X-ray diffraction pattern for monoclinic paracetamol (form I) as received.

structure of the as-received paracetamol as being that of the commercially used form I with characteristic peaks (2Θ) including 12.1 ± 0.2, 13.6 ± 0.2, 15.5 ± 0.2, 18.2 ± 0.2, 23.5 ± 0.2, 24.4 ± 0.2, and 26.6 ± 0.2.

X-ray diffraction was utilized to eliminate the possibility of a polymorphic transformation during the solubility experiments across the full range of solvents considered. High-temperature samples of solute were prepared through rapid filtration followed immediately by X-ray diffraction analysis to mitigate the risk of solvent-mediated polymorphic transformation during storage. Powder diffraction data were collected on a Philips X'Pert-MPD PRO diffractometer (PW3064 sample spinner) with nickel-filtered Cu Kα radiation (λ = 1.5418 Å) run at 40 kV and 35 mA, 2θ = 5−40°, with a step size of 0.02° 2θ and a scan speed of 0.02° s^{-1}. Samples were prepared by light grinding prior to placement in a sample holder and flattening with a glass slide. The possibility of a polymorph transformation occurring during grinding was eliminated by analyzing a sample before and after grinding. Analysis of the

Table 1. Source, Mass Fraction Purity, and Boiling Points[a] of the Materials Used in This Work

chemical name	source	mass fraction purity	analysis method	additional purification	boiling point (1 atm)/(1.25 bar(g)) (K)
methanol	Sigma-Aldrich	≥0.999	HPLC[b]	none	337.65/359.35
ethanol	Sigma-Aldrich	≥0.999	HPLC	none	351.48/373.01
1-propanol	Sigma-Aldrich	≥0.999	HPLC	none	370.35/393.25
2-propanol	Sigma-Aldrich	≥0.999	HPLC	none	355.38/376.77
1-butanol	Sigma-Aldrich	≥0.995	HPLC	none	390.65/414.82
1-pentanol	Sigma-Aldrich	≥0.99	GC[c]	none	410.65/437.67
deionized water	distilled				373.15/396.99
acetone	Sigma-Aldrich	≥0.998	HPLC	none	329.44/354.83
monoclinic paracetamol	Sigma-Aldrich	≥0.99	NMR[d]	none	

[a]All normal boiling point data listed were obtained from Sigma-Aldrich and estimated for elevated pressure using the Clausius–Clapeyron relation. [b]High-performance liquid chromatography. [c]Gas chromatography. [d]Nuclear magnetic resonance spectroscopy.

spectra indicated no polymorphic transformation with consistent peaks before and after equilibration, indicative of form I.

The eight pure solvents and one solute utilized in the present work are listed in Table 1 alongside supplier and purity data and boiling points at atmospheric pressure and under pressurization (1.25 bar(g)). The water utilized was distilled, deionized, and filtered (0.2 μm).

2.2. Pressurized-Synthetic Solubility Determination. Suitable reactors were required to perform the pressurized solubility measurements. Three glass-lined stainless steel pressure vessels were custom designed with a nominal capacity of 300 mL. These were fitted with flush mounted 15 bar pressure-rated sight glasses to facilitate the synthetic laser measurement. Vessel lids were mechanically fastened securely in place with Viton seals except in the case of acetone where EDPM seals were used for solvent chemical compatibility. The pressure vessel lids incorporated PTFE-sealed compression fittings for a PT-100 temperature gauge, chemically resistant 316 stainless steel pressure gauge, and pressure supply fittings. A pressure regulator was installed in addition to a pressure relief valve for safety reasons. For each solubility data point, predetermined masses of solvent and solute were added to, and weighed in, each reactor using an analytical balance (Mettler Toledo MS4002S, weighing capacity up to 4200 g, sensitivity ± 0.01 g).

To suspend the crystals in the solvent and further to minimize temperature and concentration gradients during the experiments, each vessel was agitated by means of a PTFE-coated magnetic stirrer bar driven by a serial magnetic stirrer plate (2mag MIXdrive 1) with 40 W stirring power and a maximum stirring volume of 3,000 mL per stir point. A maximum stir rate of 500 rpm was determined due to the potential for the formation of vortex and consequent air entrapment. A conservative 450 rpm was adopted for the solubility experiments, which is more than sufficient to ensure full suspension of paracetamol crystals.

The pressure vessels were sealed, purged, and slowly pressurized with N_2 to a pressure of 1.25 bar(g), which is sufficiently high to ensure minimal evaporation of any of the solvents used during the experiments. The pressure vessels were positioned within a thermostatic stainless steel bath (Grant GR150; 38L; stability ±0.005 K; uniformity ±0.02 K). For experiments up to 343.15 K, the bath was filled with deionized water, and the surface was covered with polypropylene spheres to avoid evaporation and minimize thermal gradients. For higher-temperature experiments, silicone oil was used to ensure appropriate thermal conduction while avoiding evaporation.

For the determination of solubility, a continuous-wave 635 nm red laser (MRL-III-635L) was utilized as is commonly adopted in laser diffraction measurements. The laser power was adjustable from 0 to 30 mW. A low power setting of 4 mW was selected to avoid excessive heating of the solution.

Silicon photodiode-based laser detectors (Coherent UV/vis Quantum Power Sensor, 325–1065 nm, calibration uncertainty ±2%, 5 μW to 100 mW wavelength dependent) were used to evaluate laser power and obscuration. The complete assembly of the apparatus used for pressurized-synthetic solubility measurements is illustrated in Figure 2.

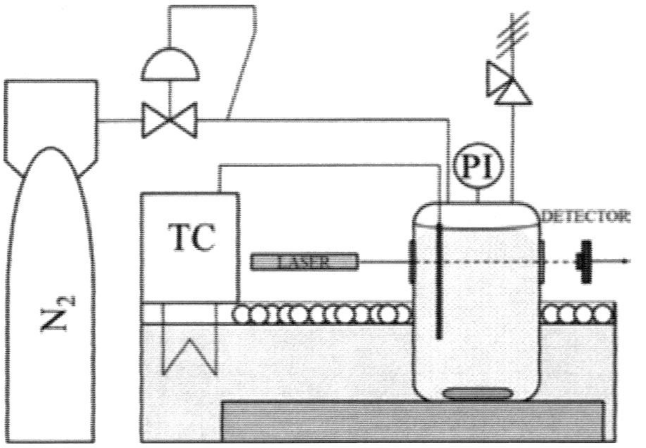

Figure 2. Schematic diagram of the experimental rig including temperature and pressure sensor, submersed magnetic stirring plate, thermostatic bath with temperature controller, 635 nm red laser, detector, magnetic stirring bar, and polypropylene spheres.

The experimental procedure adopted was broadly based on the laser monitoring approach.[10−12] This approach is also referred to as the synthetic or last crystal disappearance method.[13] The temperature of the reactors were slowly but progressively increased while monitoring obscuration of the laser beam by means of particle scattering. For each concentration of solute in solvent, a solubility temperature consistent with a sharp change in obscuration was detected as the particles dissolved. A representative sample result illustrating the sharp change in obscuration commensurate with the progressive, slow, and linear temperature increase of the reactor is shown in Figure 3 for the 0.05 molar fraction of paracetamol in 2-propanol.

Temperature profiles were implemented using the Labwise control software based on the feedback of the external

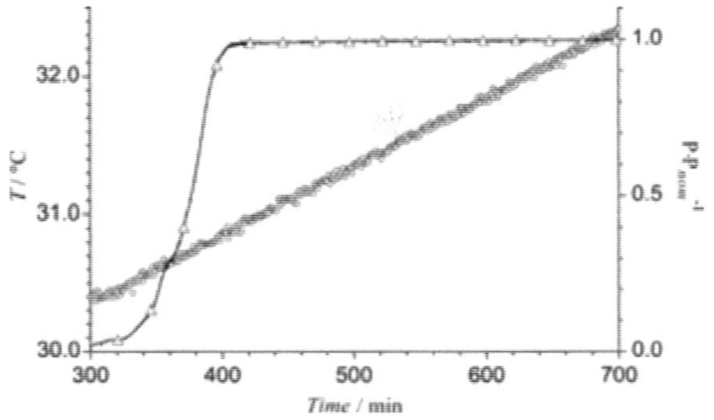

Figure 3. Pressurized-synthetic result demonstrating step change in laser obscuration as recorded for the 0.05 molar fraction of paracetamol in 2-propanol with a temperature gradient of 0.3 K/h. (○) Temperature data points and (△) normalized laser power detected.

temperature probe positioned within the reactor. A sensitivity study was performed to evaluate the effect of varying the temperature gradient, the results from which are shown in Figure 4.

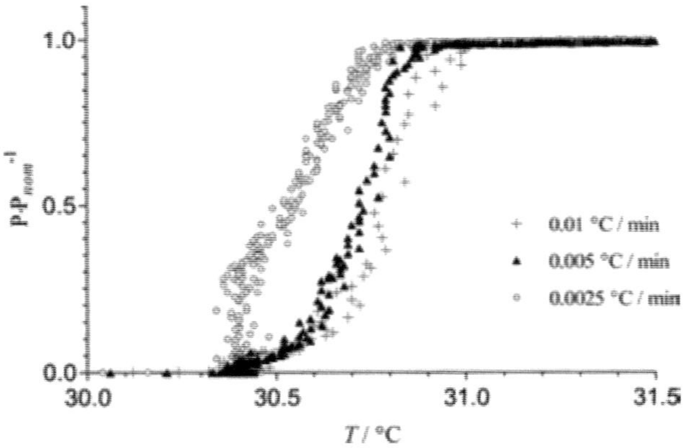

Figure 4. Sensitivity study investigating the influence of vessel temperature change gradient on the observed solubility point temperature measurements. The laser power measurement is normalized by the maximum laser power.

The results of the sensitivity study indicated a high level of consistency in the solubility temperature measurements with a standard deviation of 0.1 K. The uncertainty in the measurement is comparable with the earlier gravimetric work of Granberg and Rasmuson[5] who reported a mass-based standard deviation approximately equivalent to a temperature-based standard deviation of 0.2 at 303.15 K for paracetamol in 2-propanol.

For maximizing accuracy and ensuring sufficient time for equilibration, a conservative 0.3 K/h temperature gradient was adopted, which is significantly slower than the maximum suggested 2 K/h value[10] for synthetic measurements.

To monitor laser obscuration, data points were recorded every ten seconds with the average and standard deviation of laser power over 100 measurements returned. All experiments were conducted in the presence of a fume hood.

2.3. Gravimetric Solubility Measurements. Equilibrium solubility measurements were conducted for paracetamol in each of the eight pure solvents. The focus of the gravimetric measurements was to extend the earlier data set of Granberg and Rasmusson[5] such as to allow comparison with synthetic solubility data at more elevated temperatures. As with the pressurized-synthetic solubility determination, a Grant GR150 thermostatic stainless steel bath was used with a magnetic stainless steel stirrer plate placed on the base. Verification of temperature was obtained using a calibrated PT-100 resistance thermometer.

For experiments up to 343.15 K, the bath was filled using deionized water insulated by a layer of polypropylene balls. For higher-temperature gravimetric samples, high-temperature silicone bath oil was used. The data sets of low boiling point temperature solvents, namely acetone and methanol, were extended to 313.15 K; 2-propanol and ethanol were extended to 343.15 K, and higher boiling point temperature solvents were extended to 343.15 K in 10 K increments.

Excess solute was added to each of the solvent mixture tubes, which were then sealed to prevent solvent loss by evaporation. The solution was stirred at 500 rpm using Teflon-coated stirrer bars for a minimum of 72 h at constant temperature. Following this time, the solid residue was allowed to settle for at least 12 h, without stirring, until the saturated solution was observed to be visibly clear. This was briefly further verified using a laser light of 532 nm. The clear solution was then sampled (~10 mL) using a preheated syringe and filtered into preweighed dry glass vials using a 0.2 μm PTFE membrane (15 mm diameter) syringe filter.

The use of a cap incorporating te PTFE septum reduced the risk of solvent evaporation. The combined mass of vials and sample was measured without undue delay. The caps were then removed, and the solvents were allowed to evaporate slowly until only the solid residue of paracetamol was found to remain in the vials. In the case of volatile solvents with a boiling point of <393.15 K, drying was performed using a vacuum oven at 313.15 K for approximately 1 week with a further verification measurement performed several days later to ensure the sample was completely dry. In the case of less volatile solvents, the temperature of the vacuum oven was increased from 313.15 to 343.15 K after 3 days and sustained for approximately 1 week before the first mass measurement. A verification measurement was again performed several days later to ensure the sample was fully dry. All masses were weighed using an analytical balance (Mettler-Toledo AX054, weighing capacity up to 520 g, sensitivity ±0.1 mg).

Triplicate measurements were performed under the same conditions, and the arithmetic average value of the solubility was calculated. The estimated relative standard uncertainty based on repeat experiments (calculated as 100% standard deviation/average value) of the solubility measurement was less than 2% in each case. Negligible mass changes were noted between initial and verification measurements, thus indicating that the samples were completely dry.

3. THEORY

As noted earlier, the synthetic solubility studies conducted in this paper were carried out in pressurized vessels. The purpose of this pressurization was to control evaporation and furthermore to raise the boiling point of the solvent. It was not the intention of the study to directly affect the solubility of paracetamol in any of the pure solvents by means of the pressurization alone. Indeed, a theoretical demonstration was necessary to verify the appropriateness of negating the influence of pressure, as detailed below.

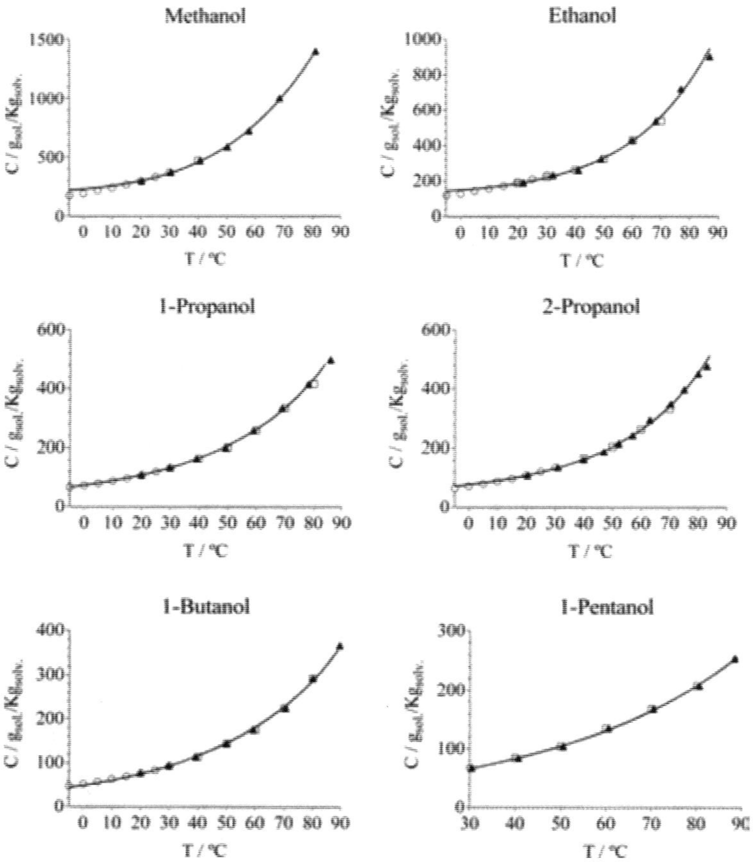

Figure 5. Solubility of paracetamol C versus temperature T in each of the alcohols as determined using (○) published literature values,[5] (▲) pressurized-synthetic data, and (□) the extended gravimetric data set. Pressurized-synthetic data are shown fitted using the NRTL model.

Early studies by Gibson[14] recognized that only in exceptional cases is the solubility of solids in liquids influenced by pressure. These exceptional cases include solutions of the carbonates, sulfates, sulfides, fluorides, and hydroxides of some alkalis, the alkaline earths, and the heavy metals in water, especially where the saturated solutions are very dilute. Pressure on the order of 1000s of atmospheres can significantly increase the solubility of these substances.

The variation of solubility with pressure for solids in pure liquids at equilibrium follows the following relationship[14]

$$\frac{dx_2}{dP} = -\frac{v_2 - v_2^S}{\left.\frac{\partial \mu_2}{\partial x_2}\right|_{P,T}} \tag{1}$$

where x_2 is the molar fraction referring to the solute, P is the pressure in pascals, v_2 is the partial volume of the solute in the solution, and $v_2{}^S$ is its specific volume in the solid phase, both expressed in m^3/mol, and $(\partial\mu_2/\partial x_2)_{P,T}$ is the rate of change of the chemical potential of the solute in the solution with a change of composition at the temperature and pressure of the experiment.

Considering the chemical potential change between solid and liquid solution and in solid phase[7]

$$\mu_2^L = \mu_2^S + kT\ln\frac{a_2}{a_{2,\mathrm{eq}}} = \mu_2^S + kT\ln\frac{x_2\gamma_2}{x_{2,\mathrm{eq}}\gamma_{2,\mathrm{eq}}} \tag{2}$$

where k represents the Boltzmann constant equal to 1.38×10^{-23} J/K, γ_2 is the activity coefficient of the solute in the liquid phase, T is the temperature in kelvin, and the subscript "eq" denotes the equilibrium conditions. Assuming the activity coefficients are approximately equal to 1 for the sake of simplicity, it is possible to evaluate the change in chemical potential with a change in solution composition

$$\left.\frac{d\mu_2}{dx_2}\right|_{P,T} = \frac{kT}{x_2} \tag{3}$$

Substituting eq 3 into eq 1 and integrating leads to the expression

$$\ln\frac{x_2}{x_2^*} = -\frac{v_2 - v_2^S}{kT}(P - P^*) \tag{4}$$

$$x_2 = \exp\left[-\frac{v_2 - v_2^S}{kT}(P - P^*)\right]x_2^* \tag{5}$$

Incorporating the pressure applied for the experiments (1.25 bar(g)) and a volume variation for polymorphic form I of paracetamol of 1.75×10^{-6} m^3/mol at the melting temperature (442.2 K),[15] it is apparent that the term inside the exponential, relative to the atmospheric conditions in eq 5, must have a magnitude order of $O(10^{-19})$ for the range of temperatures under consideration. This implies that the RHS of eq 5 must

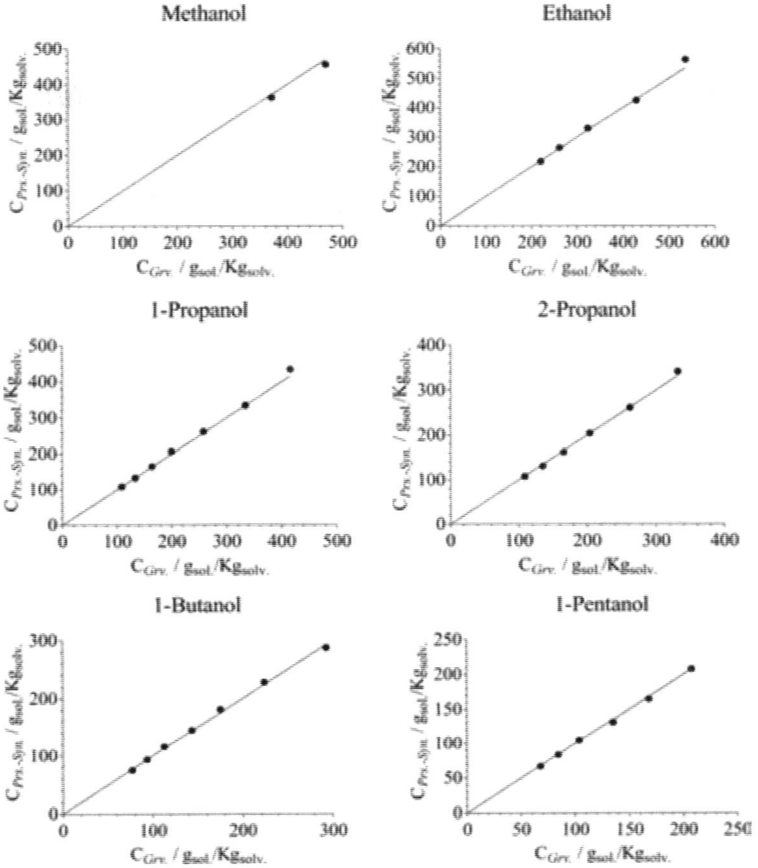

Figure 6. Comparison of gravimetric and pressurized-synthetic solubility results at temperatures up to the boiling point of each alcohol. The straight line represents an ideal correlation with gravimetric results, and the point *x* and y axis coordinates are obtained directly from gravimetric data and best-fit NRTL interpolation of synthetic solubility results, respectively.

tend to zero to balance the equation. Therefore, it can be concluded that the direct effect of pressure on solubility is negligible in the case of the present work.

4. RESULTS AND DISCUSSION

The pressurized-synthetic, as well as the extended gravimetric data from the present work and Granberg and Rasmuson[5] solubility data for paracetamol as measured in each of the different pure alcohols, is presented in Figure 5. No discernible discontinuity was observed between data obtained through the gravimetric and synthetic approaches. Each of the experimental data points represents a single concentration measurement. Given that the synthetic solubility curves were not known in advance and were estimated by extrapolation of existing data, some nonuniformity in the distribution of pressurized-synthetic data points is present. The pressurized-synthetic solubility is shown correlated using the non-random two-liquid (NRTL) model, which was found to have the best fit of the four activity coefficient models used for the thermodynamic modeling of solubility.

Using the pressurized-synthetic approach, solubility data for paracetamol in methanol was obtained up to 354.15 K, well beyond the atmospheric boiling point of methanol (337.65 K). The availability of such high-temperature solubility data overcomes the need for extrapolation with implicit error. For instance, in the case of applying the modified Apelblat correlation (eq 8) to the published data for paracetamol in methanol, an 11.2% error was calculated at the highest temperature measured (354.15 K) when compared to pressurized-synthetic measurements.

As previously noted, in the case of cooling crystallization, operations beyond the boiling up can enhance yield, throughput, and solvent usage. Consider a cooling crystallization of paracetamol in methanol; cooling from the atmospheric boiling point of methanol 337.65 K to the lowest temperature solubility point available (268.15 K) results in a theoretical yield of 80.2%. Increasing the upper temperature limit to 354.15 K facilitates an improvement in yield, which is increased to 87.5%. Indeed, further yield enhancements may be gained through extension of this method to temperatures in excess of those considered in the present work. Crystallization is a late-stage process in pharmaceutical manufacturing wherein the costs of the precursors have largely been absorbed. Therefore, the economic impact of process intensification can be very significant on an industrial scale. These benefits illustrate the clear need for high-temperature solubility data in excess of solvent atmospheric boiling points.

For validating the pressurized-synthetic method, a comparison between gravimetric- and pressurized-synthetic-derived solubility data was required. A direct comparison of the gravimetric-derived solubility data with the NRTL-modeled pressurized synthetic data is shown in Figure 6 for the alcohols.

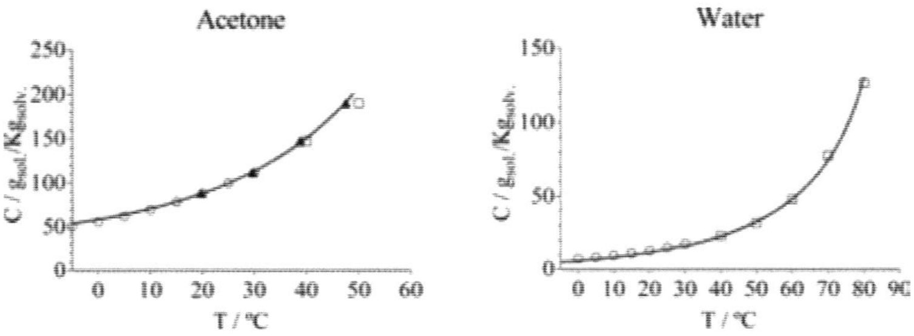

Figure 7. Solubility of paracetamol C versus temperature T in each of the nonalcohols as determined using (○) published literature values,[5] (▲) pressurized-synthetic data, and (□) the extended gravimetric data set. Pressurized-synthetic data are shown fitted using the NRTL model.

The data shown in Figure 6, in general, show excellent agreement between the solubility data obtained using the gravimetric and pressurized-synthetic methods. This is particularly true in the case of higher boiling point alcohols such as 1-butanol and 1-pentanol. Lower boiling point temperature alcohols such as ethanol, 1-propanol, and 2-propanol returned a marginally higher solubility using the pressurized-synthetic method at temperatures close to the atmospheric boiling point. This was anticipated in advance given the potential for error of the gravimetric approach at elevated temperatures.

In the case of the nonalcohols, solubility data for acetone was obtained up to a maximum of 323.15 K. The aggressive nature of hot acetone necessitated the provision of ethylene propylene diene monomer (EPDM) rather than Viton seals.

One of the drawbacks of the synthetic, and consequently the pressurized-synthetic method, is that the solute must be freely suspended within the solvent and physically obscure the laser path. The solubility of paracetamol in water is significantly less than that of all other solvents tested. Despite vigorous agitation, a significant mass of paracetamol was found to be suspended as a froth on the surface of the water. As such, it was not possible to obtain accurate solubility data for paracetamol in water using the pressurized-synthetic approach. In the case of gravimetric testing, even with limited mixing, equilibrium is ultimately reached with the opportunity to sample the clear solution between the paracetamol froth and settled crystals below. Additional gravimetric test results were taken to extend the solubility curve for paracetamol in water to 353.15 K as shown in Figure 7.

Table 2 presents the full data set of pressurized-synthetic results in both mole fraction x_2 and as $g_{solute}/kg_{solvent}$ (solute-free basis) at the measured saturation temperature under a pressure of 1.25 bar(g). Also presented is the calculated mole fraction solubility of paracetamol in each of the solvents using four activity coefficient models and the empirically modified Apelblat equation. Table 3 provides the extended set of gravimetric data for each solvent. For some data points, the mole fraction solubilities in Tables 2 and 3 are identical. In such cases, the concentration of paracetamol in pure solvent utilized for the pressurized-synthetic measurements was informed by the preceding gravimetric results.

The conversion between molar solubility x_2 and mass solubility C and vice versa was performed using the relationship

$$C = \frac{1000x_2 MW_{SOLUTE}}{MW_{SOLVENT}(1 - x_2)} \quad (6)$$

$$x_2 = \frac{\frac{C}{MW_{SOLUTE}}}{\frac{C}{MW_{SOLUTE}} + \frac{1000}{MW_{SOLVENT}}} \quad (7)$$

5. SOLUBILITY MODELING

Many methods have been proposed to describe the solubility of a solute in different pure solvents; for instance, the perturbed-chain statistical-associating fluid theory (PC-SAFT) modeling approach has been widely applied for the correlation and prediction of solubility even at elevated temperatures.[16,17] In this work, the solubility of paracetamol in the selected solvents at different temperatures is correlated with the modified Apelblat equation,[18] and Margules, Van-Laar, NRTL and Wilson activity coefficient-based models.

The modified Apelblat equation is expressed in eq 8.

$$\ln C = a_A + \frac{b_A}{T} + c_A \ln T \quad (8)$$

where C represents the equilibrium concentration in grams of solute per kilograms of solvent at temperature T expressed in kelvin; the three adjustable parameters of the empirical solubility model are indicated as a_A, b_A, and c_A. The values of these parameters were found for each solvent through regression and are summarized in Table 4. The empirical model shows good agreement with the underlying experimental data in the case of all solvents tested.

Although the correlation of Apelblat shows good agreement with the underlining experimentally derived solubility data, the correlation is empirical in nature rather than thermodynamically derived. Activity coefficient models, such as Margules, Van-Laar, Wilson and NRTL, offer a more thermodynamically rigorous approach, accounting for deviation from the ideal solubility. For calculating the binary interaction parameters utilized by these models, it was first necessary to calculate the experimental activity coefficients based on the solubility data.

The experimental activity coefficients were calculated using a nonideal solid–liquid equilibrium equation defined as[18]

$$\ln(x_2\gamma_2) = \frac{\Delta H_f|_{T_m}}{R}\left(\frac{1}{T_m} - \frac{1}{T}\right) - \frac{\Delta C_p|_{T_m}}{R}\left[\ln\left(\frac{T_m}{T}\right) - \frac{T_m}{T} + 1\right] \quad (9)$$

in which the exchange of enthalpy of fusion is represented by ΔH_f, and ΔC_p signifies the difference in heat capacity of solute between the solid and liquid states, both evaluated at the normal melting temperature T_m. Using different scanning calorimetric analyses, Hojjati and Rohani[19] measured the

Table 2. Pressurized-Synthetic and Calculated Mole Fraction Solubility of Paracetamol as Calculated Using Margules, Van-Laar, Wilson, NRTL, and Modified Apelblat Equations across the Entire Range of Solvents at Saturation Temperature T and Pressure P = 1.25 bar(g)[a]

T (K)	C (g/kg)	x^{exp}	x^{apel}	x^{Marg}	x^{VL}	x^{W}	x^{NRTL}
			Methanol				
293.45	297.81	0.0594	0.0613	0.0633	0.0587	0.0575	0.0597
303.55	371.61	0.0730	0.0658	0.0766	0.0740	0.0729	0.0725
313.95	469.93	0.0906	0.0853	0.0933	0.0925	0.0927	0.0902
323.35	586.13	0.1105	0.1099	0.1115	0.1122	0.1147	0.1111
330.95	722.80	0.1328	0.1338	0.1289	0.1308	0.1358	0.1336
341.55	1000.00	0.1749	0.1719	0.1578	0.1609	0.1705	0.1739
354.15	1400.00	0.2288	0.2228	0.2004	0.2045	0.2192	0.2291
			Ethanol				
295.15	190.61	0.0549	0.0577	0.0545	0.0525	0.0520	0.0551
305.35	232.75	0.0662	0.0596	0.0667	0.0650	0.0649	0.0654
314.15	260.93	0.0737	0.0726	0.0794	0.0777	0.0781	0.0750
322.15	323.59	0.0898	0.0894	0.0933	0.0927	0.0932	0.0896
332.95	429.13	0.1157	0.1182	0.1161	0.1177	0.1181	0.1150
341.25	536.38	0.1405	0.1445	0.1375	0.1412	0.1416	0.1402
350.05	718.80	0.1797	0.1759	0.1649	0.1719	0.1720	0.1769
360.15	902.08	0.2156	0.2154	0.2021	0.2119	0.2123	0.2205
			1-Propanol				
293.05	108.09	0.0412	0.0432	0.0406	0.0411	0.0409	0.0412
302.95	132.77	0.0501	0.0464	0.0500	0.0502	0.0501	0.0501
312.55	163.47	0.0610	0.0581	0.0613	0.0612	0.0612	0.0609
322.55	199.45	0.0735	0.0749	0.0757	0.0754	0.0755	0.0748
332.25	257.84	0.0930	0.0947	0.0934	0.0932	0.0930	0.0925
342.15	334.10	0.1173	0.1180	0.1158	0.1162	0.1153	0.1153
351.25	415.29	0.1417	0.1418	0.1411	0.1422	0.1405	0.1410
359.15	498.00	0.1653	0.1639	0.1673	0.1694	0.1667	0.1680
			2-Propanol				
293.65	108.78	0.0415	0.0434	0.0411	0.0407	0.0403	0.0415
304.05	135.00	0.0509	0.0468	0.0512	0.0505	0.0504	0.0506
313.05	161.63	0.0604	0.0579	0.0619	0.0610	0.0613	0.0605
320.25	188.25	0.0696	0.0700	0.0721	0.0712	0.0717	0.0702
325.35	214.88	0.0787	0.0799	0.0805	0.0798	0.0803	0.0788
330.25	241.50	0.0876	0.0904	0.0894	0.0889	0.0895	0.0879
336.15	294.75	0.1049	0.1042	0.1020	0.1024	0.1026	0.1022
343.45	348.00	0.1215	0.1229	0.1193	0.1207	0.1207	0.1208
353.05	450.00	0.1518	0.1501	0.1473	0.1509	0.1500	0.1524
348.15	397.21	0.1364	0.1359	0.1323	0.1348	0.1343	0.1356
356.15	476.47	0.1593	0.1594	0.1573	0.1614	0.1604	0.1631
			1-Butanol				
293.05	77.07	0.0364	0.0379	0.0341	0.0361	0.0360	0.0355
302.95	93.64	0.0439	0.0452	0.0423	0.0438	0.0438	0.0436
312.55	113.13	0.0526	0.0542	0.0522	0.0530	0.0531	0.0532
323.05	143.81	0.0659	0.0669	0.0658	0.0656	0.0658	0.0662
332.55	175.38	0.0792	0.0817	0.0811	0.0797	0.0799	0.0806
343.65	224.08	0.0990	0.1038	0.1037	0.1005	0.1006	0.1015
353.15	292.76	0.1255	0.1276	0.1289	0.1238	0.1235	0.1236
362.65	366.49	0.1523	0.1569	0.1596	0.1522	0.1514	0.1505
			1-Pentanol				
303.55	67.82	0.0380	0.0386	0.0363	0.0383	0.0381	0.0378
313.85	84.20	0.0468	0.0459	0.0458	0.0471	0.0471	0.0471
323.65	103.77	0.0571	0.0571	0.0571	0.0575	0.0577	0.0580
333.65	135.07	0.0730	0.0719	0.0720	0.0711	0.0713	0.0718
343.55	168.19	0.0893	0.0894	0.0903	0.0878	0.0880	0.0885
353.75	207.54	0.1080	0.1100	0.1139	0.1092	0.1094	0.1094
361.55	254.14	0.1291	0.1275	0.1368	0.1302	0.1298	0.1287
			Acetone				
292.95	88.09	0.0327	0.0327	0.0343	0.0325	0.0324	0.0327
302.85	111.65	0.0411	0.0410	0.0425	0.0416	0.0417	0.0413
311.95	146.94	0.0534	0.0534	0.0521	0.0533	0.0534	0.0532

Table 2. continued

T (K)	C (g/kg)	x^{exp}	x^{apel}	x^{Marg}	x^{VL}	x^{W}	x^{NRTL}
			Acetone				
320.65	189.83	0.0680	0.0678	0.0634	0.0678	0.0677	0.0682

[a]Standard uncertainties u are $u(T)$ = 1 K, $u(P)$ = 0.05 bar, and $u_r(x_2)$ = 0.02.

Table 3. Extended Gravimetric Data Set of the Solubility of Paracetamol in Pure Solvents Expressed in Both Mole Fraction x_2 and as $g_{solute}/kg_{solvent}$ (Solute-Free Basis) at Temperature T and Pressure P = 0.1 MPa[a]

T (K)	C (g/kg)	x^{exp}	C (g/kg)	x^{exp}
	Methanol		Ethanol	
293.15	297.81	0.0594	188.00	0.0542
303.15	371.61	0.0730	220.00	0.0628
313.15	469.99	0.0906	260.93	0.0737
323.15			323.59	0.0898
333.15			429.13	0.1157
343.15			536.38	0.1405
	1-Propanol		2-Propanol	
293.15	108.09	0.0412	108.78	0.0415
303.15	132.77	0.0501	135.01	0.0509
313.15	163.47	0.0610	165.54	0.0618
323.15	199.45	0.0735	203.70	0.0749
333.15	257.84	0.0930	262.13	0.0944
343.15	334.10	0.1173	331.50	0.1165
353.15	415.29	0.1417		
	1-Butanol		1-Pentanol	
293.15	77.07	0.0364	61.79	0.0348
303.15	93.64	0.0439	67.82	0.0380
313.15	113.13	0.0526	84.20	0.0468
323.15	143.81	0.0659	103.77	0.0571
333.15	175.38	0.0792	135.07	0.0730
343.15	224.08	0.0990	168.19	0.0893
353.15	292.76	0.1255	207.54	0.1080
	Acetone		Water	
293.15	88.09	0.0327	12.78	0.0015
303.15	111.65	0.0411	17.39	0.0021
313.15	146.94	0.0534	22.94	0.0027
323.15	189.83	0.0680	31.74	0.0038
333.15			48.03	0.0057
343.15			77.67	0.0092
353.15			126.48	0.0149

[a]Standard uncertainties u are $u(T)$ = 1 K, $u(P)$ = 0.05 bar, and $u_r(x_2)$ = 0.02.

Table 4. Apelblat Parameters and Root Mean Square Error (RMSE) Calculated for Each Solvent

solvent	a_A	b_A	c_A	RMSE
methanol	−193.9	7236	30.79	0.0020
ethanol	−193.3	7303	30.55	0.0034
1-propanol	−129.2	4322	20.97	0.0021
2-propanol	−156.6	5545	25.06	0.0014
1-butanol	−170.4	6393	26.93	0.0029
1-pentanol	−105	3207	17.26	0.0011
acetone	−270.4	10300	42.2	0.0128
water	−341.6	13280	52.64	3.7×10^{-5}

thermal properties of paracetamol to be T_m = 442.2 K, ΔH_f = 28.1 kJ/mol, and ΔC_p = 99.6 J mol^{-1} K^{-1} (at the fusion temperature).

Through insertion of the enthalpy, melting temperature, and heat capacity values into eq 9, and given that the experimental solubility values x_2 and temperatures are known, it is possible to evaluate the experimental activity coefficients by rewriting eq 9 for γ_2. The ideal solubility of paracetamol may also be estimated (setting $\gamma_2 = 1$).

Plots of the logarithm of experimental activity coefficients of paracetamol in alcohol- and nonalcohol-based solvents are given in Figure 8(a) and (b), respectively. Referring to eq 9, it is apparent that, at any given temperature, the activity coefficient is inversely proportional to the measured solubility. The ideal solubility is represented by the value 0 on the logarithmic activity coefficient scale with higher values relating to lower solubility. Increasing chain length in alcohols, as is shown in Figure 8(a), results in greater activity coefficient values and consequently reduced solubility. Activity coefficient trends for all solvents are shown to remain relatively flat or decrease slightly with increasing temperature.

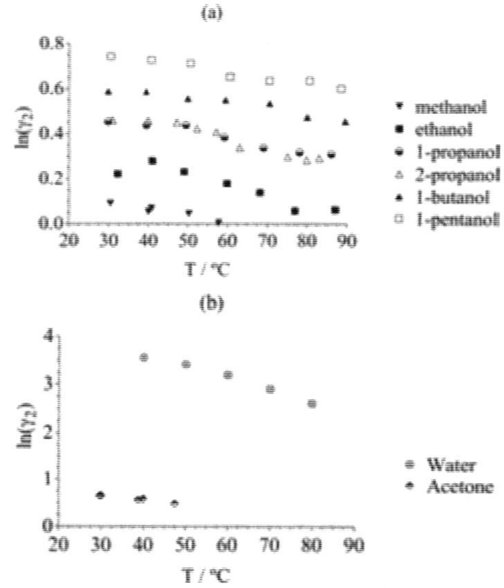

Figure 8. Logarithm of experimental activity coefficients of paracetamol at selected temperatures in (a) alcohols, (b) water, and acetone.

The highest activity coefficient values are observed with water, a highly polar solvent. Physically, this can be explained by the significant influence of the aromatic ring and methyl group in paracetamol on surrounding water molecules despite the fact that the alcohol and amide groupings in the paracetamol molecule are also polar, facilitating the formation of hydrogen bonds.

Having established the experimental activity coefficients, activity coefficient models could now be fitted to the experimental data. The four models selected are described in Table 5. Utilization of each of the models is commonly found in the literature. They can be differentiated by such factors as

Table 5. Activity Coefficient Models Used in the Present Work Complete with Their Respective Number of Parameters and Temperature Dependency Information

model	equation[a]	number of parameters	temperature dependency
Margules	$\ln \gamma_2 = \frac{A}{RT}(1 - x_2)^2$	1 $[A]$	explicit
Van-Laar	$\ln \gamma_2 = \frac{B_{VL}}{\left(1 + \frac{B_{VL}x_2}{A_{VL}x_1}\right)^2}$	2 $[A_{vl}, B_{vl}]$	implicit
Wilson	$\ln \gamma_2 = -x_2\left(\frac{\Lambda_{12}}{x_1 + \Lambda_{12}x_2} - \frac{\Lambda_{21}}{x_2 + \Lambda_{21}x_1}\right) - \ln(x_2 + \Lambda_{21}x_1)$ where $\Lambda_{ij} = \frac{v_j}{v_i}\exp\left(-\frac{\lambda_{ij}}{RT}\right)$ with $i \neq j$ and $i, j = 1, 2$	2 $[\lambda_{12}, \lambda_{21}]$	explicit
NRTL	$\ln \gamma_2 = x_1^2\left[\tau_{12}\left(\frac{G_{12}}{x_2 + x_1G_{12}}\right)^2 + \frac{\tau_{21}G_{21}}{(x_1 + x_2G_{21})^2}\right]$ where $G_{ij} = \exp(-\alpha\tau_{ij})$ and $\tau_{ij} = \frac{g_{ij}}{RT}$ with $i \neq j$ and $i, j = 1, 2$	3 $[g_{12}, g_{21}, \alpha]$	explicit

[a] R is the ideal gas constant expressed in J mol^{-1} K^{-1} (8.314 J mol^{-1} K^{-1}) and v_i or v_j are the molar volumes of the ith or jth components in cm^3/mol.

Table 6. Estimated Binary Coefficients for Paracetamol in Different Pure Solvent Using Margules, Van-Laar, Wilson, and NRTL Models

	Margules	Van-Laar		Wilson		NRTL		
solvent	A	A_{VL}	B_{VL}	λ_{12}	λ_{21}	g_{12}	g_{12}	α
methanol	127.62	17.695	−0.085	549.958	1.46×10^3	1.50×10^4	-7.77×10^3	0.1627
ethanol	624.49	0.4312	0.0903	587.877	614.984	1.844×10^4	-1.07×10^4	0.1019
1-propanol	1.28×10^3	0.5542	0.2805	1.897×10^3	−531.1754	8.86×10^3	-5.83×10^3	0.1033
2-propanol	1278.31	0.6022	0.2429	1.00015×10^3	377.5298	12597.57	−7409.39	0.1257
1-butanol	1.72×10^3	0.6497	0.6108	3.35×10^3	-7.48×10^2	1.56×10^3	−45.657	−2.94
1-pentanol	2.16×10^3	0.802	0.733	3.67×10^3	-1.61×10^2	1.67×10^3	1.49×10^2	−9.468
acetone	1.69×10^3	1.037	0.1632	−558.174	3740.595	11752.49	−5241.13	0.2406
water	8.84×10^3	3.799	0.2622	4.990×10^3	4.871×10^3	11912.37	−1477.12	0.5141

Table 7. Mean Squared Error (MSE) of the Fitted Models for Each Solvent

solvent	Margules	Van-Laar	Wilson	NRTL	MSE_{max}/MSE_{min}
methanol	7.60×10^{-3}	5.10×10^{-3}	1.50×10^{-3}	5.73×10^{-5}	132.63
ethanol	3.63×10^{-3}	1.80×10^{-3}	2.21×10^{-3}	3.00×10^{-4}	12.10
1-propanol	2.30×10^{-4}	8.98×10^{-5}	2.33×10^{-4}	2.27×10^{-4}	2.59
2-propanol	7.11×10^{-4}	3.29×10^{-4}	5.19×10^{-4}	1.70×10^{-4}	4.18
1-butanol	1.93×10^{-3}	1.28×10^{-4}	1.80×10^{-4}	5.45×10^{-4}	15.07
1-pentanol	2.21×10^{-3}	3.30×10^{-4}	2.63×10^{-4}	2.94×10^{-4}	7.58
acetone	3.76×10^{-3}	1.02×10^{-4}	1.92×10^{-4}	7.35×10^{-5}	51.15
water	5.50×10^{-2}	3.80×10^{-3}	4.21×10^{-4}	6.47×10^{-4}	130.95

complexity and the inherent nature of the temperature dependency.

The binary interaction parameters were estimated using the following mathematical optimization problem solved using the MATLAB nonlinear least-squares algorithm *nlinfit*

$$\min_{\theta \in \mathcal{R}} \sum_{i=1}^{N_p} [\ln\gamma_{2,i}|_{mod}(T, \theta) - \ln\gamma_{2,i}|_{exp}(T)]^2 \quad (10)$$

where θ is the set of binary interaction parameters to be estimated for each model and each solvent used, N_p is the number of experimental points carried out for each solvent used, and the subscripts mod and exp refer to the model and experimental activity coefficients, respectively.

Insertion of the binary coefficients from Table 6 into the activity model equations of Table 5 allows for the prediction of molar solubility, which can then be converted to mass solubility as per eq 6. Returning to Table 2, the solubility derived from each of the activity coefficient model is directly compared to the experimental data.

The selection of activity coefficient model is influenced by the quality of fit of the model to the underlying data as well as the complexity of the model. Although the quality of fit of each activity coefficient model is solvent-dependent, across all solvents it is evident that the greatest deviation between experimental and model results occurs in all plots at the extremes of the temperature measurements. Further points in this temperature space could potentially reduce these discrepancies. Indeed, the methodology adopted in the current work facilitates further extension of the temperature range, thus potentially increasing the robustness of prediction within the current temperature range.

In seven of the eight solvents examined, the Margules model is shown to have the largest mean square error (Table 7). The NRTL model has the smallest mean square error for methanol, ethanol, 2-propanol, and acetone; Wilson is smallest for 1-pentanol and water, Van-Laar is smallest for 1-propanol and 1-butanol.

6. CONCLUSIONS

Solubility measurements are essential for pharmaceutical process development. This paper utilizes the pressurized-synthetic approach to facilitate measurement of solubility data for the reference compound paracetamol at elevated temperatures, in some instances exceeding the atmospheric boiling points of the solvent. Such high-temperature data is particularly important in the case of cooling crystallization, where the difference in solubility between upper and lower temperature limits is implicitly associated with yield and industrial efficiency. The error associated with extrapolation of solubility data was demonstrated to be potentially significant (11.2% error) using the modified Apelblat model for paracetamol in methanol and extrapolating from 303.15 to 354.15 K.

The data obtained from the pressurized-synthetic measurements were validated for all solvents with the exception of water by comparison with the gravimetric data. No discernible discontinuity was observed between data obtained through the gravimetric and synthetic approaches. Excellent agreement was seen for higher boiling point solvents at elevated temperatures, whereas for more volatile solvents, the pressurized-synthetic approach indicated a marginally higher solubility when compared to that of the gravimetric approach as was expected.

The solubility data obtained for all pure solvents was modeled using the modified Apelblat empirical equation, as well as Margules, Van-Laar, NRTL, and Wilson activity coefficient models. The NRTL model was shown to have the smallest mean square error for methanol, ethanol, 2-propanol, and acetone.

The methodology presented is this paper is an important extension of the synthetic method, which allows nonintrusive access to valuable high-temperature solubility measurements. The method can be readily automated. Care must be taken when applying the method to ensure appropriate suspension of solute in solvent. When applying the methodology, one should be aware of the potential for polymorphism at elevated temperatures, and if applying to crystallization, one should carefully consider the maximum solid loading.

■ AUTHOR INFORMATION

Corresponding Author

*Tel: +353 (0)61 213134. Fax: +353 (0)61 202944. E-mail: Brian.de.souza@ul.ie. URL: http://www.ul.ie/sspc.

ORCID

Brian de Souza: 0000-0002-4469-7802

Funding

This research has been conducted as part of the Synthesis and Solid State Pharmaceutical Centre (SSPC) and funded by Science Foundation Ireland (SFI) under grant 12/RC/2275.

Notes

The authors declare no competing financial interest.

■ NOMENCLATURE

Acronyms

ACE acetone
ACN acetonitrile
API active pharmaceutical ingredient
DIW deionized water
IPA isopropyl alcohol
MSE mean square error
MSZW metastable zone width
MW molecular weight [g/mol]
PBE population balance equation
PSD particle size distribution

Variables with Units and Symbols

a_A first Apelblat parameter
a_2 Solute activity
A Margules binary interaction parameter
A_{VL} first Van-Laar interaction parameter
b_A second Apelblat parameter
B_{VL} second Van-Laar interaction parameter
c_A third Apelblat parameter
C_{eq} equilibrium concentration [g_{sol}/kg_{solv}]
ΔC_p exchange of heat capacity at constant pressure evaluated at melting point [J mol^{-1} K^{-1}]
g_{ij} NRTL binary interaction parameter
G_{ij} corrected temperature-dependent NRTL binary interaction parameter
ΔH_f exchange of enthalpy of fusion evaluated at melting point [kJ/mol]
k Boltzmann constant [J/K]
N_p number of experimental points for each solvent
P pressure [Pa]
P^* reference pressure [Pa] (atmospheric condition)
R ideal gas constant [J mol^{-1} K^{-1}]
T temperature [K]
T_m melting temperature of paracetamol [K]
v_2^S molar volume of solute in the solid state [cm^3/mol]
v_i or v_j molar volume of the ith or jth component [cm^3/mol]
x_2 solute molar fraction
x_2^* reference solute molar fraction (atmospheric conditions)

Greek Letters

α NRTL nonrandomness parameter
γ_2 activity coefficient referring to the solute
λ_{ij} Wilson binary interaction parameter
Λ_{ij} temperature-dependent Wilson binary interaction parameter
τ_{ij} temperature-dependent NRTL binary interaction parameter
θ set of binary interaction parameters for each activity equation model

■ REFERENCES

(1) Higuchi, T.; Connors, K. A. Phase solubility techniques. *Anal. Chem. Instrum* **1965**, *4*, 117–212.

(2) Worlitschek, J.; Mazzotti, M. Model-Based Optimization of Particle Size Distribution in Batch-Cooling Crystallization of Paracetamol. *Cryst. Growth Des.* **2004**, *4*, 891–903.

(3) Nagy, Z. K.; Fujiwara, M.; Woo, X. Y.; Braatz, R. D. Determination of the Kinetic Parameters for the Crystallization of Paracetamol from Water Using Metastable Zone Width Experiments. *Ind. Eng. Chem. Res.* **2008**, *47*, 1245–1252.

(4) Mitchell, N. A.; Ó'Ciardhá, C. T.; Frawley, P. J. Estimation of the growth kinetics for the cooling crystallisation of paracetamol and ethanol solutions. *J. Cryst. Growth* **2011**, *328*, 39–49.

(5) Granberg, R. A.; Rasmuson, Å. C. Solubility of Paracetamol in Pure Solvents. *J. Chem. Eng. Data* **1999**, *44*, 1391–1395.

(6) Fernandez, C. Effect of the nature of the solvent on the crystallization of paracetamol. In *Proceedings of the 14th International Symposium on Industrial Crystallisation*; Cambridge, UK, 1999.

(7) Granberg, R. A.; Ducreux, C.; Gracin, S.; Rasmuson, Å. C. Primary nucleation of paracetamol in acetone–water mixtures. *Chem. Eng. Sci.* **2001**, *56*, 2305–2313.

(8) Shakeel, F.; Alanazi, F. K.; Alsarra, I. A.; Haq, N. Solubilization Behavior of Paracetamol in Transcutol–Water Mixtures at (298.15 to 333.15) K. *J. Chem. Eng. Data* **2013**, *58*, 3551–3556.

(9) Nichols, G.; Frampton, C. S. Physicochemical characterization of the orthorhombic polymorph of paracetamol crystallized from solution. *J. Pharm. Sci.* **1998**, *87*, 684–693.

(10) Jouyban, A.; Fakhree, M. A. A. *Toxicity and Drug Testing*; Acree, B., Ed.; InTech, 2012.

(11) Sheng, X.; Wang, Q.; Chen, L.; Shen, Z.; Pei, Y. Determination and Correlation for the Solubility of Glutaric Acid in Cyclohexane + Cyclohexanol + Cyclohexanone Solvent Mixtures. *J. Chem. Eng. Data* **2016**, *61*, 3761–3769.

(12) Luo, W.; Ruan, D.; Liu, D.; Xie, K.; Li, X.; Deng, W.; Guo, C. Measurement and Correlation for Solubilities of Adipic Acid in Acetic Acid + ε-Caprolactone Mixtures and Cyclohexanone + ε-Caprolactone Mixtures. *J. Chem. Eng. Data* **2016**, *61*, 2474–2480.

(13) Hao, H.; Hou, B.; Wang, J.; Zhang, M. Solubility of Erythritol in Different Solvents. *J. Chem. Eng. Data* **2005**, *50*, 1454–1456.

(14) Gibson, R. E. General considerations of the effect of pressure on solubility. *Trans., Am. Geophys. Union* **1938**, *19*, 273.

(15) Espeau, P.; Céolin, R.; Tamarit, J.-L.; Perrin, M.-A.; Gauchi, J.-P.; Leveiller, F. Polymorphism of paracetamol: relative stabilities of the monoclinic and orthorhombic phases inferred from topological pressure-temperature and temperature-volume phase diagrams. *J. Pharm. Sci.* **2005**, *94*, 524–539.

(16) Ruether, F.; Sadowski, G. Modeling the Solubility of Pharmaceuticals in Pure Solvents and Solvent Mixtures for Drug Process Design. *J. Pharm. Sci.* **2009**, *98*, 4205–4215.

(17) Reschke, T.; Zherikova, K. V.; Verevkin, S. P.; Held, C. Benzoic Acid and Chlorobenzoic Acids: Thermodynamic Study of the Pure Compounds and Binary Mixtures With Water. *J. Pharm. Sci.* **2016**, *105*, 1050–1058.

(18) Prausnitz, J. M.; Lichtenthaler, R. N.; Azevedo, E. G. de. *Molecular Thermodynamics of Fluid-Phase Equilibria*; Pearson Education: Upper Saddle River, NJ, 1998.

(19) Hojjati, H.; Rohani, S. Measurement and Prediction of Solubility of Paracetamol in Water–Isopropanol Solution. Part 2. Prediction. *Org. Process Res. Dev.* **2006**, *10*, 1110–1118.

501976 How Raw Material Attributes and Process Parameters Affect the Global Residence Time Distribution Behavior in Continuous Direct Compression

Abstract:

Title: *How Raw Material Attributes and Process Parameters Affect the Global Residence Time Distribution Behavior in Continuous Direct Compression*

Author: Samantha Hurley

The pursuit of drug product continuous manufacturing will revolutionize the pharmaceutical industry by improving product flow in our supply chains, such as leveraging fast cycle times and flexible batch sizes, and by reducing lead times and inventory levels. A successful batch to continuous manufacturing conversion requires a very robust control scheme. One commonly described control scheme involves the use of NIR spectroscopy for blend composition monitoring. This study evaluates the use of an alternate method, a Residence Time Distribution (RTD) based process model for control of the continuous direct compression (CDC) process without direct measurement of blend uniformity. In some cases, this method is preferred over traditional spectroscopic techniques as the primary control across a wide range of throughputs, blender speeds, and material properties.

The CDC system studied consists of two continuous convection blenders in series followed by a tablet press. Each of the blenders and the tablet press are modeled as a Plug Flow Reactor (PFR) in series with two Continuously Stirred Tank Reactors (CSTRs). Since each PFR and CSTR is characterized by a single time constant, t_{Lag} and t respectively, each blending unit can be described by threeparameters. Tracer experiments were conducted by spiking API into either the inlet of blender 1 or blender 2. The effluent API concentration was measured either by blend NIR spectroscopy at the exit of blender 2 or tablet HPLC at the exit of the tablet press. The RTD model parameters were optimized by using a least squares algorithm between the blend NIR/tablet HPLC and the RTD model, previously described. The model parameters were developed across various processing conditions and material properties (e.g., bulk densities and flow indices).

The results from the tracer experiments performed show that a ratio of $t:t_{Lag}$ is constant within each blending unit across all the various conditions. This universal behavior thus allows the development of a simple empirical relationship which describes RTD model parameters across the entirety of the design space. The results indicate that the relationship between the blender 1 and blender 2 model parameters versus throughput (TP) and blender speed (BS) can be described as

where A, B, C, and D are constants. The results also indicate the relationship between the tablet press model parameters versus throughput can be described as

where A, B, and C are constants. Knowledge of this universal behavior allows interpolation of model parameters anywhere within the design space, enables simpler model validation, and significantly simplifies process model development for future products.

502628 Development of a Continuous Blending System

Abstract:

Continuous processing equipment for solid dosage manufacture at a pilot scale can be costly and consume more space than is typical for pilot plants. Merck developed its own custom continuous blending system to fit within limited space within its clinical pilot plant. This discussion will walk through the design process, key learnings, challenges and development experiences as the pilot plant team designed a system to fit inside an existing pilot plant space.

In this discussion, the team will use experiences, system design and early data to show how a clinical continuous blending system can be designed to mimic the commercial process while meeting vertical and footprint limitations.

500878 Investigating Continuous Powder Blending at Different Scales Using Residence Time Distribution Studies

Abstract:

Powder mixing or blending is an essential unit operation required to achieve the homogenization of excipients (fillers, disintegrates, lubricants, etc.) and active pharmaceutical ingredient(s). Pioneering developments in continuous drug product manufacturing have facilitated the adoption of continuous powder blending platforms.[1-3] The residence time distribution (RTD) is a probability function that describes the amount of time that a fraction of material spends within any continuous processing system. The RTD is a critical tool in continuous processing equipment and rig design, in pharmaceutical process understand and is also used to ensure the quality of the manufactured product. [4]

In this work, continuous powder lubrication performance was compared using continuous dry powder blenders at both lab scale and production scale by varying total mass flow rate and blender blade speed with a view to translating the lubrication performance between blender systems. At steady state, the RTD and residence mass hold-up in the blender were measured. The tensile strength of tablets compacted from the blend at steady state were also measured in order to characterize the lubrication response.[5] Three different regimes of blending were observed with increasing blade speed – 'mixing', 'transitioning' followed by 'conveying'. Peclet number (Pe# - ratio of axial transport to axial dispersion) was determined by fitting RTD curves using the axial dispersion model. A slower line rate was required at the lab scale in order to achieve the same Pe# at production scale. The mean residence time, τ, was found to be the key predictor of lubrication performance, independent of blender scale. For both blender systems, τ decreases with increasing mass flow rate and increasing blade speed and, where blender speed was increased, changes in τ were minimized by selecting higher mass flow rates. Longer τ produced greater variation and skewness of the RTD across each blending system which has implications for material tracking and product yield within any continuous manufacturing process. Significantly, we show that the mean residence time, τ can be calculated directly from knowledge of the residence mass hold-up and mass flow rate for both systems.

References

1. Byrn, S. *et al.* Achieving Continuous Manufacturing for Final Dosage Formation: Challenges and How to Meet Them May 20–21 2014 Continuous Manufacturing Symposium. *Journal of Pharmaceutical Sciences* **104**, 792-802, doi:http://dx.doi.org/10.1002/jps.24247 (2015).

2. Teżyk, M., Milanowski, B., Ernst, A. & Lulek, J. Recent progress in continuous and semi-continuous processing of solid oral dosage forms: a review. *Drug Development and Industrial Pharmacy* **42**, 1195-1214, doi:10.3109/03639045.2015.1122607 (2016).

3. Gao, Y., Vanarase, A., Muzzio, F. & Ierapetritou, M. Characterizing continuous powder mixing using residence time distribution. *Chemical Engineering Science* **66**, 417-425,

doi:http://dx.doi.org/10.1016/j.ces.2010.10.045 (2011).

4. Engisch, W. & Muzzio, F. Using Residence Time Distributions (RTDs) to Address the Traceability of Raw Materials in Continuous Pharmaceutical Manufacturing. *Journal of Pharmaceutical Innovation* **11**, 64-81, doi:10.1007/s12247-015-9238-1 (2016).

5. Kushner Iv, J. & Moore, F. Scale-up model describing the impact of lubrication on tablet tensile strength. *International Journal of Pharmaceutics* **399**, 19-30, doi:http://dx.doi.org/10.1016/j.ijpharm.2010.07.033 (2010).

491220 Leveraging Raw Material and Blend Properties with the Die Filling Step of the Tableting Process

Abstract:

Leveraging raw material and blend properties with the die filling step of the tableting process

V. Vanhoorne[a*], B. Van Snick[a*], W. Grymonpré[a], J. Dhondt[c], C. Segers[a], K. Van Vooren[a], T. Eerdekens[a], G. Di Pretoro[b], T. De Beer[c], C. Vervaet[a]

[a]*Laboratory of Pharmaceutical Technology, Ghent University, Ottergemsesteenweg 460, B-9000 Ghent, Belgium*

[b]*Pharmaceutical Research and Development, Division of Janssen Pharmaceutica, Johnson & Johnson, Turnhoutseweg 30, B-2340 Beerse, Belgium*

[c]*Laboratory of Pharmaceutical Process Analytical Technology, Ghent University, Ottergemsesteenweg 460, B-9000 Ghent, Belgium*

**Both authors contributed equally*

Tablets are the most widely used dosage forms for oral drug delivery, mainly due to their ease of manufacturing, accurate dosing and high patient compliance. Today, tablets still account for more than 80 % of all pharmaceutical preparations. Despite the high quantities of tablets produced daily, a lot of the established processes and formulations are the results of trial-and-error experiments. Knowledge and understanding of the influence of process parameters and formulation properties on the final product quality is of great importance for the industry, offering economic efficiency and high product quality.

Therefore, the aim of this study is to investigate the influence of a broad range of formulation properties and process parameters related to the feed frame on the tablet properties and tablet weight variation. Based on characterization of a wide range of fillers for direct compression (13 materials) and active pharmaceutical ingredients (API) (14), thirty divergent blends were composed and evaluated on the tablet press. Characterization of the individual materials included size and shape analysis (laser diffraction, static image analysis, scanning electron microscopy), specific surface analysis, tapped, bulk and true density, flowability (drained angle of repose, dynamic angle of repose, flowability index, basic flowability energy, specific energy, aeration index, permeability), compressibility, charge density, moisture content (loss-on-drying) and dynamic vapor sorption. Based on PCA analysis of the fillers and API's, 5 fillers (Pearlitol 100SD, Emcompress, Avicel PH101, Tablettose 80, Avicel PH200) and 6 API's (Caffeine, Metoprolol tartrate, Intelence, micronized, powdered and dense powder grades of Paracetamol) were selected to be combined into 30 blends (API/filler/magnesium stearate 9.93/89.33/0.75) in order to cover a broad range of formulations. Additionally, these blends were characterized. An experimental design with 6 runs (including 2 centerpoints) varying paddle speed (paddle 1: 15 – 100 rpm, paddle 2: 18 – 120 rpm) and turret speed (10 – 100 rpm) was performed on an instrumented rotary tablet press (Modul P, GEA Pharma systems). Experiments were performed at a fixed main compression force (15 kN), with a fixed pre-compression displacement (0.5 mm) and overfill (2 mm). Steady state of the tablet press was determined by evaluation of the tablet mass flow signal on a catch scale. After reaching steady state, the press was run for 3, 6 or 30 min at turret speeds of 10, 55 and 100 rpm, respectively. Samples were taken during 8, 2 and 2 s at intervals of 100, 20 and 10 s at turret speeds of 10, 55 and 100 rpm, respectively. Tablets of the individual samples were weighed and the pooled mean and pooled relative standard deviation was calculated. Additionally, the residence mass in the feed frame (RMF) was determined after each run and the bottom punch position to obtain a main compression force of 15 kN (BPMC) and ejection force were recorded. Multivariate data analysis was applied to investigate the relation between the process parameters, blend characteristics, product and process responses.

Blends with metoprolol tartrate as API and Pearlitol 100 SD or Emcompress as filler tended to stick on the punches. This behavior could be linked to the high wall friction value of metoprolol. Cellulose-based fillers did not exhibited this behavior because of their self-lubricating behavior. Additionally, high ejection forces were observed using Pearlitol SD100 as filler, resulting in picking and capping in combination with paracetamol.

Overall, the density of the blend impacted the mean tablet weight and residence mass in the feed frame. In contrast, the tablet weight variability could be related to the flowability of the blend expressed by the flow function coefficient obtained with the ring shear test. For the formulations with a well flowing, high density filler (e.g. Emcompress) no effect or a minor effect of the paddle speed and turret speed was detected on the tablet weight, tablet weight variability, RMF and BPMC. In contrast when a poorly flowing, low density filler (e.g. Avicel PH101) was used in the formulation, the tablet weight, RMF and BPMC were negatively and positively affected using a higher turret speed and paddle speed, respectively. These effects were most pronounced when the poorly flowing, low density filler was combined with an API with similar properties (e.g. Metoprolol tartrate, micronized Paracetamol). This signifies that die filling is incomplete and that the mass in the feed frame is low when poorly flowing, low density formulations are tableted. However, interestingly, the variability of the tablet weight of these formulations was affected by the turret speed or paddle speed. Higher paddle and turret speeds resulted in a reduction and increase of the tablet weight variability, respectively. The tablet weight variability of a very cohesive formulation, e.g. micronized paracetamol combined with Avicel PH101 (ffc=2.55), varied from 0.7 – 7.9% depending on the turret and paddle speed. Operating the tablet press at the highest paddle speed and lowest turret speed resulted in the lowest tablet weight variability, whereas operating the tablet press at the lowest paddle speed and highest turret speed resulted in the highest tablet weight variability. Intermediate settings of turret and paddle speed (center point conditions) resulted in a tablet weight variability of 3.6%. Therefore, it could be concluded that even poorly flowing formulations with a low density (e.g. consisting of a filler intended for wet granulation and a micronized API) could be tableted at high speed on a rotary tablet press, yielding acceptable tablet weight variability when the paddle speed is optimized.

The developed characterization database will be further expanded with more materials and additional material characteristics (e.g. compression properties as elastic recovery, plasticity constant, fragmentation) and a PLS model will be developed to link the material characteristics to process settings and responses of the tableting process. This way the model will be a valuable tool to compose formulations with new APIs during drug product development.

498979 Modelling of Pharmaceutical HME Processes

Abstract:

Modelling of Pharmaceutical HME Processes

Josip Matić*,**, Milica Stanković-Brandl**, Amrit Paudel**,Christophe Herkenne***, Jessica Lovey Martinetti***, Sophie Martel***, Johannes Khinast*,**

*Institute for Process and Particle Engineering, Graz University of Technology, Inffeldgasse 13/III, 8010 Graz, Austria

**Research Centre Pharmaceutical Engineering GmbH, Inffeldgasse 13/III, 8010 Graz, Austria

***Debiopharm Research&Manufacturing S.A., Campus "après-demain", Rue Levant 146, CP 368, 1920 Martigny, Switzerland

Email for correspondence: khinast@tugraz.at

Introduction

Hot melt extrusion (HME) is a continuous manufacturing process primarily using co-rotating intermeshing twin-screw extruders (TSE). The continuous manufacturing nature of the process allows for a steady product quality and a decrease in production costs. Furthermore, the process has the potential of increasing the bioavailability of poorly soluble drugs by enabling manufacturing of amorphous solid dispersions, making it interesting to the pharmaceutical industry. Although the process has been used in the polymer and rubber industry for years it is still not well enough understood and fairly new in the pharmaceutical industry. The HME setup for new API-polymer system and the technical transfer between different extruder scales (scale-up) are still open question.

Methodology

HME process simulations are done using fully resolved 3D simulations, or by simplified 1D simulation approaches [1-4]. The 1D mechanistic model developer in-house is able to simulate the HME process as a whole, giving information about the filling degree, pressure and temperature distribution, as well as the specific mechanical energy input (SMEC) across the screw configuration[1]. In order to setup the 1D model, the individual screw elements have to be characterized by determining the inherent conveying capacity and the pressure build up capacity[2]. The characterization is done using the mesh-free SPH simulation

approach, giving additional insight into the mixing and sheer rate distribution for every screw element, allowing for a comparison between triple- and double-flighted extruder screw element geometries [3].

Results

An HME process on a triple-flighted 16mm extruder, with two different materials, was investigated, providing recommendations for the process setup and scale-up. The validation was done by comparing the experimentally obtained torque; melt temperature and residence time distribution to the *in-silico* predictions.

Literature

[1] A. Eitzlmayr, G. Koscher, G. Reynolds, Z. Huang, J. Booth, P. Shering, and J. Khinast, "Mechanistic Modeling of Modular Co-Rotating Twin-Screw Extruders," *Int. J. Pharm.*, vol. 474, no. 1–2, pp. 157–176, 2014.

[2] K. Kohlgrüber, *Co-Rotating Twin-Screw Extruders*. Munich: Carl Hanser Publishers, 2008.

[3] A. Eitzlmayr, J. Matić, and J. Khinast, "Analysis of Flow and Mixing in Screw Elements of Co-Rotating Twin-Screw Extruders via SPH," *AIChe J.*

[4] R. Baumgartner, J. Matić, S. Schrank, S. Laske, J. G. Khinast, and E. Roblegg, "NANEX: Process Design and Optimization," *Internaitonal J. Pharm.*, vol. 506, pp. 35–45, 2016.

492270 In-Depth Experimental Analysis of Pharmaceutical Twin-Screw Wet Granulation in View of Detailed Process Understanding

Abstract:

Twin-screw wet granulation is gaining increasing interest within the pharmaceutical industry for the continuous manufacturing of solid oral dosage forms. However, limited prior fundamental physical understanding has been generated relating to the granule formation mechanisms and kinetics along the internal compartmental length of a twin-screw granulator barrel, and about how process settings, barrel screw configuration and formulation properties such as particle size, density and surface properties influence these mechanisms. One of the main reasons for this limited understanding is that experimental data is generally only collected at the exit of the twin-screw granulator barrel although the granule formation occurs spatially along the internal length of the barrel. The purpose of this study is to analyze the twin-screw wet granulation process using both hydrophilic and hydrophobic formulations, manufactured under different process settings such as liquid-to-solid ratio, mass throughput and screw speed, in such a way that the mechanisms occurring in the individual granulator barrel compartments (i.e., the wetting and different conveying and kneading compartments) and their impact upon granule formation are understood. To achieve this, a unique experimental setup was developed allowing granule characteristic data-collection such as size, shape, liquid and porosity distribution at the different compartments along the length of the granulator barrel. Moreover, granule characteristic information per granule size class was determined. The experimental results indicated that liquid-to-solid ratio is the most important factor dictating the formation of the granules and their corresponding properties, by regulating the degree of aggregation and breakage in the different compartments along the internal length of the twin-screw granulator barrel. Collecting appropriate and detailed experimental data about granule formation along the internal length of the granulator barrel is thus crucial for gaining fundamental physical understanding of the twin-screw wet granulation process.

496103 Understanding the Impact of Feedframe Design and Process Parameters on the Die-Filling Step of the Tableting Process

Abstract:

Powder compression on rotary tablet presses is widespread in various industries and for pharmaceuticals over 80 % of all oral dosage formulations are manufactured as tablets. The ease of tablet manufacturing combined with a high economic efficiency and patient compliance all contribute to the prominent role of tablets as pharmaceutical oral dosage form. With the concept of continuous processing gaining momentum in pharmaceutical manufacturing, a thorough understanding of how process and formulation parameters impact the critical quality attributes (CQA) of the end product is more than ever required. Regarding tableting processes, consistent die-filling is a crucial control variable since this will impact the final product quality (e.g. drug content and tablet weight). Although a wide range of feedframe designs is available in terms of paddle-type, chamber height, powder-flow 'enhancers' and feedframe aeration, there is limited knowledge on which specific feedframe design is the most appropriate for a specific formulation in order to obtain consistent die filling during tabletting.

The aim of this study is to investigate the impact of feedframe design and tableting process parameters on the die-filling step and other critical quality attributes (e.g. tablet tensile strength). The knowledge gained via this study will contribute to a rational selection of the feedframe design and tabletting settings in function of the formulation characteristics. Different paddle-types were selected based on their prevalence in the pharmaceutical industry and these were implemented in a standard feedframe consisting of 2 paddle wheels, whereby the feeding wheel had 8 paddles and the recovery wheel 12 paddles. The following 4 feedframe designs were evaluated: curved paddles, straight paddles, thin cylindrical paddles with varying height in the paddle wheel and curved paddle-wheels with lower thickness and a 'scraper', together with the possibility of aeration (present or not). An experimental design with 48 runs including 4x2 centerpoints for each paddle-type was conducted on a fully instrumented rotary tablet press (Modul P, GEA Pharma systems) with an additional fill-level sensor mounted on the feeding-tube which ensured a constant powder volume above the feedframe throughout all experiments. Paddle speed (15 - 100 rpm), the ratio of paddle speed 1 (i.e. feeding wheel) over paddle speed 2 (i.e. recovery wheel)(1-1.4), turret speed (10-100 rpm) and overfill-level (0-4 mm) were varied in the DOE as process parameters.

Experiments were performed at a bottom punch position generating a main compression force (15 kN), with a fixed pre-compression displacement (0.5 mm). Steady state of the tablet press was determined by monitoring the tablet mass flow signal on a catch scale. Each experimental run was performed on an empty and cleaned feedframe. After reaching steady state, tablets were sampled at different time points. The sampling duration (required to sample 20 tablets per time point) and sampling interval depended on the turret speed: a 20s sampling duration with an 80s interval at 10 rpm compared to a sampling duration of 2s with an 8s interval at 100 rpm. Tablets were analysed on their weight, tablet dimensions and hardness using a combitester (Pharmatron Smarttest 50), and the pooled mean and pooled relative standard deviation were calculated for tablet weight, tablet tensile strength and tablet porosity as responses. Additionally, the time needed to reach steady state conditions for each experimental run was calculated and the displacement signals on the pre-compression were monitored to evaluate fluctuations in the tablet weight variation. With respect to the latter, main compression forces were also recorded. Based on a PCA analysis of fillers and API's, microcrystalline cellulose (Vivapur 101) was initially selected as poorly flowing formulation for performing the experiments.

Overall, feedframe design affected the die filling and consequently the mean tablet weight: significantly higher tablet weights were obtained using straight paddles, while the feedframe with 'scraper' and lowered curved paddles resulted in significantly lower tablet weights. At some process settings with the latter paddle design, feedframe obstruction occurred due to intensive compaction inside the feedframe. However, no significant influence on the tablet weight variability was detected, indicating that the consistency of die filling seemed guaranteed for each paddle type, specifically since no interaction was observed between paddle type and overfill-level for RSD on tablet weight. The process parameters strongly impacted the tablet weight and weight variability: (a) higher turret speeds resulted in significantly lower mean tablet weight (i.e. less die-filling) and a higher tablet weight variability, (b) more overfill resulted in a higher tablet weight without an effect on tablet weight variability, (c) paddle speed 1 had a similar impact as the overfill-level on the mean tablet weight at higher paddle speeds, yielding a higher tablet weights and lower variability indicative of a more consistent die-filling at higher paddle speed 1 (i.e. feeding wheel) for this formulation. Similar trends were observed when investigating the tablet tensile strengths at each experimental run for this formulation.

The developed experimental design will be repeated with other powder formulations with varying density and flow behavior in order to investigate the robustness of this study towards other formulations. As a result, this research study will contribute to an improved approach for the set-up of future experimental tableting studies.

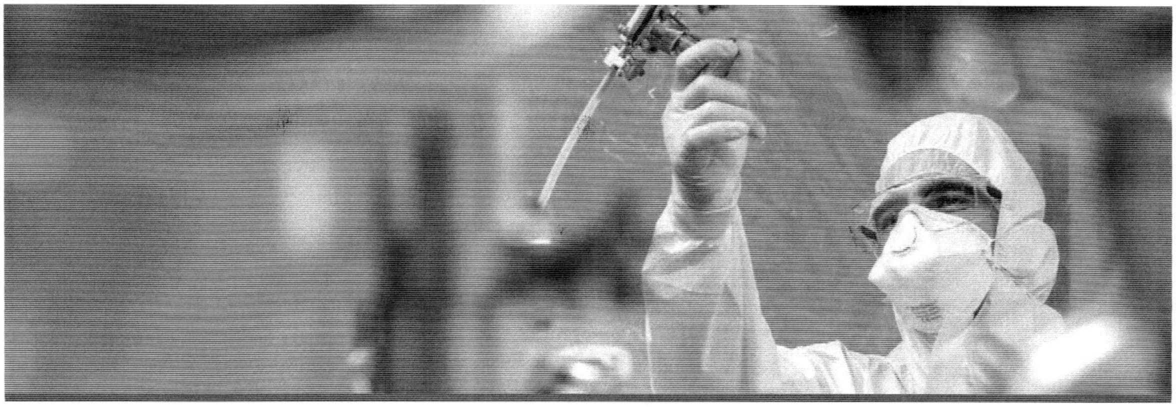

Intensification of a Continuous Process Guided by Mechanistic Modeling

Rudi Oliveira, Ana Cruz, Filipe Ataíde, Nuno Matos

October 2017

Hovione

Hovione has extensive manufacturing capacity spread over three continents

New Jersey, USA
Manufacturing facilities small volume
R&D Labs, Kilo and pilot plant
Sales and marketing for North America

Loures, Portugal
Manufacturing facilities including
R&D Labs, kilo and pilot plants

Macau, China
Manufacturing facilities

Cork, Ireland
Manufacturing facilities

Taizhou, China
Manufacturing facilities

Hovione on the Pharmaceutical Supply Chain

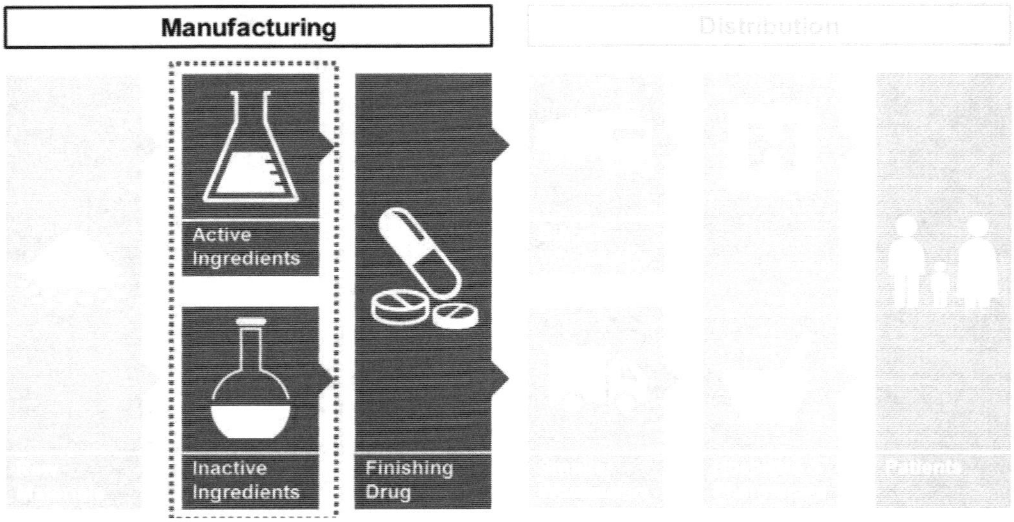

Drug Development

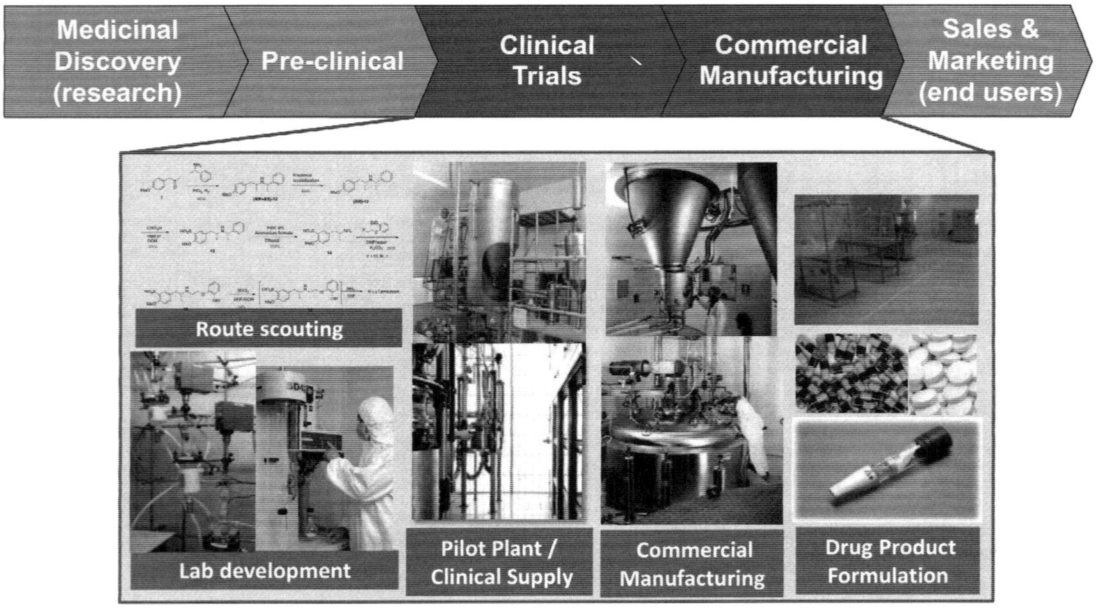

Agenda

Flow Chemistry: Features & Capabilities

Approach to Development: QbD

Feasibility Study: Batch-to-Continuous

Conclusions

Agenda

Flow Chemistry: Features & Capabilities

Approach to Development: QbD

Feasibility Study: Batch-to-Continuous

Conclusions

Batch vs Flow

Understand the Drivers

Main Features

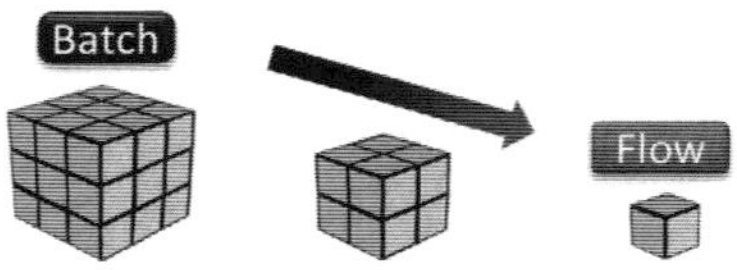

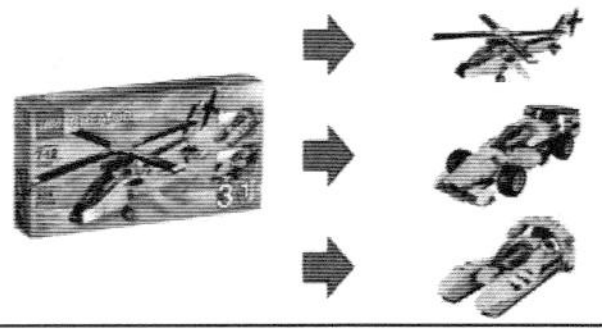

Process Drivers

Safety	Efficiency	Quality	Environmental Impact	Time to market	Batch Size Flexibility	Costs

Flow Capabilities

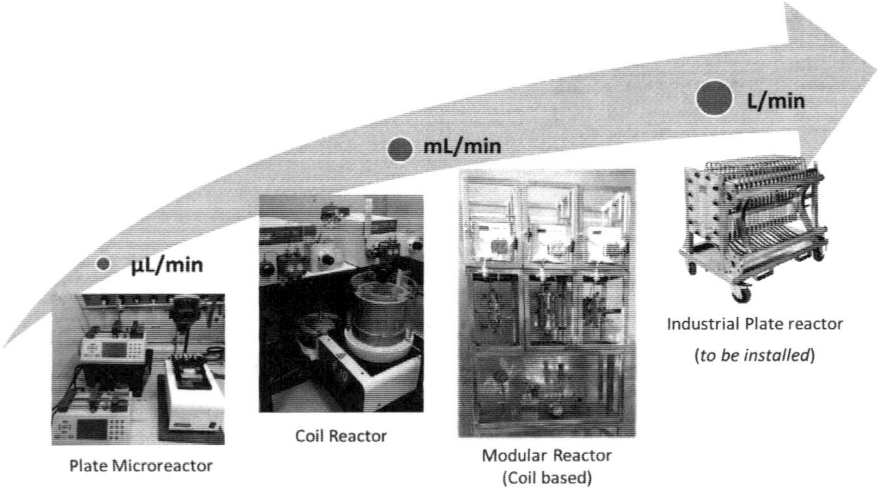

Agenda

Flow Chemistry: Features & Capabilities

Approach to Development: QbD

Feasibility Study: Batch-to-Continuous

Conclusions

Process Development at Hovione

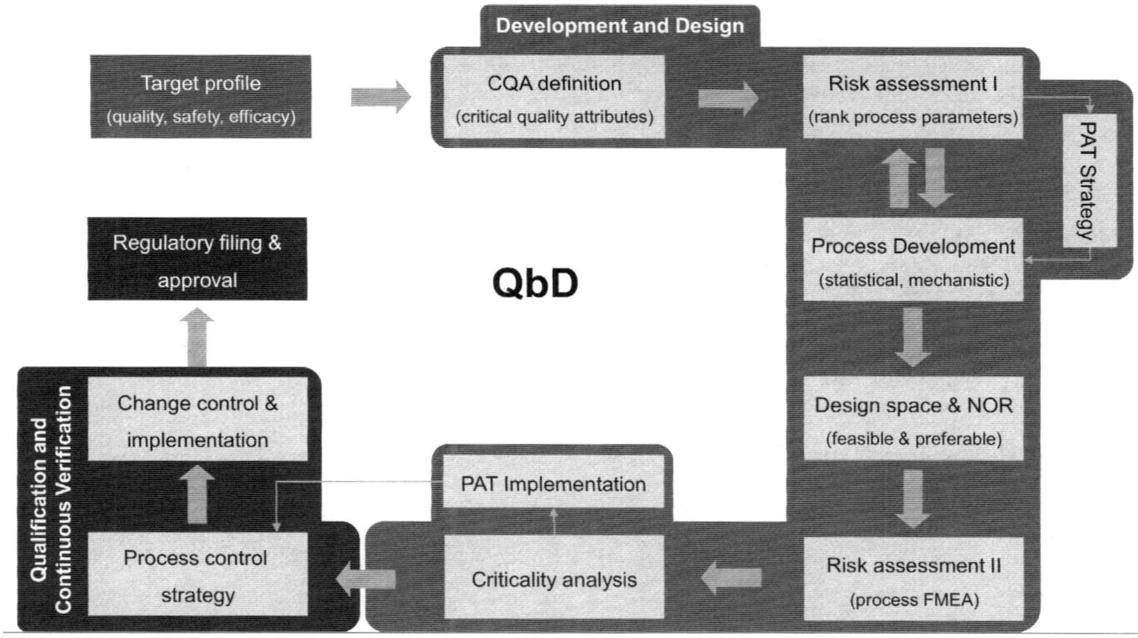

Approach to Development

Systematic workflow with gate reviews

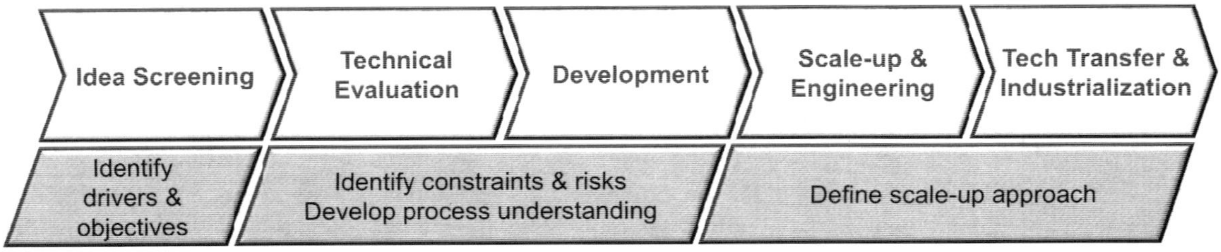

Agenda

Flow Chemistry: Features & Capabilities

Approach to Development: QbD

Feasibility Study: Batch-to-Continuous

Conclusions

Project X

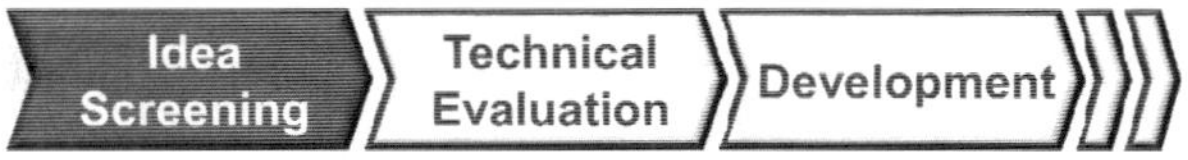

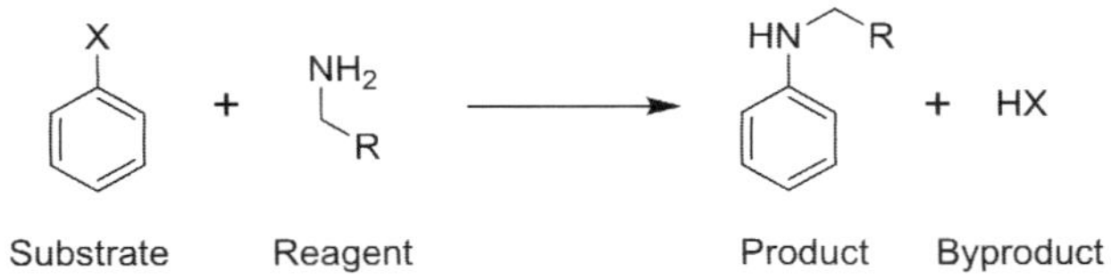

Significant exotherm at > 150ºC

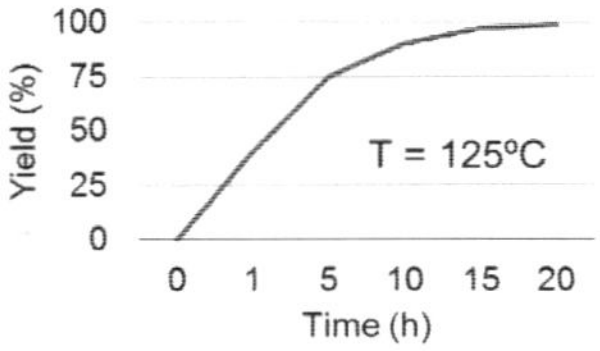

Low b.p. of solvent

Challenge
Batch-to-continuous

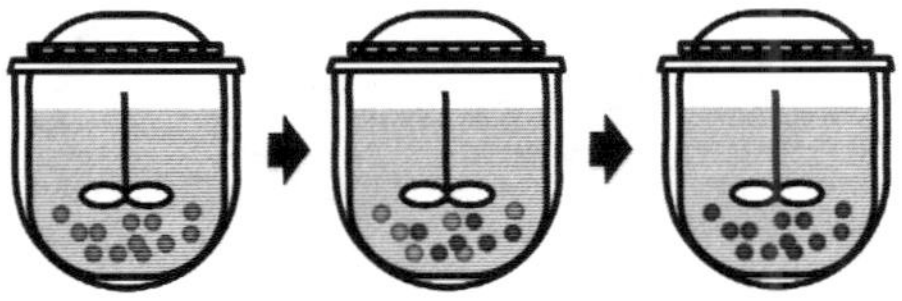

Batch Process

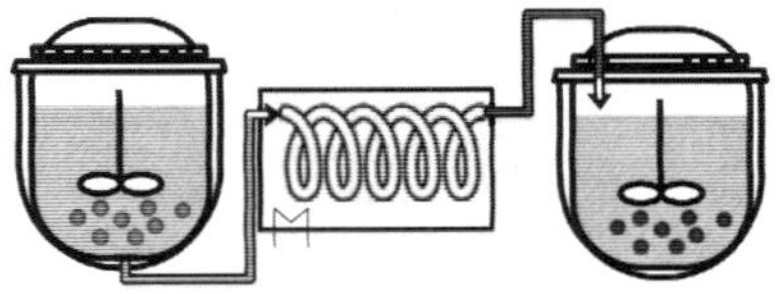

Continuous Process

Where do you start?

Modeling

Idea Screening | Technical Evaluation | Development

Batch data

DynoChem

- Design space
- Predict the relation between process parameters and relevant quality attributes

Batch Model

Idea Screening | Technical Evaluation | Development

Reaction Scheme
Substrate + Reagent ⟶ Product + Byproduct

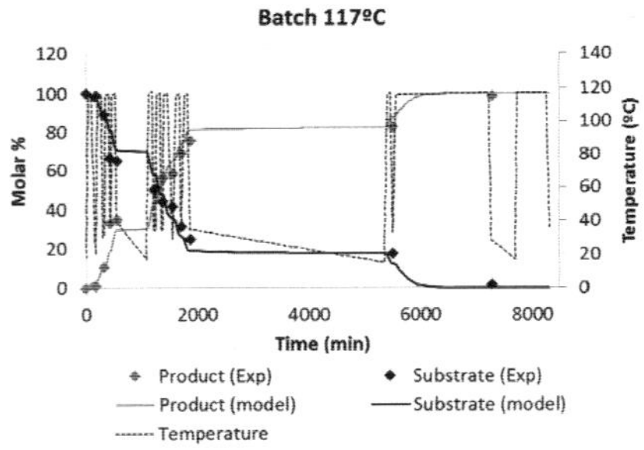

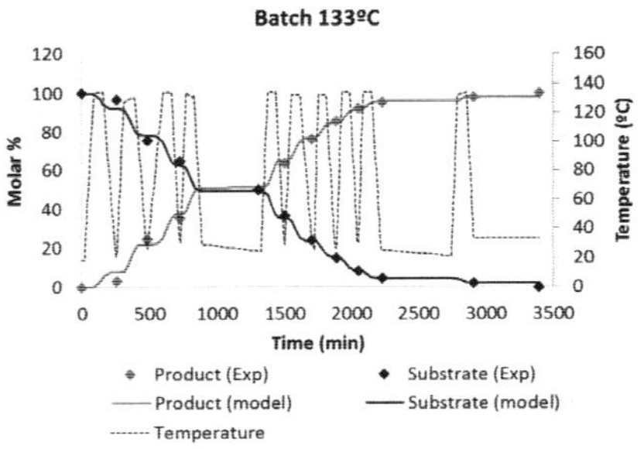

			Confidence Interval
Kinetic constant (*k* @ 117°C)	2.72 x 10^{-4}	L/mol.s	11.8%
Activation energy (E_a)	58.66	kJ/mol	11.8%

Idea Screening | Technical Evaluation | Development

Performance Estimation
under Continuous Mode

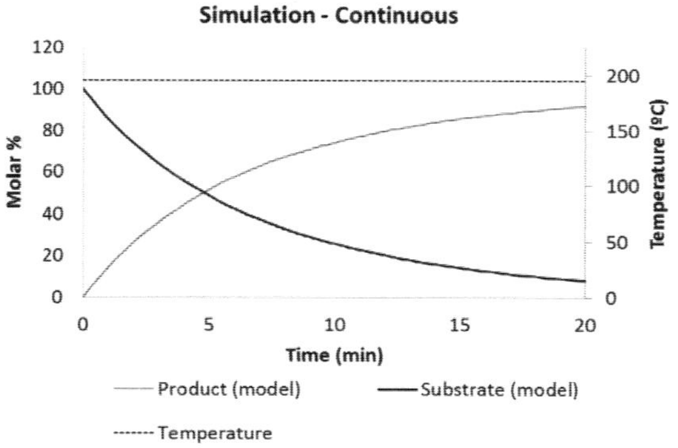

Temperature	Residence Time	Yield
195ºC	20 min	92 %

Idea Screening | Technical Evaluation | Development

Lab Experiments
under Continuous Mode

Considerations:

- Selection of the equipment:
    - Material compatibility (MoC)
    - Ranges of operation (flowrate, temperature, pressure)

Temperature	Residence Time	Yield
195ºC	20 min	92 %

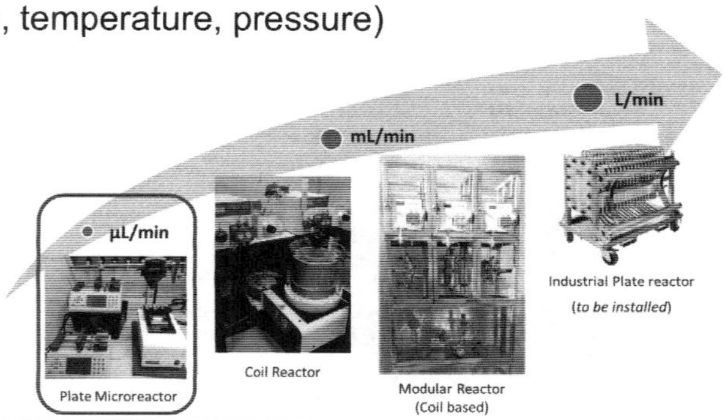

Idea Screening | Technical Evaluation | Development

Lab Experiments
under Continuous Mode

Considerations:

- Selection of the equipment;
  - Material compatibility (MoC)
  - Ranges of operation (flowrate, temperature, pressure)
- Solubility limits.

Idea Screening | Technical Evaluation | Development

Continuous Model

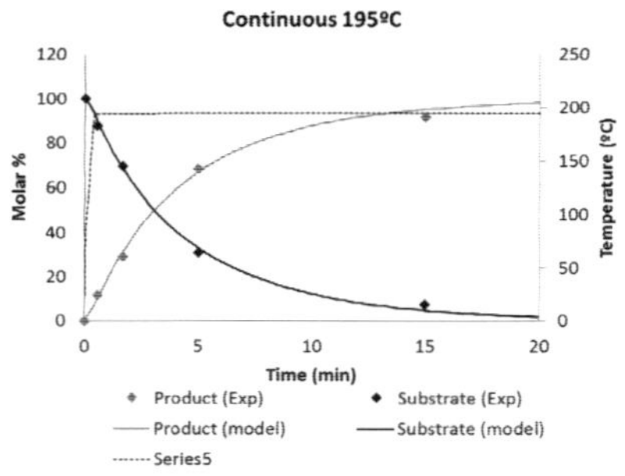

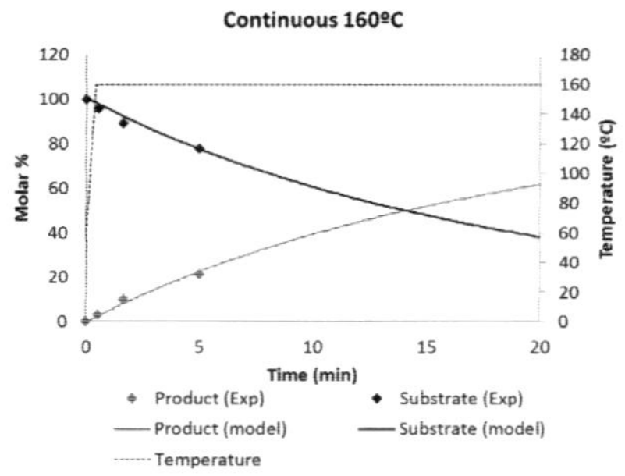

			Confidence Interval
Flow	**Kinetic constant** (k @ 195°C)	8.92×10^{-4} L/mol.s	6.9%
	Kinetic constant (k @ 117°C)	1.06×10^{-6} L/mol.s	*calculated*
	Activation energy (E_a)	76.66 kJ/mol	5.9%

			Confidence Interval
Batch	**Kinetic constant** (k @ 117°C)	2.72×10^{-4} L/mol.s	11.8%
	Activation energy (E_a)	58.66 kJ/mol	11.8%

Idea Screening | Technical Evaluation | Development

Continuous Model
Test Trials

Considerations:

- Different scale
- Different equipment

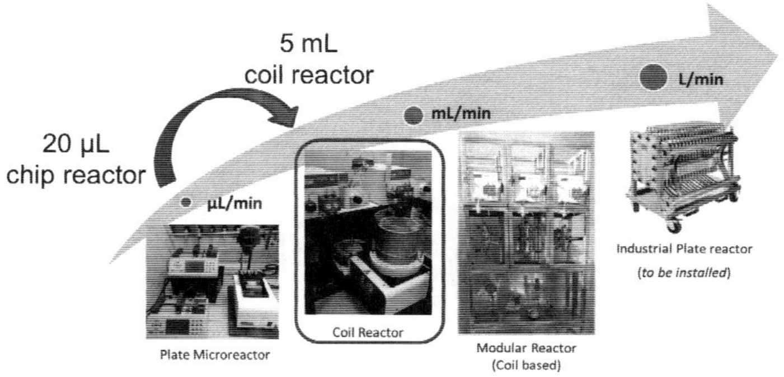

Idea Screening | Technical Evaluation | Development

Continuous Model
TestTrials

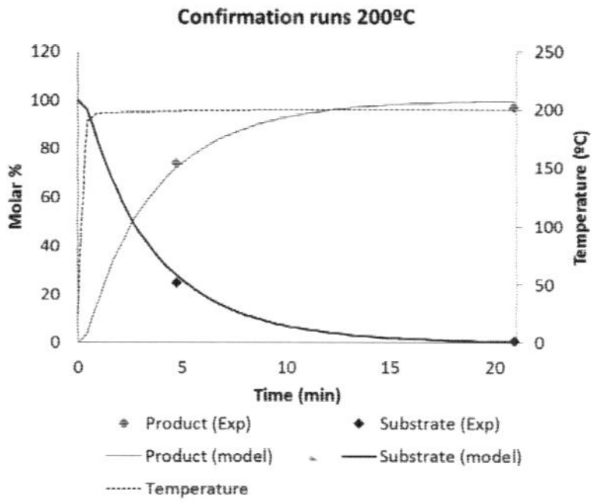

Batch performance
> 97 molar%

Temperature (ºC)	Residence time (min)	Product - model (molar%)	Product - experimental (molar%)
200	20	99.4	97.2
200	5	74.8	73.9
210	10	97.2	94.4
210	5	83.4	85.6

Stage Gate Review

Idea Screening | Technical Evaluation | Development

Process Feasible

Similar Performance to Batch

Safer scale-up

Solubility Concern

Improve Mechanistic Understanding (explore other CPPs)

Study PAT application

Agenda

Flow Chemistry: Features & Capabilities

Approach to Development: QbD

Feasibility Study: Batch-to-Continuous

Conclusions

Agenda

Flow Chemistry: Features & Capabilities

Approach to Development: QbD

Feasibility Study: Batch-to-Continuous

Conclusions

- **Development Workflow**
  - ✓ **Organize** prior knowledge
  - ✓ Plan for **faster** development
- **Modeling**
  - ✓ **Application** of prior knowledge
  - ✓ **Save material**
  - ✓ Predict process performance in a **different**: **manufacturing mode** / **scale** / **equipment**

Hovione

Thank you for your attention.

Rudi Oliveira
rsoliveira@hovione.com
hovione.com

489646 Simultaneous Estimation of Kinetic Parameters and Curve Resolution of Spectral Data -Â Applications and Extensions

Abstract:

In the context of studying reaction kinetics, nowadays a common practice is to use spectroscopic instrumentation to indirectly monitor the concentration of the species in the reaction. Recently, Chen et al[1] proposed an approach based on maximum likelihood principles for simultaneous estimation of reaction kinetics and curve resolution. This work presents further learnings and developments in the application of this technique.

Three cases studies will be presented highlighting first the criticality of pre-estimation steps, specifically: *i*) the importance of the pre-processing algorithm applied to the spectral data prior to the estimation, and *ii*) wavelength selection where a nonlinear programming (NLP) formulation is proposed as a means to efficient wavelength selection. We also present an application of mixed integer NLP (MINLP) approaches to identify which active species can be observable in the spectra (since this information is not always available). Finally, we present an Estimability Analysis (EA) that identifies the appropriate subset of model parameters to be estimated from the available spectral data. Overall, the simultaneous technique presented by Chen et al. is shown to be a reliable and flexible method to deal with the complexities of spectral data and system dynamics in a reaction system.

References

1. Chen W, Biegler LT, Muñoz SG. An approach for simultaneous estimation of reaction kinetics and curve resolution from process and spectral data. *Journal of Chemometrics.* 2016;30(9):506-522.

489834 Modeling and Monitoring of a Continuous Silylation Reaction

Abstract:

A multidisciplinary approach was utilized to develop a mechanistic kinetic model of a continuous silylation reaction within a drug substance chemical synthesis. Analysis by mid-infrared spectroscopy and ^{1}H and ^{29}Si-NMR were both valuable monitoring tools toward obtaining insight into the reaction kinetics.

A multi-step kinetic mechanism was proposed based upon the experimentation and orthogonal simulation results. The impact of reaction temperature, solvent concentration, and recipe were investigated. The model has been applied to both batch and continuous processes. Discussion will include sharing key learning from the model development and its application toward risk management and understanding the feasible operating region.

490063 Estimability Analysis for Improved Parameter Estimation in Deterministic Models: Pharmaceutical Case Studies

Abstract:

Often in the early stages of model development, one needs to estimate the model parameters using data with insufficient information. This information insufficiency in the experimental data can arise from either a poor design, or lack of measurements from the system.

This work illustrates the application of *Estimability Analysis* to aid the parameter estimation of a deterministic mathematical model. This technique has proven useful when the information content of the data available is insufficient to obtain a good estimate of all the unknown model parameters. *Estimability Analysis* ranks the parameters from the most to least estimable; taking into account the uncertainties in both the initial guesses of the parameters and the data [1]. This ranking enables the modeler to determine which parameters to estimate, and which to fix. The obtained model parameters are a good initial guess to a formal model-based experimental design exercise to determine the necessary experiments to optimally parametrize the model.

Case studies are presented from drug substance process development where a kinetic model of a reactive system is being developed; and a third case study from the modeling for a continuous direct compression line for drug product.

References

[1] Wu et al., 2011, *International Journal of Advanced Mechatronic Systems,* 3, 188-197.

490870 Solubility Data Mining and Predictive Modeling: AI+ChE

Abstract:

In the synthesis of pharmaceutical drug substances, selecting the optimal solvent system as early as possible is critical to develop efficient, commercially viable processes. Process safety, environmental impact, and yield are driven in large part by the choice of solvent system. Yet early in process development, material availability limits the amount of experimental data that can be collected. Models to predict solubility enable more focused experimentation, but with current approaches there are trade-offs associated with accuracy, solute characterization, and accessibility for process development engineers.

Performing a historical analysis of experimental data using the tools of data scientists has the potential to unlock insights and streamline future development work. Applying these tools to solubility prediction, data was mined from an internal catalog of automated solubility screening reports across dozens of projects. In all, over 64,000 solubility measurements for >700 pharmaceutically relevant organic compounds in pure and mixed solvents at various temperatures have been aggregated from these reports and analyzed. Such a data set allows for the rapid testing of hypotheses related to solvent selection, such as correlations and synergies between solvent pairs and temperature effects.

Access to this large data set enables machine learning approaches to create quantitative structure property relationship models for solubility prediction that show an improvement over the current standard approaches. Deployed web applications can allow any researcher to rapidly calculate solubility predictions by providing only a molecular structure and a single benchmark solubility measurement. This presentation aims to show how the recently available combination of algorithms (random forests, support vector machines, neural networks, etc.), data-oriented programming languages (R, Python), and cloud computing capabilities can enable chemical engineers to better utilize their data for pharmaceutical process development.

492564 Toward Efficient Development and Reliable Scale-up of Agitated Filter Drying Protocol through DEM Modeling and Simulation

Abstract:

Drying of an active pharmaceutical ingredient (API) is an important unit operation in the pharmaceutical industry as it is used to control residual solvent content, in addition to controlling appropriate physical properties. Drying operation is often time consuming, and may turn out to be a bottleneck in the overall process cycle time. Many APIs are potent compounds that require filtration and drying to be done in a single contained vessel, known as an agitated filter dryer (AFD), to minimize operator exposure.

Gradual evolution of powder bed during the drying results in continuously changing physical properties. This poses significant challenges in experimentally driven understanding of the impact of drying process parameters on final physical properties. Agglomeration and attrition, which are governed by the shear, are the major issues encountered during design and scale-up of filter drying operation. The extent of shear and its distribution changes significantly across various scales and is not amenable to experimental measurements. With advancements in computer algorithms and cheaper computational power, more fundamental insights into particle dynamics including the shear stress distribution can be assessed using Discrete Element Methods (DEM).

In this talk, we will present some of the key attributes that govern the particle dynamics calculated using DEM in the context of AFD operation. We will also provide a decision tree to guide efficient design and reliable scale-up of agitated filter drying.

497704 A Novel Microfluidic Platform for Screening of Pharmaceutical Polymorphs Under Hydrodynamically-Controlled Crystallization Conditions

Abstract:

The majority of oral drugs are delivered in solid forms such as capsules, tablets, powders or suspensions. The physicochemical properties and quality of such drugs strongly depend on their crystalline structure or polymorph, which is controlled during the crystallization process. Various factors such as solvent, temperature, co-solvent, supersaturation, and additives can affect the formation of polymorph. Screening of polymorphs with respect to these factors can be combinatorially tedious and time-consuming. High-density microtiter plates with as many as 1586 wells are commonly employed to screen ~10,000 factors per day. However, the application of microtiter plates for polymorph screening has some practical limitations as the environmental conditions in stagnant wells of the microtiter plate is much different than those in the batch crystallizers. Alternatively, the microfluidic platforms have been used to screen polymorphs using very small quantities (~100 µg) of the drugs in the early stages of developments. Such microfluidic devices locally produce a mixture of crystalline material which is not suitable for the study of nucleation and growth of polymorphs. Here we propose a novel microfluidic platform which mitigates the issues of the previous designs by facilitating nucleation and growth of single crystals. This microfluidic device can effectively screen the aforementioned factors to generate single crystals of polymorphs at the stagnation point while implementing the hydrodynamic conditions which are not considered in the microtiter plate method. In this work, we have screened polymorphs of Carbamazepine using various mole fractions of chloroform:ethanol mixtures and temperatures in the microfluidic device. Further analysis for the measurement of growth rate and morphology are also discussed. The high throughput screening system is also proposed to evaluate multiple crystallization conditions within one single device.

1

POLYMORPHISM IN DRUG COMPOUNDS BY CONTROLLING FLUID DYNAMICS DURING CRYSTALLIZATION

Stephanie Guthrie, Gaurav Giri

AICHE 2017: Solid Form Selection: Cocrystals, Salts, Solvates, Polymorphs, and Beyond

University of Virginia
Department of Chemical Engineering

2

Current drug development

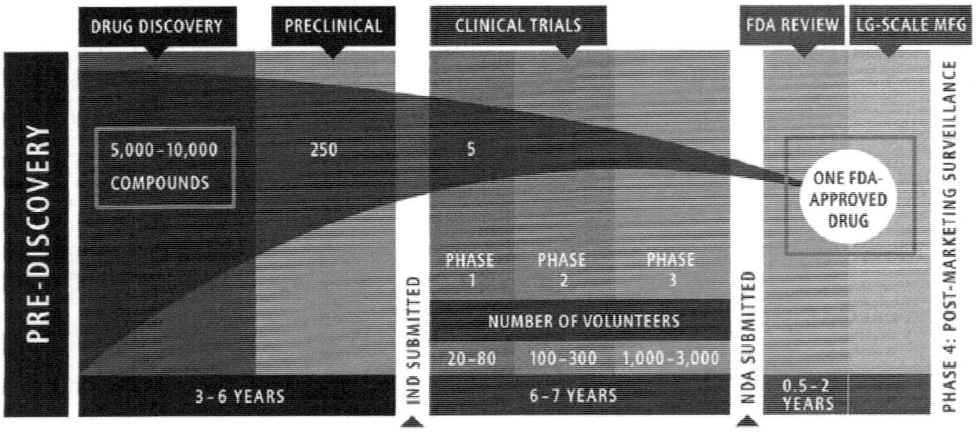

Source: Pharmaceutical Research and Manufacturers of America

3

Need to optimize polymorph detection

- Need ability to detect and isolate polymorphs
    - Polymorphs have unique physical and chemical properties
    - Color, solubility, bioavailability

ROY: 5-Methyl-2-[(2-nitrophenyl)amino]-3-thiophenecarbonitrile

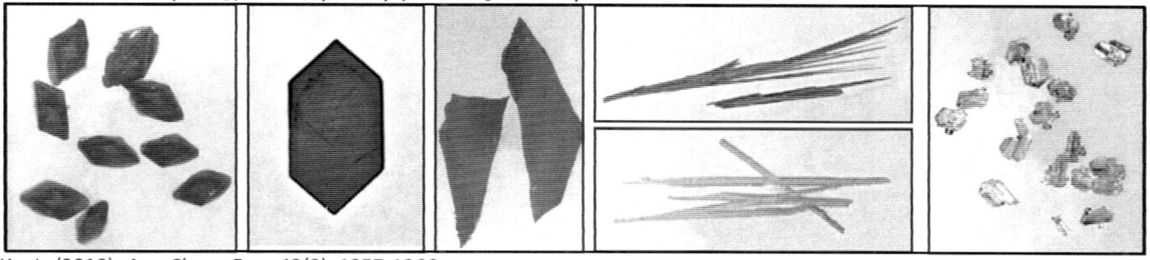

Yu, L. (2010). *Acc. Chem. Res.*, 43(9), 1257-1266.

4

Solid state changes solubility

Increased solubility

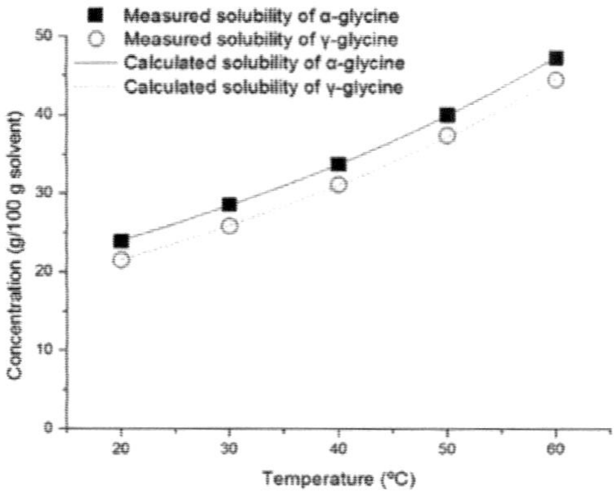

Kim, J. W., (2012). *Crystal Growth & Design*, 12(10), 4739-4744.

Enhanced dissolution rate

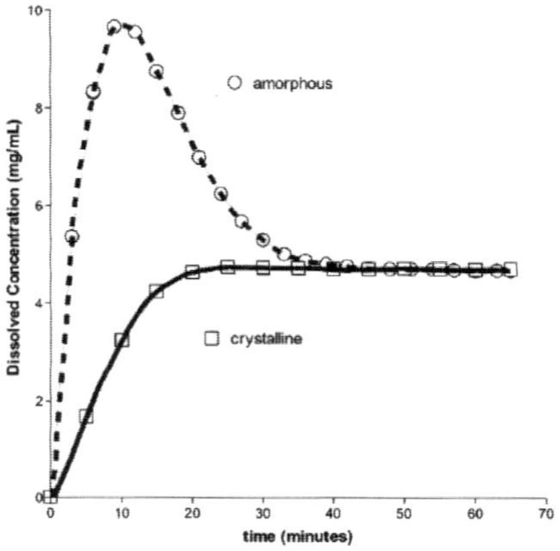

Brittain, H. (2014). *Am Pharmaceut Rev.*

Fluid dynamics can impact polymorphism

- Increased charge carrier mobility based on polymorphism in organic semiconductors

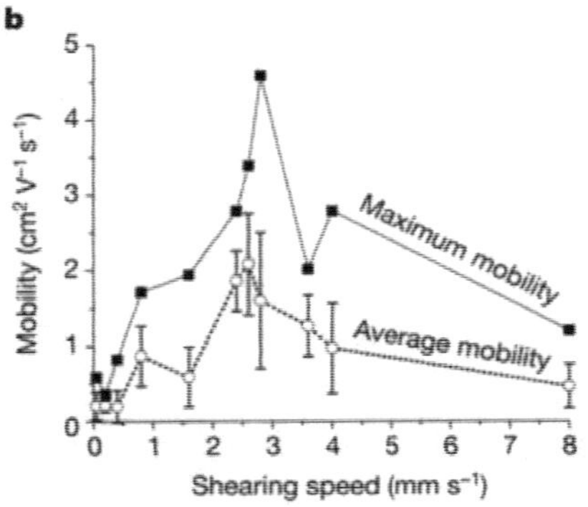

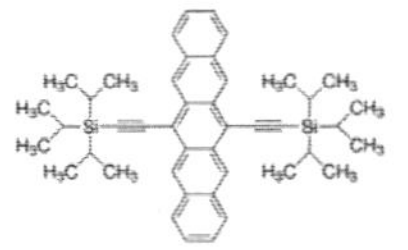

TIPS-pentacene

Giri, G. et al. (2011). *Nature* 480(7378), 504–508.

Solution shearing technique for pharmaceuticals

- Two pharmaceutically relevant molecules
  - Glycine, acetaminophen
- 10 mg/mL drug in deionized water
- Change processing parameters
  - Shearing speed: 0.01 – 3 mm s^{-1}
  - Temperature: 70-100 ˚C

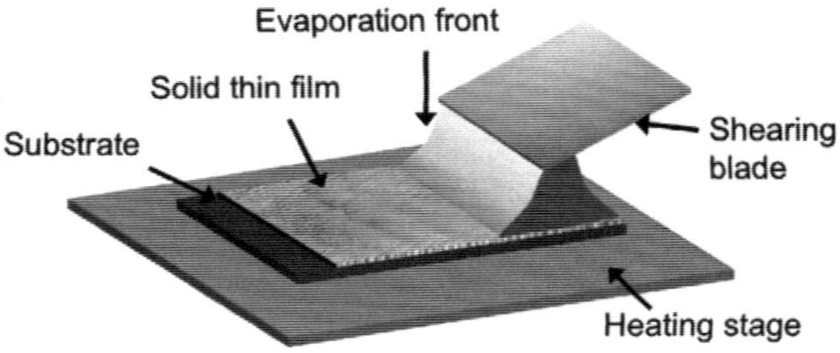

7

Shearing regimes

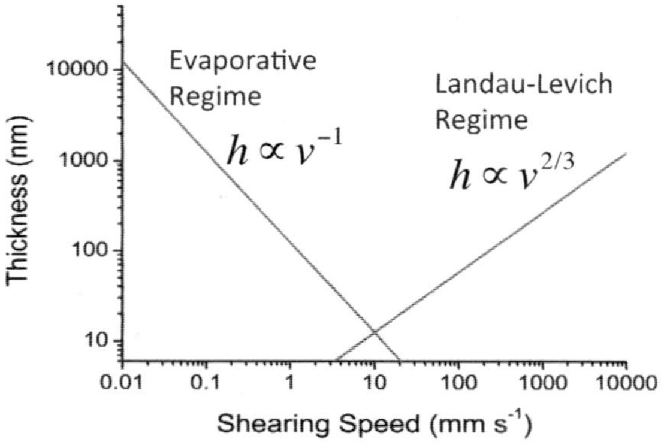

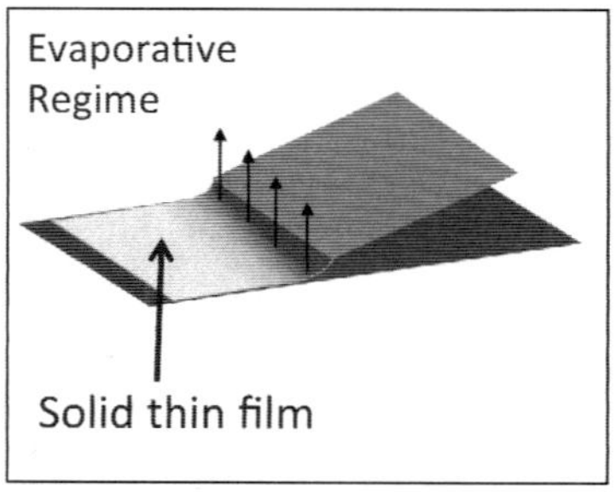

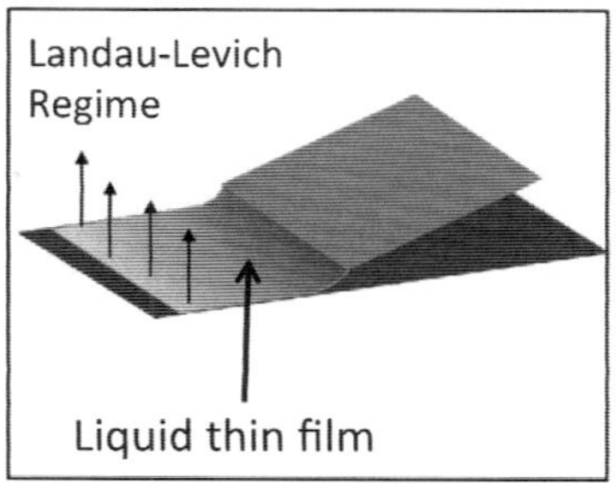

Le Berre, et al. (2009). *Langmuir*, *25*(5), 2554-2557.

8

Film thickness characterization

70°C

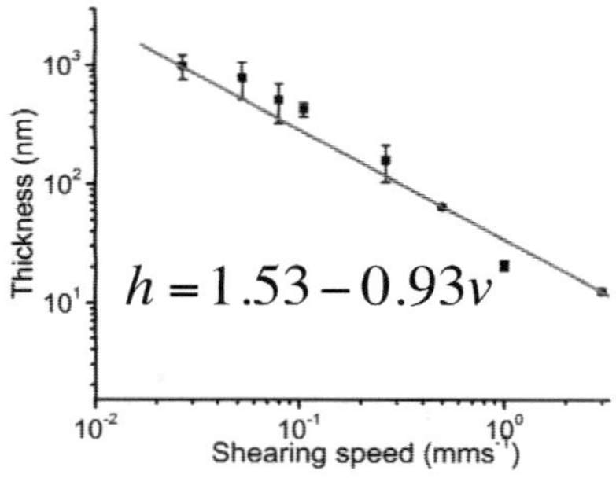

100°C

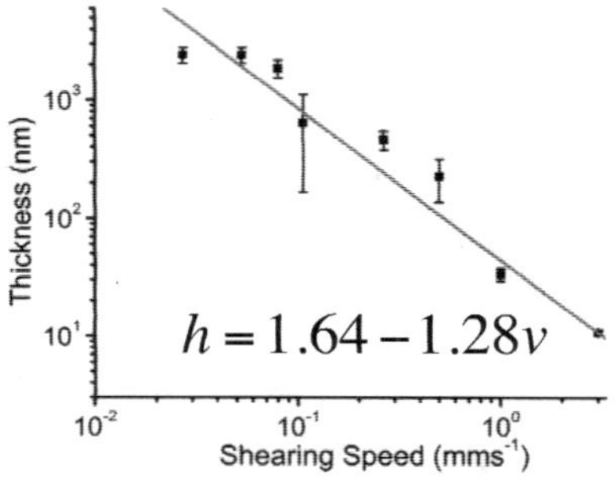

Glycine morphology

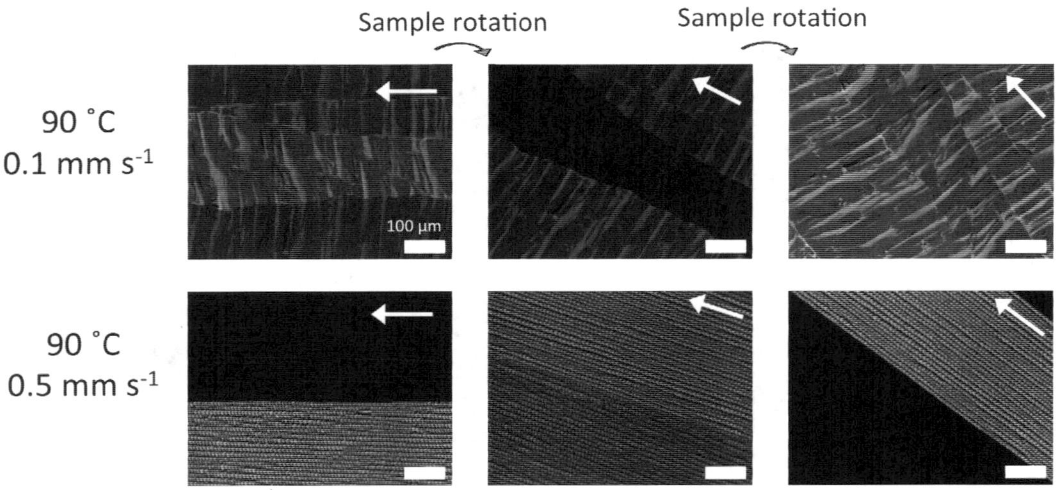

Acetaminophen morphology

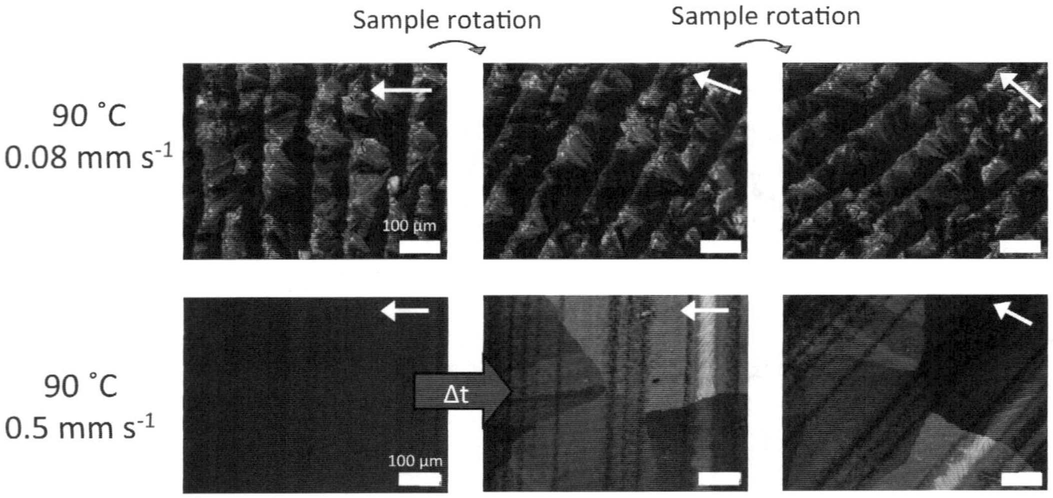

Grazing incidence X-ray diffraction (GIXD)

1. Incident X-ray beam is directed towards sample

2. X-rays are diffracted on film crystallites

3. Diffracted beam lands on area detector

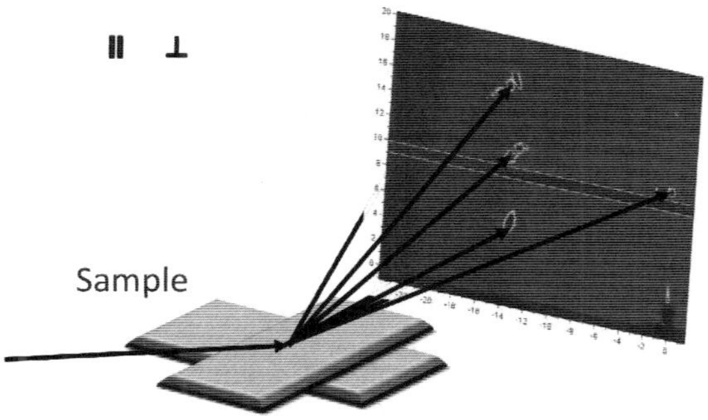

Samples are rotated parallel and perpendicular to shearing direction

- Orients new planes in the line of X-rays

Increasing order ↓

Sample type	Film Appearance	Characteristic GIXD	Description
Isotropic powder			Annular rings
Anisotropic textured film	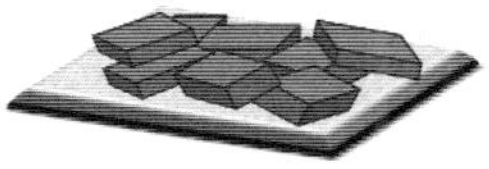	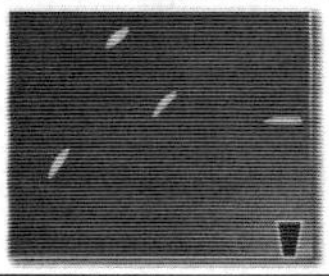	Peak smearing
Oriented film		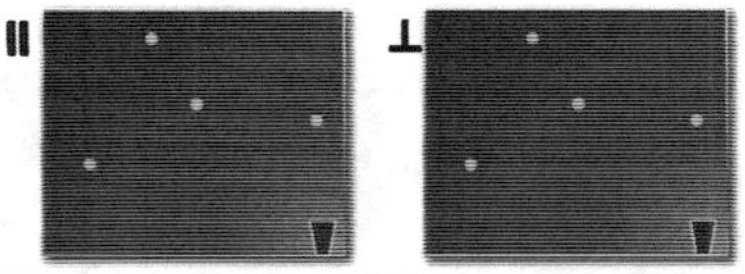	Identical peaks upon rotation
Aligned film	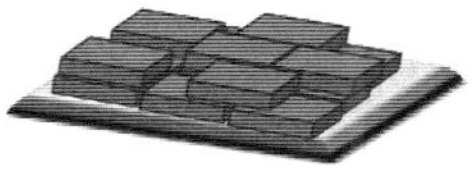	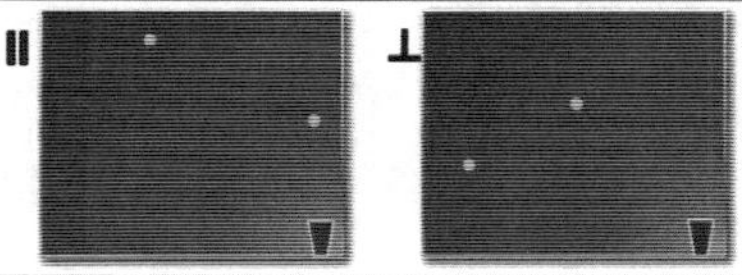	Unique peaks upon rotation

Morphological characterization for glycine

Sample type 1

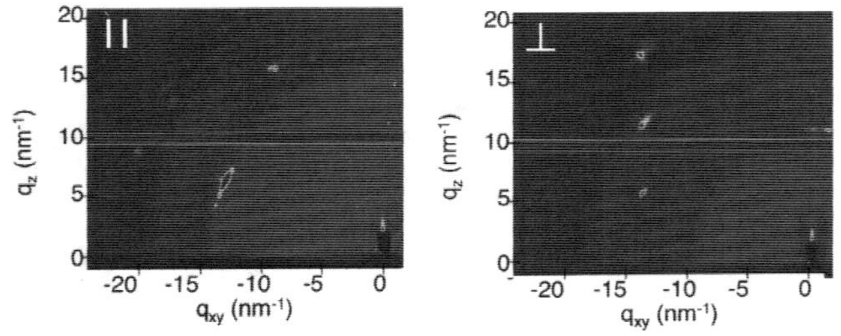

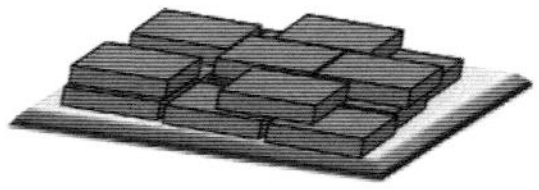

Sample type 2

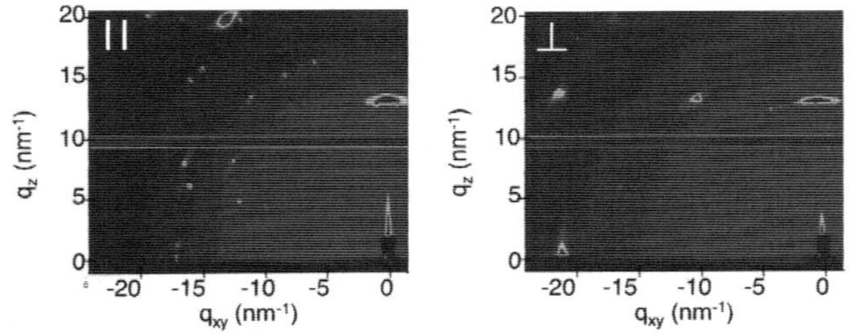

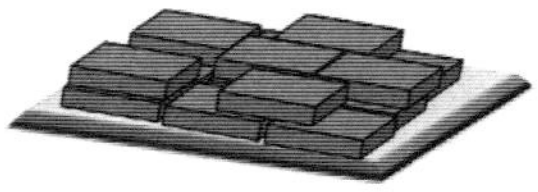

Morphology characterization for acetaminophen

Sample type 1

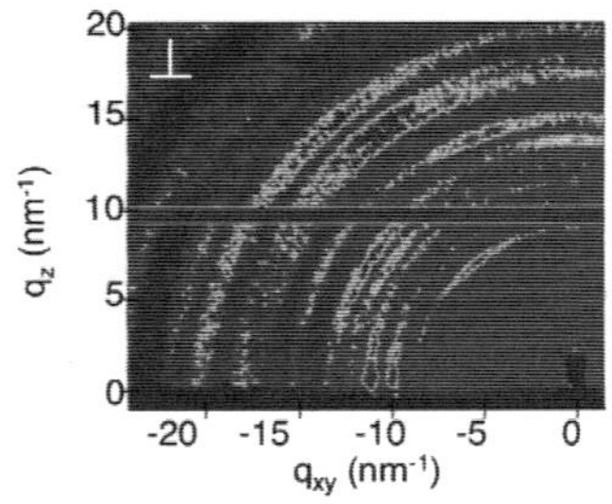

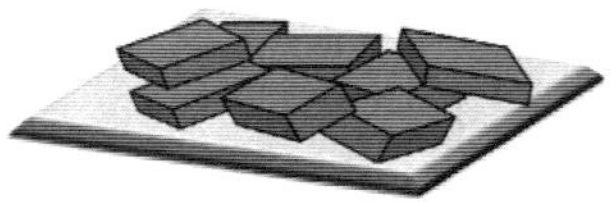

Sample type 2

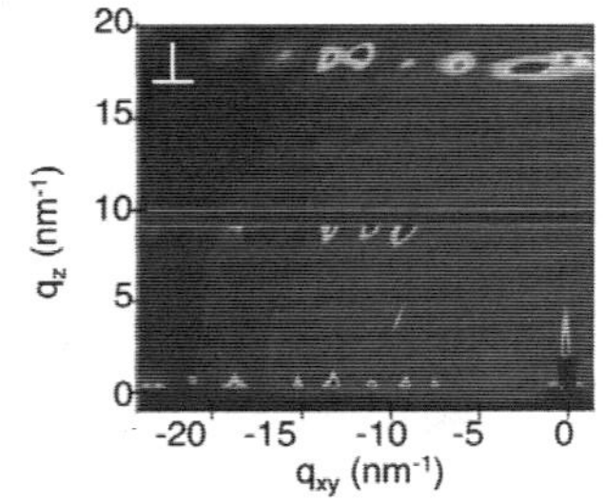

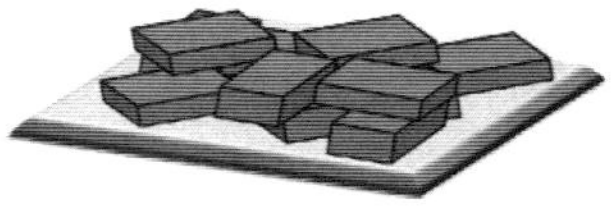

Integrated GIXD for glycine

• Confirmation of polymorphism

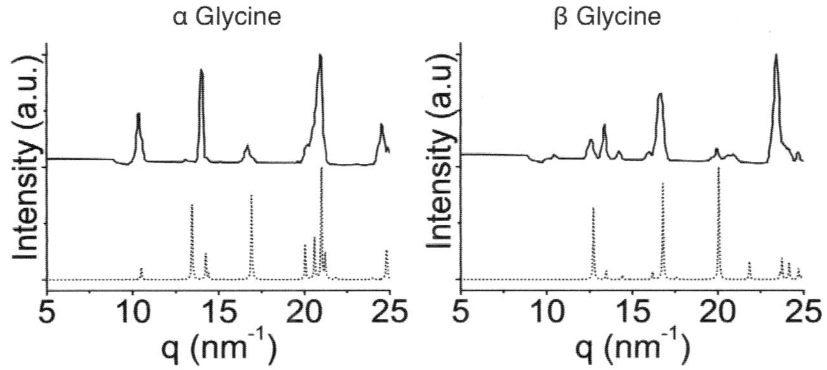

(—): Experimental (- -): Integrated from crystal structure

Integrated GIXD for acetaminophen

• Confirmation of polymorphism

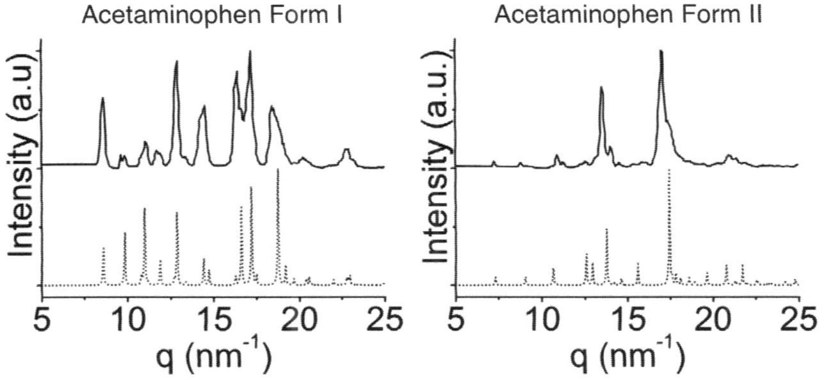

(—): Experimental (- -): Integrated from crystal structure

17

Data reduction

Acetaminophen, 0.27 mm s^{-1}, 90°C	
Sample	Polymorph assignment
1	Form II
2	Form II
3	Form I
4	Both
5	Form I

3/5 samples contain Form I
3/5 samples contain Form II

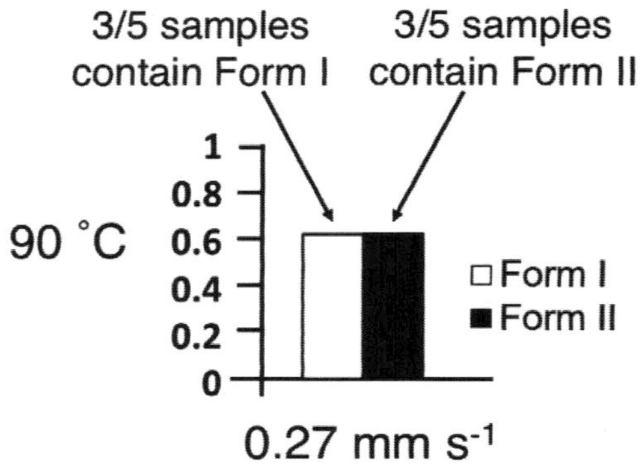

18

Tracking glycine polymorph formation

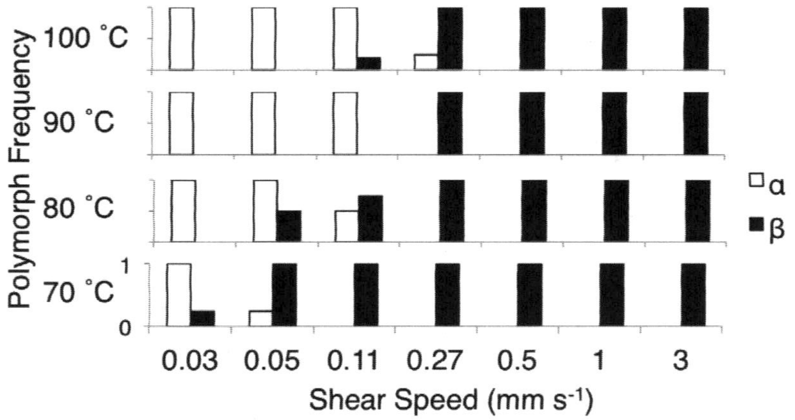

Tracking glycine polymorph formation

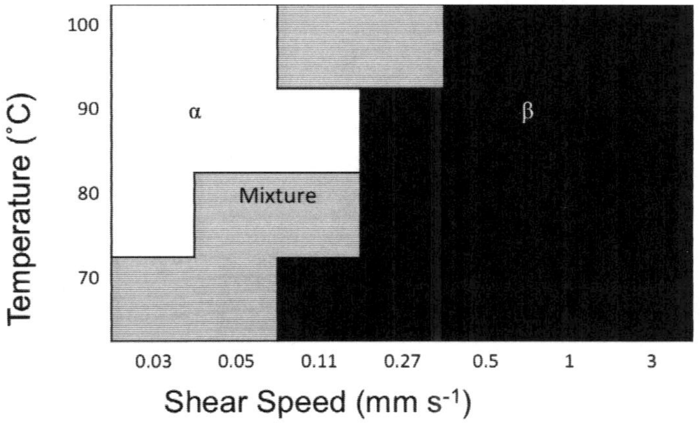

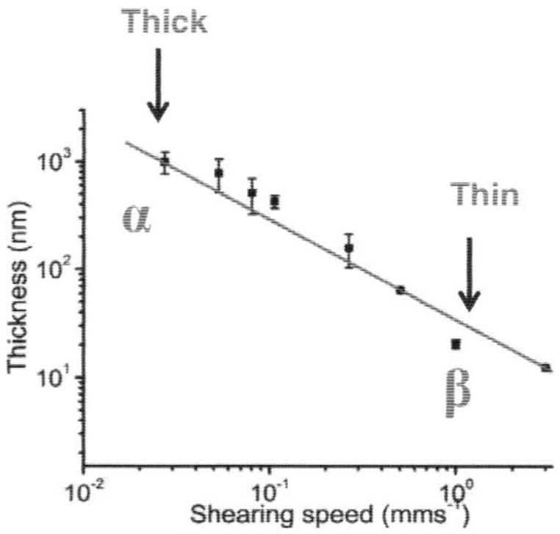

Tracking acetaminophen polymorph formation:

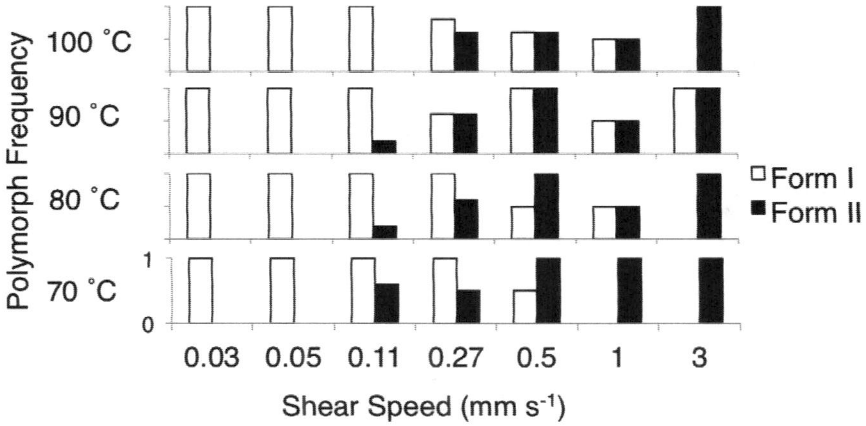

Tracking acetaminophen polymorph formation

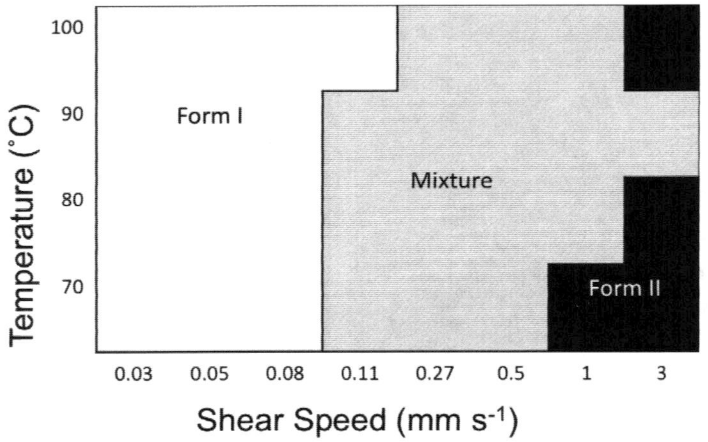

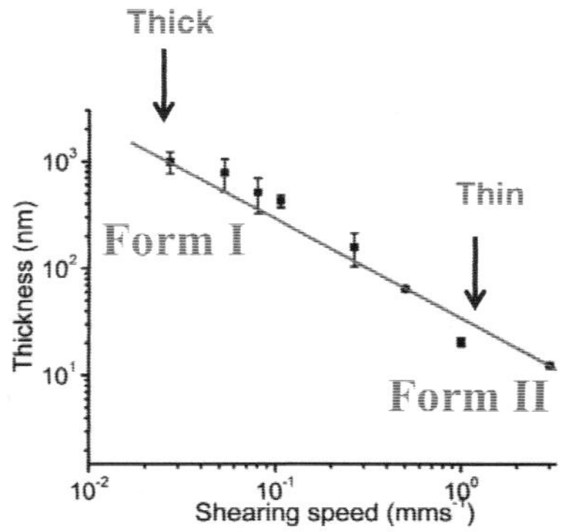

Rationale for polymorphism

- Thick films and thin films display polymorphic changes
- More stable form in thicker films
- Thin films provide a confined environment for isolation of metastable polymorphs

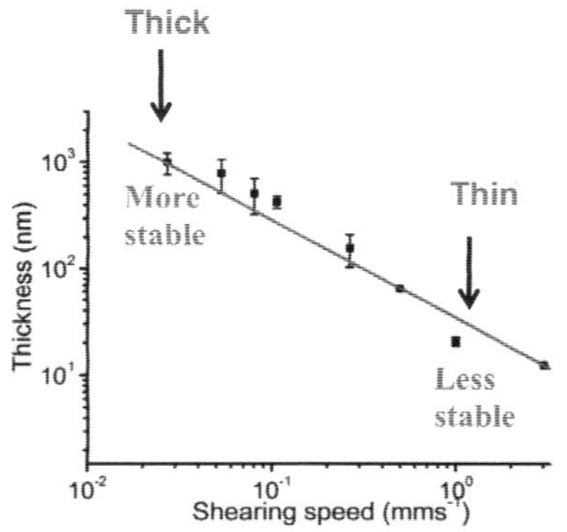

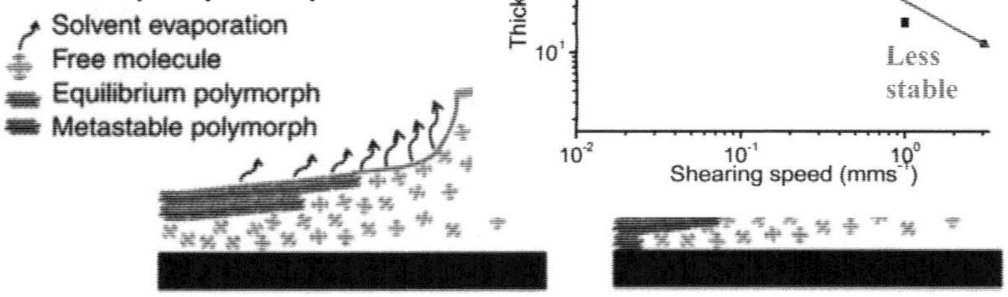

Giri, G. (2014). *Nature communications.* ***5****, 3573*

From II from amorphous transformation

- Amorphous phase isolated first
- Crystallizes over time

Acetaminophen

t =30 seconds t = 90 seconds t = 300 seconds

Carbamazepine

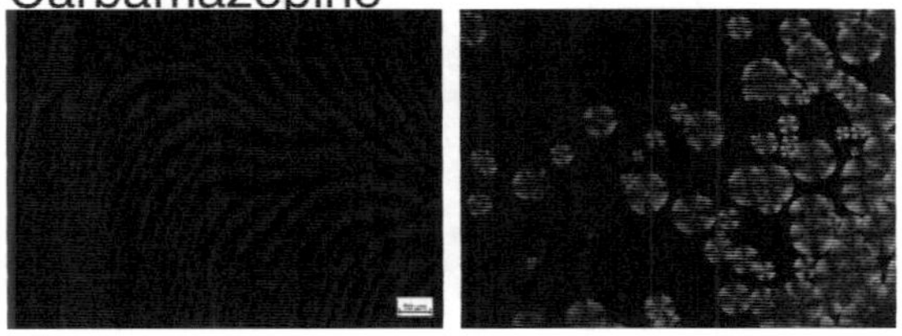

t = 0 seconds t = 10 minutes

Time temperature transformation

- Processing temperature impacts final phase formation
- Different microstructures
- Balance between nucleation and growth creates 'C' shape

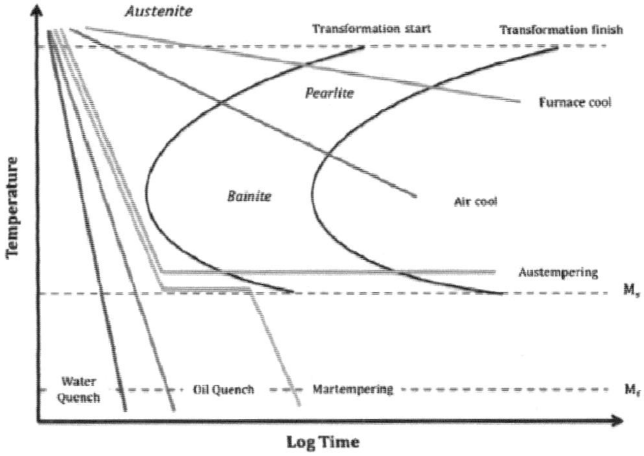

Brog, J. P (2013). *Rsc Advances*, 3(38), 16905-16931.

Crystallization kinetic diagram

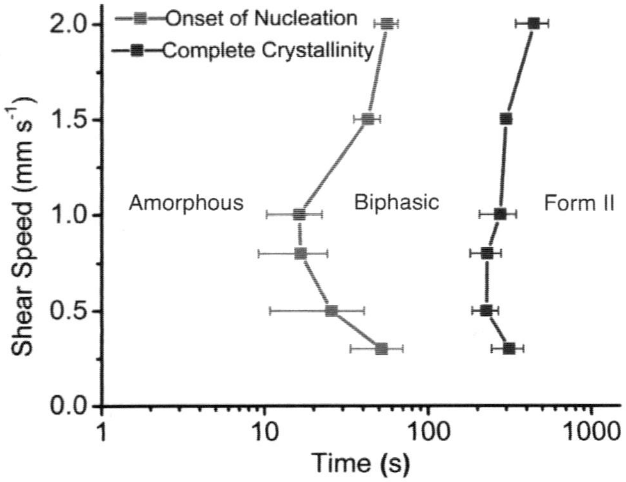

Summary

Processing techniques

Polymorphism

Processing diagram

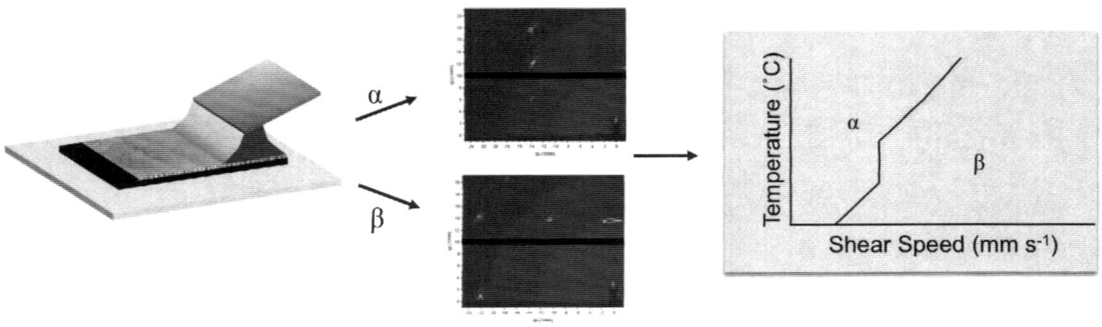

27

Acknowledgement

- Detlef Smilgies (CHESS)
- Luke Huelsenbeck
- Arian Ghorbanpour
- Giri group members

Virginia Space Grant Consortium
Aerospace Partnerships in Education, Research and Industry

Health Resources in Action®
Advancing Public Health and Medical Research

Cornell University
Cornell High Energy Synchrotron Source

28

Questions

488669 Mechanism of Selective Co-Crystallization of Cresol Isomers with Urea

Abstract:

The separation of *m*-cresol from mixtures of *m*-cresol and *p*-cresol has been a great challenge in industry since the physicochemical properties of *m*-cresol and *p*-cresol are pretty similar. In this work, co-crystallization method was used to separate *m*-cresol and *p*-cresol and the co-crystal preparation was attempted by cooling crystallization. The selective co-crystallization of *m*-cresol and *p*-cresol with urea was evaluated, based on the principle of supramolecular chemistry and crystal engineering strategy. It was found that the co-crystal former, urea, can only form co-crystal with m-cresol. One supramolecular synthon was formed between *m*-cresol and urea with the help of intermolecular mutual recognition and self-assembly process among them. Consequently, *m*-cresol can be effectively separated from *p*-cresol by the way of selective co-crystallization. The *m*-cresol _urea co-crystal (MC_U) was prepared on 1:1 molar ratio, which exhibits a regular hexagonal shape and belongs to orthorhombic system with space group P b c a and cell parameters a = 11.2502(5)Å, b = 7.1880(4) Å, c = 21.7868(15) Å. The hydroxyl group of *m*-cresol forms a H-bond with the carbonyl group in amide moiety of urea. The MC_U molecules adopt a chained layer structure. The solid-state properties of MC_U were characterized using X-ray Power Diffractometry (PXRD), Differential Scanning Calorimetry (DSC), ^{1}H nuclear magnetic resonance (^{1}H NMR) and Fourier Transformed Infrared Spectrometer (FTIR). Meantime, the selective co-crystallization process was on-line monitored by process analytical technologies/tools (PAT), such as Raman spectroscopy (Raman), Attenuated Total Reflection Fourier Transformed Infrared Spectroscopy (ATR-FTIR) and Focused Beam Reflectance Measurement (FBRM). Combined with the on-line monitoring data, ^{1}H NMR and the single crystal structure data, the mechanism of molecular recognition and self-assembly between m-cresol and urea is revealed. It was found that the whole co-crystal formation process can be divided into three steps: (1) *m*-cresol-urea dimers formation process, (2) MC_U nucleation process, and (3) MC_U growth process. And the reason for the selective co-crystallization is elucidated from the molecular level by using the corresponding calculation method. The mechanisms of selective co-crystallization as well as molecular recongnition and self-assembly between *m*-cresol and urea were proposed by analyzing the structure of *m*- and *p*-cresol, the single crystal structure data of cocrystal and the on-line monitoring data of Raman and ATR-FTIR. It was found that a chain supramolecular synthon was formed by the carbonyl oxygen (C=O) of the urea and the hydroxyl hydrogen (O-H) of the *m*-cresol relying on stronge hydrogen bonds. However, no obvious non-covalent bonds between *p*-cresol and urea molecules were observed. This failure is probably due to the position of the phenolic hydroxyl in toluene, which is affected by the methyl group. The electron cloud density of hydroxyl groups on *p*-cresol is reduced because of the presence of methyl groups, which could lead to the decrease of the electronegativity of oxygen, and thus will reduce the degree of exposure to the hydrogen atom attached to it. Obviously, *m*-cresol _urea co-crystal could be formed because of the contribution of the more exposed hydrogen atoms.

Continuous co-crystallization of benzoic acid and isonicotinamide by mixing-induced supersaturation: Exploring opportunities between reactive and antisolvent crystallization concepts

Vaclav Svoboda[1], Pól MacFhionnghaile[2], John McGinty[1], Lauren E. Connor[3], Iain D.H. Oswald[4], Jan Sefcik[2,*]

1. EPSRC Doctoral Training Centre in Continuous Manufacturing and Crystallisation, Department of Chemical and Process Engineering, University of Strathclyde, James Weir Building, 75 Montrose Street, Glasgow, G1 1XJ, United Kingdom
2. EPSRC Centre for Innovative Manufacturing in Continuous Manufacturing and Crystallisation, Department of Chemical and Process Engineering, University of Strathclyde, James Weir Building, 75 Montrose Street, Glasgow, G1 1XJ, United Kingdom
3. Collaborative International Research Programme, Strathclyde Institute of Pharmacy and Biomedical Sciences, University of Strathclyde, Technology and Innovation Centre, 99 George Street, Glasgow, G1 1RD, United Kingdom
4. Strathclyde Institute of Pharmacy and Biomedical Sciences, University of Strathclyde, 161 Cathedral Street, Glasgow, G4 0RE, United Kingdom

Corresponding Author:
Prof. Jan Sefcik
Department of Chemical and Process Engineering,
University of Strathclyde,
James Weir Building,
75 Montrose Street,
Glasgow,
G1 1XJ,
United Kingdom
jan.sefcik@strath.ac.uk

ABSTRACT

This study combines reactive and antisolvent crystallization concepts via mixing-induced supersaturation to demonstrate a wider range of options for solvent system selection in multicomponent crystallization. This approach was applied to investigate continuous crystallization of 1:1 and 2:1 co-crystals of benzoic acid and isonicotinamide. Design of Experiments was used to identify conditions where pure co-crystal phases are obtained and a continuous mixing-induced co-crystallization process was implemented to selectively produce either 1:1 or 2:1 co-crystals.

INTRODUCTION

Multicomponent crystalline materials are composed of more than one molecule in its crystal lattice. This includes co-crystals, solid solutions, solvates, and salts. The interest in multicomponent systems is a growing trend within the pharmaceutical industry[1]. Co-crystals are of benefit due to their ability to tailor physical and pharmaceutical properties of active pharmaceutical ingredients (APIs) such as solubility, bioavailability, stability, and the processability of the solid powder within industrial manufacturing processes[2]. They can also be utilized to expand IP portfolios[3]. However, co-crystallization is inherently more complex than single component crystallization as it involves an additional component with new potential solid phases. Having additional components and related process variables makes navigating the phase diagram and crystallization process more challenging.

Several methods for manufacturing co-crystals have been previously reported including both dry powder processes and solvent based processes, typically in batch processing mode. Dry powder processes are mechanochemistry based and utilize grinding at room temperature[4] or cryo-temperatures[5], polymer assisted grinding[6], high shear granulation[7] or hot melt extrusion[8]. Co-crystals from solution-based processes have been reported from evaporative and cooling[9], antisolvent[10] or reactive crystallization[11]. Cooling co-crystallization has previously been implemented in continuous platforms[12,13] and with co-crystal stoichiometry control[14]. Co-crystallization using antisolvent has been previously studied in batch processes[10], but not using continuous methods. Antisolvent crystallization of single component materials has been shown adaptable to continuous manufacturing and scale up in a number of processes[15–17]. Reactive co-crystallization mentioned in the work of Rodríguez-Hornedo et al.[11] does not involve a chemical reaction as such but a solution-based formation of a multicomponent solid phase from a mixture of two solutions undersaturated with respect to individual components using the same solvent.

Processes which rely on mixing to induce supersaturation such as reactive and antisolvent crystallizations, are well amenable to scale-up under continuous conditions[18]. Continuous manufacturing of pharmaceuticals can provide benefits such as decreased plant footprint, easier scale-up, shorter lead times and better unit performance through process intensification[19]. Since mixing can have strong impact on generation of supersaturation profile and subsequent nucleation, especially under kinetically controlled conditions[20], well-controlled mixing is essential for a control over final crystal properties, such as solid form and particle size distribution.

Antisolvent and reactive crystallization approaches can be combined in order to design a process where supersaturation is induced by mixing and is determined by the position in a quaternary phase diagram involving two crystal co-formers and two solvents. Depending on the shape of the phase diagram, which often shows a highly unsymmetrical nature[21], the co-crystal would be the thermodynamically most stable phase under some conditions. However, sometimes it may not be readily crystallized due to kinetic limitations (slow nucleation or growth) even if thermodynamically favored. Munshi and co-workers pointed out opportunities to use mixed solvents in controlling solid phase outcomes in cooling co-crystallization[22]. Solvent selection, a key design choice in crystallization, becomes more challenging in multicomponent systems. The key parameter of co-crystallization process design is the supersaturation with respect to the co-crystal phase, rather than the supersaturations of the individual co-formers. While having four components increases the complexity of mapping the phase diagram, it also allows for more options how to access solid phase regions which might not be easily accessible at a fixed solvent composition. For example, it may be possible to start with a solution of both coformers undersaturated in one solvent and add a second solvent to induce supersaturation, as in antisolvent crystallization. Alternatively, one can start with one co-former undersaturated in a given solvent mixture, and the other co-former undersaturated in the same solvent mixture and generate supersaturation by mixing these two solutions together, as in reactive crystallization. These decisions will be driven by the shape of the phase diagram and the nature of the target solid phase.

This study combines antisolvent and reactive crystallization concepts to develop a continuous co-crystallization process to produce benzoic acid (BZA) – isonicotinamide (INA) co-crystals. These molecules form two different co-crystals in 2:1 and 1:1 stoichiometric ratios which have been previously isolated and characterized from small scale cooling crystallization[23,24]. In this work, benchtop screening crystallization experiments were scaled to run in a continuous process. A Design of Experiment was carried out for the benchtop screening to better understand the effects of different process conditions on the crystallization of the two co-crystal phases. DoE screening was done in order to cut down the time and number of experiments to find suitable continuous crystallization conditions to selectively produce either 1:1 or 2:1 co-crystals by continuous crystallization.

EXPERIMENTAL SECTION

Materials. Benzoic acid (≥99.5%), isonicotinamide (99%), and ethanol (≥99.8%) were supplied by Sigma-Aldrich (Gillingham, UK). Deionised water was produced using the in-house Millipore Milli-Q system.

Batch Screening. Two sets of screening experiments were carried out. The first was an initial screening and the second was a systematic Design of Experiments (DoE) approach used for mapping of a limited region of the quaternary phase diagram (BZA, INA, Water, Ethanol) in order to identify suitable operating conditions for continuous crystallization. For both sets of screening experiments, an aqueous solution of isonicotinamide was added to an ethanolic solution of benzoic acid in a 20 mL vial, illustrated in Figure 1. Solutions were mixed at various ratios to obtain the total of 10g of solution and agitated using magnetic stirrer bar for 10 minutes past observed nucleation. All experiments were carried out at 25°C. For the initial screen, solutions of isonicotinamide in water and benzoic acid in ethanol were prepared at concentrations of 68.1 g/kg water and 213 g/kg ethanol, respectively, and mixed in ratios of 2:1, 1:1, 1:2, 1:4 and 1:9 by mass. The solid product obtained was filtered using 0.2 μm PTFE filter without washing, dried at 40°C for 24 hours and analyzed by X-ray powder diffraction (see solid characterization section). The design space for DoE was selected based on information from the initial screen. Experimental plan was created using MODDE in a 2 level full factorial design. The experimental worksheet, model fitting and results analysis was carried out in MODDE by Umetrics. Data were fitted using Partial Least Squares (PLS). The three variable parameters chosen for the DoE were: BZA-Ethanol solution concentration, INA-Water solution concentration, and the mass ratio of the two solutions mixed. Values of these three parameters combined determine the final composition of the mixture and thus the position on the phase diagram. The responses measured were induction time, solid yield, solid phases present and resulting slurry flow properties. Induction time was estimated as time after mixing until first particles were visually observed. Solid yield was taken as a percentage of total solute that was recovered as solids by filtration 10 minutes after the estimated induction time while agitating. Slurry flow was assessed by a rating system where a number was assigned on a scale from 1 to 5: 1 for an easily flowing mixture and 5 for a slurry too thick to flow with gravity. Solid phase was determined by XRPD (see solid characterization section). Based on this DoE, conditions for the production of 2:1 co-crystal were selected. Although the initial DoE was aimed to find suitable conditions for crystallization of both co-crystal forms, conditions leading to crystallization of 1:1 co-crystal resulted in solid loadings which would be too high for a continuous operation. Based on results from the initial DoE, a set of further experiments was carried out in an expanded design space to find suitable conditions for crystallization of 2:1 co-crystal. These experiments were carried out at lower solution concentrations and resulting solid phases, induction times and slurry flow ratings were determined in the same way as in the initial DoE runs.

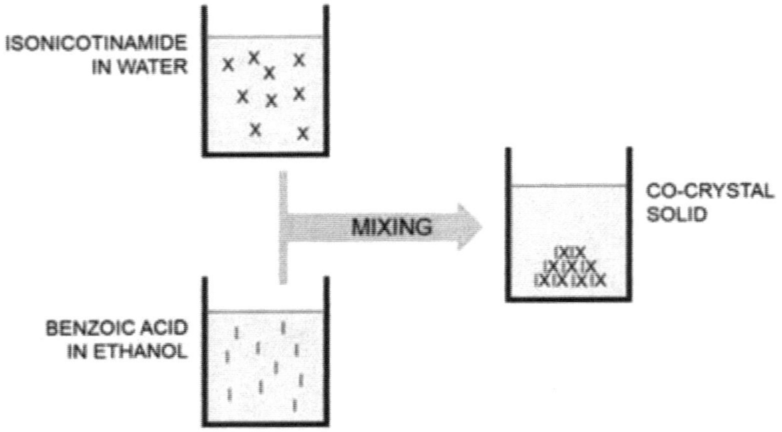

Figure 1. Solvent system used for mixing-induced co-crystallization

Solubility measurement. Solubility of the 2:1 co-crystal has been determined by a solvent addition method in the Crystalline Reactor System (Technobis), similar to a previously published procedure[25]. As compared to temperature variation or gravimetric methods, solvent addition relies on slow dilution of a suspension with a solvent (mixture) of a given composition under isothermal conditions until complete dissolution occurs, when a clear point can be detected. Solvent was added at a constant rate using PHD Ultra syringe pumps (Harvard Apparatus) to the Crystalline vials. The solubility measurement setup is shown in Figure 2. Multiple addition rates were tested (0.5 ml/hr and 0.75 ml/hr) to verify that dissolution kinetics have a negligible effect on the result at the addition rates used. Clear point was determined using both transmissivity measurement and images from the Crystalline camera. Image analysis produced more consistent results, which is in line with previous findings from Reus et al[25]. A study was also carried out to monitor the solid state transformation of the co-crystal in order to check phase stability in the same setup. Suspensions with concentrations as used for the start of solubility measurements were held for varying amount of time from 1 to 5 hours, after which the remaining solid was filtered, dried and analyzed by XRPD.

Figure 2. Technobis Crystalline and syringe pumps for solubility determination using the solvent addition method

Continuous crystallization. Continuous crystallization runs were performed using a concentric capillary mixer (Figure 3) as well as the Ehrfeld modular micro-reactor system equipped with Valve Mixer 30 module (Figure 4). The BZA-Ethanol solution was supplied to the concentric capillary mixer through the inner PEEK capillary, while the INA-Water solution was fed to the outer glass tube as shown in Figure 3. The capillary has internal and outer diameters of 0.51mm and 1.59mm, respectively. The outer glass tube has an internal diameter of 3mm. The capillary stream entered into the outer tube stream 6 cm from the T-junction. The feed solutions were pumped using Bronkhorst Mini CORI-FLOW system coupled with gear pumps allowing for accurate control of mass flow rates. The resulting slurry from the mixer was sampled 0.7m from the mixing point at the start and end of experiments. Collected slurry was filtered after a holding period in a Buchner funnel with a 0.45μm filter paper and dried overnight at 40°C. The slurry holding time without agitation was 10 minutes for 2:1 co-crystal runs, and 18 minutes for 1:1 co-crystal run. A camera was used for visual analysis of any fouling in the glass tube. Temperature of both streams was measured throughout experiments and it varied less than 2°C from 25°C. The total mass flow rates through the capillary mixer were 20, 40 and 60 g/min in 50:50 (w/w) ratio for producing the 2:1 co-crystal and 115 g/min at a 15:100 (w/w) ratio of benzoic acid solution flow rate to isonicotinamide solution flow rate for the 1:1 co-crystal. In order to demonstrate the transferability of the process to a commercial platform, the Ehrfeld modular micro-reaction system, shown in Figure 4, was used at the total mass flow rate of 20 g/min for producing the 2:1 co-crystal.

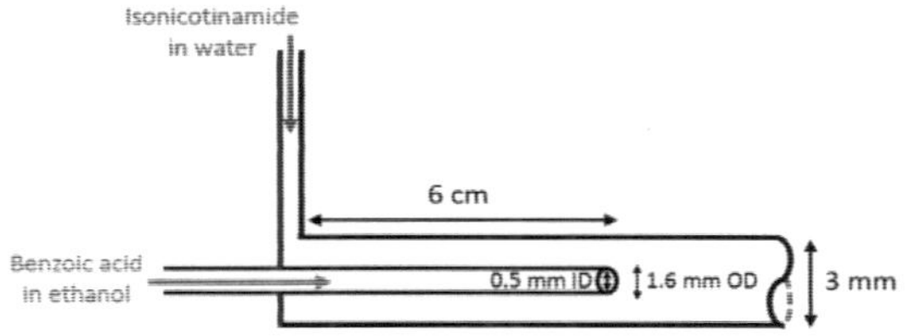

Figure 3. Concentric capillary mixer diagram with dimensions indicated

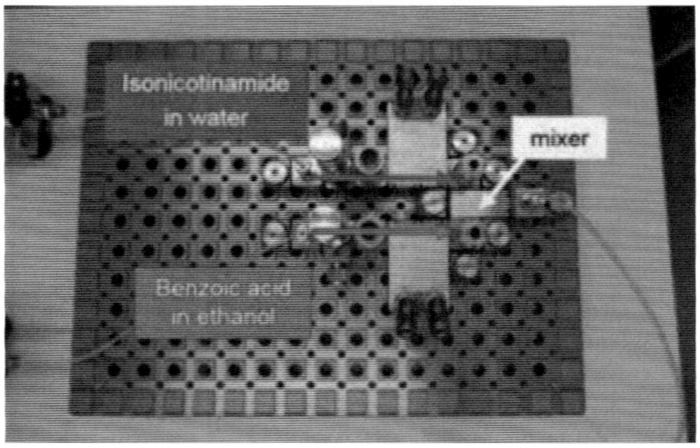

Figure 4. Ehrfeld modular micro-reaction system fitted with valve assited mixer 30

Solid characterization. Filtered and dried crystalline powder from the screening and continuous experiments was analyzed by X-ray Powder Diffraction (XRPD) and IR spectroscopy. Samples from continuous runs were further analyzed by Differential Scanning Calorimetry (DSC) paired with Thermogravimetric Analysis (TGA), and microscope image analysis for particle size measurement. Solids from continuous runs targeted to produce the 2:1 co-crystal were also analyzed by NMR. XRPD fingerprinting was performed on a sample placed in a 28 well plate, supported by Kapton film (7.5 μm thickness). Data were collected on a Bruker AXS D8 Advance transmission diffractometer equipped with θ/θ geometry, primary monochromatic radiation (Cu $K\alpha_1$ λ = 1.54056 Å), a Braun 1D position sensitive detector, and an automated multiposition x–y sample stage. Data were collected from 4 to 35° 2θ with a 0.015° 2θ step size and 1 s $step^{-1}$ count time. FT-IR measurements were taken using Bruker Tensor II, using 32 scans with 4 cm-1 resolution from 450 to 4000 cm^{-1} with diamond tip ATR sampling plate. DSC/TGA data was obtained from Netszch STA 449 F1 Jupiter. Measurements for DSC/TGA were taken from 20 to 180 °C with a ramp of 10°C/min. For Particle Size Distribution (PSD), image analysis was carried out on dry powder using Malvern Morphologi G3 using low pressure dispersion and 2.5x optics. ^1H and ^{13}C NMR analysis was carried out using Bruker Advance 3 at 400 MHz by dissolving the solid product in deuterated DMSO in 5mm NMR vials.

RESULTS and DISCUSSION

Mixing-induced supersaturation. A case study on benzoic acid – isonicotinamide co-crystallization presented here explores the space between reactive and antisolvent crystallization concepts in generating mixing-induced supersaturation to target specific solid phases of co-crystals. The illustrative phase diagram in Figure 5 is similar to the phase diagrams presented previously for the system investigated here[23]. We use it to demonstrate different modes of inducing supersaturation through mixing. The blue lines illustrate examples of reactive and antisolvent crystallization. In reactive crystallization, two undersaturated solutions, each containing a single coformer in the same solvent mixture, are mixed to supersaturate with respect to a desired co-crystal solid. In antisolvent crystallization, an undersaturated solution containing both coformers in one solvent is mixed with antisolvent to generate supersaturation. The red line illustrates a new approach used in this study, where antisolvent and reactive crystallization concepts were combined to target a specific region of the phase diagram, in this case 2:1 and 1:1 co-crystal regions. In this combined approach, two undersaturated solutions, each containing a single co-former in a single (pure) solvent, are mixed to generate supersaturated solution with respect to desired co-crystal phase in the mixed solvent.

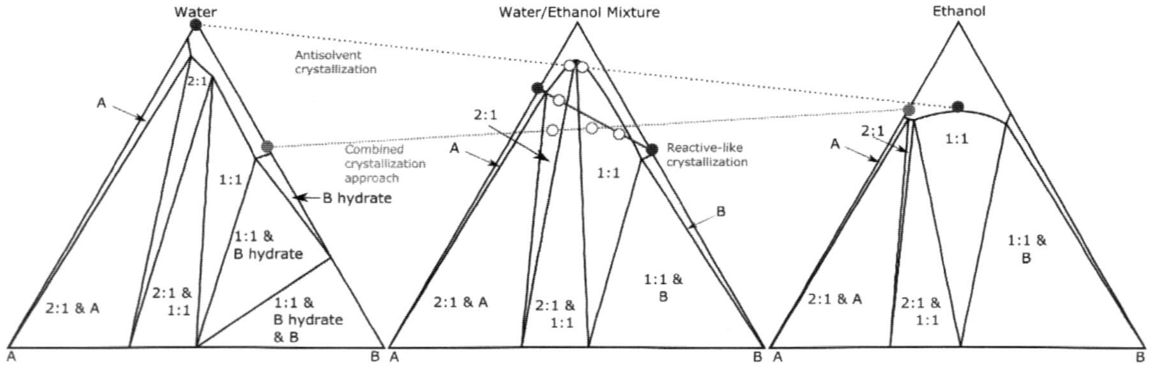

Figure 5. Illustration of three modes of inducing supersaturation through mixing: Reactive, antisolvent and combined approach. Targeted solid phases are 1:1 and 2:1 co-crystals of A and B and solvent is a mixture of water and ethanol. Dotted lines do not correspond to lever rule.

Screening. An initial screen was used to determine design space boundaries and was followed by a detailed DoE driven screen to map a design space corresponding to a region of interest in the phase diagram. XRPD from the initial screen has shown that the method of mixing solutions of isonicotinamide in water and benzoic acid in ethanol can produce both 2:1 and 1:1 co-crystals. DoE responses were chosen as key outcomes relevant for development of a continuous crystallization process. High solid loadings at the end of experiments limited solution concentrations from above, as too high solid loadings could lead to downstream blockages. On the other hand, long induction times limited solution concentrations from below, as relatively rapid nucleation is desirable. Due to the probabilistic nature of primary nucleation, induction time estimates from batch experiments are only intended to give an approximate guidance, nevertheless there was a good consistency among repeated experiments.

Despite not necessarily providing direct information on phase stability under thermodynamic equilibrium conditions, the DoE mapping provided a rapid way to find potentially suitable crystallization conditions while also giving indication about kinetics. Contour plots displayed in Figure 6 show responses at a fixed solution ratio where at higher concentrations of starting solutions the yield is higher, slurry flow rating diminishes due to higher solid loading and induction time decreases due to higher supersaturation. The co-crystal phase formed is primarily dependent on the solution flow ratio rather than the solution concentrations within the design space investigated. Contour plots for other solution ratios are located in supplementary information (ESI Figures 3-5). After targeting the desired phase by adjusting the solution flow ratio, slurry flow ratings of 2-3 were deemed as acceptable for a continuous crystallizer. The induction time model prediction is the least reliable as can be expected for a probabilistic process. The design space selected was large enough to find suitable conditions for the crystallization of the 2:1 co-crystal, while the conditions to crystallize the 1:1 co-crystal phase had to be extrapolated and subsequently validated. The extrapolation was done by choosing various concentrations of initial solutions outside of the original design space and analysing the same response parameters. The conditions from the DoE chosen as most suitable to produce the 2:1 co-crystal are 50/50 w/w solution flow ratio, with the solution concentrations of 109.4 g BZA/kg ethanol and 48.5 g INA/kg water. For the 1:1 co-crystal the optimal conditions were the solution flow ratio of 13/84 w/w (benzoic acid solution to isonicotinamide solution) with the solution concentrations of 30g BZA/kg ethanol and 40g INA/kg Water.

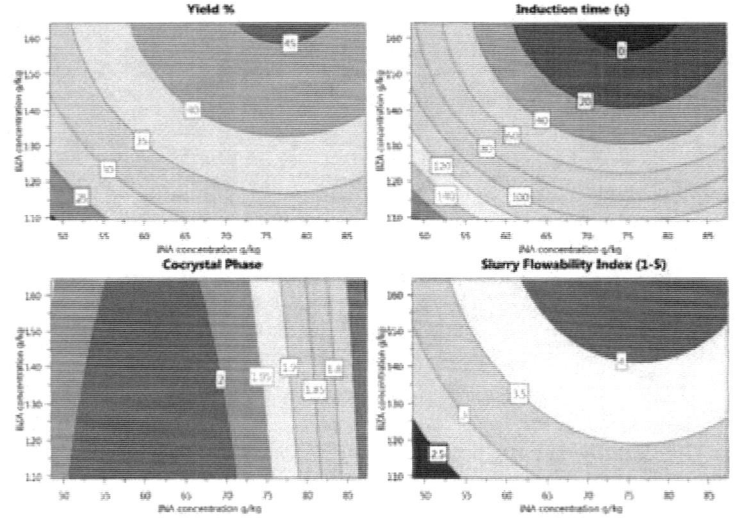

Figure 6. Contour plot showing responses from the Design of Experiments at a fixed solution ratio of 50:50 (w/w)

Solubility. During the phase stability test, the 2:1 co-crystal began to transform to the 1:1 co-crystal over time at the crystallizer outlet conditions. Transformation commenced between 158 to 264 minutes in an isothermal stirred environment as detected by XRPD, see Figure 7. This makes solubility determination more difficult as the 2:1 solid form is not thermodynamically stable at the chosen process conditions. However, the solid-liquid equilibrium of the metastable solid phase was determined for the purpose of thermodynamic yield calculations. The solubility test was timed to avoid any significant transformation. Solubility measurements were carried out in 48.8% (w/w) ethanol/water mixture, which is the same as the solvent mixture at the continuous mixer outlet during crystallization of 2:1 co-crystal. The results from six experiments suggest the solubility to be 0.0446 g/ml solvent with a high level of reproducibility based on clear points from image analysis. The spread of various measurements is illustrated in Figure 8. The high spread of clear point measurements based on transmission may be attributed to the limited ability to detect solids by transmission measurements at low concentrations, especially for thin needles. The different solvent addition rates show comparable results, thus the result is not dependent on the kinetics of crystal dissolution. The findings about the 2:1 co-crystal metastability point out that the outlet composition of the crystallizer slurry is either in the 1:1 co-crystal or a mixture of the 1:1 and 2:1 co-crystal regions on the phase diagram. The solubility of the 2:1 co-crystal has been previously reported between 0.0367 g/ml and 0.0482 g/ml for 50% water/ethanol mixture by[26] using two methods. However, the measurements reported therein have a high variability and the gravimetry method implemented did not seem to account for a possibility of a phase transformation. Therefore, it is possible that in some cases the solubility of the 1:1 co-crystal or a mixture of both co-crystal phases could have been measured. A recent study by Munshi et al.[22] shows the ternary phase diagram at 50% ethanol/water, although it was not reported whether this was 50% by mass or by volume[22]. As we have shown, solvent mediated transformation may occur in this system under conditions studied in this and previous works, and this may perhaps also help to explain the unusual shape of the phase diagram reported in the work of Munshi et al.

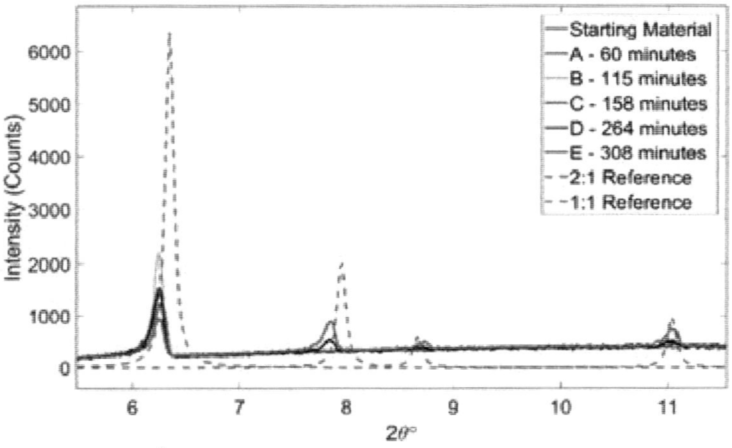

Figure 7. XRPD from suspension stability study of 2:1 co-crystal in 48.8% ethanol – water (w/w)

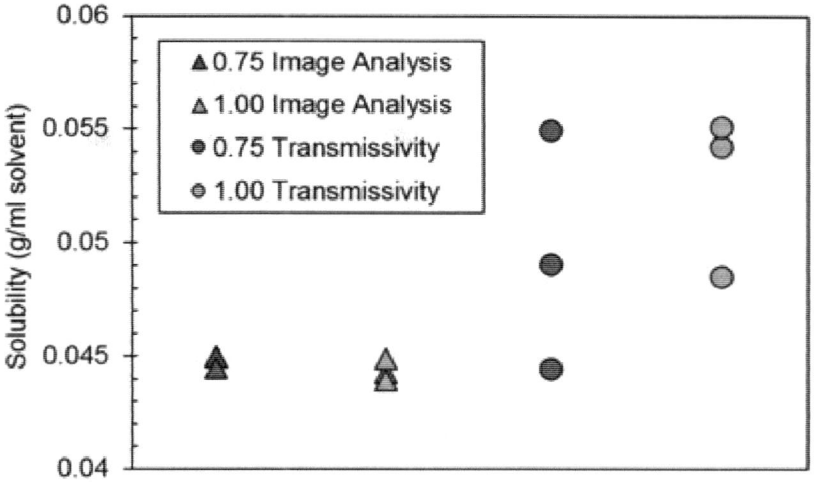

Figure 8. Solubility of 2:1 co-crystal in 48.8% ethanol – water (w/w) at 25 °C

Continuous crystallization. Moving from benchtop vial-based screening to a continuous platform introduces new process parameters and challenges. The overall flow rate was found to have a strong impact on the mixing regime and thus final particle properties. Three different overall flow rates were examined with the aim of producing the 2:1 co-crystal: 60 g/min, 40 g/min and 20 g/min at 50:50 ratio (w/w) of solution streams. The continuous runs revealed fouling and subsequent blockages to be a challenge, however blockage issues decreased with lower flowrates. Antifouling measures such as low power sonication could be used to mitigate this issue[27]. The solid output was 0.507 g/min, 0.342 g/min and 0.0259 g/min for overall flowrates of 60 g/min, 40 g/min and 20 g/min, respectively. The L50 of particles obtained at overall flow rates of 60 g/min, 40 g/min and 20 g/min was 361μm, 317μm and 238μm, respectively. Despite producing the desired solid form (Figure 10) from all conditions, mixing performance drastically reduced at the overall flow rate of 20 g/min resulting in a lower solid yield. The low velocity difference between the capillary jet and the outer flow did not provide sufficient mixing to rapidly achieve the desired supersaturation. The same process at the overall flow rate of 20 g/min resulted in 0.0484 g/min solid output in the Ehrfeld platform. Clearly, mixing conditions influence nucleation kinetics of co-crystal as the solid yield and resulting particle size distribution (see ESI Figure 6) varies with overall flowrate although the overall solution composition is the same in all cases. In order to achieve a higher yield in the crystallization process, increasing of residence time downstream from the mixing point combined with additional cooling stages should be considered. Further process optimization would benefit from more detailed knowledge of the relevant phase diagram and additional experiments in an expanded design space.

A continuous crystallization was also carried out to produce the 1:1 co-crystal, using flow rates of 15 g/min of benzoic acid solution through the capillary and 100 g/min of isonicotinamide solution through the outer tube. For this experiment the solutions were less concentrated to avoid blocking due to potentially high solid loading of the resulting slurry. The induction time at these concentrations was relatively long compared to the 2:1 co-crystal experiments and particles were not immediately visible in the mixer outlet. The solid output after holding the slurry for 18 minutes was 0.0726 g/min.

Solid characterization. Solid state characterization techniques were selected in order to establish solid phase identity and purity. Crystal unit cell parameters were previously documented for both co-crystal phases which makes powder X-ray diffraction suitable for phase verification. The co-crystal phases also have unique melting points verifiable by DSC. NMR was also used to independently verify the stoichiometry of co-formers in the crystalline sample. The morphological differences in crystals for the 2:1 and 1:1 forms were minor. The needle shaped morphology of both co-crystal forms can be seen in Figure 9. Results from Malvern Morphologi G3 are shown in supplementary information and similar aspect ratio distributions were found for both co-crystal forms. Analysis by XRPD on the dry samples from the capillary mixer is shown in Figure 10 (XRPD on samples from the Ehrfeld platform are in supplementary information ESI Figure 9). Lines A, B and C are samples from continuous runs designed to produce the 2:1 co-crystal. Simulated patterns are generated from unit cell data available from the Cambridge Structural Database under structure labels: BUDWEC[28] and MOVTOH[23] for the 1:1 and 2:1 co-crystals respectively. Comparison with D – the simulated pattern suggests that a pure 2:1 co-crystal phase has been produced by all runs. Similarly, line E shows pure 1:1 co-crystal by comparison to the 1:1 co-crystal simulated pattern – line F. This is further supported by results from DSC shown in Figure 11. DSC for the 2:1 co-crystal samples shows a sharp thermal event previously reported at 142 – 143°C for the 2:1 co-crystal[29]. The TGA curves also show a decrease in mass from 148-150°C onwards indicating sample decomposition. Similarly, DSC for 1:1 sample shows a unique endotherm above 160 °C, at the expected melting point of the 1:1 co-crystal[29] with subsequent decomposition.

For additional confirmation, the samples were analyzed using ATR-IR. The absorbance spectra are in Figure 12. The figures show a spectrum for each starting material as well as samples from continuous experiments. It is clear that the co-crystal phases have a well-defined unique spectra and it is possible to distinguish the 2:1 and 1:1 co-crystals using IR spectrometry.

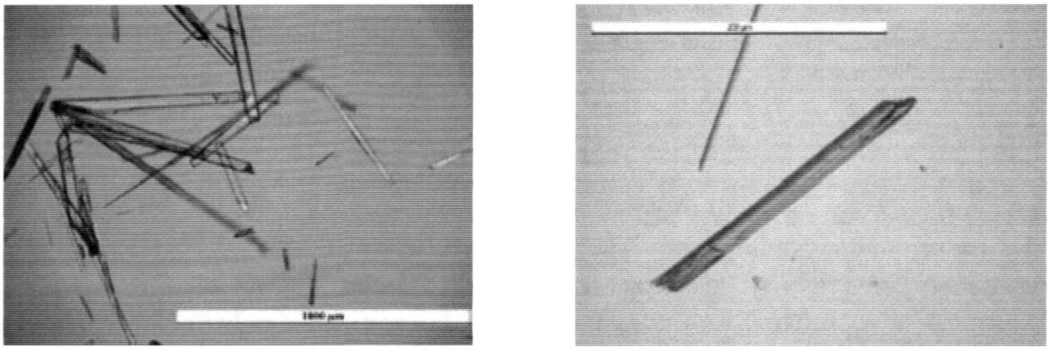

Figure 9. Optical microscope images: 2:1 co-crystal (Left) and 1:1 co-crystal (Right)

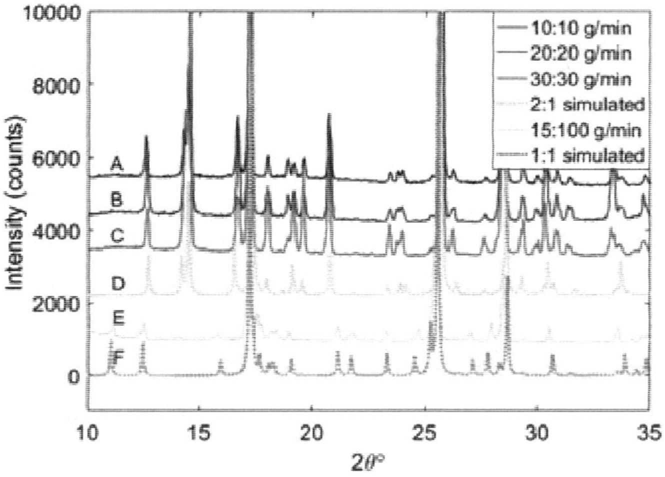

Figure 10. XRPD for samples from all continuous runs together with calculated patterns

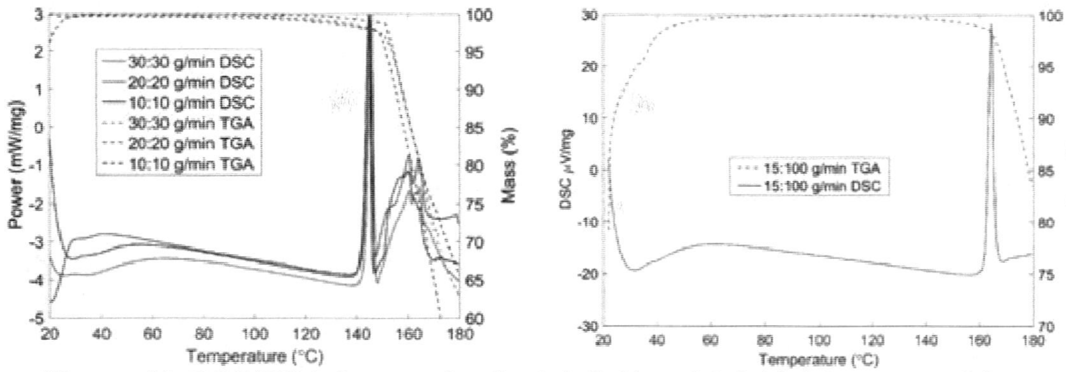

Figure 11. DSC/TGA for samples for 2:1 (left) and 1:1 (right) co-crystal from continuous runs

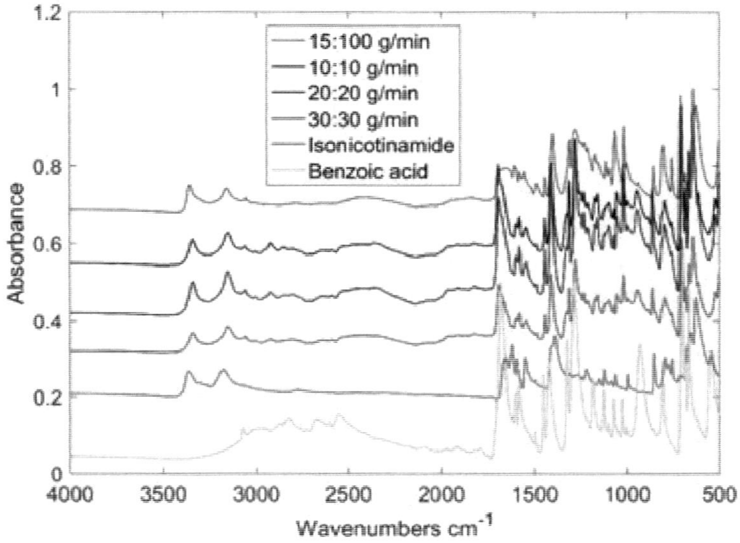

Figure 12. ATR-IR spectra for starting material and samples from continuous runs

NMR analyses on solid samples dissolved in deuterated DMSO from each continuous run confirmed (within 2% error) a 2:1 ratio for each sample from the continuous runs aimed to produce 2:1 co-crystal (ESI Figure 8). Continuous experiments using 40 g/min and 60 g/min flow rates produced particles of very similar particle size and shape characteristics. Results from 20 g/min continuous experiment show significantly smaller particles at much lower solid yield (ESI Figure 6). The aspect ratios for the particles from 20 g/min run are more needle-like with a narrower distribution.

We showed that both 2:1 and 1:1 pure forms of the co-crystal can be produced using the combined approach method based on mixing-induced supersaturation in a continuous crystallization process. The combined antisolvent and reactive crystallization approach allows for additional flexibility in reaching specific phase regions in multicomponent systems using multiple solvents.

CONCLUSIONS

We have used Design of Experiments to find conditions suitable for formation of 2:1 and 1:1 co-crystals of benzoic acid and isonicotinamide by mixing ethanolic solutions of benzoic acid and aqueous solutions of isonicotinamide in isothermal conditions. The input parameters were the two solution concentrations and the mass ratio of the two solutions, while the output variables were the solid form produced, solid yield, induction time and slurry flow rating. Optimal conditions were then identified for both 2:1 and 1:1 co-crystal formation and continuous crystallization experiments were performed at several total flow rates. Both solid forms were selectively produced in continuous crystallization as confirmed by XPRD, IR, DSC and NMR. Mixing conditions were found to influence the solid yield and particle size distribution of crystal produced although not the solid form.

The combined antisolvent and reactive crystallization approach can be a useful tool for reaching solid phases that are otherwise difficult to access through mixing-induced supersaturation. Under conditions of thermodynamic control, it may be possible to reach certain regions of phase diagrams that would be not accessible by other approaches due to non-ideal shapes of some phase diagrams. Under conditions of kinetic control, it may be possible to influence nucleation kinetics through the initial mixing of two solutions which would depend on compositions of both solutions as well as mixing conditions applied.

ACKNOWLEDGMENT

We would like to thank Maria Briuglia for assistance with Technobis Crystalline. The authors would like to thank EPSRC and the Doctoral Training Centre in Continuous Manufacturing and Crystallisation (Grant Ref: EP/K503289/1) for funding Vaclav Svoboda and John McGinty. The authors would like to thank EPSRC and Centre for Innovative Manufacturing in Continuous Manufacturing and Crystallisation (Grant Ref EP/I033459/1) for funding Pól MacFhionnghaile. The authors would like to thank the International Strategic Partnership between University of Strathclyde and Nanyang Technological University Singapore for funding Lauren E. Connor. The authors would like to thank EPSRC for Fellowship (Grant Ref EP/N015401/1) for funding Iain D.H. Oswald. The authors would like to acknowledge that this work was carried out in the CMAC National Facility supported by UKRPIF (UK Research Partnership Fund) award from the Higher Education Funding Council for England (HEFCE) (Grant ref HH13054).

REFERENCES

(1) Domingos, S.; André, V.; Quaresma, S.; Martins, I. C. B.; Minas da Piedade, M. F.; Duarte, M. T. J. Pharm. Pharmacol. 2015, 67, 830–846.
(2) Elder, D. P.; Holm, R.; Diego, H. L. de. Int. J. Pharm. 2013, 453, 88–100.
(3) Trask, A. V. Mol. Pharm. 2007, 4, 301–309.
(4) Friščić, T.; Jones, W. Cryst. Growth Des. 2009, 9, 1621–1637.
(5) Macfhionnghaile, P.; Hu, Y.; Gniado, K.; Curran, S.; Mcardle, P.; Erxleben, A. J. Pharm. Sci. 2014, 103, 1766–1778.
(6) Hasa, D.; Carlino, E.; Jones, W. Cryst. Growth Des. 2016, 16, 1772–1779.
(7) Rehder, S.; Christensen, N. P. A.; Rantanen, J.; Rades, T.; Leopold, C. S. Eur. J. Pharm. Biopharm. 2013, 85, 1019–1030.
(8) Moradiya, H. G.; Islam, M. T.; Halsey, S.; Maniruzzaman, M.; Chowdhry, B. Z.; Snowden, M. J.; Douroumis, D. CrystEngComm 2014, 16, 3573.
(9) Wittering, K.; King, J.; Thomas, L.; Wilson, C. Crystals 2014, 4, 123–140.
(10) Lee, M.-J.; Chun, N.-H.; Wang, I.-C.; Liu, J. J.; Jeong, M.-Y.; Choi, G. J. Cryst. Growth Des. 2013, 13, 2067–2074.
(11) Rodríguez-Hornedo, N.; Nehm, S. J.; Seefeldt, K. F.; Pagán-Torres, Y.; Falkiewicz, C. J. Mol. Pharm. 2006, 3, 362–367.
(12) Zhao, L.; Raval, V.; Briggs, N. E. B.; Bhardwaj, R. M.; McGlone, T.; Oswald, I. D. H.; Florence, A. J. CrystEngComm 2014, 16, 5769.
(13) Powell, K. A.; Bartolini, G.; Wittering, K. E.; Saleemi, A. N.; Wilson, C. C.; Rielly, C. D.; Nagy, Z. K. Cryst. Growth Des. 2015, 15, 4821–4836.
(14) Powell, K. A.; Croker, D. M.; Rielly, C. D.; Nagy, Z. K. Chem. Eng. Sci. 2016, 152, 95–108.
(15) Alvarez, A. J.; Myerson, A. S. Cryst. Growth Des. 2010, 10, 2219–2228.
(16) Brown, C. J.; Adelakun, J. A.; Ni, X. Chem. Eng. Process. Process Intensif. 2015, 97, 180–186.
(17) Ferguson, S.; Morris, G.; Hao, H.; Barrett, M.; Glennon, B. Chem. Eng. Sci. 2013, 104, 44–54.
(18) Lindenberg, C.; Schöll, J.; Vicum, L.; Mazzotti, M.; Brozio, J. Chem. Eng. Sci. 2008, 63, 4135–4149.
(19) Baxendale, I. R.; Braatz, R. D.; Hodnett, B. K.; Jensen, K. F.; Johnson, M. D.; Sharratt, P.; Sherlock, J.-P.; Florence, A. J. J. Pharm. Sci. 2015, 104, 781–791.
(20) Bałdyga, J.; Makowski, ł.; Orciuch, W. Chem. Eng. Res. Des. 2007, 85, 745–752.
(21) Leyssens, T.; Springuel, G.; Montis, R.; Candoni, N.; Veesler, S. Cryst. Growth Des. 2012, 12, 1520–1530.
(22) Munshi, T.; Redha, B.; Feeder, N.; Meenan, P.; Blagden, N. Cryst. Growth Des. 2016, 16, 1817–1823.
(23) Seaton, C. C.; Parkin, A.; Wilson, C. C.; Blagden, N. Cryst. Growth Des. 2008, 9 (1), 47–56.
(24) Boyd, S.; Back, K.; Chadwick, K.; Davey, R. J.; Seaton, C. C. J. Pharm. Sci. 2010, 99, 3779–3786.
(25) Reus, M. A.; van der Heijden, A. E. D. M.; ter Horst, J. H. Org. Process Res. Dev. 2015, 19, 1004–1011.
(26) Redha, B. H. Impact of mixed solvent on co-crystal solubility, ternary diagrams and crystallisation scale-up. Crystallisations of Isonicotinamide- Benzoic Acid Co-crystals from Ethanol-Water Co-solvent System., University of Bradford, 2013.
(27) Al Nasser, W. N.; Pitt, K.; Hounslow, M. J.; Salman, A. D. Powder Technol. 2013, 238, 151–160.
(28) Aakeröy, C. B.; Beatty, A. M.; Helfrich, B. A. Angew. Chem. Int. Ed. 2001, 40, 3240–3242.
(29) Buanz, A.; Prior, T. J.; Burley, J. C.; Raimi-Abraham, B. T.; Telford, R.; Hart, M.; Seaton, C. C.; Davies, P. J.; Scowen, I. J.; Gaisford, S.; Williams, G. R. Cryst. Growth Des. 2015, 15, 3249–3256.

497864 A Multiscale Computational Method for Prediction of Polymorphs

Abstract:

Physical properties of solid pharmaceutical drugs stem from their crystalline structure or polymorph. Pharmaceutical molecules can undergo polymorphism to attain more than one crystalline structure depending on the crystallization conditions. Experimental screening of polymorphs is expensive and time-consuming. Therefore, the accurate methods to predict polymorphs have become essential to build in quality into the pharmaceutical products. The most prominent method to predict crystalline polymorph is the lattice energy minimization technique, which uses appropriate sampling and optimization algorithms to identify space groups and lattice energy. This approach has limitations in treating the solute-solvent interactions and capturing the effect of additives, solvents, co-solvents and hydrodynamic conditions of the crystallizer. Therefore, there is a need to develop a model which considers these effects involved in multiple length and time scales during the crystallization process. Here we propose a consistent multiscale approach to predict dynamics of polymorph formation, which includes hierarchical simulation of sequential events â€“ i) lattice formation via self-assembly of molecules, ii) growth of lattice and formation of molecular clusters, and iii) development of faces and evolution of faceted crystal.Â The effect of the solute-solvent interactions, additives, solvents, co-solvents and hydrodynamic conditions are embedded into the calculation of free energy barriers for molecular attachment/detachment in these three events. An example of polymorphism in Glutamic acid crystals using the multiscale method will be shown.

499540 Polymorph Selection By Continuous Precipitation

Abstract:

Due to polymorphic variation in mechanical, electrical, stability, bioavailability and physicochemical properties, there always exists an optimal polymorph for a given application. Sometimes, a thermodynamically metastable form is preferred. In these cases, intelligent process design must be utilized to produce the desired polymorph. By judiciously selecting a solvent, temperature, inlet supersaturation, and residence time, one can typically produce a pure, preferred polymorph in a continuous crystallization process regardless of the polymorphâ€™s thermodynamic stability.Â Design rules exist to guide the process engineer for systems in which heterogeneous nucleation and size independent linear crystal growth are the dominant mechanisms of solid deposition from solution. [1] Here, we hypothesize that these findings remain qualitatively true in precipitative systems in which homogeneous nucleation and/or agglomeration are also important crystal growth mechanisms. This hypothesis is supported by our recent experiments in which we continuously produce metastable forms of agglomerative $CaCO_3$. We have also extended the population balance analysis to include systems in which agglomeration is significant. Here we report these experiments and calculations and explain the implications for general polymorphic systems that exhibit significant agglomeration.

[1] Farmer, T. C., Carpenter, C. L., & Doherty, M. F. (2016). Polymorph selection by continuous crystallization. *AIChE Journal*, *62*(9), 3505-3514.

489642 Assessing the Adequacy of a Sampling Frequency in the Application of PAT to a Continuous Manufacturing Process for Drug Product

Abstract:

The implementation of spectroscopic instruments to determine the concentration of active pharmaceutical ingredient (API) in powder samples is vast and well documented in the scientific literature. The introduction of continuous manufacturing lines for solid oral drug product has triggered increased interest in the application of such tools to determine the concentration of API in moving streams of powder. A key question in the development of such methods is the determination of an adequate frequency of sampling for the PAT devices, as it pertains the overall dynamics of the system

Our contribution will discuss a hybrid approach that uses information from a deterministic model, and the variability in the NIR spectra, along with time-series analysis techniques (such as ARX models, partial auto correlation coefficients, and cross-correlation coefficients) and the Shannon theorem of sampling to assess the adequacy of proposed frequencies of sampling and averaging in the spectrometer located in the feed-frame of the tablet press.

492244 Continuous Production Technology for Pharmaceutical Semi-Solid and Liquid Formulations: Processability and Influence of Process Parameters on the Product Quality

Abstract:

N. Bostijn[1,*], T. De Beer[1], W. Dhondt[2], J. Van Renterghem[1], C. Vervaet[3]

[1]Laboratory of Pharmaceutical Process Analytical Technology, Ghent University, Ottergemsesteenweg 460, 9000 Ghent, Belgium

[2]Janssen Global Technical Operations, Pharmaceutical Mfg (PM) Platform, Turnhoutseweg 30, 2340 Beerse, Belgium

[3]Laboratory of Pharmaceutical Technology, Ghent University, Ottergemsesteenweg 460, 9000 Ghent, Belgium

INTRODUCTION

Pharmaceutical production today mainly consists of batch processes. These are time-consuming, expensive and sub-optimal processes. Continuous processes on the other hand, allow for more flexibility, an easier scale-up and better quality control. The aim of this study is to investigate whether the Contimix system, a modular continuous production system, initially developed for the continuous production of semi-solid and liquid cosmetics, can be employed for pharmaceutical applications as well. A commercially available pharmaceutical cream, which is originally produced via a batch process, was manufactured with the Contimix system. In a first step, the processability of the cream with the Contimix system was investigated. Furthermore, the influence of the process parameters on the product quality was studied through statistical Design of Experiments (DoE), where the different process parameters were varied simultaneously.

MATERIALS AND METHODS

Materials

A pharmaceutical cream (Janssen Pharmaceutica N.V., Beerse, Belgium) was selected as model formulation.

Methods

All creams were produced with a continuous production system (Contimix, Hebold systems, Cuxhaven, Germany). A two level fractional factorial design was applied to study the influence of five process parameters on the product quality: process temperature, throughput, mixing speed in mixing vessel 1, mixing speed in mixing vessel 2 and water phase temperature. The studied quality attributes of the cream (DoE responses) were Active Pharmaceutical Ingredient (API) concentration (at the beginning, middle and end of a DoE run), yield point, viscosity of the yield point, viscosity (at low, intermediate and high shear rate) and pH. The data was analyzed using the MODDE Pro 11.0 software (MKS Umetrics, Umeå, Sweden). The API concentration of the creams was measured using a Raman Rxn2 spectrometer (Kaiser Optical Systems, Ann Arbor, MI, USA), equipped with a fiber-optic PhAT probe. The laser wavelength was 785 nm. Spectra were collected off-line with an exposure time of 20 s and a resolution of 5 cm^{-1}. The spectral range of the system was from 150-1890 cm^{-1}. Rheological measurements were conducted with a Haake Mars III rheometer (Thermo Fisher Scientific, Karlsruhe, Germany). The experiments were carried out at a fixed temperature (25°C), using a 60 mm parallel plate geometry. A stress ramp and viscosity curve were obtained from each produced cream. The pH of the creams was measured with a glass HI 3220 probe (Hanna instruments, Temse, Belgium).

RESULTS AND DISCUSSION

Prior to the experiments, the raw materials of the cream were divided into three pre-phases, which were batch mixed before the actual continuous production. This additional step was necessary since the Contimix system does not feature as many pumps as there are components in the cream. Furthermore, there is no possibility of introducing powders directly into the process. The results of the DoE show that the influence of the process parameters on the pH was negligible. The API concentration at the three time points between the experimental runs was not influenced by the process parameters. However, when the API concentration within the runs was evaluated, a downward trend in API concentration was detected for runs with a high throughput. This was caused by an inhomogeneity of pre-phase 3, which contained the API. Even for runs produced at a lower throughput, a similar trend was observed, suggesting that the homogenization of pre-phase 3 was not optimal, leading to an inhomogeneity of the API in the final product. The stress ramp experiments showed an influence of the process temperature. A higher process temperature was linked with a lower viscosity. The viscosity curve experiments showed an influence of the process temperature and throughput, but both influences were minimal. However, when the viscosity curves were investigated, a difference between the runs at low and high process temperature was visible. A possible explanation for the influence of the process temperature on the yield point and viscosity can be found in the short time between manufacturing the creams and conducting the rheological experiments. To draw conclusions about this effect, a number of runs at low and high process temperature should be executed and the viscosity of these creams should be measured directly (1 day) and a longer period (4 weeks) after production. Despite achieving processability, further improvements are necessary to transfer this continuous production technology to the pharmaceutical industry. The batch pre-step should be replaced by a continuous step. In order to achieve this, a way of feeding powder directly into the system is necessary. This would make the production system completely continuous, with all the advantages this entails. The original batch produced cream was measured with the same techniques as the creams of the DoE runs and compared with a center point run (run 18). The pH and API concentration were similar for both creams. From the stress ramp experiment, it can be concluded that the yield point of run 18 was slightly higher than that of the original product, indicating that both creams have similar rheological properties. This was confirmed by the viscosity curves, where the viscosity decrease with increasing shear rate was similar.

CONCLUSION

Processability of a pharmaceutical cream with the Contimix, a continuous production system, was achieved. From the screening design, it can be concluded that the process temperature had a significant influence on the viscosity of the creams, i.e. higher process temperature resulted in creams with a lower viscosity. Changing the other process parameters had no significant impact on the measured quality attributes. Furthermore, the Contimix was capable of producing a commercially available pharmaceutical cream in a continuous way with a similar quality as the original formulation.

494641 Modeling of Residence Time Distribution of a Continuous Dry Granulation Tableting Line

Abstract:

In recent time academia and industry have shown the economic and process related benefits of continuous pharmaceutical manufacturing. Regulatory authorities support strongly this approach of processing as well. This trend leads to a need for advanced control strategies, which enable automated operation of integrated production lines. With the first continuous plants being implemented for continuous production of solid oral dosage forms, out-of-spec material discharge strategies and process control with the aim to ensure robust product quality and to save resources have to be developed. Malfunctions of process equipment, unexpected process events and start-up or shut-down phases can lead to out-of-spec material. Due to material holdup and back mixing effects in specific unit operations, the out-of-spec material distribution can spread over process time. Therefore, knowledge of the residence time distribution (RTD) is essential not only for batch definition and material tracking aspects. For the development of cost-effective discharge strategies a fundamental tracking of the material through the process based on RTD is required.

In this work a dry granulation line, as one of the common ways of solid dosage form manufacturing, was investigated. The line consists of a feeding, blending, dry granulation, pneumatic transport and tableting step. For the determination of the residence time distribution in the considered unit operations as well as in the entire line a step change of API was performed at different total throughputs. As a system response the concentration of API was measured via in-line NIR spectroscopy at the outlet of blender, granulation and tablet press. Additionally, samples of powder, granules and tablets were drawn and analyzed off-line by means of UV/Vis as a reference.

Based on the experimental RTD data a transfer function was calculated and the transfer behavior of single unit operations modeled. The RTDs for single unit operations were interlinked and the entire process simulated with customized models in the process simulation software gSolids. The selected approach allows the prediction of out-of-spec material amount and the discharge time for different throughputs. This work shows promising results for both the inter- and extrapolation of RTDs for various throughputs and comparison with validation runs of the entire line.

494709 Robust State Estimation of Feeder and Blender Systems in Continuous Pharmaceutical Manufacturing Systems

Abstract:

Abstract

The development and application of advanced monitoring, operational- and design-making strategies for continuous pharmaceutical manufacturing require good understanding of the current state of the system. Moving horizon estimation (MHE) can estimate the current system state by solving a well-defined optimization problem where process complexities are explicitly taken into consideration as constraints. Traditional MHE techniques assume random measurement noise governed by some normal distributions. However, this assumption is not always valid since the actual distributions are often unknown or exhibiting departures from normal form. In addition, conventional state estimators fail to consider gross or systematic errors, such as bias, outliers, and drifts. The presence of gross errors can contaminate the state estimates and the reconciled measurements, causing undesirable â€œsmearingâ€ effects [1]. As a result, research on state estimation is closely related to topics of data reconciliation (DR) and gross error detection (GED).

Robust estimators, also known as the maximum likelihood estimators or M-estimators, are commonly used in reconciling measurements contaminated with gross errors. Robust estimators have been implemented in many research areas, including state estimation, data reconciliation, and parameter estimation [2, 3, 4]. Systematical comparisons between various robust estimators can be found in literature [1, 5]. To improve the accuracy and robustness of state estimation, we incorporate robust estimators within the standard MHE skeleton, leading to an extended MHE for dynamic state estimation with gross errors.

In continuous tablet or capsule production lines, the optimal operation of the feeder-blender system (FBS) is crucial in order to meet the critical quality attributes of the final product. Advanced monitoring and control strategies of FBS require fast and accurate estimates of the current process state. In this work, the proposed MHE framework is implemented on a simple FBS configuration which includes two loss-in-weight (LIW) feeders and one continuous blender. Based on our experiment results, we propose a first-principle dynamic model for screw feeders and a reduced-order, compartment-based model for continuous blenders. Both dynamic models are used in our MHE framework, to result in a dynamic optimization (DO) problem constrained by differential algebraic equations (DAEs).

The proposed MHE strategy is first tested on simulation runs of the FBS where measurements are contaminated with gross errors such as outliers and drifts. Numerical results show significant benefits in applying the extended MHE technique with robust estimators. In addition, we implement our MHE framework in a pilot plant FBS, which is part of a continuous dry granulation line, and demonstrate on-line dynamic state estimation and data reconciliation. Experimental results confirm that the proposed MHE framework is robust to gross errors and can provide accuracy state estimates. Moreover, the efficient solution of the MHE realized in this work, suggests feasible application of on-line state estimation on more complex continuous pharmaceutical processes.

Bibliography

[1] Özyurt, Derya B., and Ralph W. Pike. "Theory and practice of simultaneous data reconciliation and gross error detection for chemical processes." Computers & chemical engineering 28.3 (2004): 381-402.

[2] Tjoa, I. B., and L. T. Biegler. "Simultaneous strategies for data reconciliation and gross error detection of nonlinear systems." Computers & chemical engineering 15.10 (1991): 679-690.

[3] Arora, Nikhil, and Lorenz T. Biegler. "Redescending estimators for data reconciliation and parameter estimation." Computers & Chemical Engineering 25.11 (2001): 1585-1599.

[4] Nicholson, Bethany, Rodrigo López-Negrete, and Lorenz T. Biegler. "On-line state estimation of nonlinear dynamic systems with gross errors." Computers & Chemical Engineering 70 (2014): 149-159.

[5] Prata, Diego Martinez, José Carlos Pinto, and Enrique Luis Lima. "Comparative analysis of robust estimators on nonlinear dynamic data reconciliation." Computer Aided Chemical Engineering 25 (2008): 501-506.

496016 Large-Scale Experimental Comparison of Batch and Continuous Technologies for Pharmaceutical Tablet Manufacturing

Abstract:

This work presents experimental comparison of batch and continuous technologies for pharmaceutical tablet manufacturing at a scale of 100 kg. The investigation covered the units of mixing, granulation, drying, blending, and compression, for which three processes were established based on batch and continuous technologies. For the batch process, conventional facilities were used, with employing fluidized bed granulation and high shear granulation as the granulation unit. For the continuous process, a recently developed facility was used, which interconnects all the units and is capable of producing tablets at the speed of 25 kg/hr. Common to these three processes, a 100 kg of raw materials were provided that contained ethenzamide as an active pharmaceutical ingredient (API), and mannitol and crystalline cellulose as excipients. As a preparation, experiments at a small scale were performed in order to determine process parameters in the experiments at a scale of 100 kg. During or after the production, samples were taken for investigating the following three items. First was the quality of granules as intermediate products, e.g., particle size distribution, mean diameter, and loose bulk density. Second was the quality of the tablets, e.g., tablet hardness, content of API, and dissolution profiles. Third was the yield of the overall process considering various causes of losses, such as stabilization of the continuous process, sticking to the equipment, or condition setting of compression. For the continuous process, stability of product quality during the process operation was also investigated because the total run time can be up to hours.

The experimental result indicated that the product quality of all processes fulfilled the standard values which are often used in the industry. Granules made by fluidized bed granulation were porous, while granules made by high shear granulation and the continuous process were spherical. Although hardness of tablets made by the high shear granulation process was lower than that by the other two processes, there are no significant differences of dissolution profiles among the three processes. In the continuous process, product quality measured by mean particle diameter of granules was stable for the entire run time except for the initial nine minutes. Because of this reason, yield of the continuous process was lower than those of the other two processes, which indicates that stabilization would be one of the important challenges in continuous technology. In conclusion, the three investigated processes using continuous, fluidized bed, and high sheer granulation showed similarity in terms of product quality. However, intermediate products showed different quality characteristics such as size/shape, and especially in continuous technology, yield was found to be an improvement opportunity. The actual decisions should be based on various aspects such as quality, economy or safety, and thus, development of multiobjective design methodology is desired in future.

496075 A Cost-Benefit Analysis of Continuous Manufacturing Using Discrete Event Simulation

Abstract:

In the pharmaceutical industry, the demand for small volume, market specific products has increased in recent years. However, current batch processes do not allow the flexibility to meet smaller demands without a high cost of inventory and risk of discards. Continuous processing may offer advantages to traditional batch processing including shorter lead times, reductions in inventory, decreases in management overhead, and fewer discards or loss of tenders. However, the specific conditions which make continuous manufacturing beneficial are currently not well understood. To quantify the benefit and determine the conditions where continuous manufacturing is an attractive alternative for products, a discrete event simulation model was developed. The model compares continuous to batch tablet formulation processes under numerous different scenarios including; various market conditions, production planning strategies, capacity constraints, and production volumes. The model uses a product currently produced at a Merck production facility as a base case. The outputs from the model define the product, market, and production conditions best suited for a continuous manufacturing platform. This ultimately enables a mechanism for identifying products that are a good fit for a continuous manufacturing platform.

496960 Real-Time Monitoring and Control of API Concentration in a Tablet Press for Continuous Manufacturing of Tablets

Abstract:

Active pharmaceutical ingredient (API) concentration in powder blends is a critical material attribute (CMA) that is considered one of the most important product critical quality attributes (CQAs) for their final product (tablets). To determine the API concentration before tableting, real-time monitoring of API concentration before and after compaction has been performed for a variety of batch and continuous processes via Near-Infrared (NIR) spectroscopy [1, 2]. This analytical technology, which has been highly encouraged within the Quality by Design (QbD) and Process Analytical Technology (PAT) frameworks, has shown to be a valuable for detecting issues in CMAs during the process and has become a critical in-process sensor in the pharmaceutical industry [3].

This work presents a feed-forward controller for continuous manufacturing by using a NIR-based real-time monitoring technology at a tablet press feed frame. The basic idea behind feed forward control is to measure important process variables and take corrective action before the disturbances upset the process. The feed forward controller therefore needs to receive information about the known and predictable effects of process disturbances before those occur [4]. The proposed methodology involves measuring the inherent variability in the API concentration of material coming from the process and adjusting tablet weight to maintain overall API potency in tablets. We measured the residence time distribution (RTD) information from the rotary tablet press together with the NIR from the feed frame to develop the proposed framework as follows.

We evaluated the RTD of our Fette 1200i tablet press unit. A step change of API concentration of powder blend was given before the feed frame and the API concentration by the feed-frame NIR and compressed tablets were measured [5]. Based on these profiles, RTD of the feed frame was shown to have a similar RTD profile as a continuous stirred tank reactor (CSTR). RTD of the turret was found to be negligibly short and no difference was observed between the profiles of the feed-frame and tablets. The estimated RTDs are useful for prediction of the CQA and its controls. The formulation for our case study consisted of micronized acetaminophen, microcrystalline cellulose (Avicel® PH102), and magnesium stearate at a ratio of 15:84:1 by weight, respectively. A portable JDSU NIR probe was attached in the feed frame, sampling the powder spectra through one of the unit's glass windows. Five (5) different blends ranging from 10 – 20% API concentration were made as calibrations for the NIR model. NIR spectra were measured in-line using the 5 powders blends to build the calibration model *in situ*. A calibration model with small calibration error was constructed by partial least squares (PLS) analysis. Validation experiments with four different acetaminophen concentrations demonstrated that the constructed model has high accuracy during long-term operation.

Overall, based on the feed-frame's RTD profile and the information from the NIR sensor's model, a feed-forward controller was developed to account for changes in powder blend API concentration and adjust tablet weight to decrease variability of API potency in tablets. It was verified by both software simulations and experiments that API potency could be maintained at the target value by the feed-forward control compared to the open-loop process.

References

1. Vanarase, A.U., et al., *Real-time monitoring of drug concentration in a continuous powder mixing process using NIR spectroscopy.* Chemical Engineering Science, 2010. **65**(21): p. 5728-5733.

2. Jarvinen, K., et al., *In-line monitoring of the drug content of powder mixtures and tablets by near-infrared spectroscopy during the continuous direct compression tableting process.* Eur J Pharm Sci, 2013. **48**(4-5): p. 680-8.

3. USFDA, *Guidance for Industry: PAT — A Framework for Innovative Pharmaceutical Development, Manufacturing, and Quality Assurance*, U.D.o.H.a.H. Services, Editor. 2004: Rockville, MD.

4. Myerson, A.S., et al., *Control Systems Engineering in Continuous Pharmaceutical Manufacturing. J. Pharm. Sci.*, 2015. **104**: p. 832-839.

5. Ward, H.W., et al., *Monitoring blend potency in a tablet press feed frame using near infrared spectroscopy.* Journal of Pharmaceutical and Biomedical Analysis, 2013. **80**: p. 18-23.

489666 Use of Model Discrimination Method in Drug Substance Process Development

Abstract:

Drug substance process development often includes several reaction steps to produce the Active Pharmaceutical Ingredient (API), where each step leads to a desired API intermediate. Mathematical models with different structures can be built to represent the proposed reaction mechanisms. It is common for multiple model constructs to equally explain the data gathered at this early stage. At this point it is important to select a model structure among the multiple candidates proposed. This work highlights the application of a model discrimination method to achieve this goal.

The model discrimination method was used to identify a discriminative experimental recipe for a reaction system of interest [1]. The experiment was then conducted to generate data that was qualitatively compared against the model predictions. The model that better explains the trends of the experimental data is deemed to be the best representation of the system among the candidates. To illustrate the use of this method, case studies for different reactions in drug substance process development will be given.

Once a model is discriminated, further investigation can be performed to optimally parameterize the model by following a model based experimental design method. A representative model can help the process development team determine safer and more efficient operating conditions.

Reference

[1] Espie, D. and Macchietto, S. (1989), The optimal design of dynamic experiments. AIChE J., 35: 223â€“229. doi:10.1002/aic.690350206.

489846 Model-Aided Development for a Mixed Anhydride Formation in a Drug Substance Manufacturing Process

Abstract:

Mathematical models are valuable tools in process development to determine acceptable parameter ranges and control strategies for manufacturing. Some of the benefits of modeling in pharmaceutical development are its ability to reduce material costs and resource needs, while simultaneously increasing process understanding and quality control.

A case study is presented here that demonstrates the use of automated experimentation to gather high-value reaction data for kinetic parameter estimation of a six reaction step mechanism with nine components. The model was used to simulate kinetic and thermal profiles for varying addition strategies of a key reagent. The resulting model was used to define parameter targets and ensure process robustness in preparation for a successful registration stability campaign.

494892 Design of Efficient Metal Nanocatalysts for Cross Coupling Reactions Using Finite Difference Time-Domain Simulations

Abstract:

Design of Efficient Metal Nanocatalysts for Cross Coupling Reactions using Finite Difference Time-Domain Simulations

Andishaeh Dadgar, Farshid Mohammadparast, Marimuthu Andiappan, School of Chemical Engineering, Oklahoma State University, Stillwater, OK, USA.

Cross coupling reactions (e.g., Suzuki and Heck couplings) have been traditionally carried out using homogeneous palladium (Pd) based catalysts, and in recent years, the use of heterogeneous metal nanocatalysts has been of great interest. However, the metal nanocatalysts (e.g., Pd, Au) require more drastic reaction conditions (e.g., high temperatures) because of their low catalytic activity. The operation of cross couplings under these drastic reaction conditions compromises long-term stability of the nanocatalysts due to leaching of metal atoms. Therefore, there remains a critical need to develop a heterogeneous metal nanocatalyst with high activity and long-term stability for cross couplings.

In this contribution, we show that finite-difference time-domain (FDTD) simulations can be used as platform to design efficient heterogeneous metal nanocatalysts. Our FDTD simulations predict that the UV-Vis surface plasmon extinction spectra of the metal nanoparticles are very sensitive to the local environment, and therefore, changes in even a small fraction of the surface atoms (e.g., leaching of the surface atoms) can be identified from the changes in the extinction spectra. We will show that a combination of FDTD simulations and a simple UV-Vis surface plasmon spectroscopic technique can be used to develop efficient heterogeneous metal nanocatalysts for a variety of cross coupling reactions.

495908 Nonlinear Optimisation of Continuous Artemisinin Crystallisation with Explicit NRTL Model-Based Solubility Prediction

Abstract:

Continuous processing is studied for use in pharmaceutical production to overcome the inherent drawbacks of batch processing, which is the current paradigm in the pharmaceutical and fine chemicals industry. Developments in small-scale flow technologies (and in microreactors in particular) have accelerated this effort; flow chemistry has not only allowed higher efficiencies but also enabled access to reaction pathways that have not been previously available for batch production (Jensen, 2017). This sustained interest in Continuous Pharmaceutical Manufacturing (CPM), along with the development of Process Analytical Technology (PAT) explains the strong interest towards Quality by Design (QbD) for pharmaceutical production processes, which enjoys significant regulatory body encouragement and support (Lee et al., 2015).

One crucial aspect of continuous processing is the continuous separation of Active Pharmaceutical Ingredient (API) products, often at low scales of process throughput and strict purity requirements. Research in product separation, including downstream drug product formulation, is a vibrant area (Ierapetritou et al., 2016; Vural Gürsel et al., 2017; Yabuta et al., 2017). The recent technological development of plug flow crystallisers and Continuous Oscillatory Baffled Crystallisers (COBC, which operate in an oscillatory plug flow regime) offers production advantages, including enhanced heat and mass transfer and improved profile temperature control (Brown and Ni, 2011; McGlone et al., 2015).

Our recent publications explore process modelling, simulation and optimisation of the continuous synthesis and separation of ibuprofen (Jolliffe and Gerogiorgis, 2017a) and artemisinin (Jolliffe and Gerogiorgis, 2017b). By using systematic kinetic parameter estimation, solubility prediction via the UNIFAC method, and comprehensive economic evaluation, optimal total costs have been determined.

Artemisinin is a key antimalarial, and its continuous synthesis has recently been demonstrated (Kopetzki et al., 2013; Gilmore et al., 2014); product recovery and purity enhancement via crystallisation is a viable technique (Malwade et al., 2016). The present paper investigates further the modelling and systematic optimisation of continuous artemisinin crystallisation incorporating explicit solubility predictions via the Non-Random Two-Liquid (NRTL) model. Modelling a plug-flow COBC helps ensure high heat and mass transfer rates, with minimal impact of temperature and concentration gradients on the API product. Toluene, used in a promising CPM process for artemisinin (Horváth et al., 2015), is assumed to be the process solvent; the upstream portion of the flowsheet (reactor) performance is analysed on the basis of previous work (Jolliffe and Gerogiorgis, 2016). The antisolvents considered encompass three different binary mixtures (ethanol and ethyl acetate, ethanol and acetone, or acetone and ethyl acetate) chosen due to to their pharmaceutical processing suitability and the availability of solubility data and NRTL parameters (Nti-Gyabaah et al., 2010). The ratio of antisolvents in the binary mixture ranges continuously between 1:9 to 9:1 by weight. Similarly, the ratio of process solvent (toluene) to binary antisolvent mixture ranges between 1:1 to 1:5 by weight. A continuous crystalliser operating temperature range (between 40 °C and 5 °C) has been considered. Crystal growth and productivity have been modelled on the basis of similar systems (Horváth et al., 2015), and impurities have been assumed to not co-crystallise with the API product.

The nonlinear optimisation (NLP) formulation encompasses the optimal choice of antisolvent, rate of antisolvent addition and the operating temperature which is preferable for continuous artemisinin crystallization. Our results include objective function (minimal cost) surfaces for a range of key variables, such as API recovery; insightful three-dimensional illustrations are obtained for process cost as a function of decision variables, e.g. crystallisation operating temperature and ratio of binary antisolvent mixture addition over incoming reactor effluent. The systematic optimisation of technoeconomic indices under explicit process equilibria and thermodynamic solubility constraints offers unprecedented insight into fine-tuning Continuous Pharmaceutical Manufacturing (CPM) process design and operation.

LITERATURE REFERENCES

Brown, C.J., Ni, X., 2011. Online evaluation of paracetamol antisolvent crystallization growth rate with video imaging in an oscillatory baffled crystallizer. Cryst. Growth Des. 11, 719–725.

Gilmore, K., Kopetzki, D., Lee, J.W., Horváth, Z., McQuade, D.T., Seidel-Morgenstern, A., Seeberger, P.H., 2014. Continuous synthesis of artemisinin-derived medicines. Chem. Commun. 50, 12652–12655.

Horváth, Z., Horosanskaia, E., Lee, J.W., Lorenz, H., Gilmore, K., Seeberger, P.H., Seidel-Morgenstern, A., 2015. Recovery of artemisinin from a complex reaction mixture using continuous chromatography and crystallization. Org. Process Res. Dev. 19, 624–634.

Ierapetritou, M., Muzzio, F., Reklaitis, G., 2016. Perspectives on the continuous manufacturing of powder-based pharmaceutical processes. AIChE J. 62, 1846–1862.

Jensen, K.F., 2017. Flow chemistry—Microreaction technology comes of age. AIChE J. 63, 858–869.

Jolliffe, H.G., Gerogiorgis, D.I., 2017a. Technoeconomic optimisation of a conceptual flowsheet for continuous separation of an analgaesic Active Pharmaceutical Ingredient (API). Ind. Eng. Chem. Res., in press.

Jolliffe, H.G., Gerogiorgis, D.I., 2017b. Technoeconomic optimisation and comparative environmental evaluation of continuous crystallisation and antisolvent selection for artemisinin recovery. Comput. Chem. Eng., in press.

Jolliffe, H.G., Gerogiorgis, D.I., 2016. Process modelling and simulation for continuous pharmaceutical manufacturing of artemisinin. Chem. Eng. Res. Des. 112, 310–325.

Kopetzki, D., Lévesque, F., Seeberger, P.H., 2013. A continuous-flow process for the synthesis of artemisinin. Chem.–Eur. J. 19, 5450–5456.

Lee, S.L., O'Connor, T.F., Yang, X., Cruz, C.N., Chatterjee, S., Madurawe, R.D., Moore, C.M.V., Yu, L.X., Woodcock, J., 2015. Modernizing pharmaceutical manufacturing: from batch to continuous production. J. Pharm. Innov. 1–9.

Malwade, C.R., Buchholz, H., Rong, B.-G., Qu, H., Christensen, L.P., Lorenz, H., Seidel-Morgenstern, A., 2016. Crystallization of artemisinin from chromatography fractions of Artemisia annua extract. Org. Process Res. Dev. 20, 646–652.

McGlone, T., Briggs, N.E.B., Clark, C.A., Brown, C.J., Sefcik, J., Florence, A.J., 2015. Oscillatory Flow Reactors (OFRs) for continuous manufacturing and crystallization. Org. Process Res. Dev. 19, 1186–1202.

Nti-Gyabaah, J., Gbewonyo, K., Chiew, Y.C., 2010. Solubility of artemisinin in different single and binary solvent mixtures between (284.15 and 323.15) K and NRTL Interaction Parameters. J. Chem. Eng. Data 55,

3356–3363.

Vural Gürsel, I., Kockmann, N., Hessel, V., 2017. Fluidic separation in microstructured devices – Concepts and their Integration into process flow networks. Chem. Eng. Sci.

Yabuta, K., Hirao, M., Sugiyama, H., 2017. Process model for enhancing yield in sterile drug product manufacturing. J. Pharm. Innov. 1–12.

499096 Multi-Layered Modelling Techniques for the Development of Continuous Manufacturing Processes

Abstract:

Traditionally the majority of active pharmaceutical compounds have been manufactured at large plant scales using batch methodologies. Continuous manufacturing is becoming an increasing attractive alternative however due to apparent benefits such as smaller manufacturing footprints, increased process safety, lower manufacturing costs and increased supply-chain flexibility. Realising all of the above benefits using conventional continuous equipment is however quite challenging and sometimes not achievable. In this project, a new approach to the design of continuous manufacturing platforms is being presented which is not restricted to using such conventional equipment but which is based around the bespoke design and manufacture of all unit operations, optimised to fit the specific process being examined. Processes are quantified in detail using first principle models which in turn are used to inform computational fluid dynamics (CFD) simulations leading to tailored equipment designs which can be manufactured with high precision using metallic 3D printing.

This presentation will focus on a number of case studies which demonstrate the linking of these various modelling techniques for several different unit operations required for the continuous manufacturing of pharmaceutical compounds including liquid-liquid separations, biphasic reactors and homogenous reactors. In particular however, the talk will focus on the development of a novel distillation column methodology, employing an inert gas to enable the concentration of solvent streams at low temperatures (required by temperature sensitive APIs). First principles theoretical models were used in order model the behaviour of such gas-liquid systems in bubble columns, enabling verification with experimental data. However as a bubble column does not represent the optimum design of such a gas-liquid contacting system, CFD was then employed to examine possible designs concepts suitable for 3D printing and to quantify evaporation rates in such possible configurations.

Following the modelling processes, these unit operations are 3D printed from stainless steel to enable their use as part of manufacturing processes. This approach leads to significant process intensification as all unit operations are designed specifically for the process being examined and are hence manufactured to ensure the processâ€™s optimum performance under the required conditions. Utilising 3D printing informed by process models to manufacture the process train also leads to significant cost reductions as unit operations become essentially disposable, reducing both the upfront capital costs and also essentially eliminating ongoing cleaning costs.

Several results demonstrating this approach are presented, including a biphasic reactor which is able to reduce the required reaction residence time by approximately 100 fold compared to an agitated batch reactor and a continuous homogenous reactor with dimensions of 15cm x 7cm x 3cm which is capable of achieving a throughput of 580,000 kg/L/day, representative of manufacturing output levels for many modern pharmaceutical compounds.

495661 A Refined Non-Random Two-Liquid Segment Activity Coefficient Model for Solubility Modeling

Abstract:

Accurate correlation and robust prediction for complex molecule solubility or solid-liquid equilibrium (SLE) in solvents and solvent mixtures have drawn much interest recently[1]. Solubility models such as the Non-Random Two Liquid Segment Activity Coefficient (NRTL-SAC) model[2] and its variations such as UNIQUAC-SAC[3] and COSMO-SAC with apparent sigma profiles[4] are now routinely used in the industry. This paper presents a refinement to the successful NRTL-SAC model in several perspectives. Specifically, the model is refined by modifying the Flory-Huggins expression in the entropy calculation. Additionally, a new conceptual segment has been proposed to better characterize polar components. At last, temperature dependency of activity coefficient is addressed in the refined model. We present the rationale for the refined model, along with the updated conceptual segment numbers for common solvents. Both the original and the refined NRTL-SAC models are employed to calculate vapor-liquid and liquid-liquid equilibrium for binary solvent systems and solubility of several drug molecules in solvents and solvent mixtures. The refined model provides improved prediction performance when compared with the original model, especially for systems containing larger size molecules. The refined NRTL-SAC model will be further extended to provide predictive capability for complex molecule solubility in excipients and polymers.

1. Tanveer, S.; Hao, Y.; Chen, C.-C., Introduction to solid-fluid equilibrium modeling. *Chemical Engineering Progress* **2014,** *110* (9), 37-47.

2. Chen, C.-C.; Song, Y., Solubility Modeling with a nonrandom two-liquid segment activity coefficient model. *Industrial & Engineering Chemistry Research* **2004,** *43* (26), 8354-8362.

3. Haghtalab, A.; Yousefi Seyf, J., Vaporâ€"liquid and solidâ€"liquid modeling with a universal quasichemical segment-based activity coefficient model. *Industrial & Engineering Chemistry Research* **2015,** *54* (34), 8611-8623.

4. Islam, M. R.; Chen, C.-C., COSMO-SAC sigma profile generation with conceptual segment concept. *Industrial & Engineering Chemistry Research* **2014,** *54* (16), 4441-4454.

490465 Overcoming Challenges for Scale-up of Continuous Pharmaceutical Drug Substance Processes

Abstract:

The area of continuous technology and reactor design applied to pharmaceutical drug substance processes has advanced significantly over the last decade. Primary drivers of flow versus batch operations are enhanced safety, greater process control, increased productivity, and better operational stability. However, several obstacles need to be overcome prior to successfully scaling-up flow processes from laboratory to manufacturing. Some examples are solid fouling, stability process hold, and mapping feasible operational space for process perturbations. This talk will focus on two case studies related to a: 1) Grignard reaction carried out in a continuous stirred tank reactor (CSTR) and, 2) hydrogenation reaction run in a pipes-in-series plug flow reactor (PFR). The investigated case studies will illustrate methodology and use of experimental tools in conjunction with mathematical models to overcome the above challenges. Specifically, the Grignard case study involved issues with instable process holds and solid fouling issues (affecting downstream performance) that were overcome by mechanistically probing the system with careful experimental design and analytical techniques. On the other hand, mathematical models were used for the hydrogenation case study to demonstrate their power in mapping out the feasible operational space. These efforts enabled scaling up these reactions to successfully run in various pivotal and validation campaigns.

485306 Preparation of Hazardous "on-Demand" Reagents Using Continuous Processing

Abstract:

Hazardous and unstable reagents are often forgone for use during manufacturing due to storage and safety concerns. These reagent are often replaced with more expensive alternatives. Continuous processing allows for the preparation of hazardous and/or unstable reagents that can be used directly in the production of pharmaceuticals. This talk will focus on using the Corning ® Advanced Flow Reactor™ (AFR) for the continuous preparation and use of several hazardous and unstable reagents.

489921 Model-Aided Development for a Continuous Amide Bond Formation in a Drug Substance Manufacturing Process

Abstract:

Mathematical models are valuable tools in development to determine acceptable parameter ranges and control strategies for continuous manufacturing. Some of the benefits of modeling in pharmaceutical development are its ability to reduce material costs and resource needs, while simultaneously increasing process understanding and quality control.

A case study is presented here that demonstrates the use of NMR reaction monitoring (ReactNMR), to gather high-value reaction data in batch mode for kinetic parameter estimation. The model describes the amide bond formation between an aniline and a mixed anhydride. The kinetic model was then employed to describe a continuous process that includes a plug flow reactor (PFR), a delay coil, two mixed-suspension, mixed-product-removal (MSMPR) crystallizers in series, and a filter/dissolution vessel. The resulting model was used to enable risk assessment, test failure modes, and build decision trees for use in the event of a process upset at manufacturing scale.

492008 Effects of Catalyst and Metals Pretreatment for the Continuous Hydrogenation of a Halogenated Aromatic Nitro Compound

Abstract:

Continuous processing for the synthesis of active pharmaceutical ingredients (APIs) continues to be a growing area of interest and investment. Continuous heterogeneous catalysis offers both safety and capital cost advantages over a traditional batch autoclave.

This investigation explored impurity control for the continuous hydrogenation of a halogenated aromatic nitro compound. The hydrogenation utilized a palladium catalyst for the reduction of an aromatic nitro compound to an aniline. In particular, the work involved development of a packed bed reactor for the continuous heterogeneous catalysis. Several challenges appeared in transitioning the chemistry from batch to continuous processing. These challenges were characterized using advanced monitoring techniques (flow-NMR, online HPLC, etc.) and investigated using a combination of batch and flow processing. As part of that effort, a catalyst pretreatment was investigated which had been demonstrated as effective in batch mode for suppressing an impurity. Parameters investigated included duration and temperature of pretreatment, as well as liquid flow profile. Also investigated was the effect of residual metals in the feed solution entering from a previous synthetic step. Feed pretreatment and its impact on reaction impurity profile and bed life were evaluated.

492261 Large Scale Production ofÂ Pharmaceutical Intermediates in the Flow Mode

Abstract:

In recent years, there have been significant efforts in the Pharmaceutical Industry to adopt continuous processing technologies to support the chemical/unit operation steps. Major drivers for developing flow processes and continuous manufacturing for specific steps include process safety issues, poor heat transfer, poor mass transfer in multiple phase reactions, presence of unstable intermediates, extreme reaction conditions requirements, and the need for short residence time. Since 2015, STA Pharmaceutical, a WuXi AppTec company, has established a leading continuous processing technology platform to support process R&D as well as plant scale production of pharmaceutical intermediates. The reaction types supported to date include cryogenic and room temperature lithiations, deprotection, photochemistry, ozonolysis, oxidations, dehydrations, coupling, cyclization, amination, and azide chemistries among other reactions. Flow steps included the above chemical steps as well as unit operations including phase separation, crystallization, filtration, and dissolution. The scale of the operations ranged from a few hundred grams to a few hundred kilograms under GMP and non-GMP modes. This presentation gives an overview of the continuous processing platform in STA followed by case studies illustrating the key benefits of this technology platform.

494952 Plasmonic Metal Nanocatalysts As Platform for Continuous Synthesis of Drug Substances

Abstract:

Plasmonic Metal Nanocatalysts as Platform for Continuous Synthesis of Drug Substances

Farshid Mohammadparast, Andishaeh Dadgar, Marimuthu Andiappan, School of Chemical Engineering, Oklahoma State University, Stillwater, OK, USA

Approximately 25% of all the reactions performed in the pharmaceutical industry consist of cross-coupling reactions. The cross-couplings have been traditionally carried out by homogeneous palladium (Pd) catalyzed batch processes. The National Science and Technology Council (NSTC) has recently identified the continuous manufacturing of pharmaceuticals as one of five manufacturing areas of emerging priority in the USA, revealing that continuous manufacturing of pharmaceuticals has the potential to reduce manufacturing cost by up to 40-50%. Given these advantages, in recent years, metal nanoparticle catalyzed cross-couplings have been of great interest since they can be used as a platform to switch cross-couplings from traditionally used batch mode to continuous flow mode. Additionally, heterogeneous metal nanocatalysts can be immobilized on a rigid support in the form of a packed-bed reactor, and therefore, can avoid the need for subsequent difficult steps to separate the catalyst and product.

In this contribution, we show that heterogeneous plasmonic metal nanocatalysts can be used as efficient catalysts to drive cross couplings. Specifically, we demonstrate that visible light can be used as a powerful tool to make plasmonic metal nanocatalysts as efficient photo-catalysts with activity and stability superior to that of traditionally used homogeneous Pd based catalysts. These plasmonic metal nanocatalysts can be used to design efficient cross-coupling flow processes to produce drug substances.

Abstract, AIChE 2017

Developing Scale-Up Approach for Fast Chemical Reactions in Continuous Flow

Because of potential high API demand for a particular Merck product currently under development, a new chemical synthesis was developed with improved yield. One of the chemical transformations in the synthesis was developed as a continuous flow step, improving reaction conversion by over 15% compared from batch mode. The reaction exhibits very fast kinetics (in the order of ms) requiring very good mixing. Successful scaling up of such fast reactions requires two important pieces of information:

1. Understanding of how reaction performance is affected by mixing.

2. A reliable way of measuring/scaling up mixing across different scales/geometries of static mixers.

In order to gain understanding on the effect of mixing, the continuous reaction was ran at various flow rates, static mixer geometries/sizes and temperatures. The collected data for reaction conversion was analyzed statistically in order to obtain a mathematical model for its dependence on the above process parameters. The model had acceptable predictive capabilities and confirmed an expected trend: for each mixer, the conversion was increased with increasing flow rate (improved mixing) and decreasing temperature (slower kinetics, allowing for better mixing). The drawback of this approach was that all these models were static mixer specific. If a different static mixer was to be used, a new study was needed in order link the flow rates to reaction conversion. A different, more generalized method was needed to measure/characterize mixing, independent of the mixer type.

Many research attempts to characterize and scale-up continuous mixing in static mixers usually involve correlations using scale-up parameters such as Reynolds number (Re), residence time (τ) or turbulent energy rate (ε). To link all these scale up parameters to the mixing intensity, a method known as the Fourth Bourne Reaction is usually employed. Some of these parameters usually work well, however they may fail to capture any differences in the internal geometry of the static mixer. Instead of using the above scale-up parameters, the conversion from the Fourth Bourne Reaction was used directly as a surrogate process parameter. This direct method of measuring mixing is more accurate as it will capture any internal changes to the mixing pattern (such as back mixing), which may not be captured if the other scale-up approaches are used. It can be seen as a snapshot of the mixing pattern inside the static mixer.

In order to show the applicability of this method, the Fourth Bourne Reaction tests were performed on the same mixers used in the continuous reaction, within the same flow rate ranges. For each mixer type/flow rate combination, the Bourne reaction conversion (BRC) number was assigned and used as a surrogate process parameter. Then the BRC and temperature only were used in the same statistical analysis. Both BRC and temperature were found to be statistically significant and the overall model had the same predictive capabilities. The advantage of this method was the ability to substitute the mixer type/flow rate combinations with the single BRC parameter. This allowed the three model plots produced previously for each of the mixers to be reduced to only one: BRC and temperature only. This methodology also makes scale-up much more straight-forward: one need to show only that the BRC is the same across scales in order to ensure identical mixing (provided the BRC test is executed identically).

501505 A Method to Reduce DimensionalityÂ of Powder Flow Characterization

Abstract:

Background: Material flow properties often have substantial impact on the performance of the manufacturing processes. For solids materials, the flow behavior is a multi-dimensional characteristic. Currently, a practical way is to measure powder properties using several standardized lab scale tests so that the flow behavior of a material can be described using a large group of indices. However, this approach is often time-consuming, and some of the indices obtained may be redundant. **Purpose:** The purpose of this study is to examine the intrinsic relations between different flow indices in a multivariate approach, and to develop a methodology that identifies an optimal set of measurements depending on the availability of time, materials, and instrument. **Method:** The proposed methodology includes techniques to characterize material flow properties and multivariate analysis. Pharmaceutical materials with varying flow properties were firstly characterized. The flow properties of each material was represented by multiple flow indices. To reduce the dimensionality of the database, the mathematical correlations between flow indices within each test were firstly examined. Computational iterations of principal component analysis was then performed. Depending on availability of instruments, the amount of materials, and time needed for each test, each flow index can be associated with an assigned weight. The criteria of selecting the optimal set was based on correlation coefficient of the Weighted Euclidean distance. **Results:** Results suggested that by assigning different weights to each flow index depending on the availability of time, materials, and instrument, different optimal sets of characterization measurements can be identified. The optimal set significantly reduced the cost of measurements. In addition, compared to the original dataset, the optimal set was able to capture maximal information of the material flow behavior. The optimal set of measurements can also be used prior to establishing the predictive correlations between material flow properties and process performance. **Conclusion:** The work presented here has shown an efficient approach to reduce dimensionality of material flow testing. This approach is especially powerful when the amount of time, instrument, or a material is limited, or when latent variables representing material flow properties are needed to develop process control strategy.

Using Mechanistic Modeling of Chromatography to Increase Process Understanding

Tim Fattor, Steve Hunt*, Jonathan Rocher*, and Bob Todd**
**KBI Biopharma*

Although the past decade has seen a rise in the use of mechanistic modeling of chromatography unit operations to increase process understanding, the biopharmaceutical industry's preferred approach to characterization of these processes is still the resource-intensive and experimentally-based design of experiments (DoE). Although this traditional approach does increase process understanding to a limited degree, resource pressures invariably limit the size (number of experiments) and scope (number of parameters investigated) of these studies and therefore result in a diminished level of process understanding. Due to inherent limitations, DoE based process characterization studies often do not adequately address important sources of process variability, including the impact of: feed stream variability, difficult to control parameters (e.g. column bed height, resin-lot), and variability in buffer preparation. Further, DoE studies are not well suited to identify the edges-of-failure for a process. As mechanistic modeling of chromatography unit operations is uniquely suited to address all of these limitations, we have modeled an intermediate cation exchange purification step using a targeted set of calibration experiments. Using this model, we demonstrate the increased process understanding that can be achieved through mechanistic modeling of chromatography unit operations.

OUTLINE

1. INTRODUCTION
   - Continuous processes: set up, monitoring and control
   - Multivariate monitoring
2. METHODOLOGY
   - Construction of the multivariate model
   - Informatics set up
   - Monitoring
3. RESULTS
   - R&D site: real time monitoring and retrospective analysis
   - Manufacturing site: real time monitoring
4. IMPLEMENTATION
5. FUTURE PERSPECTIVES
6. CONCLUSIONS

Multivariate monitoring of a continuous manufacturing process for API: enhancing the power of real time data 2

1. INTRODUCTION

Continuous processing in pharmaceutical industry

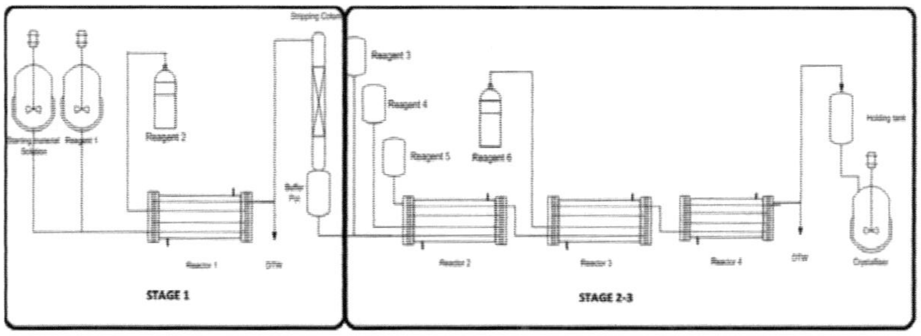

- Smaller footprint
- No inventory
- Flexible batch size
- Lower solvent and energy consumption
- Consistent quality

- Sensitive to solid formation
- Little regulatory precedents

1. INTRODUCTION

Set up of continuous API synthesis: process and monitoring

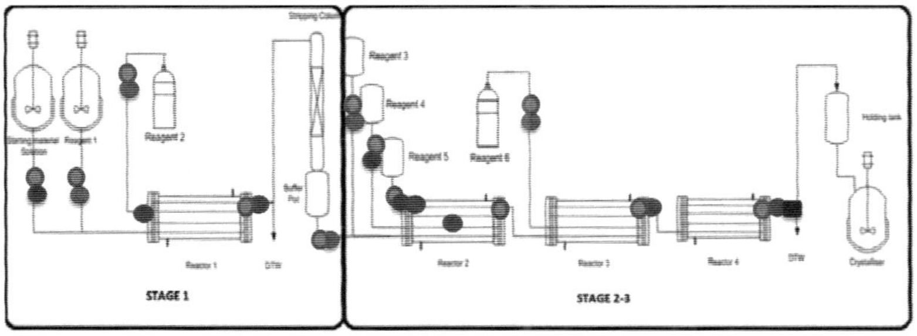

- Conductivity sensors
- Pressure sensors (each pump, reagent and reactor)
- Flow rate sensor (each pump and reagent)
- Pump speed sensor (each pump: not shown)
- Temperature sensors across reactors and heat transfer fluid (not shown)
- Online HPLC

R&D: 30 sensors
Manufacturing: 29 sensors

R&D: 40 sensors
Manufacturing: 52 sensors

1. INTRODUCTION

Set up of continuous synthesis: monitoring and control

- Temperature and flow rate controlled within predefined ranges in order to control product quality (feedback loops) → actively monitored: excursion triggers alarm and divert to waste
- All other sensors → passively monitored: investigated for RCA, process change etc.

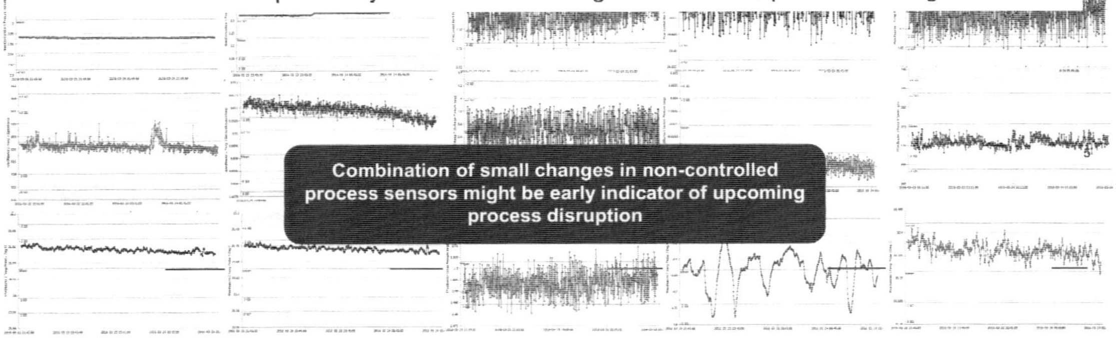

1. INTRODUCTION

Multivariate monitoring

- MSPC model can summarize variability of multiple process sensors readings over time
- Summarizes main trends and correlation patterns in a new trend and calculates a limit based on the 95% confidence interval of normal distributed data.
  - Simplified visualization of process performance
  - Understand natural versus impactful process variation
  - Enhanced process control resulting in more consistent quality
  - Early drift detection can trigger correction actions and prevent forced shut down
  - Facilitates root cause analysis

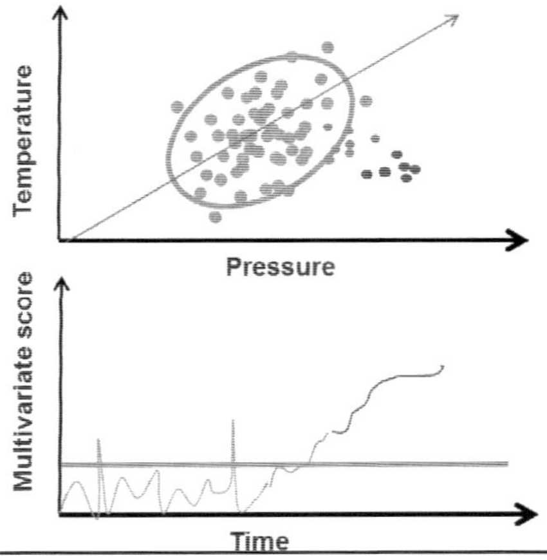

2. METHODOLOGY

1. Construction of the multivariate model

- Data selection
  - State-of-control measurements only (↑sensitivity)
  - Fixed process settings (↑sensitivity)
  - Sufficient process length to ensure representative for future processes (↑robustness), e.g. include full feed bottle cycle

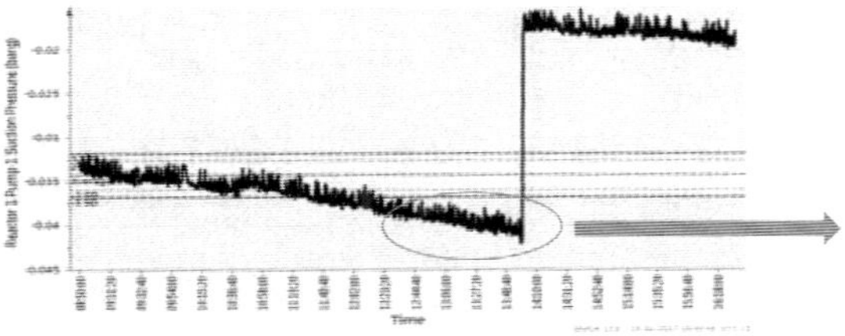

Triggers false positive alarm

2. METHODOLOGY

1. Construction of multivariate model

- Pre-processing
  - Process sensors different scales and ranges → scale to unit variance (divide by SD)
  - Controlled sensors: observed standard deviation <<< accepted standard deviation → down weight
- Model selection
  - Cross validation prone to over fit model due to collinearity of data (i.e. consecutive time points not independent) → check validity of selected model components (e.g. >5%)

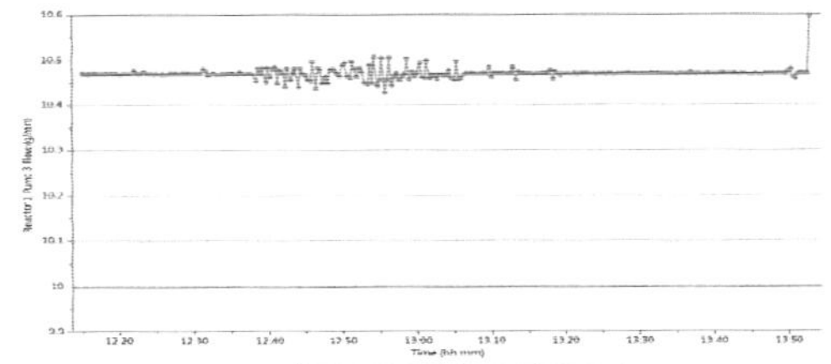

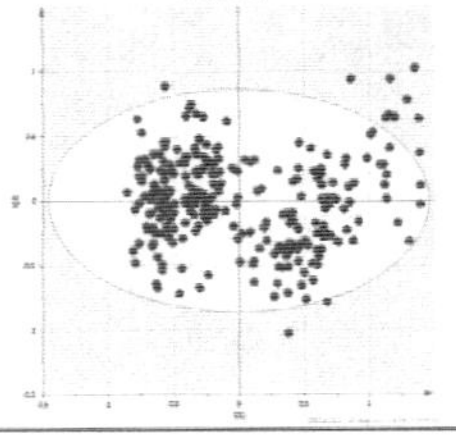

2. METHODOLOGY

Informatics set up

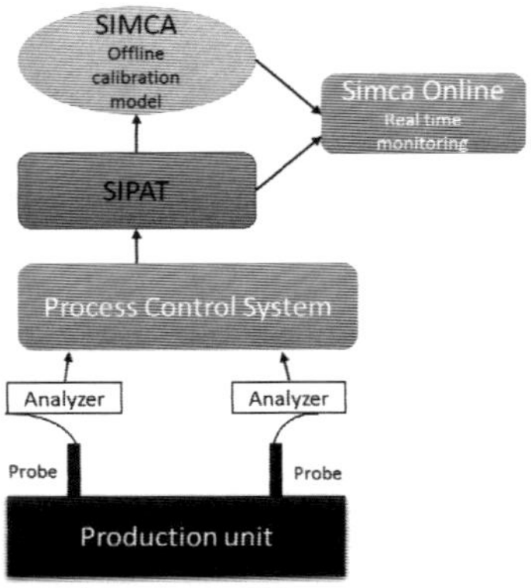

- Sensor readings from probe every second
- Sensor readings collected by SIPAT every 20 seconds
  - Statistical analysis confirmed no information lost by decreased frequency
- Historical data extracted via SIPAT and imported to Simca to build model defining operational space off line
- Multivariate model imported to Simca Online and set up sensor and alarm configuration
- Monitoring in real time started via Simca Online

2. METHODOLOGY

3. Monitoring

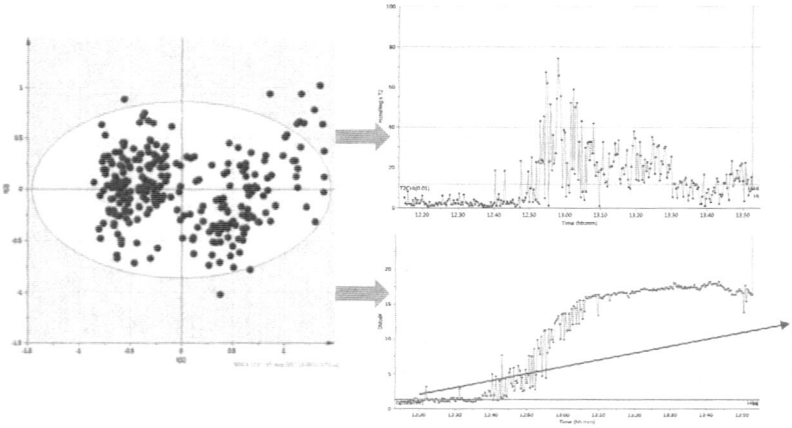

By default multivariate alarm flagged if:

- Three consecutive measurements exceed 99% statistical limit
- 10 out of 100 consecutive measurements exceed 99% statistical limit

Needs optimizing to reduce false positive alarms, whilst maintaining sensitivity for relevant process changes

3.RESULTS

1. R&D site monitoring: evaluation during development campaign

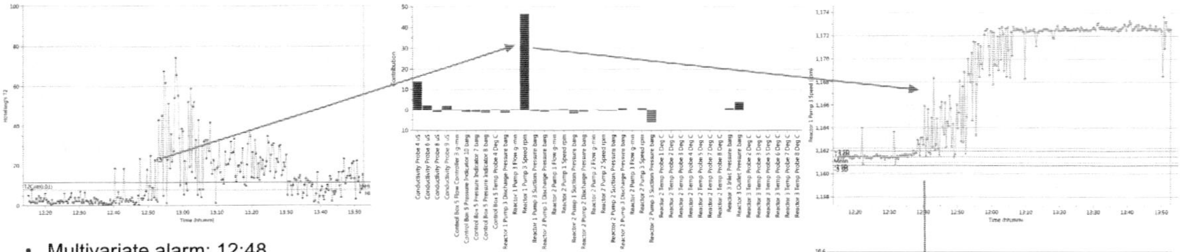

- Multivariate alarm: 12:48
- Root cause multivariate alarm: 15 rpm change in pump speed
- Univariate alarm: 13:52
- Action: pump switch at 13:52
- Increase in pump speed detected one hour before the univariate flow rate alarm → pump switch based on multivariate alarm could prevent flow rate excursion and divert to waste

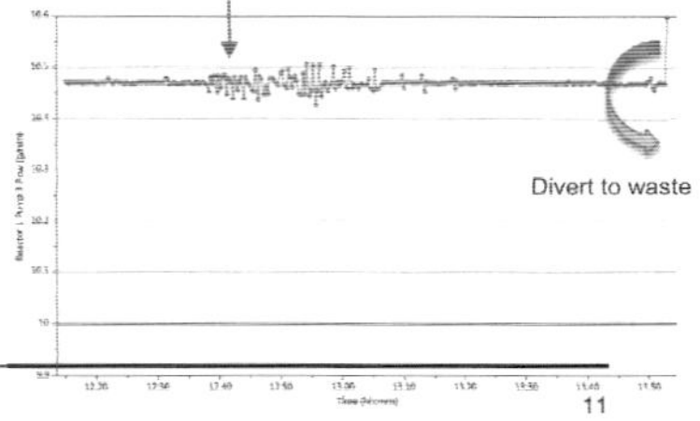

Multivariate monitoring of a continuous manufacturing process for API: enhancing the power of real time data

11

3. RESULTS

1. R&D site monitoring: evaluation during development campaign

Observed process failure	Multivariate detection	Process change	Univariate Alarm	Action
Stage 2 feed pump failure	12:48	15 rpm increase pump speed	13:52	Switch pump
Reagent 2 flow failure	13:18	0.6 barg increase upstream p	15:28	Increase pressure on reagent + divert to waste
7 more similar flow failures for reagent 2 flow				
Reagent 3 pump failure	02:19	12 rpm increase in pump speed	06:38	None (naturally restored)

Multivariate monitoring of a continuous manufacturing process for API: enhancing the power of real time data

12

3. RESULTS

1. R&D site: retrospective RCA for recurrent reagent stream blockage

Largest process changes before blockage:

- Controller differential pressure
- R3 downstream pressure
- Column pressure

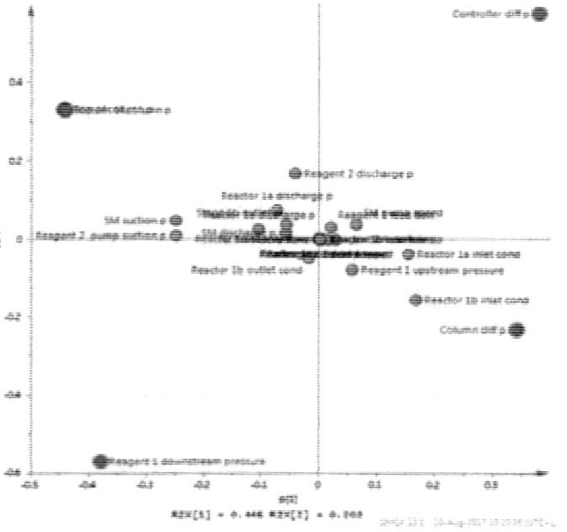

- MVA confirmed root cause blockage
  - Originates from reagent stream
  - Does not start in the reactor

Wealth of real time process data facilitates retrospective root cause analysis

3. RESULTS

2. Manufacturing site monitoring: evaluation during development campaign

Observed process failure	Multivariate detection	Process change	Univariate Alarm	Action
Reagent 2 flow failure	14:53	0.35 barg increase p after flow controller	14:05 (24 h later)	Divert to waste
2 more reagent 2 flow failures as described above				
Reagent 3 pump failure	12:16	Spikes (2 g/min increase) flow rate	12:36	Switch pump
Reagent 1 pump failure	06:06	0.015 bar increase supply filter differential p	06:23	Switch pump
Reagent 6 depletion	23:14	3 % increase flow control output	23:27	Divert to waste and top up reagent 6

4. IMPLEMENTATION

Plan: routine implementation for predictive maintenance by operators

- Procedural flow and decision tree
- List with specific actions in case of detected drift, e.g. switch pump in case of 5 consecutive pump pressure excursions

Aim: detect early onset process failures enabling actions to maintain consistent process performance and avoid divert to waste

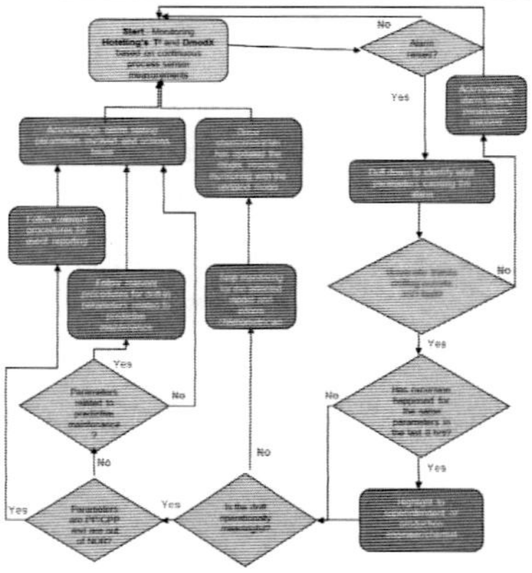

5. FUTURE PERSPECTIVES

Learning

- Increased understanding model sensitivity/robustness and improve implementation strategy
- Combine process sensor data with quality data (e.g. online HPLC): enable detailed quality information at high frequency
- Combine system models and multivariate process monitoring to investigate opportunities for process control

Implementation

- Implement for all high volume processes (API and drug product) containing multiple continuous synthesis stages or unit operations as part of the control strategy (quality assurance and batch release)
- Develop model validation and maintenance plan compliant with regulatory requirements and accommodating high flexibility (e.g. frequent model updates to enable set point changes within the design space)

Advanced Control

- Real-time control of product quality using feedback and feedforward loops

6. CONCLUSIONS

- Multivariate monitoring enabled early detection onset process failure
  - Prolonged time window from detection till process excursion
  - Enables preventative actions to ensure process performance and avoid divert to waste
- Right balance between model sensitivity and model robustness (minimized false alarms) is critical:
  - Optimize data selection to define operational space
  - Optimize monitoring diagnostics limits and alarm settings based on manufacturing experience
- Routine implementation for predictive maintenance by operators
  - Standardized procedures (prevent subjective decisions, actions changing product quality, e.g. set point change)
  - Rigorous yet agile plan for model validation and maintenance accommodating frequent updates
- Future: embed more quality aspects or achieve real time optimization and control

Multivariate monitoring of a continuous manufacturing process for API: enhancing the power of real time data 17

6. ACKNOWLEGEMENTS

- Malcolm Berry: project lead
- Hannah Robinson, Peter Clements for engineering advice
- Peter Shapland for chemistry advice
- Peter Hamilton and Christian Airiau: platforms support and vision
- Ian Barylski: SIPAT and SOL support

Multivariate monitoring of a continuous manufacturing process for API: enhancing the power of real time data 18

500198 Global System Analysis of Interconnected Flow Sheet Models for Drug Product Manufacturing to Performance

Abstract:

Flowsheet models developed using gPROMS platform presents a holistic modeling and simulation approach to design drug product manufacturing processes. Often, these models involve estimation of several parameters using experimental data and measurements. Global Systems Analysis (GSA) is a powerful tool which can be used to identify the cardinal factors of the different parts of the process and their effect on critical quality attributes. In this presentation, we will share our recent experience using GSA through two case studies. First case study will focus on the development of API dry milling model and application of GSA in identifying inputs which are most influential for the model predictions. In the second case study, a comprehensive flowsheet that combines API crystallization, downstream solids operations and drug performance is thoroughly analysed using GSA.

499041 Closed Loop Dynamics of Ribbon Density in a Dry Granulation Process

Abstract:

End to end continuous processing is currently a key focus area in pharmaceutical manufacturing. The increase in research and development is evident from recent FDA approvals of drug products manufactured using continuous downstream processes [1]. Dry granulation is one of the unit operations added for improving material handling and blend uniformity in the continuous tableting line.

Dry granulation is an inherently continuous process. The particle size distribution and bulk density of the granules are critical quality attributes of the process, which depend primarily on the ribbon relative density. The ribbon density is hence a crucial intermediate quality attribute of the unit operation. It is thereby valuable to monitor and control the ribbon density for ensuring desired granule qualities.

Assurance of product quality in real-time requires satisfactory process models with a robust sensor network and control architecture. A dynamic, simplified process model along with real-time monitoring capability is important to the implementation of supervisory control of the process to achieve Quality by Control.

The Johanson model [2] describes the fundamental solid mechanics of roller compaction as a standalone steady state process. FEM analyses have suggested modifications to the model for improved prediction of ribbon density [3]. A process scale model, however, needs to consider the production rate and requires integration of previous unit operations that feed into the roller compactor, in addition to the solid mechanics of the roller compactor.

The process model by Reynolds et al [4] integrates the screw conveyor with the roller compaction system, describing the effect of screw speed on the ribbon density and production rate at steady state. A dynamic process model is, however, essential for the implementation of a control system, as proposed by Hsu et al [5]. The model did not include the flow rates through the system and lacked experimental validation.

This work focuses on a modified process model incorporating the solid mechanics, process dynamics and throughput. The model combines the dynamics of material flow from Hsu et al (2010a) and Reynolds et al (2010) and solid mechanics suggested by Liu and Wassgren (2016). Demonstration of real-time ribbon density monitoring using NIR and microwave sensors [6,7] enable the validation of such a dynamic model. Additionally, the model evaluates the integrated system of a hopper and screw conveyor feeding into the roller compactor.

Blends comprising of acetaminophen and microcrystalline cellulose at inlet flow rates of 8-12 kg/h and varying roll pressures are used for real-time monitoring of ribbon density. The solid mechanics material parameters are estimated using ribbons. The process variables and ribbon density are recorded and monitored using DeltaV DCS.

This work progresses to demonstrate the implementation of ribbon density as proposed by Hsu et al. The integrated process model and real time ribbon density monitoring enable the development of control strategies for the desired ribbon density and eventually the granule properties.

References:

1. https://blogs.fda.gov/fdavoice/index.php/2016/04/continuous-manufacturing-has-a-strong-impact-on-drug-quality/
2. Johanson JR. A Rolling Theory for Granular Solids. *J Appl Mech*. 1965;32(4):842-848
3. Liu Y, Wassgren C. Modifications to Johansonâ€™s roll compaction model for improved relative density predictions. *Powder Technol*. 2016; 297:294-302.
4. Reynolds G, Ingale R, Roberts R, Kothari S, Gururajan B. Practical application of roller compaction process modeling. *Comput Chem Eng*. 2010;34(7):1049-1057
5. Hsu S-H, Reklaitis G V., Venkatasubramanian V. Modeling and Control of Roller Compaction for Pharmaceutical Manufacturing. Part I & II: Process Dynamics and Control Framework. *J Pharm Innov*. 2010;5(1-2):14-36
6. Gupta A, Austin J, Davis S, Harris M, Reklaitis G. A Novel Microwave Sensor for Real-Time Online Monitoring of Roll Compacts of Pharmaceutical Powders Online-A Comparative Case Study with NIR. *J Pharm Sci*. 2015;104(5):1787-1794.
7. McAuliffe MAP, Oâ€™Mahony GE, Blackshields CA, et al. The Use of PAT and Off-line Methods for Monitoring of Roller Compacted Ribbon and Granule Properties with a View to Continuous Processing. *Org Process Res Dev*. 2015;19(1):158-166.Â

493080 Integrated Control and Data Management System for Continuous Pharmaceutical Manufacturing Process

Abstract:

Continuous manufacturing (CM) is evolving as a preferred platform for pharmaceutical products involving solid dosages forms. One advantages of CM among others is that the product quality can be controlled in real time and thereby its open the possibilities of achieving Quality by Control (QbC) and real time release (RTR) paradigm. However, there are different levels of challenges associated with design and implementation of an efficient control system into pharmaceutical manufacturing plant. Identification of right control strategies such as feedforward (FF), feedback (FB), or combined FF/FB[1] as well as right control algorithms such as PID, MPC, or hybrid MPC-PID is often a tough decision that need to be made prior to implementation of integrated control system into the manufacturing plant[2]. Selection of a right control strategy and algorithm is mandatory task both for efficient manufacturing as well as regulatory point of view. Therefore, the different control options need to be evaluated and the best integrated control system needs to be implemented into the pilot-plant. The integrated control system is then need to be coupled with material traceability framework for real time identification of batches so that the product can be recall in case if needed. The continuous manufacturing is a data rich process and therefore a systematic data management framework is also desired to be implemented to collect, store and manage all the relevant data in real time.

In this work, different control strategies for direct compaction continuous tablet manufacturing process has been developed and implemented into our pilot-plant facility using industry standard control platform. The critical control variables are drug concertation, powder level, main compression force, tablet weight and hardness. Both, feedback (FB) as well as feedforward (FF) control strategies have been implemented into the pilot-plant. Feedback (FB) control strategy has been compared with a combined feedforward/feedback (FF/FB) control strategy. A novel method for real time monitoring of powder density has been developed and integrated with feedforward control loop. The feedforward control loop takes the powder bulk density signal as an input and mitigate the effect of variations in powder bulk density on CQAâ€™s proactively and thereby reduces the rejections significantly in compare to only feedback control strategy. Three control algorithms (PID, MPC, hybrid MPC-PID) have been also implemented and compared for feedback loops. The flexible control system is designed in such a way that any control scheme can be easily selected. The control system is also modular in nature that can be easily transferred to any other control platform and manufacturing plant. The continuous tablet manufacturing pilot-plant runs under each control scheme and their performance are evaluated and thereby a best control system has been identified. A systematic methodology for material traceability has been also developed and coupled with integrated control system. The methodology starts with material and process specifications followed by scanning the powder drums and transmitting the material and time information to the control platform. The control platform then utilizes the information stored in the database, generated through residence time distribution model and information collected through real time monitoring to track the material during continuous manufacturing. All the relevant data generated during continuous manufacturing has been systematically collected, stored and organized in a data hub (OSI PI). The PAT data has been collected through PAT data management tool (synTQ) and then transferred to the PI system while the other operational data has been collected in control platform (DeltaV) and send to PI system via OPC communication. Finally, the data has been transferred to the cloud system for enterprise wise efficient sharing and applications.

The objective of this presentation is two-fold: first to evaluate the performance of integrated control system of continuous pharmaceutical tablet manufacturing process and then to highlight the material traceability and data management framework.

References

1. Singh, R., Muzzio, F., Ierapetritou, M., Ramachandran, R. (2015). A combined feed-forward/feed-back control system for a QbD based continuous tablet manufacturing process. PROCESSES Journal, 3, 339-356.
2. Haas, N. T., Ierapetritou, M., Singh, R. (2017). Advanced model predictive feedforward/feedback control of a tablet press. Journal of Pharmaceutical Innovation, 1-14, DOI 10.1007/s12247-017-9276-y.

497291 *In silico* Process Characterization of Drug Product Filling Operations: Predicting Quality Attributes from First Principles

Abstract:

Process characterization of drug product filling operations is typically a time- and labor-intensive process requiring substantial manufacturing line down-time due to the large number of process inputs and lack of mechanistic understanding of the fluid dynamics and filler systemsâ€™ sensitivities to drug product properties (e.g. viscosity, density, surface tension), or variations in site, equipment type, settings, and ambient conditions.

As an alternative to empirical process characterization, we have developed a computational model which predicts process quality attributes â€“ for instance flow rate, fill volume/weight, cavitation, droplet formation, and product concentration â€“ from first principles of product dynamic properties, fluid dynamics, structural mechanics and control theory. Formulation of the model couples finite element and finite volume solutions to structural deformation and fluid dynamics in pinch valves, and leverages analytical solutions to ordinary differential and algebraic equations accounting for orifice and pipe flow as well as equipment control laws.

We present an application of the model to a Time/Pressure filling operation, including formulation, deployment as a software application, and development of a workflow for in-silico process characterization utilizing design space response studies, local sensitivity studies, and uncertainty analysis.

497879 Pinducer Analysis for the Design of a Nucleation Subsystem for Continuous Slug-Flow Crystallization

Abstract:

The reliable and convenient control of crystallization processes has received increased academic and industrial interest. Ultrasonication has been used to aid crystallization, such as accelerating nucleation, reducing crystal size, reducing aggregation, or changing polymorphic ratios [1]–[3]. Most bench-scale continuous sonocrystallization studies are based on configurations involving an ultrasonic probe directly inserted into the solution/slurry or a tube placed within a bath. An alternative multiphase millifluidic tubular crystallizer design combined the advantages of using a probe (focused energy) and a bath (indirect ultrasonication) for continuous nucleation, with the nuclei size tuning by changing the ultrasonic amplitude [4].

To better understand and improve the design of a continuous crystallization process, we employ a needle-tip pressure transducer ("pinducer") to measure and analyze the "primary effect" of ultrasonication, which is the generation of pressure waves (the term "primary effect" is used as in Ref. [5]). This analysis is used to gain understanding on the crystallization outcome (the "secondary effect"). This separation of the analysis of the primary effect from the secondary effect provides greater insight and understanding than past studies that only reported the crystallization results of applying ultrasonication. The pressure waves of ultrasonication are measured for different equipment configurations, and different positions and orientations of the pinducer relative to the ultrasonication probe. The quantification of the pressure waves as a function of frequency provides insights into bubble dynamics and is used to guide the design of the nucleation subsystem and its positioning relative to other subsystems of the continuous crystallization apparatus. The analysis of the spatial localization of energy dissipation that occurs during ultrasonication informs the design of the focused indirect ultrasonication zone to enhance reproducibility and the control of nuclei size distribution [6]. Although the focus of this study was on a particular continuous crystallizer design, the same procedures for the analysis of physical phenomena via pinducers employed in this study could be applied to provide insights and guidance into other ultrasonication-facilitated crystallization designs.

References

[1] J. R. G. Sander, B. W. Zeiger, and K. S. Suslick, "Sonocrystallization and sonofragmentation," *Ultrason. Sonochem.*, vol. 21, no. 6, pp. 1908–1915, 2014.

[2] O. Narducci, A. G. Jones, and E. Kougoulos, "Continuous crystallization of adipic acid with ultrasound," *Chem. Eng. Sci.*, vol. 66, no. 6, pp. 1069–1076, 2011.

[3] R. J. P. Eder, S. Schrank, M. O. Besenhard, E. Roblegg, H. Gruber-Woelfler, and J. G. Khinast, "Continuous sonocrystallization of acetylsalicylic acid (ASA): Control of crystal size," *Cryst. Growth Des.*, vol. 12, no. 10, pp. 4733–4738, 2012.

[4] M. Jiang, C. D. Papageorgiou, J. Waetzig, A. Hardy, M. Langston, and R. D. Braatz, "Indirect ultrasonication in continuous slug-flow crystallization," *Cryst. Growth Des.*, vol. 15, no. 5, pp. 2486–2492, 2015.

[5] V. S. Sutkar and P. R. Gogate, "Design aspects of sonochemical reactors: Techniques for understanding cavitational activity distribution and effect of operating parameters," *Chem. Eng. J.*, vol. 155, no. 1–2, pp. 26–36, 2009.

[6] M. Jiang, C. Gu, and R. D. Braatz., "Analysis of focused indirect ultrasound via high-speed spatially localized pressure sensing and its consequences on nucleation," submitted.

499788 Understanding Metastable Zone Width without Nucleation Rate

Abstract:

Traditionally, the metastable zone width (MZW) in solution crystallization is thought to originate from the path-dependence of the steady-state nucleation rate of a solute. However, new experimental findings appear to go against this popular notion. In this work, we propose an alternate viewpoint to understand the MZW, and show that MZW may not necessarily be a function of the nucleation rate as conventionally understood.

Our work presents experimental data on the MZWs of various solutes and show that these data can be explained without invoking the concept of a nucleation rate. We use data from our own work as well as that from the recent literature. The applicability of the new concept of MZW to both evaporative and cooling crystallizations is shown. The stochastic view point of nucleation is contrasted against a more â€˜deterministicâ€™ view point of the classical nucleation theory. The pitfalls of obtaining nucleation kinetics from MZW experiments are discussed. Both induction time and MZW measurements are recommended as complementary experimental techniques to characterize the kinetics of nucleation in solution crystallization.

499208 Molecular Interactions between Nanocellulose and Crystallizing Pharmaceuticals

Abstract:

Cellulose is a highly abundant natural polymer that, being a plant-based material, exhibits renewability and biodegradability. It is frequently used in the form of microcrystalline cellulose as a filler and binder in pharmaceutical tablets. A nanoscale form of cellulose can be prepared by removing much of the amorphous cellulose material and leaving small crystalline particles with a large concentration of hydrogen bonding groups on the surface. Due to its strong hydrogen bonding tendency and the presence of charges on the nanocellulose surface, it has the potential to bind with drug molecules and act as a release inhibitor, a stabilizer or a site for heterogeneous crystallization of the drug itself. Beyond hydrogen bonding and charge, however, the nanocellulose surface can be readily altered to display a variety of functional groups, leading to flexibility for rational design of nanocellulose-drug composites. We use nanocellulose functionalization, such as carboxylation and sulfonation, to design composites containing nanocellulose and crystallized pharmaceutical products, with an aim towards using the molecular interactions between nanocellulose and drugs to direct formation of desired polymorphs and crystal sizes. The high surface area of the nanocellulose provides a high concentration of interaction sites and the small size of the nanocellulose excipient particles has interesting performance in promoting assembly and packing of the composite in drug products such as compressed tablets and dense free standing films.

500635 A New Model for Solubility Prediction to Guide Solvent Selection for Process Development

Abstract:

Solvent selection to obtain adequate yield and impurity rejection is a critical activity in the development of crystallization processes for pharmaceutical molecules. This activity requires the determination of solubilities for a solute molecule (moreover a polymorph of a solute molecule) in a variety of different solvents and solvent combinations. This is referred to as solvent screening. There are roughly fifty different solvents commonly used in synthetic chemical manufacturing. Although experimentally determining the solubility of a single solute in each of these solvents is an approachable (be it impractical) problem, experimental screening becomes unapproachable for mixtures of two or more solvents. Conventionally, intuition has driven the reduction of the complete set of binary blends to a more tractable set of mixed solvent systems on a case-by-case basis. As such the majority of solvent combinations remaining unexplored for a given molecule. A predictive model was developed to better facilitate the exploration of single solvent systems and blends.

This model was developed starting from a mechanistic perspective for crystal growth and dissolution, which is most applicable at temperatures far from the melt. The model can be implemented based on either the regression of a limited set of solubility data and the subsequent prediction of solute characteristics (magnitude of polar, dispersive and hydrogen bonding interactions, size) or using molecular modelling to approximate these characteristics directly. The regression approach has proven to be efficient (requiring minutes) and accurate for single solvent systems. This model has been extended to mixtures of two solvents, and coupled with crystallization process design calculations to predict yield, and provide a rough process outline to the end user. The intention being that once a â€œgoodâ€ system has been found (determined using the model) it can be verified experimentally. This model has been applied to several systems (~20 solutes and 200 solute/solvent combinations) including the modelling of a solubility data set for paracetamol and caffeine collected within the Enabling Technologies Consortium.

Heterogeneous Nucleation of Small Molecule Crystals in the Presence of Nanoparticles

Lucrèce H. Nicoud, Leia M. Dwyer, Allan S. Myerson*

Chemical Engineering Department, Massachusetts Institute of Technology, Cambridge, MA, U. S. A.
***Corresponding author**: myerson@mit.edu

Abstract
In this study, we evaluated the possibility to use nanoparticles as a means to obtain fundamental insights on the heterogeneous nucleation of organic molecules. To do so, we investigated the crystallization of diphenhydramine hydrochloride in ethanol in the presence of silica coated magnetic nanoparticles. We performed crystallization experiments at two different scales, namely in 1 mL vials and in 100 mL vessels. At both scales, crystallization was induced by two different ways, either by cooling or by anti-solvent (heptane) addition. It was found that the presence of nanoparticles promotes crystallization, in particular at low supersaturation. However, the high variability of induction times and the aggregation of the nanoparticles makes it difficult to extract quantitative information regarding the impact of the nanoparticles on the crystallization kinetics.

Introduction

Surfaces are ubiquitous in crystallizers, in particular due to the walls of the vessel, the impeller or the presence of dust. Surfaces generally lower the energy barrier to nucleation, so that heterogeneous nucleation commonly dominates over primary homogeneous nucleation in most laboratory or plant settings. However, the physical mechanisms underlying heterogeneous nucleation remain poorly understood. Because the early stages of solution crystallization play a decisive role in determining the final crystal properties, such as the crystal size, shape and polymorphism, a fundamental understanding of nucleation is a key requisite to improve control over crystallization processes [1, 2].

Experimental studies aiming at investigating crystallization in the presence of surfaces employed various types of templates including crystals [3], polymers [4] and self-assembled monolayers (SAMs) [5, 6]. The latter are highly ordered and oriented molecular assemblies that form spontaneously by the adsorption of a surfactant on a solid surface [7]. SAMs have been reported to influence crystal size [8], enantiopurity [9], morphology [10], and polymorphism [11].

In this work, we evaluate the possibility to use nanoparticles to investigate the mechanisms of heterogeneous nucleation. Nanoparticles seem particularly suitable for this purpose due to their large surface area and tunable properties. Nanoparticles are indeed available in a wide range of sizes (from few nanometers to few micrometers), shapes (most often spheres, but also cubes, rods and nanostars) and functionalities (mainly thanks to the large portfolio of thiols and silanes, which allow the surface functionalization of gold and silica nanoparticles, respectively). One may therefore envision performing studies with various types of well characterized nanoparticles in an attempt to understand how the uncontrolled presence of foreign objects such as dust may impact nucleation in regular crystallizers. Nanoparticles have been shown to promote the formation of protein crystals at the microliter scale [12, 13, 14], but their quantitative impact on the crystallization kinetics still needs to be elucidated.

In this context, we investigate the impact of silica coated magnetic nanoparticles on the crystallization behaviour of diphenhydramine hydrochloride (DPH), which is a commercial pharmaceutical ingredient. Silica coated magnetic nanoparticles were selected as they combine two main advantages: *(i)* the magnetic core allows performing magnetic decantation, which is very practical for several preparation steps (e.g. to wash, concentrate) and for a potential reuse after crystallization, *(ii)* the silica shell allows functionalizing the surface with a wide variety of silanes. We start our study with experiments performed at the milliliter scale and then increase the crystallization volume to 100 mL. At both scales we induce crystallization by two different ways, namely anti-solvent (heptane) addition and cooling.

1. Materials and Methods

1.1 Materials

1.1.1 Suppliers

DPH was purchased from BeanTown Chemical at a purity $\geq 99\%$. Anhydrous heptane 99% was purchased from Sigma-Aldrich and 200 proof ethanol from Deacon Labs. Silica coated magnetic nanoparticles with a total diameter of 200 nm were purchased from Spherotech Inc (catalog number SIM-025-10H). N-decyltriethoxysilane was purchased from Gelest (catalog number SID2665.0) to perform functionalization of the silica surface of the nanoparticles with decane.

1.1.2 DPH solution preparation

DPH solutions were prepared by dissolving DPH in ethanol at 35 °C under magnetic stirring. The solutions were then filtered with a PTFE syringe filter prior to performing crystallization experiments. Filters with a 0.2 μm and 0.45 μm cut-off were used for crystallization studies at the small scale (1 mL) and large scale (100 mL), respectively. DPH solutions were prepared at a concentration of 315.8 g/L^{EtOH} so as to reach 300 g/L^{EtOH} after addition of the nanoparticle solution (95 % DPH solution and 5 % nanoparticle solution in volume). For control experiments performed in the absence of nanoparticles, the corresponding quantity of pure ethanol was added to reach the same final DPH concentration. In the following, the nanoparticle concentrations are given in grams of nanoparticles per liter of DPH solution (i.e. DPH dissolved in ethanol before a potential addition of heptane). The units are abbreviated as g/L^{DPH} for the sake of simplicity.

1.1.3 Nanoparticles solution preparation

Before usage, the nanoparticles (Nps) were washed, first with water and then with ethanol. To do so, a given volume of nanoparticle solution was placed in a tared vial on top of a magnet. After magnetic decantation, the solvent was removed and a volume of water equal to twice the initial volume of the nanoparticle solution was placed on top of the nanoparticles. The nanoparticles were then redispersed by sonicating the solution for 2 min (Branson 3510 ultrasonicator bath). This wash procedure was then repeated with ethanol. After magnetic decantation and solvent removal, the last traces of ethanol were eliminated by drying the nanoparticles in a vacuum oven at around 60 °C. The vial was then weighted to determine the mass of nanoparticles and an appropriate volume of ethanol was added to reach a target nanoparticle concentration. The nanoparticles were redispered by placing the vial in the bath sonicator for 10 min.

It was checked with dynamic light scattering (DLS) that the nanoparticles are properly dispersed after sonication. Measurements were performed with a DynaPro NanoStar instrument (Wyatt Technology) at a 90° angle using a laser whose wavelength was 658 nm.

1.2 Nanoparticles functionalization

Before functionalization, the silica coated magnetic nanoparticles were washed as described in the previous section. A 5 g/L solution of nanoparticles in a 95/5 % v/v ethanol/water mixture was prepared. The water was acidified with acetic acid at a concentration of 10^{-2} M. The silane was added to the solution to reach a 2 % v/v concentration. The solution was sonicated in the bath sonicator during 2 min. The functionalization reaction was then performed at 25 °C under shaking at 500 rpm during 24 h. The temperature was subsequently increased to 70 °C during 2 h. The solvent was removed by magnetic decantation and the nanoparticles were dried in the vacuum oven at 90 °C during 4 h. Finally, the nanoparticles were washed twice with ethanol using sonication to redisperse the nanoparticles at each wash step.

The extend of functionalization was assessed by Fourier transform infrared (FTIR) spectroscopy (IdentifyIR Smiths). To do so, few microliters of a concentrated nanoparticle suspension were deposited on the sample holder and 256 scans were recorded after ethanol evaporation.

1.3 Solubility

Saturated solutions of DPH were prepared by magnetically stirring an excess of DPH crystals in contact with a selected solvent (either pure ethanol or a mixture of ethanol and heptane) during 24 h. The temperature was maintained at a set value thanks to a water bath. After 24 h, the solution was filtered with a 0.2 μm PTFE syringe filter and the filtrate was placed in tared vials. The vials were weighted immediately after introducing the filtrate and after evaporation of the solvent. The solvent was first evaporated under the hood and the last traces were removed by drying the crystals in a vacuum oven at around 60 °C. Once the mass of the samples was constant over time, the solubility was computed as the mass of crystals divided by the mass of solvent that had evaporated.

1.4 Crystallization likelihood (1 mL)

The tendency of a solution to crystallize was assessed by preparing at least 5 samples per investigated condition and visually inspecting after 24 h whether crystals formed. To do so, 1 mL samples were placed in glass vials (ThermoScientific, C4000-1W) incubated in a thermoshaker (MultiTherm, heat/cool, Benchmark Scientific) at a selected temperature and under 500 rpm shaking. For anti-solvent crystallization, the temperature was set to 25 °C. The formed crystals were observed with an optical microscope (Nikon Eclipse ME 600).

1.5 Crystallization kinetics (100 mL)

Batch crystallization kinetic experiments were performed using an EasyMax 102 worksation (Mettler Toledo). The solution was stirred at 250 rpm with a pitched-blade impeller (Alloy-C22, 38 mm diameter, downwards). The total solution volume was 100 mL. For anti-solvent crystallization, the reactor temperature was maintained at 25 °C and heptane was poured at once on top of the DPH solution to reach a heptane concentration of 33 % v/v^{tot}. For cooling crystallization, the reactor temperature was decreased from 25 °C to 15 °C at a rate of 1 °C/min.

The crystallization kinetics was monitored by focused beam reflectance measurements (FBRM) using a lasentec probe (S400Q).

In few experiments, particle vision measurements (PVM) were performed to visually observe the nanoparticle aggregates and the crystals (ParticleView V19, Mettler Toledo).

2. Results and Discussion

2.1 Choice of nanoparticles

The core-shell nanoparticles selected for this study combine the advantages of a practical handling and of a possible functionalization thanks to the magnetic core and to the silica surface, respectively. Indeed, the magnetic behaviour of the nanoparticles is very convenient for all the preparation steps such as washing, concentration and surface functionalization. The silica shell instead prevents the nanoparticles from extensive aggregation and allows the grafting of a wide variety of silanes, which then confer various surface properties to the nanoparticles.

A transmission electron microscopy picture of the nanoparticles is shown in Figure 1, where it is seen that the magnetic core is around 160 nm and the silica shell approximately 20 nm thick. The total diameter of 200 nm was confirmed by DLS analysis.

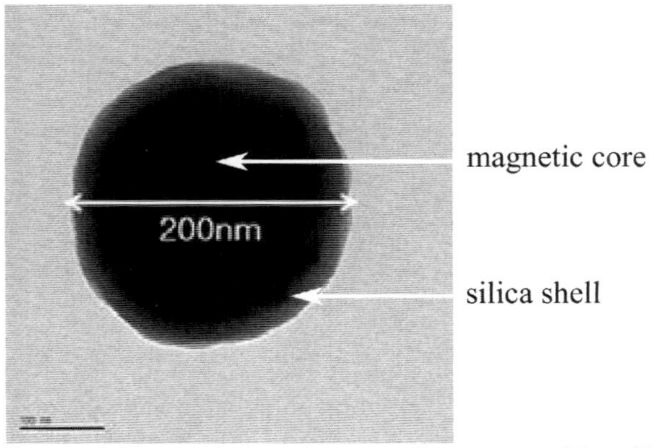

Figure 1. Transmission electron microscopy of the silica coated magnetic nanoparticles. Provided by Spherotech Inc.

To give an order of magnitude, a 1 g/L solution of such nanoparticles corresponds to a concentration of 6×10^{13} nanoparticles per liter and to a volume fraction of 0.03 %. Therefore, even a quite dilute solution of nanoparticles provides a large number of foreign objects, which are as many potential nucleation sites.

2.2 Nanoparticle functionalization

The silica shell of the nanoparticles allows functionalizing the surface with various silanes. As an example, the FTIR spectrum of nanoparticles functionalized with n-decyltriethoxysilane, which provides a functionalization with decane molecules, is shown in Figure 2. As a comparison, the spectra of the non-functionalized (bare) nanoparticles and of the silane are also shown. In contrast to the bare nanoparticles, the functionalized nanoparticles exhibit three small additional peaks in the range from 2800 to 3000 cm^{-1}. These peaks coincide with those of the silane, thus suggesting that the functionalization reaction was successful. The low intensity of the peaks corresponding to the grafted decane as compared to the pure silane can be attributed to the small mass of the surface groups with respect to the bulk of the nanoparticles.

2.3 Solubility

Figures 3(a) and (b) show experimental data of DPH solubility as a function of the heptane volumetric fraction and temperature, respectively. The solid line in Figure 3(a) represents the dilution line upon heptane addition starting from a DPH solution at 300 mg/mL^{EtOH} = 382.2 mg/g^{EtOH}. It is observed that crystallization is expected to occur from a thermodynamic point of view at heptane concentrations above approximately 10 % in volume. In the case of cooling, the DPH concentration is kept constant, as indicated by the flat solid line in Figure 3(b). It is observed that crystallization is expected to occur at temperatures roughly below 23 °C.

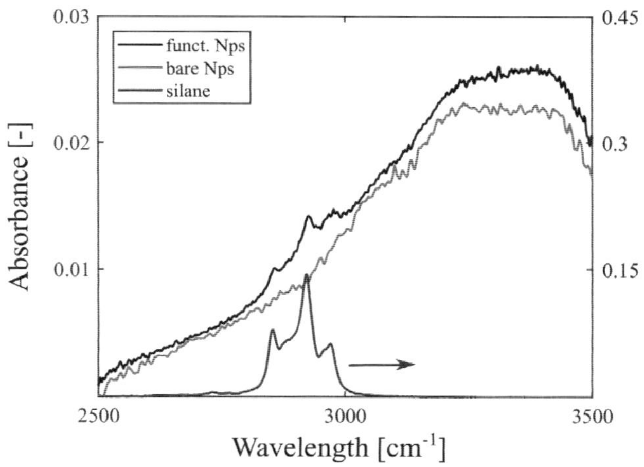

Figure 2. FTIR spectrum of nanoparticles functionalized with decane as compared to the spectra of the bare nanoparticles and of the corresponding silane (whose intensity is read on the right axis).

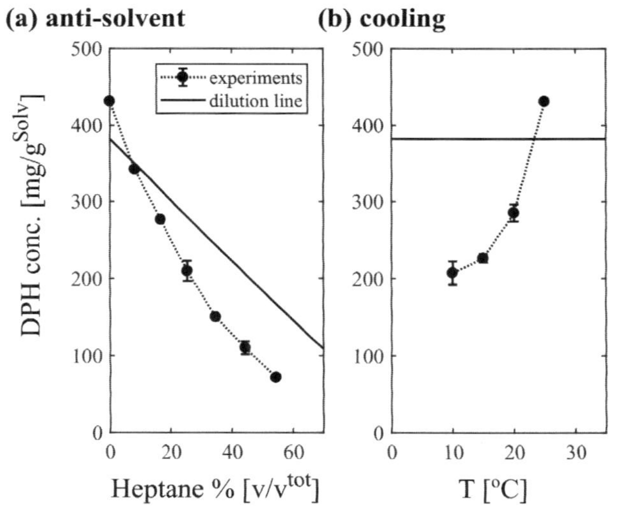

Figure 3. DPH solubility as a function of (a) heptane volumetric fraction and (b) temperature. Symbols correspond to experimental data obtained by gravimetry, whereas solid lines indicate the operating conditions, i.e. a DPH solution at 300 mg/mL^{EtOH} = 382.2 mg/g^{EtOH} diluted with heptane or cooled down.

2.4 Crystallization likelihood (1 mL)

2.4.1 Non-functionalized nanoparticles

In a first set of experiments, the impact of the nanoparticles (Nps) on the crystallization behaviour of DPH was investigated at the milliliter scale in glass vials. For each condition under investigation, at least five samples were prepared and visually inspected after 24 h of shaking at a target temperature to determine whether crystals formed during the experiment time frame. The results for anti-solvent crystallization and cooling crystallization are shown in Figure 4(a) and (b) respectively.

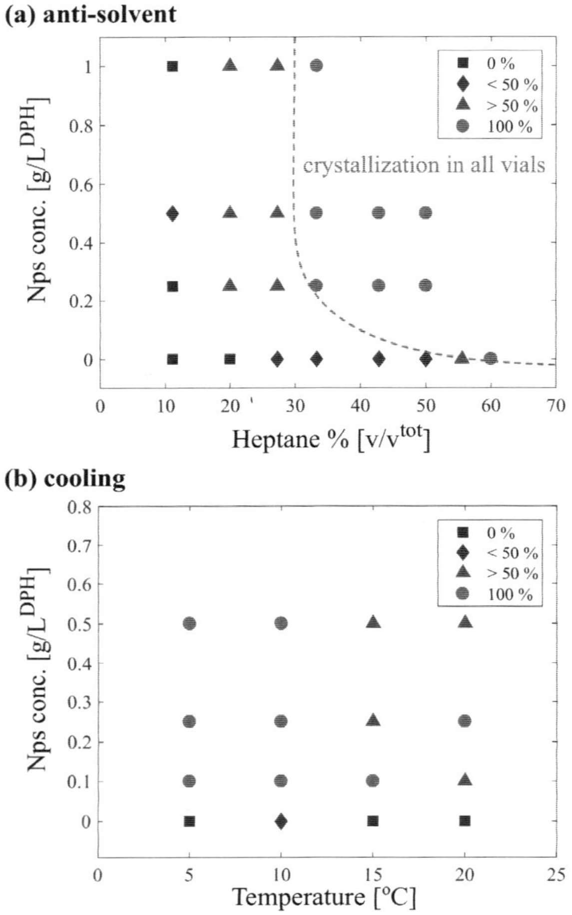

Figure 4. Crystallization likelihood at the milliliter scale in (a) anti-solvent crystallization and (b) cooling crystallization. For each condition, at least five samples were prepared and visually inspected after 24 h of shaking to determine whether crystals formed. The percentages in the caption indicate the fraction of samples that crystallized. In the case of anti-solvent crystallization, the temperature was set to 25 °C. Non-functionalized nanoparticles were used.

Let us first consider the samples without the nanoparticles, i.e. the row of symbols at the bottom of both diagrams. According to the solubility data shown in Figure 3, the selected conditions of heptane concentration and temperature were expected to induce crystallization in all samples. However, it is observed in Figure 4 that most of the samples did not crystallize within 24 h. For example, none of the samples whose heptane concentration was below 20 % crystallized (blue squares). At heptane concentrations between 25 and 55 %, only some of the samples crystallized (purple symbols) and it was necessary to reach 60 % heptane so as to observe crystallization in all the vials (red circles). To verify that there was no mistake in sample preparation, a seed was added in the samples that did not crystallize within 24 h. This resulted in rapid crystallization, thus confirming that all the solutions were supersaturated. This means that, although all the samples were expected to crystallize from a thermodynamic point of view, a significant kinetic barrier prevented crystallization to occur during the time frame of the experiment.

Interestingly, it is observed in Figure 4 that the addition of non-functionalized nanoparticles resulted in the crystallization of many more samples as compared to the situation without nanoparticles. For instance, all the samples at 33 % heptane crystallized in the presence of nanoparticles, whereas less than half crystallized in the absence of nanoparticles. Similarly, none of the samples at 15 °C crystallized in the absence of nanoparticles whereas the vast majority of the samples with the nanoparticles crystallized.

Overall, these data show that nanoparticles promote crystal formation, in particular at low supersaturation values, i.e. in the metastable zone.

2.4.2 Functionalized nanoparticles

Following the idea of a previous study [15], crystallization experiments were performed with nanoparticles functionalized with decane, which is an anti-solvent for DPH. As a reference, the addition of 50 % decane in volume reduces the solubility of DPH from 432 to 137 mg/g^{Solv}. The functionalization of the silica coated magnetic nanoparticles with decane was discussed in Section 2.2.

According to the original work, the addition of nanoparticles functionalized with anti-solvent molecules was able to induce the crystallization of slightly under-saturated solutions of sodium chloride, D-mannitol and fenofibrate. Here, we investigated whether the addition of decane functionalized nanoparticles could induce the crystallization of DPH solutions at 25 °C with heptane concentrations varying from 0 to 10 % in volume. According to the solubility data shown in Figure 3(a), such conditions lead from slightly under-saturated up to saturated solutions. Several nanoparticle concentrations ranging from 0.25 g/L^{DPH} to 1 g/L^{DPH} were tested and several repetitions were performed for each condition. However, none of the samples crystallized within 24 h.

After observing that decane functionalized nanoparticles were not able to induce the crystallization of DPH solutions in the range 0-10 % heptane, larger heptane volume fractions were investigated. As in the case of the bare nanoparticles, crystallization was observed in all the samples whose heptane concentration was above 33 %. Nevertheless, one noteworthy difference was observed in terms of crystal color. Figures 5(a) and (b) show pictures of the crystals formed with non-functionalized nanoparticles and with decane functionalized nanoparticles, respectively. The top pictures were taken with a regular camera, whereas the bottom pictures were taken with an optical microscope. In both samples, the heptane concentration was 43 % v/v^{tot} and the nanoparticle concentration 0.25 g/L^{DPH}. Due to the red/brown colour of the nanoparticles, it is possible to visually assess the entrapment of nanoparticles in the crystal lattice. It is seen in Figure 5 that the crystals formed in the presence of the bare nanoparticles are darker than those formed in the presence of the nanoparticles functionalized with decane. This might suggest that the surface functionalization with decane reduced nanoparticle entrapment.

Figure 5. Pictures of DPH crystals formed in the presence of (a) bare and (b) decane functionalized nanoparticles. The heptane concentration was 43 % v/v^{tot} and the nanoparticle concentration 0.25 g/L^{DPH}. After crystallization, the crystals were washed with heptane. The top pictures were taken with a regular camera, whereas the bottom pictures were taken with an optical microscope.

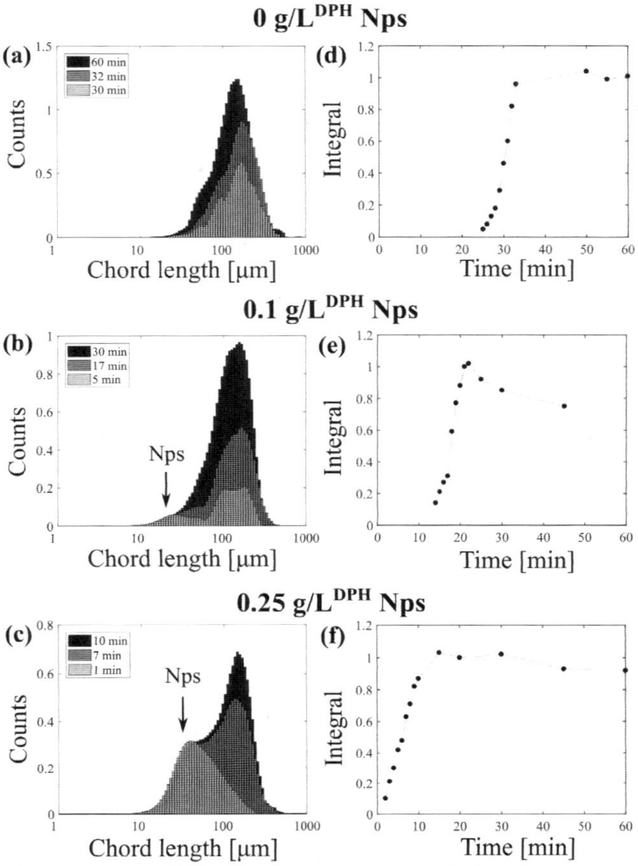

Figure 6. (a-c) Cube weighted chord length distributions monitored by FBRM at different time points and nanoparticles concentrations, as indicated in the legend. (d-f) Integral of the chord length distribution as a function of time for the same three nanoparticle concentrations. The peak corresponding to the nanoparticles (centered at around 20 µm) was substracted before computing the integral and the curves were normalized between 0 and 1. Bare nanoparticles were used.

2.5 Crystallization kinetics (100 mL)

After investigating the crystallization likelihood at the milliliter scale, batch crystallization kinetics were performed in a 100 mL stirred vessel equipped with a FBRM probe. For anti-solvent crystallization, a heptane concentration of 33 % v/v^{tot} was selected, whereas for cooling crystallization a temperature of 15 °C was chosen.

Figure 6(a) shows cube weighted chord length distributions recorded in the absence of nanoparticles at three different time points during anti-solvent crystallization. It is observed that the crystal size is in the order of 100 µm. The number of crystals increases with time, but their size remains relatively constant. This indicates that crystal growth is significantly faster than crystal nucleation.

Figures 6(b) and (c) show the chord length distributions at three different time points in the presence of 0.1 g/L^{DPH} and 0.25 g/L^{DPH} bare silica coated magnetic nanoparticles, respectively. As compared to the situation without the nanoparticles, an additional peak centered at around 20 µm is observed. This peak does not correspond to crystals but is due to the presence of nanoparticles. This was evidenced by an experiment performed below the saturation (25 °C, no heptane), during which a similar peak was observed, as further discussed in Section 2.6.

In order to evaluate the mass of crystals formed as a function of time, the peak corresponding to the nanoparticles was substracted and the integral of the cube weighted chord length distribution was computed. The results are presented in Figures 6(d), (e) and (f) for 0, 0.1 and 0.25 g/L^{DPH} nanoparticles, respectively. In the absence of nanoparticles, a lag time of approximately 25 min is observed before crystallization starts. Then, the mass of crystals increases sharply until it reaches a plateau. In few experiments, a slight decrease in the number of counts was observed at large reaction times (see Figures 6(e)). This might be due to crystals intermittently sticked to the FBRM probe falling off the probe window.

Several repetitions of these anti-solvent experiments were performed and the results are reported in Figure 7(a) in terms of lag times. It is seen that the average lag time decreases with the concentration of nanoparticles.

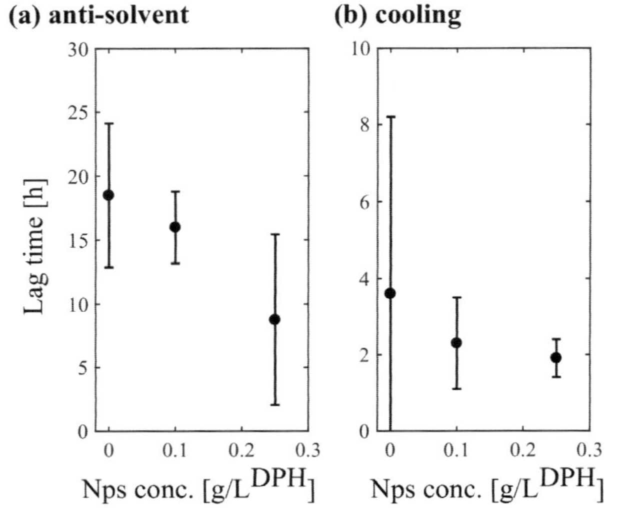

Figure 7. Lag time as a function of the concentration of nanoparticles in the case of (a) anti-solvent and (b) cooling crystallization. Non-functionalized nanoparticles were used.

Similar studies were carried out in the case of cooling crystallization. Chord length distributions comparable to those obtained during anti-solvent crystallization were obtained. The lag time is shown as a function of the nanoparticle concentration in Figure 7. Once more, it is seen that the average lag time decreases upon nanoparticle addition. However, the poor reproducibility makes it difficult to properly quantify the impact of the nanoparticles.

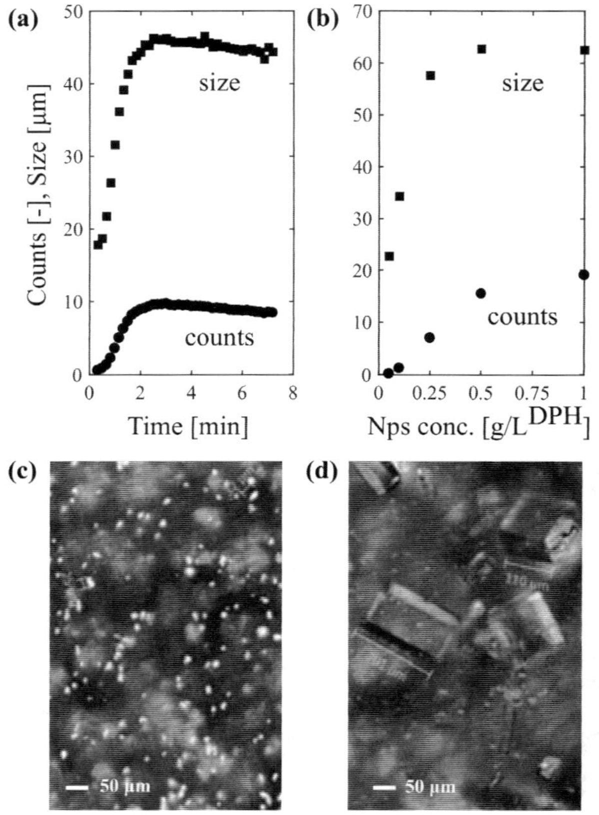

Figure 8. (a) Total number of counts (circles) and average size (squares) of the cube weighted distribution measured with FBRM as a function of time in a 300 g/L^{EtOH} DPH solution at 25 °C in the presence of 0.25 g/L^{DPH} of bare nanoparticles. (b) Total number of counts (circles) and average size (squares) of the cube weighted distribution measured with FBRM at the early stages of an anti-solvent crystallization experiment. (c) PVM images recorded after 8 min of the experiment reported in panel (a). (d) PVM images recorded after 3 h of cooling at 15 °C following the experiment reported in panel (a).

2.6 Nanoparticle aggregation

In this section, we investigate in more details the aggregation behaviour of the nanoparticles. Figure 8(a) shows the total number of counts (circles) and the average size (squares) of the cube weighted distribution measured with FBRM in an experiment performed in the presence of 0.25 g/L^{DPH} of bare nanoparticles below the supersaturation, i.e. where no crystals can form. In this experiment, the temperature was maintained at 25 °C and no heptane was added. It is seen that both the number of counts and the average size increase during the first two minutes and then become almost constant with time. Note that an average size close to 45 μm is reached, which is more than two order of magnitude larger than the initial nanoparticle diameter (200 nm as measured by DLS).

The nanoparticle aggregates were visualized by PVM and a representative picture recorded after 8 min is shown in Figure 8(c). It is seen that particulates in the order of few tens of microns are observed, in agreement with the FBRM data shown in Figure 8(a).

After performing particle vision measurements, the temperature was decreased to 15 °C so as to induce crystal formation. Figure 8(d) shows a representative picture recorded after 3 h of cooling, where both nanoparticle aggregates and crystals are observed. The crystals appear much larger than the nanoparticle aggregates, in line with the bimodal distribution shown by the FBRM data in Figure 6.

The impact of the concentration of nanoparticles on their aggregation behaviour was studied under supersaturated conditions in the case of anti-solvent crystallization. It was already observed in the context of Figure 6 that an increase in the nanoparticle concentration from 0.1 to 0.25 g/L^{DPH} leads to an increase in the height of the peak corresponding to the nanoparticles. Figure 8(b) shows how the total number of counts and the average size related to the peak of the nanoparticles increase as a function of the nanoparticle concentration. A sharp rise in the size of the nanoparticle aggreates is observed at nanoparticle concentrations below

0.25 g/L^{DPH}. Above that value, the nanoparticle aggregates reach an average size of around 60 μm.

Nanoparticle aggregation represents a major issue in this study. Indeed, aggregation reduces both the the surface area and the number of particles in solution. Assuming that the aggregates are compact spheres with a uniform size distribution, the surface area and the number of particles are inversely proportional to the aggregate diameter and to the cube of their diameter, respectively. Considering that in the present case aggregates are approximately 100 times larger than the primary particles, aggregation roughly reduces the surface area by a factor 100 and the number of particles by 10^6.

Besides reducing the surface area and the number of particles, aggregation also complicates the interpretation of experimental data. Indeed, aggregates are non regular structures, which are difficult to characterize with accuracy in terms of shape and size distribution. In addition, aggregation is affected by a number of parameters, such as time and nanoparticle concentration as shown in Figure 8, but also potentially temperature, solvent type, solute concentration, nanoparticle functionalization, etc. It is therefore difficult to rationalize the quantitative impact of nanoparticles on the crystallization kinetics monitored under various experimental conditions. For instance, the observation that the aggregate size increases with nanoparticle concentration prevents a detailed analysis of the impact of the nanoparticles concentration on the kinetics of crystallization. Similarly, studies aiming at investigating different surface functionalization would require the decoupling between the impact of surface functionalization on the aggregation of nanoparticles and on the kinetics of crystallization.

3. Conclusion

In this study, we assessed the possibility to use nanoparticles to gain fundamental insights into heterogeneous nucleation mechanisms. This was motivated by the fact that nanoparticles offer a large surface area and are tunable in terms of various properties, such as size, shape and surface chemistry.

At the milliliter scale, it was shown that silica coated magnetic nanoparticles promote the crystallization of diphenhydramine, which was induced either by heptane addition or cooling. Then, the crystallization kinetics was monitored with FBRM in 100 mL stirred reaction vessels and it was found that the presence of nanoparticles reduces the lag time before crystallization occurs. However, the effect of the nanoparticles seems less pronounced than at the milliliter scale and the large variability in the lag time measurements makes it difficult to draw solid conclusions.

Finally, it was shown that the nanoparticles significantly aggregate under the considered experimental conditions, which reduces the surface area and the number of particles in solution. This complicates the interpretation of experimental results and hinders the quantitative analysis of the impact of the nanoparticles on the crystallization kinetics, at least for the system under investigation.

Acknowledgments

Financial support from the Swiss National Foundation is gratefully acknowledged (grant P2EZP2_168909). The support of DARPA for a portion of this project is gratefully acknowledged.

References

[1] A. S. Myerson. *Hanbook of industrial crystallization, 2nd Edition*. Butterworth-Heinemann, 2002.

[2] P. G. Vekilov. Nucleation. *Cryst Growth Des*, 10(12):5007–5019, 2010.

[3] M. D. Ward. Bulk crystals to surfaces: Combining x-ray diffraction and atomic force microscopy to probe the structure and formation of crystal interfaces. *Chem Rev*, 101(6):1697–1725, 2001.

[4] C. P. Price, Ad. L. Grzesiak, and A. J. Matzger. Crystalline polymorph selection and discovery with polymer heteronuclei. *J Am Chem Soc*, 127(15):5512–5517, 2005.

[5] A. Singh, I. S. Lee, K. Kim, and A. S. Myerson. Crystal growth on self-assembled monolayers. *CrystEngComm*, 13(1):24–32, 2011.

[6] J. Aizenberg, A. J. Black, and G. M. Whitesides. Control of crystal nucleation by patterned self-assembled monolayers. *Nature*, 398(6727):495–498, 1999.

[7] J. C. Love, L. A. Estroff, J. K. Kriebel, R. G. Nuzzo, and G. M. Whitesides. Self-assembled monolayers of thiolates on metals as a form of nanotechnology. *Chem Rev*, 105(4):1103–1170, 2005.

[8] A. Y. Lee, I. S. Lee, S. S. Dette, J. Boerner, and A. S. Myerson. Crystallization on confined engineered surfaces: A method to control crystal size and generate different polymorphs. *J Am Chem Soc*, 127(43):14982–14983, 2005.

[9] J. Chen and A. S. Myerson. Pasteur revisited: Chiral separation by crystallization on self-assembled monolayers. *CrystEngComm*, 14(24):8326–8329, 2012.

[10] A. Y. Lee, A. Ulman, and A. S. Myerson. Crystallization of amino acids on self-assembled monolayers of rigid thiols on gold. *Langmuir*, 18(15):5886–5898, 2002.

[11] R. Hiremath, J. A. Basile, S. W. Varney, and J. A. Swift. Controlling molecular crystal polymorphism with self-assembled monolayer templates. *J Am Chem Soc*, 127(51):18321–18327, 2005.

[12] Y. Takeda and F. Mafuné. Induction of protein crystallization by platinum nanoparticles. *Chem Phys Lett*, 647(Supplement C):181–184, 2016.

[13] S. Ko, H. Y. Kim, I. Choi, and J. Choe. Gold nanoparticles as nucleation-inducing reagents for protein crystallization. *Cryst Growth Des*, 17(2):497–503, 2017.

[14] F. Hodzhaoglu, F. Kurniawan, V. Mirsky, and C. Nanev. Gold nanoparticles induce protein crystallization. *Cryst Res Technol*, 43(6):588–593, 2008.

[15] S. A. Kulkarni and A. S. Myerson. Reversible control of solubility using functionalized nanoparticles. *Chem Comm*, 53(8):1429–1432, 2017.

487832 The Influence of Solution Conditions on the Self-Assembly of Pre-Nucleation Clusters

Abstract:

The concept of pre-nucleation clusters has been established to, successfully, explain phenomena, in nucleation kinetics, that they were inconsistent with the classical nucleation theory.[1] The key idea is that pre-nucleation clusters are formed providing a comfortable niche for nuclei to be formed.[2] Phenomenological theories exist describing this two-step nucleation process.[3] However, there is a lack of qualitative understanding of the fundamental phenomena underpinning the formation of pre-nucleation clusters both in a supersaturated and an understaturated environment.

Using the models proposed in more recent works is being shown that bigger pre-nucleation clusters provides an environment giving more chances of survival to the nucleus inside it. These findings were tested experimentally using a number of industrially relevant molecules (arginine, glycine, glutamic acid). These molecules are particularly attractive as they have relatively large nucleation times. In addition the formation of pre-nucleation clusters as nucleation precursors for them has been reported in literature. Using DLS it was shown that in aqueous systems, the size of the pre-nucleation clusters scales with the supersaturation. These findings are in accordance with the findings proposed from previous works. Upon filtration, both in a supersaturated and an undersaturated regime, the pre-nucleation clusters recover their initial size in a few hours, in the absence of any agitation.

Experiments were conducted in systems containing water and ethanol to investigate the influence of solutionâ€™s chemical potential on the size of the pre-nucleation clusters. It is shown that, in both the undersaturated and the supersaturated regime, the formation of the pre-nucleation clusters follows the fundamentals of self-assembly in solution.[4] The findings suggest that the influence of the solvent can overcome that of supersaturation, for example an aqueous system containing arginine at Sâ‰ˆ0.5 (0.8 g of L-Arginine) gives bigger pre-nucleation clusters than a solution containing 30 % v/v ethanol at Sâ‰ˆ0.9.

The findings of this work verify the majority of the findings reported in literature. However, it reports different kinetics of recovery of the pre-nucleation clusters size upon filtration. It highlights the applicability of the fundamentals of self-assembly in solution. The findings suggest that in depth understanding of the self-assembly phenomena are required for the development of efficient strategies for the control of crystallization kinetics via stirring[5] and for the formulation of amorphous solid dispersions.

References

1. P.G. Vekilov. Nucleation. *Crystal Growth & Design*. **2010**, 10, 5007.
2. B. Peters. On the coupling between slow diffusion transport and barrier crossing in nucleation. *The Journal of Chemical Physics*. **2011**, 135, 044107.
3. W. Pan, A.B. Kolomeisky, P.G. Vekilov. Nucleation of ordered solid phases of proteins via a disordered high-density state: Phenomenological approach. *The Journal of Chemical Physics*. **2005**, 122, 174905.
4. J.N. Israelachvili. Intermolecular and Surface Forces. 3rd ed. Oxford, Academic Press; **2011**.
5. A. Jawor-Baczynska, B.D. Moore, H.S. Lee, A.V. McCormick, J. Sefcick. Population and size distribution of solute-rich mesospecies within mesostructured aqueous amino acid solutions. *Faraday Discussions*, **2013**, 167, 425.]
6. M. Diwan, H. Li, F. Nordstrom. Primary Nucleation Under High Shear Conditions. *AICHE Conference Proceedings*, **2015**

490516 Simulations and Experiments Validation on the Membrane Based Response Technology for Nucleation Detection and Crystallization Control

Abstract:

Simulations and Experiments validation on the Membrane based Response Technology for Nucleation Detection and Crystallization Control

Xiaobin Jiang, Wu Xiao, Dapeng Lu, Rui Zhao, Gaohong He*

State Key Laboratory of Fine Chemicals, School of Chemical Engineering, Petrochemical Energy-efficient Separation Technology Engineering Lab of Liaoning Province, Dalian University of Technology, Dalian, P. R. China

*Corresponding author: Email: hgaohong@dlut.edu.cn

Abstract

The mass transfer flux decline and the crystallization on the membrane interface is often considered as a fouling behavior [1,2]. For now, non-attention had been paid on applying this sensitive nucleation response phenomenon for the crystal nucleus detection. While, the inevitable nucleation on the membrane interface under certain operational conditions inspired us that the membrane interface could perform as the response medium of the nucleation detection. The responding mechanism is constructed base on the chain effect of ¡®nucleation on the pores of the membrane interface-transmembrane flux sharply decrease¡¯. The inflection point of the transmembrane flux confirms the nucleation timing on the membrane interface. The accurate recording of the transmembrane flux data can be utilized to calculate the solution concentration to determine the metastable zone width (MSZW), which is key data to determine the crystallizaiton operative window [3,4].

Obviously, without relying on the optical signal, membrane based nucleation response method would be easy and more meaningful in the case of opaque solutions that laser intensity response method is invalid [5]. This advantage will improve the controlling accuracy and flexibility during MSZW measurement. Moreover, with the proposed method, it is easy to enhance control over the process. The real operative conditions during crystallizaiton can be achieved in this membrane based technology. Detailed simulations and experiments on the typical binary aqueous solution systems were implemented to represent the role of T_f, T_p and A/V_0 on the supersaturation degree adjustment.

In addition, it also reported that the polymer interface with nanometer scale pores had significant impact of on inducing crystallization and polymorphism control [6,7]. The foundational work can promote the theoretical research and application of the MDC based nucleation response and crystallization control by providing the qualified microporous membrane as desired nucleation detection and response interface.

Acknowledgment

We acknowledge financial contribution from from National Natural Science Foundation of China (Grant No. 21527812, 21676043, U1663223), Changjiang Scholars Program (T2012049), the Fundamental Research Funds for the Central Universities (DUT16TD19) and Education Department of the Liaoning Province of China (No. LT2015007).

References

[1] Meng, S.; Ye, Y.; Mansouri, J.; Chen, V., Fouling and crystallisation behaviour of superhydrophobic nano-composite PVDF membranes in direct contact membrane distillation. *Journal of Membrane Science* **2014,** 463, 102-112.

[2] Chen, G.; Lu, Y.; Krantz, W. B.; Wang, R.; Fane, A. G., Optimization of operating conditions for a continuous membrane distillation crystallization process with zero salty water discharge. *Journal of Membrane Science* **2014,** 450, 1-11.

[3] Ulrich, J.; Jones, M. J., Industrial Crystallization. *Chemical Engineering Research and Design* **2004,** 82, (12), 1567-1570.

[4] Ulrich, J.; Strege, C., Some aspects of the importance of metastable zone width and nucleation in industrial crystallizers. *Journal of Crystal Growth* **2002,** 237, 2130-2135.

[5] Jiang, X.; Ruan, X.; Xiao, W.; Lu, D.; He, G., A novel membrane distillation response technology for nucleation detection, metastable zone width measurement and analysis. *Chemical Engineering Science* **2015,** 134, 671-680.

[6] Curcio, E.; L¨®pez-Mej¨ªas, V.; Di Profio, G.; Fontananova, E.; Drioli, E.; Trout, B. L.; Myerson, A. S., Regulating Nucleation Kinetics through Molecular Interactions at the Polymer¨CSolute Interface. *Crystal Growth & Design* **2014,** 14, (2), 678-686.

[7] Diao, Y.; Helgeson, M. E.; Siam, Z. A.; Doyle, P. S.; Myerson, A. S.; Hatton, T. A.; Trout, B. L., Nucleation under Soft Confinement: Role of Polymer¨CSolute Interactions. *Crystal Growth & Design* **2012,** 12, (1), 508-517.

495067 Niimbl: The National Institute for Innovation in Manufacturing Biopharmaceuticals

Abstract:

Technology innovation, workforce development, and regulatory advancement have been a hallmark of the industry for the past several decades. As the industry continues to expand, mature and evolve, there are new opportunities for shared risk in technology innovation, for partnerships to facilitate training, and for new approaches to engage with health authorities. It can be argued that collaborative efforts in this space, done appropriately, can reduce the burden and investment that would otherwise be required by individual organizations to advance their technology needs and address issues with speed, quality, and cost. At the same time, there is significant diversity in the industry in product types as well as projected needs which may mitigate some of the benefits for collaborative efforts moving forward. This presentation is intended to catalyze a discussion regarding the opportunities and challenges associated with collaborative technology development and will include a discussion about the ways that health authorities and federal agencies can foster collaboration among biopharmaceutical manufacturers, suppliers, government, and academic partners through a Manufacturing USA institute.

498214 The Transcelerate Comparator Network: Transforming the Supply Chain for Clinical Comparator Medicines

Abstract:

The procurement of comparator medicines required for use in clinical trials for the development of new medicines, poses a formidable challenge for supply chain professionals within biopharmaceutical research and development. This case study describes the synthesis and evolution of the TransCelerate Comparator Network, a biopharmaceutical industry forum that facilitates the procurement of comparator medicines and exchange of information between participating member companies. The goal of this network is to enable member companies to conduct clinical trials more efficiently and with enhanced integrity.

The TransCelerate Comparator Network marks a transformative shift in the manner in which biopharmaceutical companies have historically secured comparator medicines to conduct clinical trials. It is a game changer in that it has led to the creation of a platform that provides participating member companies access to clinical medicines and related documentation which facilitates their ability to conduct clinical trials more efficiently and mitigates risk to the continuity of clinical trials. Additionally this network has fostered an environment where member companies can collaborate and develop community based knowledge and learn from the experiences of others.

The TransCelerate Comparator Network is a great example of how multiple companies across an industry can come together and collaborate in the precompetitive space to develop creative solutions that address long standing challenges and in the process simplify the supply chain to increase efficiency, boost resilience, enhance integrity and reduce operating costs.

490317 Accelerating Pharmaceutical Development through Precompetitive Collaboration: The Enabling Technologies Consortium

Abstract:

The Enabling Technologies Consortium (ETC), which originated as a Working Group within the International Consortium for Innovation and Quality in Pharmaceutical Development (IQ Consortium), is a consortium of pharmaceutical and biotech companies focused on the optimization of technologies that enable pharmaceutical development. The goal of ETC is to simplify and accelerate the process for establishing collaborations by standardizing the procedures and agreements under which they operate; to proactively identify high-value opportunities to deliver innovative technologies where the business case is compelling and collaboration with the broader external community (vendor, academic, government) is required. The current scope of the ETC is in technology within the Chemistry, Manufacturing and Controls (CMC) of drug substance development. This presentation provides an update on the progress of the overall â€˜Enabling Technologiesâ€™ effort since its inception in August 2015 using the drug substance drying working group as a case study

502398 Cmac: A Partnership Approach to Precompetitive Collaboration for Pharmaceutical Manufacturing Research

Abstract:

Modern research increasingly demands multidisciplinary approaches with many national funders increasingly recognizing the potential benefits for supporting the development of close relationships between industry and academia. CMAC is a large collaborative UK based manufacturing research centre that is focused on advanced crystallization and accelerating the adoption of continuous manufacturing to enable future medicines supply chains. Led from the University of Strathclyde, CMAC brings together a multidisciplinary academic team from across several UK universities (Bath, Cambridge, Imperial College, Loughborough, Leeds and Sheffield) as well as a range of active international academic partnerships. Now in its second phase and starting a new 7 year, £50M programme, CMAC was co-created with industry partners to ensure activities and outputs align with areas of existing and emerging industrial need. Our members include GSK, AZ, Novartis, Bayer, Eli Lilly, Takeda and Roche as well as over 20 technology providers and SMEs. This scale and breadth of expertise brings excellent opportunities to inform research direction and inspire researchers however it also demands that all aspects of the collaborations are well co-ordinated and supported. The collaborative approach to establishing this critical mass activity has also attracted significant investments from UK research councils & agencies (EPSRC, SFC/HEFCE) as well as charities and pharmaceutical industry and has stimulated a number of subsequent initiatives aimed at transforming the UK research and innovation system. In addition to the manufacturing research programme, CMAC has active programmes in other key areas, namely: skills and training, facilities development and industrial translation, aimed at maximisingthe benefits to all of our partners.

This presentation will review the model and different approaches used across CMACâ€™s precompetitive programme and will highlight challenges and lessons from such large collaborative activities involving academia and industry.

488365 A Perspective on a Pre-Competitive NSF Engineering Research Center Focused on Pharmaceutical Manufacturing: C-Sops

Abstract:

Throughout the last ten years, The Center for Structured Organic Particulate Systems (C-SOPS) has existed as a National Science Foundation ERC focused on the pre-competitive development of pharmaceutical manufacturing products and processes. During peak levels of activity, six universities and over fifty corporations were involved in the program. A major focus of the program throughout and even today is continuous solid dose manufacturing. Through this program, much of the groundwork resulting in todayâ€™s commercialized continuous solid dose manufacturing approaches by member organizations such as Vertex, Janssen, Eli Lilly and others was developed pre-competitively.Â Today, C-SOPS continues to work closely with top pharmaceutical companies adopting advanced manufacturing technology such as continuous.Â In this talk, a brief history of C-SOPS will be presented along with lessons learned on pre-competitive efforts and where the program is going today.

496512 The Development of an Interactive Knowlege Epicenter for the Advancement of Continuous Manufacturing

Abstract:

Continuous manufacturing has increased in popularity amongst pharmaceutical organisations in recent times, and much research is currently being conducted both in academic and industrial settings. However a large portion of the research which is being conducted may be classified as precompetitive and thus a significant opportunity exists for the industry as a whole to augment knowledge accumulation rates by working together and developing one common, central space where information can be easily presented, located and interacted with. The Continuous Manufacturing Repository is being developed in order to achieve this objective. It is an easily accessible online environment in which participating companies can selectively share their publically available presentations and/or publications, creating the 'one-stop-shop' for continuous manufacturing knowledge. This presentation will focus on how the CM repository works, how companies can get involved and will demonstrate the long term vision for the repository.

492264 Scale-up Modelling and Analysis of a Pharmaceutical Crystallization Process

Abstract:

One of the great challenges in the design of industrial crystallization processes is to predict the influence of crystallizer scale on the process behavior and product quality attributes. Industrial crystallizers demonstrate far from well-mixed behavior. Local variations of critical process variables such as supersaturation, particle concentration, temperature profiles, etc. exist within industrial crystallizers. Consequently, crystallization process models that are based on the assumption of well-mixed systems are not fit for use in supporting optimal scale-up and design of industrial crystallizers. A more detailed insight into the mixing behavior and its consequences on local crystallization phenomena â€"such as a distributed parameter model- is needed in order to achieve reliable design, including an appropriate process scale-up.

This work presents an approach to develop a predictive model of a pharmaceutical crystallization process scale-up based on compartmental modelling. While full blown CFD models could be used to better describe and predict mixing characteristics in reactors as opposed to compartmental modeling, they are not practical for industrial applications due to heavy computational burden. Compartmental modelling is such that a trade-off approach is used to overcome the limitations of well-mixed models, by considering localized mixing, heat transfer and fluid hydrodynamics separately from crystallization kinetics within a crystallizer. While compartmental modelling is a well-known and successfully applied technique in other areas of process engineering, such as characterizing mass transfer for oxygen in industrial fermentation, there is a lack of comprehensive studies on the application of compartmental modeling approach to the scale-up of industrial pharmaceutical crystallization.

In order to apply compartmental modelling to pharmaceutical batch cooling crystallization, the crystallizer is divided into finite number of compartment volumes in a way that every compartmental zone possesses minimized or negligible gradients in profiles, such as crystal distribution, super-saturation and energy dissipation within their individual volumes. The same set of model equations and model parameters are defined in order to solve the balance equations for the conservation of population, mass and energy for every compartment. However, spatial variations in the crystallizer are modelled by different rates of nucleation, growth, dissolution and attrition in the individual compartments. The compartment modelling is implemented in MATLAB/Simulink.

The process behavior of the pharmaceutical batch cooling crystallization is analyzed with respect to crystallizer scale, process variables and operation conditions. The influence of the non-uniformly distribution of the related process variables is shown by comparing the multi-compartment model with a single-compartment well-mixed model.

Acknowledgment

We would like to thank the Danish Council for Independent Research (DFF) for financing the project with grant ID: DFF-6111600077B.

499257 Removal of Impurity from API and Reduction of Agglomeration during Drying

Abstract:

In this work we describe removal of an impurity from an API which was not detected in HPLC and can be observed only in NMR in specific solvents. A kinetic model was developed for conversion of this impurity to API using online NMR. Based on the prediction from the model the impurity was converted into API and isolated after crystallization. There were challenges in milling of the API due to large chunks (about 1 -2 cm) which were formed due to agglomeration during filtration and drying. This issue was resolved by using filter dryer, application controlled vacuum and application of agitation in the filter dryer based on LOD of the API during drying.

491244 Viscoelastic Properties and Dissolution Kinetics of Amorphous Drug Dispersion Films, Utilizing the Quartz Crystal Microbalance

Abstract:

Amorphous drug dispersions are becoming more and more utilized within the pharmaceutical industry, along with their academic study[1-2]. Their primary use is in facilitating the intake of poorly water soluble drugs by patients[3-4]. Plenty of research has been performed on understanding their physiochemical characteristics and stability in a wide range of conditions[2, 5]. Of the research published, little has gone into real time studies of the physiochemical properties of such films as they degrade. Our team, sought to utilize the Quartz Crystal Microbalance (QCM) to visualize how these drug blended films had altered physiochemical properties from unloaded ones in a variety of different environments[6].

Carbamazepine and Copovidone were the compounds used to form amorphous drug films that were spun cast onto the QCM sensors. Solution blends in Methanol of 17% Copovidone weight by volume and Carbamazepine drug loadings of 35 to 5 mg/mL were used and spun cast. The lack of crystallinity was proven by a combination of several instrumental techniques including optical microscopy, Powder X-ray Diffraction (PXRD), and Differential Scanning Calorimetry (DSC), and the presence of Carbamazepine was shown via Fourier transform infrared spectroscopy (FTIR).

Quantitative data, including the thickness, density, and viscoelasticity, on the structural properties of the film was obtained by utilizing both the Sauerbrey and Viscoelastic models for the films. The films were found to be better modelled as soft rather than rigid, and the Sauerbrey model was not as effective. The data obtained from the Viscoelastic model was corroborated with ellipsometry to obtain a thickness. Within this body of work, the films at varying temperatures and on different surfaces to analyse their effects on the physiochemical properties of the films.

Furthermore, the dissolution kinetics of these films when submerged in water and ethanol were investigated in real time. In water, depending on the drug loading, this led to the nucleation and crystallization of the Carbamazepine in the excipient film, where the contact of the particles on the sensors are seen. The particle sizing was performed using Dynamic Light Scattering (DLS) to obtain size measurements of the Carbamazepine powder that formed.

All these works combined, showcase the strength of the QCM as a surface specific tool and its potential within the pharmaceutical formulations industry in helping to optimize the combinations of excipients and active pharmaceutical ingredients. Additionally, it offers strong insight into how the properties of the amorphous drug films change with temperature. Future studies could utilize alternative sensors and cells to further expand our knowledge and understanding of these amorphous drug films.

References:

1. Baghel, S.; Cathcart, H.; O'Reilly, N. J. Polymeric Amorphous Solid Dispersions: A Review of Amorphization, Crystallization, Stabilization, Solid-State Characterization, and Aqueous Solubilization of Biopharmaceutical Classification System Class II Drugs. *Journal of Pharmaceutical Sciences* **2016,** *105*, 2527-2544.

2. Douroumis, D.; Bouropoulos, N.; Fahr, A. Physicochemical characterization of solid dispersions of three antiepileptic drugs prepared by solvent evaporation method. *Journal of Pharmacy and Pharmacology* **2007,** *59*, 645-653.

3. Alonzo, D. E.; Zhang, G. G. Z.; Zhou, D.; Gao, Y.; Taylor, L. S. Understanding the Behavior of Amorphous Pharmaceutical Systems during Dissolution. *Pharmaceutical Research* **2010,** *27*, 608-618.

4. Ehmann, H. M. A.; Winter, S.; Griesser, T.; Keimel, R.; Schrank, S.; Zimmer, A.; Werzer, O. Dissolution Testing of Hardly Soluble Materials by Surface Sensitive Techniques: Clotrimazole from an Insoluble Matrix. *Pharmaceutical Research* **2014,** *31*, 2708-2715.

5. Weuts, I.; Van Dycke, F.; Voorspoels, J.; De Cort, S.; Stokbroekx, S.; Leemans, R.; Brewster, M. E.; Xu, D.; Segmuller, B.; Turner, Y. T. A.; Roberts, C. J.; Davies, M. C.; Qi, S.; Craig, D. Q. M.; Reading, M. Physicochemical Properties of the Amorphous Drug, Cast Films, and Spray Dried Powders to Predict Formulation Probability of Success for Solid Dispersions: Etravirine. *Journal of Pharmaceutical Sciences* **2011,** *100*, 260-274.

6. Reviakine, I.; Johannsmann, D.; Richter, R. P. Hearing What You Cannot See and Visualizing What You Hear: Interpreting Quartz Crystal Microbalance Data from Solvated Interfaces. *Analytical Chemistry* **2011,** *83*, 8838-8848.

496430 Improving the Dissolution of Ibuprofen By Co-Milling of Ibuprofen and MCC Mixture in a Ball Mill

Abstract:

In this work, we investigated the improvement of dissolution of ibuprofen by co-milling the ibuprofen with microcrystalline cellulose, MCC (hydrophilic excipient) in dry mode. Ibuprofen and MCC were taken in 1:1 weight ratio and co-milled for different times in a planetary ball mill. Co-milled, v-blended and as received samples were characterized by different techniques. Particle size distribution measured using laser diffraction method exhibited a shift to lower particle size. To understand the morphology, shape and surface changes, optical microscopy and Field Emission Scanning Electron Microscopy (FE-SEM) were performed. While the crystallinity and crystal level changes (such as defects, strain energy and crystallite size etc.) were measured using Powder X-Ray Diffraction (P-XRD), the presence of different functional groups on the surface and the interactions between drug and excipient; and hydrogen bond formation was evaluated using Fourier Transform Infrared Spectroscopy (FTIR). The changes in the crystallinity due to co-milling were further confirmed using Differential Scanning Calorimetry (DSC) where the shift of melting endotherm to lower temperature was observed. Acidic and basic nature of the surface was evaluated by Inverse Gas Chromatography (IGC) measurements by acidic and basic parameters. The wettability measurements performed by the sessile drop method shows lower contact angle for co-milled samples demonstrating improvement in wettability of the ibuprofen. This is also confirmed by the dissolution test in USP Paddle II apparatus. The study shows that co-milling imparts surface hydrophilicity apart from size reduction by generating defects, dislocations and decreased crystallite size by the impact of balls on the surface. Increased interactions between carboxylic acid group and hydroxyl group result in hydrogen bond formation thereby improving the wettability and dissolution. Increase in polar surface energies and increase in the ratio of basic to acidic constant indicates hydrophilic nature of the surface which contributes to enhancing the dissolution of ibuprofen. Thus, co-milling can be an alternate solventless method to improve dissolution of BCS Class II drugs.

489947 Scientific Considerations on Continuous Crystallization and Its PAT Method Validation

Abstract:

Continuous manufacturing (CM) is an emerging technology in the pharmaceutical manufacturing sector. As the final purification/isolation step, crystallization can significantly affect the drug substance critical quality attributes (CQAs) and is one of the critical steps in an integrated continuous manufacturing (CM) process. Continuous crystallization has the potential to improve manufacturing efficiency and product quality, but both industry and FDA currently lack commercial experience with the technology. Here, we focus on the application of quality risk management to the development of a continuous crystallization system. Specifically, we designed and built an automated two-stage Mixed Suspension Mixed Product Removal (MSMPR) crystallization platform for a model compound (Carbamazepine, CBZ) that exhibits multiple polymorphs. The crystallization process includes integration of Process Analytical Technology (PAT) tools (on-line Raman microscopy and Focused Beam Reflectance Microscopy, FBRM) for real-time monitoring. Material transport, process control, and product collect approaches were investigated as the three key areas for risk identification and mitigation during continuous crystallization process development. Our proof of concept continuous crystallization system uses feedback controls to achieve constant levels in crystallizers, a centralized automation program coded in LabView, and PAT monitoring for polymorphs and particle size distribution (Raman and FBRM). We developed a PAT method utilizing Raman Spectroscopy to monitor solute concentration and polymorphs of CBZ. A calibration model was developed and validated; and the limit of detection for a metastable polymorphic form was quantified. The impact of multiple factors, such as solid content, and temperature, on the model prediction was studied. The model was validated following the principles described in USP <1225>. The results demonstrated that a linear model can predict the solute concentration with less than 10 percent error. The repeatability and intermediate precision was evaluated and the relative standard deviation was below 5 percent. The limit of detection and quantification were found to be 3.8 and 7.4 mg/ml, respectively. The limit of detection for the metastable form was determined by monitoring the ratio of characteristic peaks when increasing the percentage of the metastable form in relation to the total amount of crystals in the solution. A statistically significant change in the peak ratio was observed after the concentration of the metastable form reached 25 percent.

499363 An Engineering Approach to Concentration of Temperature-Sensitive Pharmaceutical Process Streams for Continuous Crystallization of an API

Abstract:

An Engineering Approach to Concentration of Temperature-Sensitive Pharmaceutical Process Streams for Continuous Crystallization of an API

P. Roche[1], G.K. Pascual[1], R. C. Jones*[1], P. Donnellan*[1], and B. Glennon[1]

[1] *Synthesis and Solid State Pharmaceutical Centre (SSPC), School of Chemical and Bioprocess Engineering, University College Dublin, Belfield, Dublin 4, Ireland*

Keywords: Solution Concentration, Continuous Processing, Bubble Column, Crystallization, MSMPR, PAT Technologies

Introduction

Traditionally, most active pharmaceutical compounds (API) have been manufactured at large plant scales using batch methodologies. Continuous manufacturing is becoming an increasing attractive alternative due to apparent benefits such as smaller manufacturing footprints, increased process safety, lower manufacturing costs and increased supply-chain flexibility. [1] A common problem faced in small molecule API production is the necessity for concentration steps between unit operations i.e. where the upstream process (e.g. reaction) is fully developed at a given concentration of solution; while the downstream process (e.g. crystallization) is operated at a different concentration range, driven by the metastable limits of the crystallisation process. To date, concentration steps have been carried out as evaporative crystallisation, i.e. boiling off a known amount of solvent until a desired concentration is reached. [2] However, some APIs have shown signs of degradation at high temperatures. [1]This study aims to provide a low temperature method under ambient pressure for concentration of API solvent solutions.

Section 1 â€“ Solution Concentration: A Novel Approach

This section of work focuses on employing a simple, lab-scale bubble column as a gas-liquid contacting system to achieve a desired removal rate of solvent. A model-based, mass transfer rate prediction is shown and compared with experiments performed for several common solvents used in the pharmaceutical industry. Thermodynamic predictions of Vapour-Liquid Equilibria were used in conjunction with mass transfer coefficients to accurately predict the rate of removal of solvent for the system and verified experimentally. [2]

For this work, a solution of Paracetamol in Methanol was chosen as the process, due to the well-researched saturation limits of paracetamol [3]and the reasonable results obtained from previously studied trials with methanol solutions in the column. The solution was reduced to an optimum concentration for the proceeding step from a dilute feed, while assuring that nucleation would not occur in any of the transfer lines by operating safely within the saturation limits.

Section 2 â€“ Continuous Crystallization Optimization

Following the establishment of model predictions, the column was employed immediately upstream of the feed of a continuously operated mixed suspension, mixed product removal (MSMPR) crystallizer with intermittent withdrawal via dip pipe [4, 5]using an automated pressure supply for the cooling crystallization of paracetamol in methanol. In-situ Focused Beam Reflectance Measurement (FBRM G400, Mettler Toledo) was used for real-time monitoring of chord length distribution and to accurately ascertain the point at which steady state operation had been reached; while in-situ PVM (ParticleView V819, Mettler Toledo) provides microscopy quality images in real time to monitor the crystal habit.

The overall objective of the work was to enable the adjustment of crystallization parameters, based on the optimum concentration provided by the gas-liquid contacting system, and to provide robust prediction of the crystallizer performance. Simple extraction of growth and nucleation kinetics from the MSMPR at steady state was accomplished using the population balance model.

References

1. Rawat, T. and I.P. Pandey, *Forced Degradation Studies for Drug Substances and Drug Products - Scientific and Regulatory Considerations.* Journal of Pharmaceutical Sciences and Research, 2015. **7**(5): p. 238-241.

2. Sandler, S., *Other Types of Phase Equilibria in Fluid Mixtures*, in *Chemical, Biochemical and Engineering Thermodynamics*. 2006, John Wiley & Sons, Inc. p. 575-593.

3. Granberg, R.A. and A.C. Rasmuson, *Solubility of Paracetamol in Pure Solvents.* Journal of Chemical Engineering Data, 1999. **44**: p. 1391-1395.

4. Power, G., et al., *Design and Optimization of a Multistage Continuous Cooling Mixed Suspension, Mixed Product Removal Crystallizer.* Chemical Engineering Science, 2015. **133**: p. 125-139.

5. Morris, G., et al., *Estimation of Nucleation and Growth Kinetics of Benzoic Acid by Population Balance Modeling of a Continuous Cooling Mixed Suspension, Mixed Product Removal Crystallizer.* Organic Process Research & Development, 2015. **19**(12): p. 1891-1902.

497499 The Effect of Axial Dispersion on Crystal Size Distribution in a Meso-Scale Continuous Oscillatory Baffled Crystallizer

Abstract:

The effect of axial dispersion on crystal size distribution in a meso-scale continuous oscillatory baffled crystallizer

Iyke Onyemelukwe[1], Qinglin Su[2], Zoltan K. Nagy[1,2], Chris D. Rielly[1]

[1] Department of Chemical Engineering, Loughborough University, Loughborough, LE11 3TU, UK

[2] Davidson School of Chemical Engineering, Purdue University, West Lafayette, IN 47906, USA

Abstract

Tubular crystallizers, specifically the continuous oscillatory baffled crystallizer (COBC), for continuous manufacturing have been an active research area in recent years, e.g. investigating how axial dispersion is affected by flow regime, operating conditions and the geometry of the periodic constrictions [1] and how the resulting residence time distribution (RTD) influences the supersaturation distribution and the resulting product crystal size distribution [2]. In this study, an experimental setup of a meso-scale COBC was developed to study the effects of axial dispersion on a cooling crystallization process of the glycine in water system. First, the axial dispersion effect, characterized by the Peclet number, $\mathbf{P_e}$, was measured experimentally by using an imperfect pulse method, tracking the concentration-time history of a tracer at two axial locations after the tracer input for both the liquid and particle phases. Dispersion models for the tracer concentration in each phase were then fitted to calculate the dispersion parameters. The method used a fast Fourier transformation (FFT) to convolute the measured input concentration profile with the parameterized model RTD to almost perfectly match the measured output profile; The effects of the oscillation frequency and amplitude on the dispersion coefficients were studied.

A seeded cooling crystallization of glycine and water system was run in the meso-COBC under the same range of through-flow and oscillatory operating conditions. Broad crystal size distributions of final crystals were observed. Furthermore, a steady-state mathematical model of COBC, consisting of population balance model of crystals, mass balance equation of solute, and energy balance equation for heat transfer, was developed to provide an in-depth understanding of the dispersion effect on crystal size distribution. It was found that the axial dispersion had a less significant effect on the crystal size distribution under the current operating conditions, whereas the fast cooling rate along the tube had a far greater impact, which results in extensive primary nucleation. An optimal trajectory of temperature along the tube was important in controlling the supersaturation and the crystal size distribution [3].

However, the nature of the counter-current heat transfer, which leads to a series of natural cooling curves, one for each change in jacket temperature, makes difficult the control of temperature trajectory in a limited number of tube segments. Perspectives on using the meso-scale COBC for crystallization kinetics screening for continuous manufacturing and for scaling up to larger COBC facilities will be given.

References

[1] L. N. Ejim, S. Yerdelen, T. McGlone, I. Onyemelukwe, B. Johnston, A. J. Florence and N. M. Reis, "A factorial approach to understanding the effect of inner geometry of baffled meso-scale tubes on solid suspension and axial dispersion in continuous, oscillatory liquid-solid plug flows," *Chemical Engineering Journal,* vol. 308, pp. 669-682, 2017.

[2] R. Kacker, S. I. Regensburg and H. J. Kramer, "Residence time distribution of dispersed liquid and solid phase in a continuous oscillatory flow baffled crystallizer," *Chemical Engineering Journal,* vol. 317, pp. 413-423, 2017.

[3] Q. Su, B. Benyahia, Z. K. Nagy and C. D. Rielly, "Mathematical modeling, design, and optimization of a multisegment multiaddition plug-flow crystallizer for antisolvent crystallizations," *Organic Process Research Development,* vol. 19, pp. 1859-1870, 2015.

495126 Continuous Heterogeneous Crystallization and the New Method of Making Tablets

Abstract:

A novel continuous heterogeneous crystallization process was developed, which the active pharmaceutical ingredient is crystallized directly on the surface of an excipient within the crystallizer. The product is then directly formed into tablets without the need for complex downstream processing steps such as milling, sieving, granulation, and blending. The innovative process has no particle processing and handling unit operation. The APIs and excipients systems selected by study of molecular interaction mechanisms, and surface functionality of substrate. The effect of various process parameters on the process stability and drug loading were studied.

Within the range of our analysis, proportional drug loading can be easily tailored for a wide range depending on required drug dosage per tablet. The fully continuous processes, and eliminating the downstream steps result in producing crystalline compounds and final form (tablets) in significantly faster, more efficient, with lower footprint, and more economical. This novel approach eliminates almost all the conventional unit operations in the drug product process, by connecting the API purification and crystallization step directly to the tablet making final stage.

KEYWORDS: Continuous manufacturing, Heterogeneous nucleation, Surface facilitated nucleation, Drug product

491965 A Comparative Study of Continuous Crystallization in an Oscillatory Baffled Crystallizer and a Mixed-Suspension-Mixed-Product-Removal Crystallizer

Abstract:

Crystallization has become a predominant technique in particle design technology and it is widely applied in the pharmaceutical industry. It directly dictates downstream processes and heavily influences overall drug properties such as bioavailability and dissolution rate. Traditionally crystallization has been carried out in batch mode where operating conditions vary with time, often resulting in inconsistent products.[1] Continuous crystallization, on the other hand, has been identified to improve reproducibility, reduce operating cost and simplify scale-up.[2] A well-studied system for continuous crystallization operation is Mixed-Suspension-Mixed-Product-Removal (MSMPR) crystallizer where an agitator is commonly used to provide mixing. There are challenges associated with stirred tank systems such as poor local mixing, low heat and mass transfer, and varying shear rate.[3] To address these challenges, a novel commercial oscillatory baffled reactor (OBR) has been studied. It consists of a reactor tank and oscillating â€˜donutâ€™ shaped baffles to provide more uniform mixing.[4] Oscillatory systems have been well studied for reactions and synthesis because of their improved mixing. Interest in the use of OBRs for crystallization processes has increased due to enhanced mass and heat transfer capabilities and reduced shear imposed on crystals.[5]

In the work proposed here, a comparison between continuous crystallization in an MSMPR versus an OBR is carried out to assess the differences in final product properties. Continuous operation in both vessels is achieved using peristaltic pumps and vacuum for slurry removal. The residence time distribution (RTD) is studied in each system using homogeneous and heterogenous tracers at different mixing conditions. Continuous cooling crystallizations of paracetamol is then evaluated in both crystallizers at the same feed concentration and mean residence time with varying power density levels. At each power density level, the start-up time and steady state crystal properties (i.e. mean size, CSD) are analyzed to determine the effect of hydrodynamics. A rigid and robust online FTIR probe, capable of correcting for signal interference caused by oscillations, is used to monitor the solute concentration during operation. This study will improve the understanding of oscillatory systems as an alternative crystallization vessel and the understanding of the impact of different mixing mechanisms on the final crystal size distribution.

1. Nagy ZK, Aamir E. Systematic design of supersaturation controlled crystallization processes for shaping the crystal size distribution using an analytical estimator. *Chem Eng Sci*. 2012;84:656-670. doi:10.1016/j.ces.2012.08.048.

2. Alvarez AJ, Singh A, Myerson AS. Crystallization of Cyclosporine in a Multistage Continuous MSMPR Crystallizer. *Cryst Growth Des*. 2011;11(10):4392-4400. doi:10.1021/cg200546g.

3. Kacker R, Regensburg SI, Kramer HJM. Residence Time Distribution of Dispersed Liquid and Solid Phase in a Continuous Oscillatory Flow Baffled Crystallizer. *Chem Eng J*. 2017;317:413-423. doi:10.1016/j.cej.2017.02.007.

4. Hewgill MR, Mackley MR, Pandit AB, Pannu SS. Enhancement of gas-liquid mass transfer using oscillatory flow in a baffled tube. *Chem Eng Sci*. 1993;48(4):799-809. doi:10.1016/0009-2509(93)80145-G.

5. Ni X, Mackley MR, Harvey AP, Stonestreet P, Baird MHI, Rama Rao NV. Mixing Through Oscillations and Pulsationsâ€"A Guide to Achieving Process Enhancements in the Chemical and Process Industries. *Chem Eng Res Des*. 2003;81(3):373-383. doi:10.1205/02638760360596928.

490309 Continuous Drying of Pharmaceuticals

Abstract:

Continuous Drying of Pharmaceuticals

M. Kreimer[1], I. Aigner[1], S. Sacher[1], M. Krumme[2], T. Mannschott[2], P. van der Wel[3], A. Kaptein[3], J.G. Khinast[1,4]

[1] Research Center Pharmaceutical Engineering (RCPE) GmbH, Inffeldgasse 13, 8010 Graz, Austria

[2] Novartis Pharma AG, Novartis Campus, 4056 Basel, Switzerland

[3] Hosokawa Micron B.V., Gildenstraat 26, 7005 BL Doetinchem, Netherlands

[4] Graz University of Technology, Institute for Process and Particle Engineering, Inffeldgasse 13, 8010 Graz, Austria

Traditional production of pharmaceutical products has been mostly conducted via batch processing during the last decades. This kind of production is inflexible and involves cost-intensive process development, especially during up-scaling within phases II and III of drug product development. A lot of effort has been made from academia, industry and regulators to overcome these challenges by development of fully continuous manufacturing processes for pharmaceutical goods. This paradigm shift from batch to continuous processing is demanded as more and more NME (new molecule entities) are entering the market and demanding flexible production processes. Furthermore, continuous pharmaceutical manufacturing facilities allow 100% real-time quality assurance and process control. A complete manufacturing line may include very different unit operations starting with API synthesis and ending up with compaction of final tablets. Although a range of manufacturing equipment is available for various unit operations, there is still a gap in linking primary and secondary manufacturing.

Continuous drying of APIâ€™s is one of those challenging process steps. Consistent quality of the dried solid product is crucial and has to be maintained after crystallization, washing and filtration. During these processes extensive effort is taken to optimize particle properties such as particle morphology, size or bulk flowabiltiy. Therefore, the particle properties of the intermediates defined and tuned in previous steps have to be at least unaltered during drying or even optimized. This raises several challenges during the drying process of wet powders, because agglomeration and attrition are present permanently as counteracting forces due to shear movement.

Besides matching of specified product quality, the key to successful continuous drying of powders is a robust process design and consistent process ability. These process design considerations include, amongst others, avoiding of dead zones and a narrow residence time distribution. The latter is favorable to avoid mixing of particles with different moisture content and to diminish the tendency of agglomeration or attrition. One approach to avoid those issues is the application of forced-feed.

In this work experiments were conducted on two different drying platform technologies with different model substances, including APIâ€™s and excipients with particle sizes below 100Â Âµm. The two investigated drying platforms are based on extrusion and a contact-convective drying technology. Different process settings were investigated and successful configurations achieved. Initial moisture content of the feed material was varied to analyze the drying rate and capacity of the dryers. Measurements were performed for residual moisture content, particle size distribution and flowability and results were compared between raw, wet and dried material. Residual moisture levels below 0.1% were reached. No agglomeration or attrition was observed during the drying procedure as particle size distribution was unchanged for the dried product. Furthermore, similar results were obtained for powder flowability prior and after drying.

496616 Continuous Fluidized Bed Drying of Pharmaceutical Materials

Abstract:

Recent years, the continuous manufacturing has become a trending technology in the pharmaceutical industries. Traditional batch manufacturing requires long processing times and consumes a tremendous amount of materials during process development. The US FDA is encouraging the companies to step forward into cleaner, flexible, more efficient continuous manufacturing. Although there are no regulatory hurdles for implementing continuous manufacturing, the lack of understanding of the process is limiting the integration of unit operations. Take the fluidized bed as an example; batch fluidized bed drying has been a traditional way of removing the moisture from the pharmaceutical ingredients due to its well-mixing and gentle drying behavior. The flow properties and the resulting tablet properties will be affected because of the change of moisture and particle size during the drying process. When a continuous process is adopted, the material property distribution can spread over process time. Therefore, understanding the residence time distribution is crucial to the process control.

In this study, experiments were conducted in a modified Glatt GPCG-2 continuous fluidized bed dryer/granulator. A twin-screw feeder was used to feed the wet materials. The residence time distributions for a pharmaceutical excipient (Fujicalin) and a placebo granule (Lactose/Avcil) were investigated using the colored tracer. Fujicalin is a free flowing porous material, and the Lactose/Avcil granule has very different flow properties before and after drying. The tracer concentration was calibrated and quantified with a color spectrophotometer that measures percent reflectance as a function of wavelength.[1] The residence time distribution curve was fitted by a dispersion model. The material properties were examined before and after the process. The effect of the operating conditions - material feed rate, air flow rate and initial loading were studied. The results showed that the Glatt continuous fluidized bed dryer behaved like a CSTR. The residence time could be reduced by applying a high material feed rate. The air flow rate and the initial loading did not have a significant effect on residence time for the conditions that we have investigated. As we previously found that the zero-order kinetic is dominant during the drying process, the outlet material moisture can be easily predicted by combining the RTD model and the drying kinetic.[2]

[1] Emady, H.N.; Wittman, M.; Koynov, S. *et al.* A simple color concentration measurement technique for powders. Powder Technology 2015, 286, 392-400.

[2] Chen, H.; Liu, X.; Bishop, C. *et al.* Fluidized bed drying of a pharmaceutical powder: A parametric investigation of drying of dibasic calcium phosphate. Drying Technology 2016, null-null.

486763 Protein and Gold Nanoparticle Based Radiation Sensor

Abstract:

High energy X-ray radiation has been used for ablation of malignant tumors for quite some time. Nevertheless, medical practitioners are unable to predict with high accuracy the amount of radiation reaching the point of interest below the tissue level. To help address this issue, we are developing a colorimetric sensor for radiation using the gold nanoparticle forming property of bio-molecules (particularly Maltose Binding Protein (MBP)) in presence of ionizing radiation. As gold nanoparticles have distinct color, gold nanoparticles formed on irradiation changes the color of the solution and thus can be used as colorimetric sensor. The color change is observed for doses of X-ray radiation as low as 2 grays (1 gray (Gy) = absorbance of 1 joule of radiation energy per kilogram of matter). The complete procedure for cloning and synthesizing MBP is also described. The system also shows a radiation dose dependent color shift (from 3 Gy to 15 Gy), which can be used to discriminate between different delivered doses.

498352 Direct Detection of Nucleic Acids without Amplification

Abstract:

Rapidly detecting the source of an infection remains challenging. Techniques using nucleic acid amplification have promise in providing answers quickly, but remain vulnerable to contamination and lack thermal stability because of their reliance on enzymatic proteins. We are developing a sensitive, inexpensive and portable instrument to automate the direct detection and identification of pathogenic nucleic acids, which will enable health professionals to rapidly identify infections in humans and subsequently accelerate the rational treatment of such infections. The current prototype performing the assay produces results from non-invasive samples in one hour and requires only minimally-processed crude samples. Our method promises to be robust for field and physician office applications, allowing distributed healthcare decisions outside of traditional laboratories.

All bacteria, viruses and fungi have within them nucleic acids (DNA, RNA) that determine their properties (benign or pathogenic) – and detecting the presence of specific nucleic acids is the fastest and the most optimal method to detect the source of an infection. Our methodology, called CAPTURE™ (Confirming Active Pathogens Through Unamplified RNA Expression), is unique in that it works in the presence of unknown amounts of unrelated nucleic acids, proteins and cellular debris. The design of the custom DNA 'capture' probes used in the assay is arguably the most integral step in our process, and as such, requires the most detailed method. Through a combination of locally compiled databases and proprietary software, ideal candidate sequences from pathogen genomes can be determined which exhibit the appropriate thermodynamic behavior required of the panel. The CAPTURE™ method is currently being used to identify several dozen specific pathogens and also to quickly determine the presence of any bacteria in a patient sample. The ability to determine universal bacterial signatures in a sample is a powerful tool for combating antimicrobial resistance, as it has the potential to limit the prescription of unnecessary antibiotics.

We have developed a CAPTURE™ panel for the analysis of urinary tract infections (UTI) that correctly identifies and differentiates the gram-negative bacteria *Escherichia coli* (which cause over 85% of UTI), *Klebsiella pneumoniae, Proteus mirabilis* and *Pseudomonas aeruginosa*; the gram-positive bacteria *Enterococcus faecalis* and *Staphylococcus aureus*; and fungal *Candida* species. The use of multiple captors for each pathogen allows for a statistical cluster analysis and introduces systematic redundancy to ensure efficacy. Our CAPTURE™ technique offers health care professionals a reliable and a quick way to screen for pathogens in clinical samples.

499935 PhotoacousticÂ Imaging to SimultaneouslyÂ Detect the Accumulation of Multiple Contrast Agents within Tumors

Abstract:

Modern cancer therapeutics have shifted focus from generic therapeutics to developing personalized treatment plans for greater safety and effectiveness. Since drugs are rapidly being developed specifically for each cancer phenotype, there is an increased demand for fast and inexpensive methods to determine cancer phenotypes and morphologies. Current diagnostic imaging modalities utilizing X-ray CT, MRI, and PET scans are limited to black-white images that cannot be used to differentiate multiple disease marker contrast agents at a time. In addition, the high costs of capital equipment and dangers of excessive imaging associated with these techniques inhibit their widespread use in diagnostics. Photoacoustic (PA) imaging, a hybrid light and sounds imaging technique, has shown to be a safe and inexpensive diagnostic technique with high spatial resolution in 3D. Traditional PA contrast agents, however, tend to have broad absorption peaks in the NIR range which renders it difficult to simultaneously image more than one signal at a time in deep tissue. Here we present the formulation of a series of PA active NPs with sharp and separable absorbance profiles in the NIR range for simultaneous multiplexed imaging **(Fig. 1)**. PA dyes are encapsulated inside NPs using the controlled self-assembly mechanism, Flash Nanoprecipitation (FNP). Four new contrast agents, with sharp absorbance maxima between 600-900 nm, were created by encapsulating a variety of phthalocyanine derivatives. We were able to concurrently detect the concentrations of contrast agents mixed together with >95% deconvolution efficacy. Using these dye particles, we demonstrated that information on tumor morphology and surrounding blood flow can be acquired simultaneously **(Fig. 2)**. As proof of concept, we co-injected RGD modified NPs and non-modified NPs with different labeling agents and tracked NP biodistributions for both particles simultaneously. Using this technology, we accessed the effect of NP ligand modification on both targeting efficacy onto the tumors and off-targeting accumulation in the liver using a single animal model. Over modification of the NPs resulted in rapid liver clearance and poor accumulation in the tumor; at low modifications, the tumor to liver accumulation ratio is 9.9 ± 4.2, while at high RGD modifications the tumor to liver accumulation ratio is 52 ± 22. Our particles enable new quantitative and rapid methods of assessing nanoparticle biodistributions.

500760 DNA Methylation Detection with an Engineered Protein That Binds Hemi-Methylated DNA

Abstract:

DNA methylation, specifically the methylation of cytosine bases, is an important biomarker, as abnormal DNA methylation patterns are found in many different types of cancer. Currently, a small number of cancer hospitals evaluate the methylation status of the MGMT gene promoter to determine the best course of treatment for patients with glioblastoma. However, improved methylation detection techniques are required in order to expand the availability of such testing to more patients. Methyl-CpG-binding domain (MBD) proteins bind specifically to methylated DNA sequences, and many assays have been developed that use these proteins in methylation profiling of DNA. The wild-type proteins in the MBD family bind specifically to symmetrically methylated CpG dinucleotides. Here, we describe the engineering of a new MBD that binds to hemi-methylated DNA but not unmethylated DNA, which allows for the detection of a methylated target sequence hybridized to an unmethylated DNA probe. The binding affinity of this engineered protein toward a DNA sequence with a single hemi-methylated CpG dinucleotide was evaluated, and a dissociation constant of 5.6 ± 1.4 nM was achieved. The protein was expressed in soluble form as a fusion protein with green fluorescent protein, and performance in distinguishing between hemi-methylated (methylated target DNA bound to an unmethylated probe) and unmethylated (unmethylated target DNA bound to an unmethylated probe) DNA in an interfacial binding assay was demonstrated. Using this protein, an assay was developed and optimized to determine the methylation status of the MGMT gene promoter by hybridizing the target DNA to a probe on a surface and determining the methylation status of the captured target based on the amount of MBD bound. Such an assay is useful in guiding treatment plans for glioblastoma patients because MGMT promoter methylation is correlated with better response to alkylating chemotherapeutics such as temozolomide.

501477 Effective Physiological and Anatomical Parameters on Fractional Fluid Reserve (FFR) of Coronary Artery Stenosis

Abstract:

Coronary stenosis is an abnormal narrowing in a coronary artery that is caused by an atherosclerotic lesion and reduces lumen space. FFR is a gold standard method to assess the hemodynamic significance of coronary stenosis. FFR is the ratio of maximum blood flow distal to a stenotic lesion to normal maximum flow in the same artery. Normal FFR is 1 and an FFR < 0.8 is considered hemodynamically significant. FFR is typically measured invasively in the cardiac catheterization laboratory using a pressure sensor wire in the coronary artery. Recently, investigators have applied computational fluid dynamics (CFD) methods to determine FFR using coronary images obtained from different modalities such as angiograms or CT scans. The purpose of this study was to determine the effect of eleven physiological and anatomical factors on the accuracy of computing FFR. In this study, blood flow was modeled as a single phase Newtonian fluid. The physiological factors were pressure as inlet boundary condition, mass flow rate as outlet boundary condition, heart rate, blood viscosity and blood density. Anatomical factors were the diameter of coronary artery, length of coronary segment modeled, percent stenosis, stenosis length, the position of stenosis relative to coronary segment modeled, and stenosis shape (concentric vs. eccentric stenoses). The parameters that affected FFR the most were determined by Plackett–Burman statistical analysis. Analysis of variance (ANOVA) showed a high variance coefficient (R^2) value of 0.928. Our analysis demonstrated that percent stenosis, diameter of coronary artery, mass flow rate, and stenosis shape had the most significant effect on FFR ($p < 0.05$). Increase in mass flow rate decreased FFR significantly, while the diameter of the coronary artery was directly proportional to FFR. As for stenosis shape, FFR decreased more significantly due to concentric stenosis than eccentric stenosis. CFD provided analysis of coronary physiology that would have been difficult or impractical to do experimentally. Accurate anatomical reconstruction is critical for computationally deriving FFR.

Keywords: Fractional Fluid Reserve; Coronary Artery Stenosis; Plackett–Burman Method; Computational Fluid Dynamics;

501819 Analysis of Carbonyl Compounds in Exhaled Breath for Identificationf of LUNG Cancer Biomarkers

Abstract:

Some carbonyl compounds have been identified as biomarkers of lung cancer. We report a fabricated reactive chip with thousands of micropillars in a microfluidic channel that have been engineered to selectively preconcentrate trace volatile ketones and aldehydes in exhaled breath. These carbonyl compounds are directly related to lung cancer induced metabolic dysfunction and oxidative stress. The chips are fabricated using deep reactive ion etching (DRIE) of silicon. The micropillar surface is functionalized with a reactive quaternary ammonium aminooxy salt for trapping trace carbonyl compounds by means of an oximation reaction. The chips are used for directly preconcentrating carbonyl compounds in breath. The concentrated ketone- and aldehyde-aminooxy adducts are analyzed by ultra-high performance liquid-chromatograh-mass spectrometry (MS). Some carbonyl compounds including hydroxyl-acetaldehyde, 2-butanone, 4-hydroxy-hexenal are identified as lung cancer biomarkers by statistical data analysis of healthy control group breath samples and lung cancer patient breath samples.

498490 Residence Time Distribution and Segregation Studies Trough Real Time Measurements By Near Infrared Spectroscopy

Abstract:

Currently, pharmaceutical companies are moving from batch to continuous manufacturing. Blender is a critical unit operation of continuous pharmaceutical manufacturing process that assures the blend uniformity and thereby drug content in final product [1]. The near infrared (NIR) spectroscopy is an important tool to measure the drug concentration at blender outlet and therefore to characterize the continuous blender [2-4]. The characterization of a continuous blender was achieved through residence time distribution (RTD) analysis and segregation studies by using a set of pharmaceutical excipients and active ingredients with material properties that defines a principal components analysis score plot.

RTD characterization was performed through in-line measurements by near infrared (NIR) spectroscopy of changes in composition adding a pulse of tracer into the system. Each material used was analyzed by using a continuous blending setup consisting in one feeder connected to a GEA continuous blender through a hopper. At the blender outlet, a chute was connected with a lateral window to control the level of powder with a rotary valve. Perpendicular to the powder flow, a near infrared probe with the light source was connected for spectral acquisition in diffuse reflectance mode. The reflected light is transferred to the interferometer for processing through fiber optic. Six experiments for each material has been performed at tow flow rates and 3 impeller speeds (revolution per minute (RPM)). A second feeder was necessary in cases where low density exceeds the gear capacity of a single feeder setup. Results of in-line RTD measurements demonstrated the selectivity of calibration models and low limit of detection required for analysis of tracers providing critical information for blender characterization.

Segregation studies were conducted for five formulations in continuous blending. For each configuration in blender RPM, samples were collected for further analysis in an ASTM D 6940-04 segregation tester. Near infrared prediction of ten samples collected for each experiment were performed by using 1-Dimentional sampling approach. Calibration models for each formulation were prepared and implemented. Results indicates that each blender configuration has a different impact in the segregation behavior of the API concentration and the NIR method provided a fast and reliable results.

Reference

[1]. Yijie Gao, Aditya Vanarase, Fernando Muzzio, Marianthi Ierapetritou, Characterizing continuous powder mixing using residence time distribution, Chemical Engineering Science, Volume 66, Issue 3, 1 February 2011, Pages 417-425, ISSN 0009-2509, http://doi.org/10.1016/j.ces.2010.10.045.

[2]. Singh, R.; Sahay, A.; Karry, K. M.; Muzzio, F.; Ierapetritou, M.; Ramachandran, R., Implementation of an advanced hybrid MPCâ€"PID control system using PAT tools into a direct compaction continuous pharmaceutical tablet manufacturing pilot plant. International journal of pharmaceutics 2014, 473 (1â€"2), 38-54.

[3]. ColÃ³n, Y. M.; Florian, M. A.; Acevedo, D.; Méndez, R.; Romañach, R. J., Near Infrared Method Development for a Continuous Manufacturing Blending Process. Journal of Pharmaceutical Innovation 2014, 9 (4), 291-301.

[4]. Cárdenas, V.; Cordobés, M.; Blanco, M.; Alcalà, M., Strategy for design NIR calibration sets based on process spectrum and model space: An innovative approach for process analytical technology. Journal of Pharmaceutical and Biomedical Analysis 2015, 114, 28-33.

499494 Use of Process Analytical Technology (PAT) in Advanced Manufacturing Process Monitoring and Control: Scientific Considerations and Regulatory Challenges

Abstract:

Significant scientific advancement made in PAT tool design and process analysis has resulted in the increasing use of PAT during process development. However, the actual implementation of PAT in a manufacturing setting has lagged despite its potential for advanced manufacturing process monitoring and control. In this presentation, we will discuss some of the scientific considerations and regulatory challenges when evaluating a PAT system for process monitoring and control for the manufacturing setting. Some aspects will be discussed are:

1) Identification of *critical process parameters or critical quality attributes* for monitoring and control purpose;

2) Risk-based strategy for identifying *measurement interface* between process analyzer and process stream and data acquisition frequency;

3) Model building framework, pros and cons of *each modeling methods* for process monitoring and control 4) Incorporating the PAT monitoring and control as part of *control strategy* for manufacturing.

Several examples will be included to highlight the use of PAT for advanced manufacturing process monitoring and control, in the spirit of facilitating broad adaptation of PAT in advanced manufacturing for robust manufacturing and product quality.

491877 3D Raman Imaging: A Method to Study the Effects of Lubrication on the Microstructure of Tablets

Abstract:

3D Raman Imaging: A Method to Study the Effects of Lubrication on the Microstructure of Tablets

Shashwat Gupta, Savitha Panikar, Fernando Muzzio

Rutgers- The State University of New Jersey

When it comes to pharmaceutical tablets, there are known knowledge gaps in the current manufacturing processes, particularly regarding the relationship between material properties of ingredients, processing conditions, and resulting tablet attributes. These three aspects can be linked through a common feature of understanding the microstructure of a sample. A better understanding of tablet microstructure has the potential to unlock the black box between process parameters and performance attributes. Product structure understanding of this nature can aid in prediction of product performance, process development as well as deviation investigations. Understanding the microstructure of a tablet requires knowing the spatial distribution of the different ingredients and void space, which can be enabled by Chemical Imaging (CI).

Raman imaging is a powerful tool for studying tablet structure, with superior chemical specificity than NIR and is also less affected by physical characteristics of the sample.

This study focuses on using 3D Raman Imaging system (H2Optx), as an off-line measurement system, to bridge this gap for tablets made using a Direct Compression (DC) continuous line, for the lubrication process. For pharmaceutical operations, lubrication is essential in order to reduce the friction between the surfaces of manufacturing equipment and that of organic solids as well as to ensure the continuation of an operation. The most common pharmaceutical lubricant used is Magnesium Stearate (MgSt). Till now, extensive structural studies have not been conducted to study the effects of lubricant(s) on tablet properties.

In this work, we investigate the particle size distribution, spatial variability (Variograms) of MgSt, using the 3D Raman Images. These parameters are then correlated with product performance (tablet hardness and dissolution profile) as well as formulation (amount of MgSt) and process parameters (Total exposure to shear and Residence Time), with the help of Principle Component Analysis.

Variographic data helped to quantify the distribution of MgSt in tablets of experimental conditions. On the other hand, to discern the variability in the distribution of MgSt within the layers of a tablet, we needed PCA score plots of the variographic data obtained from the 3D Raman images. They revealed a stark difference in the clustering of the layers for the tablets with different process and formulation parameters. The hardness of the tablets from the two experimental conditions agree with the expected trend. A wider distribution of MgSt resulted in tablets with a lower tablet hardness. However, there is not a clear trend in the dissolution profile with respect to the distribution of MgSt.

An example of a comparison between 3D Raman Images of tablets elucidating the effect of total exposure to shear is shown below:

(a)

(b)

Figure 1: 3D Raman Images (a) Lower exposure to shear: MgSt, green – particles <50 microns, red – particles from 50 to 250 microns and (b) Higher exposure to shear: MgSt, green – particles <50 microns, red – particles from 50 to 250 microns

The future research will involve studying how the distribution of MgSt changes in a tablet, as we move ahead in a continuous line i.e. comparison between post blending and post tableting samples. All the work will be used as a platform for an on-line measurement system.

493745 Development and Validation of an in-Line NIR Spectroscopic Method for Continuous Blend Potency Determination in the Feed Frame of a Tablet Press

Abstract:

INTRODUCTION

The desire within the pharmaceutical industry to shift towards continuous processing strengthens the need to invest in process analytical technology (PAT), because continuous real-time quality monitoring and control is essential for continuous production. Due to its fast, non-invasive and non-destructive character, near-infrared (NIR) spectroscopy is one of the most widely used PAT tools for the analysis of pharmaceutical tablets. The fact that NIR can be implemented into a process environment via fiber optics makes it an excellent in-line PAT tool.

Nowadays, NIR spectroscopy for potency analysis is one of the most widely used PAT applications for tablets [1–4]. This is often achieved by on-line or off-line NIR examination of tablet cores, resulting in a considerable lag time between the actual tablet sampling and the subsequent quality analysis.

To avoid this time lag, in this study, a measurement set
measurement of the circulating blend inside the feed frame of a tablet press (MODUL™P, GEA Process Engineering n.v., Halle, Belgium).

Secondly, an NIR calibration model was developed and validated, enabling in-line and real-time blend potency measurements within the optimised interfacing.

Next, the residence time distribution (RTD) inside the feed frame was studied to determine the potential lag time between the measured concentrations in the feed frame of the tablet press and the measured concentrations in the resulting tablets. Finally, influences of the process parameters on the in-line and off-line RTD were investigated.

MATERIALS

During the set-up optimisation and quantification model development study, a formulation was used consisting of sodium saccharine (JMC corporation, South Korea), microcrystalline cellulose (Avicel® PH 102, FMC biopolymer, Ireland), dibasic calcium phosphate (Emcompress™, JRS Pharma, Germany) and magnesium stearate (Mallinckrodt, Ireland)(0.5% w/w) as a lubricant for tableting.

For the RTD studies, a blend consisting of lactose (FastFlo 316, Foremost) and magnesium stearate (Mallinckrodt, Ireland)(0.5% w/w) was employed. Spiking was performed using a formulation consisting of lactose and sodium saccharine (in equal parts), and magnesium stearate (0.5% w/w) was added for lubrication.

NIR ACQUISITION

The in-line NIR spectra were collected in reflection mode using a diode array spectrometer: SentroPAT FO (Sentronic GmbH, Germany). A SentroProbe DL RS NIR (Sentronic GmbH, Germany) connected to the SentroPAT system, through a fiber-optic cable, was mounted inside the feed frame. Spectra were acquired at a speed of 15 spectra per second. A spectral range of 1150-2200 nm was covered with a spectral resolution of 2 nm.

The sampled tablets were inspected off-line in transmission mode using an FT-NIR analyser (Antaris II, Thermo Fisher Scientific, USA), furnished with a tablet analyser module. In transmission mode, a larger sample volume is measured, since the light penetrating through the sample is measured. This measurement mode is not suitable in-line because of the longer integration time and the need to be able to measure through the sample, which is impossible in the current design of the feed frame.

DEVELOPMENT AND OPTIMISATION OF SET-UP FOR IN-LINE MEASUREMENTS INSIDE THE FEED FRAME

An NIR probe was mounted on top of the feed frame in such a position that it measured the circulating blend just before compression. In order to ensure a reproducible and robust implementation of the NIR probe in the feed frame, a micrometer was installed in the tablet press, allowing a precise positioning of the probe. During tableting, NIR spectra were continuously acquired.

Initially, disturbances of the NIR signal, caused by the rotating paddle wheels of the feed frame were observed. The use of mathematical spectral filters to reduce spectral noise can be very complex and can result in deletion of important physical information about the powder blend within the feed frame [5]. Therefore small notches were created inside the paddle wheel fingers of the tablet feed frame, avoiding disturbances in the spectra and hence eliminating the need for mathematical or spectral filters.

The effect of paddle wheel type, probe height and paddle wheel speed on the stability of the NIR signal was investigated by means of a full factorial experimental design, developed in MODDE 10 (Umetrics, Umeå, Sweden). Optimal process settings, resulting in most stable NIR spectra, were used to make NIR calibration models.

DEVELOPMENT AND VALIDATION OF IN-LINE NIR CALIBRATION MODELS

Calibration models were developed by regressing the in-line NIR spectra against the true sodium saccharine concentration of the tablets (based on known weighed masses). In-line spectra were acquired during tableting of five blends with different concentrations. Models were calculated for the prediction of two different API target concentrations: 5% and 20% (w/w). Several calibration models for the prediction of this API content were developed, comparing the use of single spectra or averaged spectra, as well as using partial least squares (PLS) regression versus ratio models. The best models were selected and validated by means of accuracy profiles [6–8], by processing physical mixtures with four different API concentrations (validation set).

RTD DETERMINATION OF THE FEED FRAME

In order to determine the RTD of the feed frame, initially, the feed frame was filled with the lactose formulation. On top of this lactose blend, a pulse of 50 g of the spiking blend (saccharine/lactose) was introduced and on top of the spiking blend, 2500 g of lactose blend was added. During tableting, in-line NIR spectra were continuously acquired and tablets were sampled for off-line potency analysis. This experiment was repeated using different tableting parameters, according to a full factorial experimental design.

The mean RTD was calculated via the approach used by Kumar et al.[9]. The influence of the tableting parameters on the RTDs was evaluated using MODDE 10. In addition, the difference and link between in-line measured RTDs and off-line measured RTDs was investigated to gain a better understanding of the mechanisms inside the feed frame.

CONCLUSION

In this work, an in-line NIR method with high speed acquisition of spectral information to allow real-time release (RTR) was developed for use in the feed frame of a tablet press.

An in-line NIR spectroscopic model was developed and validated for the monitoring of API concentration of a powder blend in the feed frame during a tableting process.

Finally, RTD of the feed frame was determined in-line and off-line, helping to gain a better understanding of the flow/mixing processes going on inside the feed frame.

This study showed that in-line NIR measurements in the feed frame of a high speed rotary tablet press can be applied as a method for powder composition monitoring during the tableting process. When coupled with existing tablet press control systems, this method represents, together with the knowledge of the RTD of the feed frame, a big step towards real time release testing and real time release.

REFERENCES

[1] W. Li, A. Bashai-Woldu, J. Ballard, M. Johnson, M. Agresta, H. Rasmussen, et al., Applications of NIR in early stage formulation development. Part I. Semi-quantitative blend uniformity and content uniformity analyses by reflectance NIR without calibration models, Int. J. Pharm. 340 (2007) 97–103. doi:10.1016/j.ijpharm.2007.03.040.

[2] W. Li, L. Bagnol, M. Berman, R. a. Chiarella, M. Gerber, Applications of NIR in early stage formulation development. Part II. Content uniformity evaluation of low dose tablets by principal component analysis, Int. J. Pharm. 380 (2009) 49–54. doi:10.1016/j.ijpharm.2009.06.032.

[3] J. Gottfries, H. Depui, M. Fransson, M. Jongeneelen, M. Josefson, F.W. Langkilde, et al., Vibrational spectrometry for the assessment of active substance in metoprolol tablets: A comparison between transmission and diffuse reflectance near-infrared spectrometry, J. Pharm. Biomed. Anal. 14 (1996) 1495–1503. doi:10.1016/0731-7085(96)01800-6.

[4] E. Peeters, A.F. Tavares da Silva, M. Toiviainen, J. Van Renterghem, J. Vercruysse, M. Juuti, et al., Assessment and prediction of tablet properties using transmission and backscattering raman spectroscopy and transmission NIR spectroscopy, Asian J. Pharm. Sci. 11 (2016) 547–558. doi:10.1016/j.ajps.2016.04.004.

[5] A.J. O'Neil, R.D. Jee, A.C. Moffat, The application of multiple linear regression to the measurement of the median particle size of drugs and pharmaceutical excipients by near-infrared spectroscopy., Analyst. 123 (1998) 2297–2302. doi:10.1039/a806001k.

[6] P. Hubert, J.J. Nguyen-Huu, B. Boulanger, E. Chapuzet, P. Chiap, N. Cohen, et al., Harmonization of strategies for the validation of quantitative analytical procedures: A SFSTP proposal - Part I, J. Pharm. Biomed. Anal. 36 (2004) 579–586. doi:10.1016/j.jpba.2004.07.027.

[7] P. Hubert, J.J. Nguyen-Huu, B. Boulanger, E. Chapuzet, P. Chiap, N. Cohen, et al., Harmonization of strategies for the validation of quantitative analytical procedures A SFSTP proposal – Part II, J. Pharm. Biomed. Anal. 45 (2007) 70–81. doi:10.1016/j.jpba.2004.07.027.

[8] P. Hubert, J.J. Nguyen-Huu, B. Boulanger, E. Chapuzet, P. Chiap, N. Cohen, et al., Harmonization of strategies for the validation of quantitative analytical procedures: A SFSTP proposal–Part III, J. Pharm. Biomed. Anal. 45 (2007) 82–96. doi:10.1016/j.jpba.2007.06.032.

[9] A. Kumar, J. Vercruysse, M. Toiviainen, P.E. Panouillot, M. Juuti, V. Vanhoorne, et al., Mixing and transport during pharmaceutical twin-screw wet granulation: Experimental analysis via chemical imaging, Eur. J. Pharm. Biopharm. 87 (2014) 279–289. doi:10.1016/j.ejpb.2014.04.004.

500787 Identification of Key/Non-Key Process and Model Parameters of Polishing Chromatography, Primary Containers and Autoinjectors Using Mechanistic Models

Abstract:

Identifying key/non-key process parameters is an essential step in the Quality-by-Design methodology. It demonstrates true process understanding and it enables risk-based decision making. Typically, this is done by FMEA and DOE studies which have some important limitations. First-principles and mechanistic models are being used to advance the vision of rational drug substance development via a rigorous and quantitative Quality-by-Design approach. In this work, we use three mechanistic models to study the factor/response of drug/process/device systems of practical interest to industry. We employ well-established global sensitivity analysis techniques (Sobol indices) to address the so-called factor screening, factor prioritization and factor fixing setting [Saltelli et al., 2008]. This enables a rigorous analysis of key/non-key process and model parameters to accelerate design iterations, manage/reduce risk, and ultimately assure quality via a probabilistic in-silico QbD design space. This new methodology can work as a complement (or a replacement) of more traditional and experiential-based FMEA approaches. This is the first time such a rigorous analysis has been published for a polishing chromatography system. The application to the primary container case-study reveals that only half of the factors of this process are non-essential. For the autoinjector device two sub-components happen to be the most relevant to system variability.

大连理工大学
Dalian University Of Technology

In-Situ Agglomeration Measurement during KDP Crystallization Based on Double-View Image Analysis

Yan Huo, Tao Liu*, Cai Y. Ma, Xue Z. Wang

Institute of Advanced Control Technology
Dalian University of Technology, Dalian, China

1

Institute of Advanced Control Technology

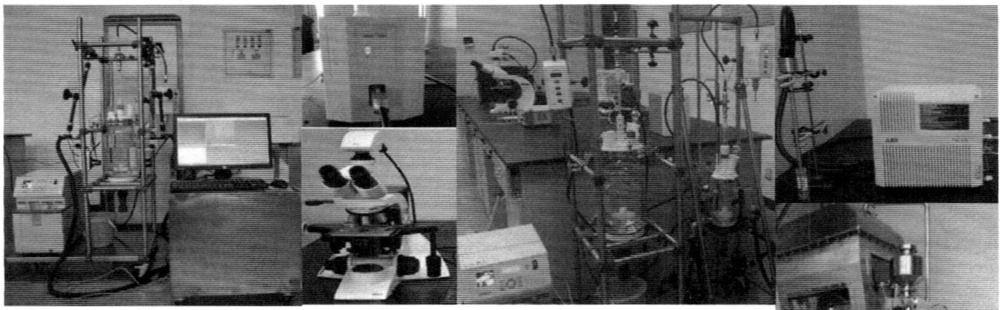

Research interests:

- In-situ crystallization & drying process monitoring using ATR-FTIR, FT-NIR, 2D/3D imaging systems;
- PBM and CFD modeling of crystallization processes;
- Online model predictive control and optimization of crystallization & drying processes;
- Design of in-situ measurement and control devices.

2

Outline

大连理工大学
Dalian University Of Technology

1 Introduction

2 Problem statement

3 Experimental setup

4 Agglomeration measurement

5 Conclusions & Outlook

1. Introduction

大连理工大学
Dalian University Of Technology

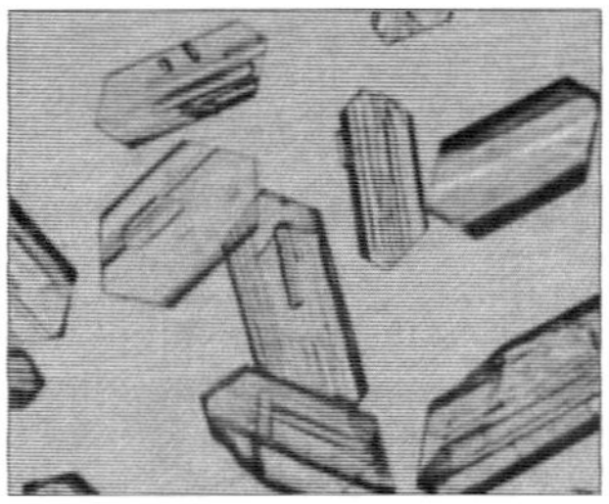

Particle shape

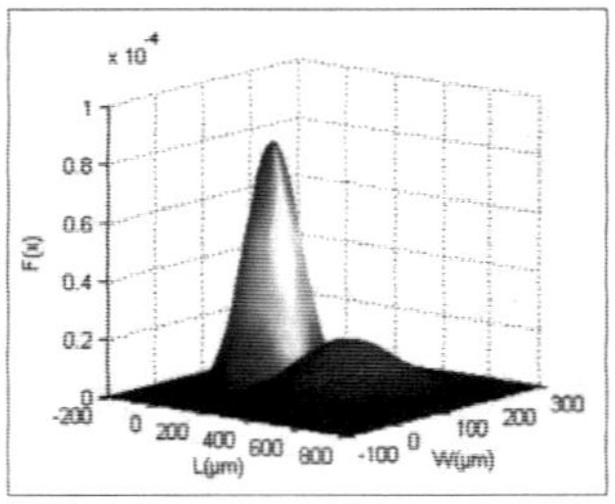

Particle size distribution

Particle agglomeration

On-line imaging systems have been increasingly applied to monitor industrial crystallization processes at the micro-scale in the past two decades, offering on-line observation of crystal shapes, the crystal size distribution, and particle agglomeration.

2. Problem statement

Pseudo agglomeration problem

Overlapping?

Agglomeration?

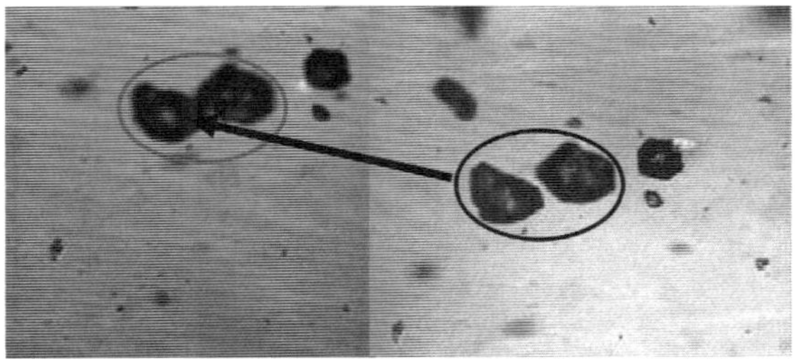

Left view Right view

3. Experimental setup

Online monitoring of crystallization process

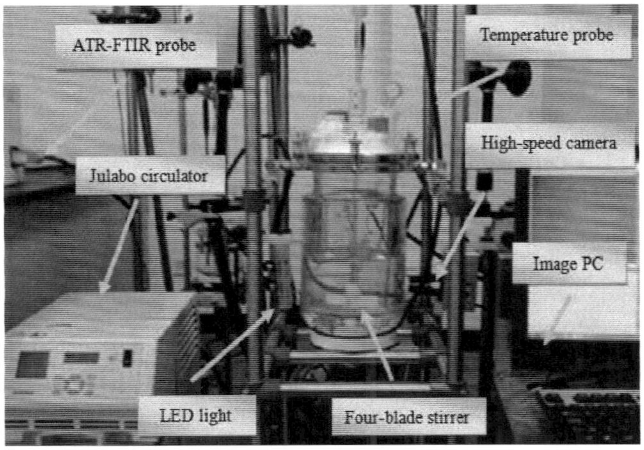

No	Device
1	4L glass vessel
2	Temperature probe
3	Julabo circulator
4	4-paddle agitator
5	Image PC
6	High-speed cameras
7	ATR-FTIR

Two cameras (UI-2280SE-C-HQ) with CCD sensors and USB Video Class standard were made by IDS Imaging Development Systems GmbH.

3. Experimental setup

大连理工大学 Dalian University Of Technology

Online monitoring of crystallization process

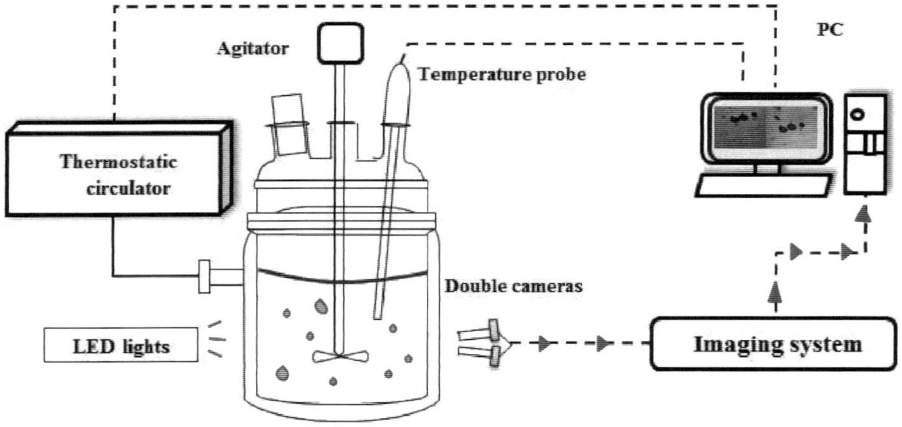

4. Agglomeration measurement

大连理工大学 Dalian University Of Technology

Image processing for agglomeration measurement

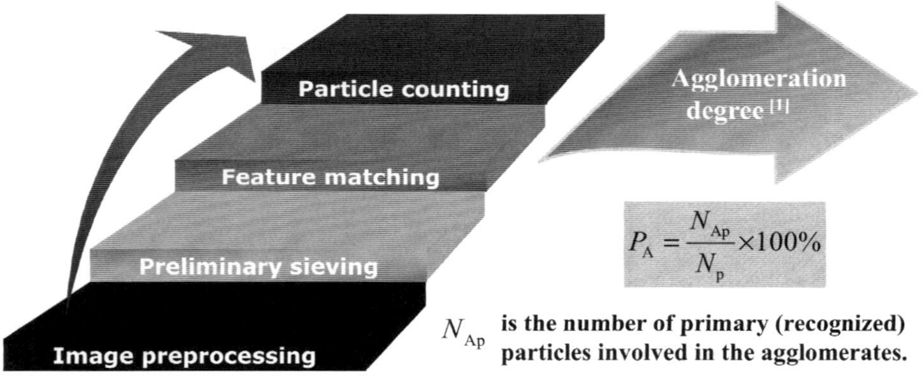

$$P_A = \frac{N_{Ap}}{N_p} \times 100\%$$

N_{Ap} **is the number of primary (recognized) particles involved in the agglomerates.**

N_p **is the total number of valid particles counted in the image.**

[1] Nývlt, J.; Karel, M., Crystal agglomeration. *Crystal Research and Technology,* 1985.

4. Agglomeration measurement

Dalian University Of Technology

Step 1: Image preprocessing
(including image compression and segmentation)

Image compression

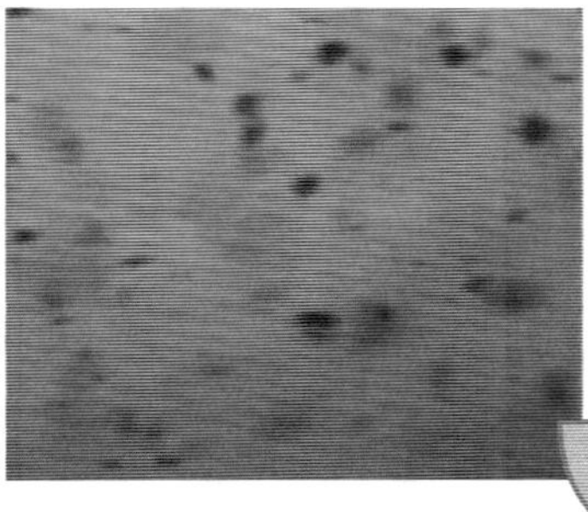

Original image

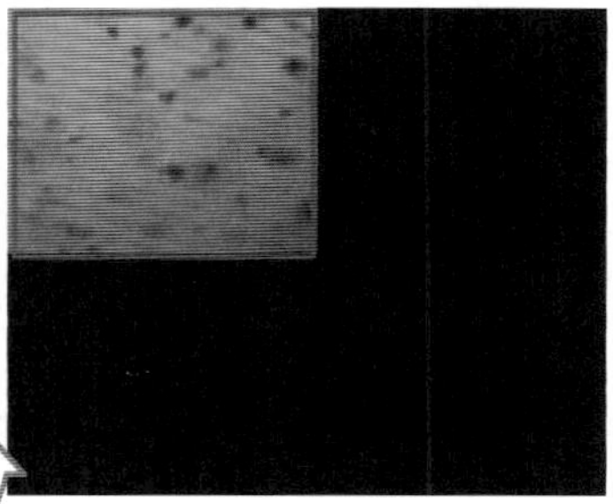

Compressed image

Wavelet transform

Objective: reduce the raw image size to one quarter for saving computation effort.

9

4. Agglomeration measurement

Dalian University Of Technology

Image segmentation

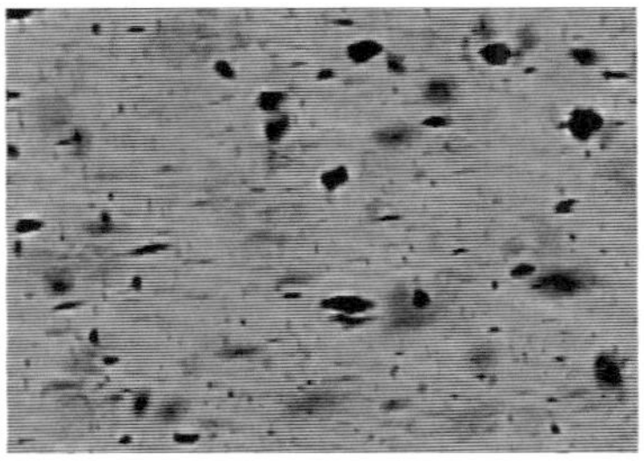

Reduced image

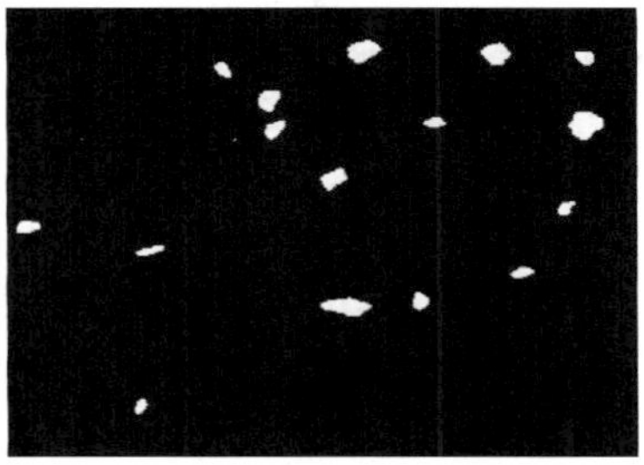

Segmented image

Task: perform image sementation to extract the particle shapes from the image.

Solution: the uneven background is extracted by using an opening operation algorithm [2]; the segmentation is computed by the Otsu thresholding function[3]

[2] Chen, J.; Wang, X. Z., A wavelet method for analysis of droplet and particle images for monitoring heterogeneous processes. Chemical Engineering Communications 2005, 192, (4), 499-515.
[3] Gonzales, R. C.; Woods, R. E.; Eddins, S. L., Digital image processing using MATLAB. Pearson Prentice Hall: Upper Saddle River, NJ, 2004.

10

4. Agglomeration measurement

Dalian University Of Technology

Step 2: Preliminary sieving

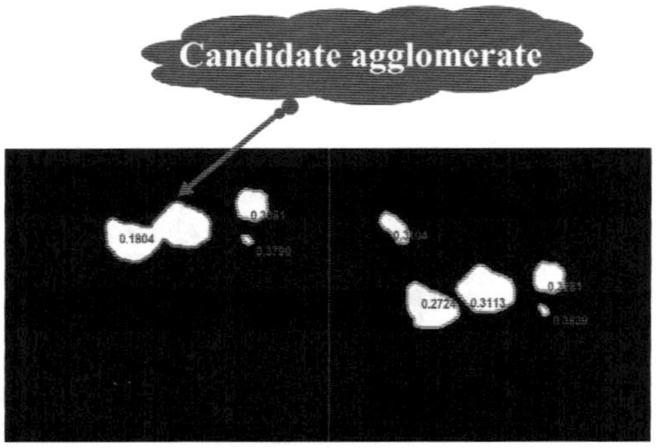

Key idea: sieve out candidate agglomerates by defining the particle shape feature.

Inner distance based curvature

$$\mathrm{IDC} = \frac{\sum_{n=1}^{N} c_n}{N}$$

$$x_c = \frac{1}{N}\sum_{n=0}^{N-1} x_n \qquad y_c = \frac{1}{N}\sum_{n=0}^{N-1} y_n$$

$$d_n = \sqrt{(x_c - x_n)^2 + (y_c - y_n)^2}$$

$$\rho_n = d_n - \bar{d} \qquad c_n = \frac{\rho_n''}{(1+\rho_n'^2)^{3/2}}$$

N is the number of particle boundary.

c_n s the curvature of the nth deviation distance.

11

4. Agglomeration measurement

Dalian University Of Technology

Step 3: Feature matching

Interest point location

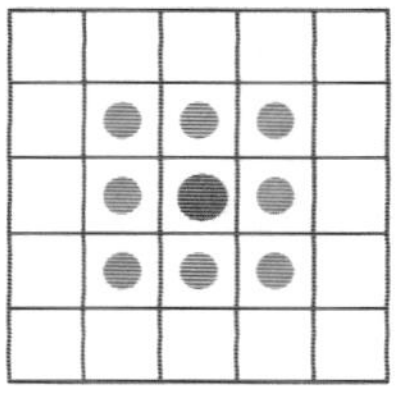

Objective: find the extremum value in a local region

The Hessian matrix for approximation is adopted to determine the extremum point

$$\mathrm{Det}(H) = D_{xx}D_{yy} - (0.9D_{xy})^2$$

D_{xx}, D_{yy} , D_{xy} are the approximations of the Gaussian second order derivatives for the image by using the box filters [4], respectively.

[4] Lowe, D. G., Distinctive image features from scale-invariant keypoints. *International Journal of Computer Vision*, 2004.

12

Step 3: Feature matching

Feature descriptor

Improved texture descriptors

(to cope with image noise and uneven illumination background)

The local binary pattern (LBP)

DCLBP (deviation compensated LBP)

$$DCLBP_{P,R}(i,j)=\sum_{p=1}^{P}s(g_p-g_a)2^p$$

where g_p is the gray values, g_a is a deviation location parameter

$$s(g_p-g_a)=\begin{cases}1, & if\ g_p-g_a\geq T;\\ 0, & if\ g_p-g_a<T.\end{cases}$$

MDLBP (magnitude difference based LBP)

$$MDLBP_{P,R}(i,j)=\sum_{p=1}^{P}s(m_p-\overline{m_p})2^p$$

where $m_p=|g_c-g_p|$

$\overline{m_p}$ is the averaged value for $[m_0,m_1,...,m_{P-1}]$

$$s(m_p-\overline{m_p})=\begin{cases}1, & if\ m_p-\overline{m_p}\geq T\\ 0, & if\ m_p-\overline{m_p}<T\end{cases}$$

[5] Ojala, T.; Pietikäinen, M.; Mäenpää, T., Multiresolution gray-scale and rotation invariant texture classification with local binary patterns. *IEEE Transactions on Pattern Analysis & Machine Intelligence*, 2002, 24, (7), 971-987.

Numerical example: three texture patterns with different features

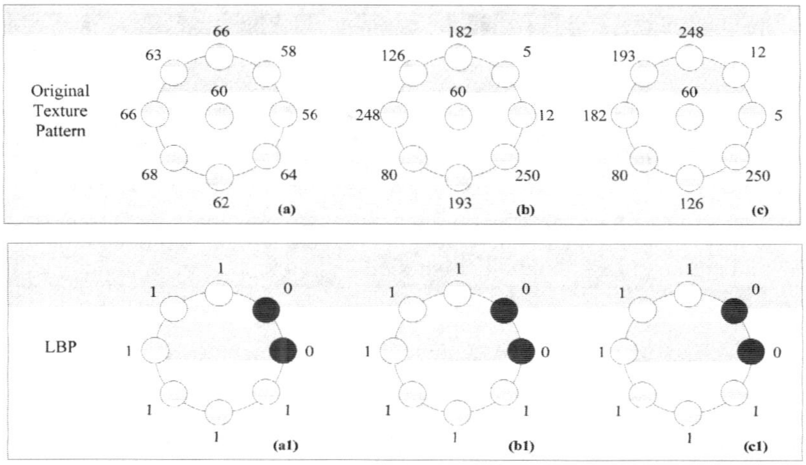

These texture patterns are classified into the same pattern by using LBP!

4. Agglomeration measurement

大连理工大学
Dalian University Of Technology

Effectively recognized by DCLBP and MDLBP

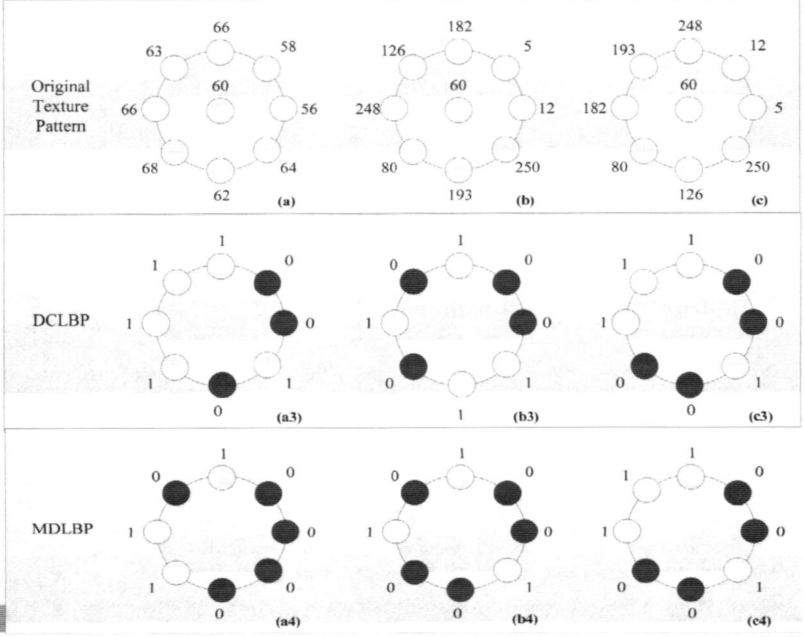

4. Agglomeration measurement

大连理工大学
Dalian University Of Technology

Descriptor matching

Exclude pseudo agglomerates due to overlapping

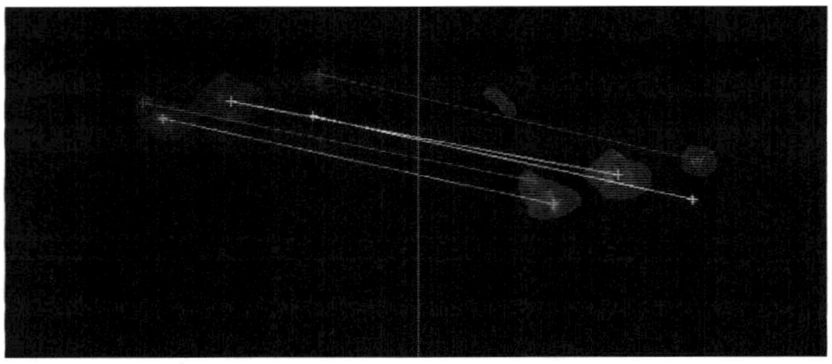

Left view Right view

Task: match the interest points between double-view images in real time.

Solution: the approximately nearest neighbor searching algorithm via K-dimension tree[6].

[6] Arya, S.; Mount, D. M.; Netanyahu, N. S.; Silverman, R.; Wu, A. An optimal algorithm for approximate nearest neighbor searching, *Acm-Siam Symposium on Discrete Algorithms*. 1994, Arlington, Virginia, 1994; 1994; pp 573-582.

16

Comparison of matching results by using different algorithms

$$P_{pr} = \frac{N_{pa}}{N_{pa} + N_{pm}} \qquad P_{pm} = \frac{N_a}{N_t}$$

Methods	Particle recall ratio (%)	Point matching accuracy (%)	Averaged computation time per frame (s)
Proposed	97.96	92.01	0.9347
SIFT [7]	94.90	91.83	1.3703
ASIFT [8]	93.88	91.65	1.8061
SURF [9]	95.92	88.47	0.8925
ORB [10]	90.82	83.95	0.4126

Higher　Higher　Faster

References:

[7] Lowe, D. G., Distinctive image features from scale-invariant keypoints. *International Journal of Computer Vision,* 2004, 60, (2), 91-110.

[8] Morel, J. M.; Yu, G., ASIFT: A new framework for fully affine invariant image comparison. *Siam Journal on Imaging Sciences*, 2009, 2, (2), 438-469.

[9] Bay, H.; Ess, A.; Tuytelaars, T.; Gool, L. V., Speeded-up robust features (SURF). *Computer Vision & Image Understanding*, 2008, 110, (3), 346-359.

[10] Rublee, E.; Rabaud, V.; Konolige, K.; Bradski, G. In ORB: An efficient alternative to SIFT or SURF, *International Conference on Computer Vision*, 2011, 2564-2571.

Step 4: Particle counting

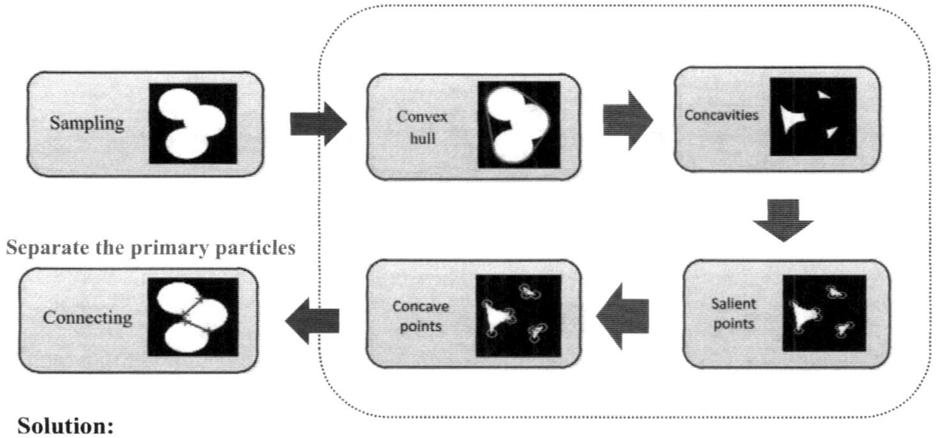

Solution:

A concave point on the agglomerate boundary is determined if all the connecting lines between the concave point and its adjacent salient points are inside the agglomerate.

4. Agglomeration measurement

大连理工大学
Dalian University Of Technology

Experiment: recognized results for two different cases

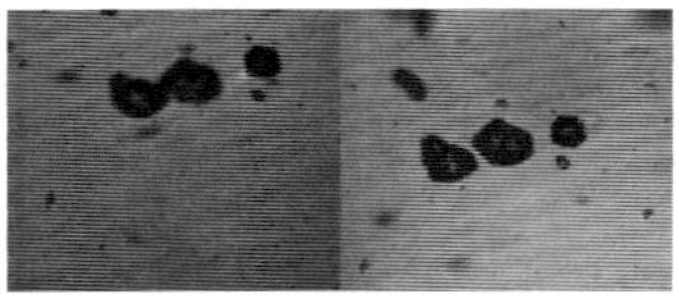

Case 1

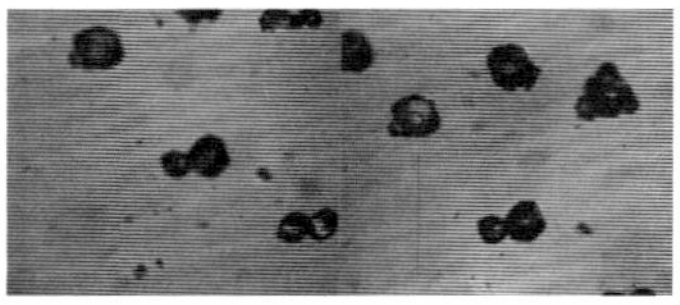

Case 2

Target parameters in the double-view images	Case 1	Case 2
The number of pseudo agglomerates	1	0
The number of true agglomerates	0	2
The number of unagglomerated particles	5	3
The number of primary particles involved in the agglomerates	0	4
The total number of all the primary particles	5	7
The agglomeration degree	0%	57.14%

19

4. Agglomeration measurement

大连理工大学
Dalian University Of Technology

Microscopic double-view images captured during the KDP (KH_2PO4) crystallization process

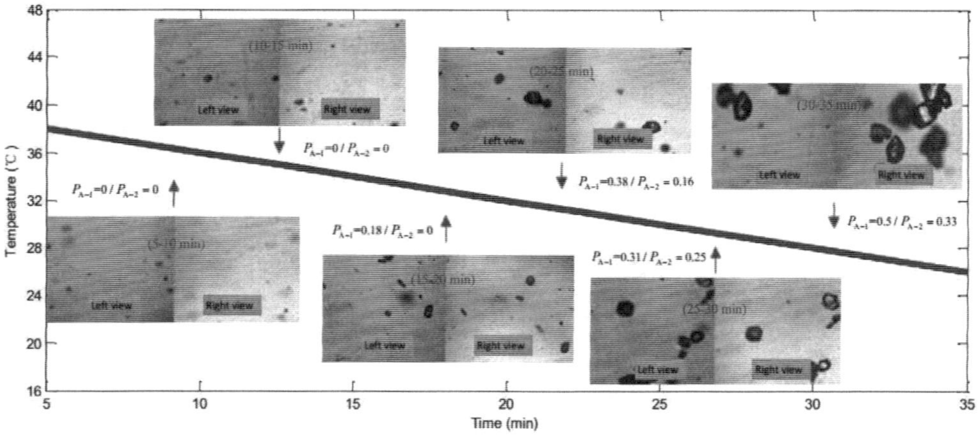

20

5. Conclusions & Outlook

大连理工大学 Dalian University Of Technology

- A synthetic image analysis method is proposed for monitoring particle agglomeration during a crystallization process.
- By defining an IDC index to detect possible agglomeration, a preliminary sieving algorithm is given to sieve out candidate agglomerates.
- Two texture descriptors, DCLBP and MDLBP, are proposed to recognize pseudo agglomerates with good accuracy.
- An efficient particle counting algorithm is established to assess the agglomeration degree.
- **It is interesting to develop a fast three-dimensional reconstruction method based on double-view images for CSD estimation in the future work.**

Yan Huo, Tao Liu*, Xue Z. Wang, Cai Y. Ma, Xiongwei Ni. On-line detection of particle agglomeration during solution crystallization by microscopic double-view image analysis. *Ind. Eng. Chem. Res.*, 2017, accepted, DOI: 10.1021/acs.iecr.7b02439.

21

Acknowledgement

Funding support

- **NSF China Grant (61633006)**
- **National Thousand Talents Program of China**

22

Institute of Advanced Control Technology

大连理工大学
Dalian University Of Technology

Thanks for your attention & comments!

Lab website: http://act.dlut.edu.cn/

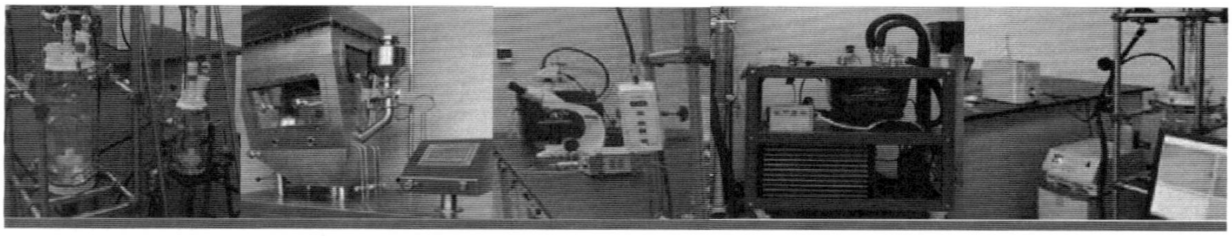

491432 Progress in Industrial Implementation of Optical Coherence Tomography (OCT) for Real-Time Evaluation of Pharmaceutical Coating Processes

Abstract:

The coating thickness of modified release coatings, enteric release coatings or coatings containing active pharmaceutical ingredients (APIs) is of biopharmaceutical importance. For modified release coating the coating thickness directly affects the dissolution profile. Thus, methods for in-line measurement of coating thickness can bring significant benefits with regards to end-point determination and monitoring of process trajectories.

Typical established methods to measure the coating thickness are weight and diameter gain of a limited number of drawn samples. In-line measurements mainly focus on spectroscopic techniques (NIR and Raman) or particle size distribution (PSD) analysis of beads. An emerging technology to measure the coating thickness is optical coherence tomography (OCT). OCT has several advantages compared to the state-of-the-art as it provides a direct measurement of coating thickness, only based on refractive index, not depending on chemometrics or on PSD statistical analysis. Additionally, OCT can deliver information of mean coating thickness and uniquely its variation within single samples and between the samples of a batch.

Optical coherence tomography has been presented as a proof-of-concept prototype at several conferences and papers are dealing with the data evaluation algorithms. This talk will focus on the application of an OCT system ready for industrial implementation in drum coating of tablets and fluid-bed coating of beads. Challenges in an industrial setting are for example the eccentricity of the drum, compensation of dispersion by thick process windows and the vast variation of coating materials, leading to different appearances of the OCT image.

The potential of OCT for monitoring and control of pharmaceutical coating processes is shown by in-line and at-line data for tablets and beads.

The authors would also like to acknowledge the contributions of Manuel Zettl, Philipp Jocham, Gregor Hauseder, Lukas Pichler and Michael Weissensteiner for experimental work and software development.

490572 Advanced Control Strategy Approaches for Continuous Pharmaceutical Manufacturing: The Regulatory Landscape

Abstract:

The FDA believes that strategic efforts to encourage and sustain improvements in pharmaceutical manufacturing quality are critical to the FDA mission of making safe, effective and high-quality drug products consistently available to the patients who need them. Over the past decade since the publication of the PAT guidance, there has been significant progress in modernizing the pharmaceutical manufacturing sector. There has been a general move away from fixed uni-variate parameter specifications towards a multivariate design space approach and the adoption of in-process measurement systems to aid process development. At the same time industry and regulators are being confronted with and increasing complex challenges including unprecedented drug recalls and shortages. The implementation of advanced control strategies can help address these challenges by increasing the efficiency, flexibility, and robustness of manufacturing processes. In FDAâ€™s experience to date, continuous pharmaceutical manufacturing is serving as a driving force for modernizing control strategy approaches including model based control, multivariate monitoring, analysis of large of data sets, and Real Time Release Testing (RTRT). Process models that can capture the dynamic relationships between process parameters, raw materials, and product attributes in a quantitative and predictive manner support the development and implementation of these advanced control strategy approaches. While there are no regulatory barriers for adopting advanced control strategy approaches, there still may be hurdles encountered when fitting these approaches into the existing regulatory framework. In this presentation, scientific and regulatory considerations for advanced control approaches designed to mitigate identified risk areas are discussed including performance criteria and lifecycle considerations. Continuous drug product and drug substance manufacturing case studies are utilized to highlight these considerations.

495150 Pharmacy-on-Demand: A Continuous Revolution in Pharmaceutical Manufacturing?

Abstract:

Unlike its counterparts in the petrochemical, food and drink industries; pharmaceutical manufacturing still applies a non-continuous or â€œbatch-wiseâ€ approach. Thus, current pharmaceutical processes are ill-equipped to react to market forces or surges in demand due to humanitarian outbreaks. Active pharmaceutical ingredients (API) are world travelers with complex supply chains and long batch syntheses, meaning it can take months to manufacture and deliver the final drug product to the patient. This enormous space/time demand can result in reduced profits and inflated capital and operating expenditure.

With the era of blockbuster drugs apparently over and Pharmaceutical companies racing to the bottom of the pricing market in recent times, a paradigm shift towards continuous manufacturing may provide the answer. The benefits of flow chemistry in the synthesis of API has now been successfully demonstrated on a number of occasions. Likewise, many Drug Product unit operations are currently conducted in a continuous manner. The present challenge is now the implementation of an end-to-end manufacturing strategy that can allow these methodologies to work in concert, in a streamlined manner that enables drug substance and drug product production in one manufacturing platform.

Pharmacy on Demand (POD) seeks to address these manufacturing inefficiencies through the use of a compact and reconfigurable platform, capable of multi-step syntheses for a catalogue of API. Substantial advances in complex reaction telescoping, real time formulation, reaction engineering and pumping technologies have enabled the design, construction and implementation of the platform. Tight control of process inputs and the implementation of in-line process analytical technologies enables high volumes of APIs (Atropine, Benadryl, Ciprofloxacin, Diazepam, Doxycycline, Lidocaine and Ibuprofen) to be manufactured over a three-week campaign period. This timeframe involved the synthesis, recrystallization and formulation of each API to their correct dosage form. These solid or liquid formulated Drug Products are produced to a standard that passes all USP specifications.

496111 Platforms for Integrated Continuous Drug Subtance Design and Manufacture

Abstract:

As pressure grows on pharmaceutical companies to deliver increasing levels of process performance, reduced capital and manufacturing costs the industry is looking more and more to continuous processing to help tackle some or all of these challenges.

Within this presentation case studies will be presented where unique continuous processing hybrid platforms have been utilised for the development and operation of continuous processes. These processes included heterogeneous and homogenous chemistries, work-up/distillations, crystallization and isolation. Specifically within the case studies presented details will be outlined, via PAT integration, engineering kinetic and statistical models, experimental and lab rig design via 3D printing, required to enable the design, optimisation and operation of these platforms . During the presentation the design considerations over CSTRs, PFRs, hybrid or novel 3D printed reactor design configurations will be presented and how such design selections can impact or tailor the design properties of the intermediates and importantly the final API solid state characteristics.

496380 Residence Time Distribution and Material Traceability for Continuous Drug Substance Processes

Abstract:

Title: Residence Time Distribution and Material Traceability for Continuous Drug Substance Processes

Authors: Christopher Polster[1], Venkata Ramana Reddy[3], Carla Luciani[1], Stephen Jeffery[2], Martin Johnson[1], Kevin Chinn[4], Hod Finkelstein[4]

[1] *Eli Lilly & Co., Indianapolis, IN-46221, USA*

[2] *Eli Lilly S.A. Irish Branch, Dunderrow, Kinsale, Co. Cork, Ireland, P17 NY71*

[3] *Lilly Capability Center India, Bangalore, India*

[4] *Trutag Technologies, Inc., 2045 Lauwiliwili St, Unit 301, Kapolei, HI 96744*

Abstract

The concept of residence time distribution (RTD) to analyze chemical reactor performances appeared earlier 1930â €™s.[1] Danckwerts materialized the field by defining the most common RTDs.[2] Currently, the use of RTDs to understand mixing characteristics of the units/trains as well as material traceability and batch definition is an established approach.

With the rapidly growing adoption of continuous manufacturing processes in the pharmaceutical industry, the understanding of RTD through modeling and characterization is imperative to support both R&D activities and long term commercial manufacturing of both drug substance and drug product.[3],[4],[5]

In this work, some of the techniques used to measure RTD are described. Also, libraries were developed to provide a tool for R&D and manufacturing scientists to develop and simulate plant flowsheets that can be used for material traceability, determining diversion boundaries, and guide decision trees for diversion scenarios. Specific case studies will be shared.

---

[1] MacMillin, R.B., Weber, M. The Theory of Short-Circuiting in Continuous-Flow Mixing Vessels in Series and the Kinetics of Chemical Reactions in such Systems. *Transactions of American Institute of Chemical Engineers* **1935**, 31,

409-458.

[2] Danckwerts, P.V. Continuous Flow Systems. Distribution of Residence Times. Chem Eng. Sci. 1953, 2, 1-13.

[3] Manley, L. Manufacturing Platformâ€™s Residence Time Distribution Using Major and Minor Unit Formula Excipient Step Changes and Soft Sensors. *2016 AIChE Annual Meeting* (342c).

[4] Garcia-Muñoz, S., Luciani, C., Shi, Z., Roth, W., Butterbaugh, A. Hanson, J., Manley, L., Barnes, L., Leavesley, I., Almaya, A. Model-Aided Development of a Continuous Drug Product Manufacturing Process.

[5] Engisch, W., Muzzio, F. Using Residence Time Distribution (RTDs) to Address the Traceability of Raw Materials in Continuous Pharmaceutical Manufacturing. *J. Pharm Innovation* **2016**, 11, 64-81.

494238 A Miniature CSTR Platform for Continuous Processing of Multiphase Systems

Abstract:

Continuous manufacturing has emerged as a powerful tool for producing fine chemicals (e.g. active pharmaceutical ingredients (APIs)) over the past decade. However, numerous important reactions in pharmaceutical industry involve multiphase interactions, which create challenges for flow chemistry to gain a wider acceptance. For example, the presence of stoichiometric amount of solids as reagents, by-products, or products leads to the issue of blocking a flow reactor, and the problem is more severe for the milli-scale or micro-scale flow systems. Besides, the mass transfer resistance at the phase boundaries sometimes becomes the rate-limiting step. Considering current widely used tubular reactors, only providing limited interfacial interactions imposed by the laminar flow pattern, a more universal and versatile reactor design is desired for handling solids in flow, generating well-controlled mixing performance, and at the same time, offer compatibility to withstand harsh chemical environment.

Here, we present a new miniaturized continuously stirred-tank reactor (CSTR) for continuous processing of multiphase systems. The design, with the feature of plug-and-play, takes efficient mass and heat transfer, chemical compatibility, and ease to fabricate and clean into account. Single-phase residence time distribution (RTD) measurements of the CSTR unit reveal nearly ideal CSTR mixing. It also has the capability to tune the RTD by connecting several units in series to achieve approximate plug flow reactor (PFR) RTD, which gives the flexibility to optimize the selectivity based on the reaction kinetics.

Two versions of miniature CSTRs were developed for: (i) solid handling in flow, and (ii) high-performance liquid-liquid mixing.

Solid handling CSTR: employing the non-stick polytetrafluoroethylene (PTFE) as the reactor body material with minimized inter-unit connection enables the CSTR to continuously handle high-concentrated slurries for hours without significant signs of clogging. Two solid-forming reactions, (i) glyoxal reacting with cyclohexylamine to form N,Nâ€²-dicyclohexylethylenediimine, and (ii) sulfonylation of 2-octanol with methanesulfonyl chloride, demonstrate its practical usefulness as a powerful for flow chemistry field.

High-performance liquid-liquid CSTR: incorporating the magnetic-coupling as the driving force to provide external agitation inside the confined reactor space gives the possibility to generate extraordinary two-phase mixing performance without losing the capability to seal the reactor. We demonstrate the reactor with a comparable liquid-liquid mass transfer coefficient to Corning Advance-flow Reactor, which is the state-of-the-art among commercial reactors. More importantly, it gets around the problem of coupling of flowrate and mixing, and is capable to accommodate reaction kinetics from seconds to hours. Phase transfer catalyzed reactions were demonstrated in this newly designed CSTR, with the corresponding kinetic model developed for Makosza phase-transfer mechanism. In addition, the success of carrying out biphasic Suzuki-Miyaura reactions further validates its suitability for the pharmaceutical applications.

Grams of API to Commercial-Scale Continuous GMP Drug Product Manufacture in Less Than a Year

Carl Hartmann

Lilly

Phase I Clinical Supply Options

Phase I Clinical Demand: <u>initial</u> forecast **<u>200,000+</u>** doses

Drug-in-capsule
Output: 200 capsules / hr
Consider: RBA study, low throughput

Small-Scale Development
Req: 250 g API

Continuous Tablet Dev
Req: under 2 kg API
Output: 50k-100k /hr

CT Manufacture Decision

?

Time

2

Small-Scale Development Defined

- Dispense, Blend, Clean

Lubrication Scale Model

3+ hrs and 200-10,000 g API → 30 min and 1-20 g API

- Setup, Compress, Clean

Intrinsic Material Properties Models

3+ days → 1 hr

3

Small-Scale, but Big Data

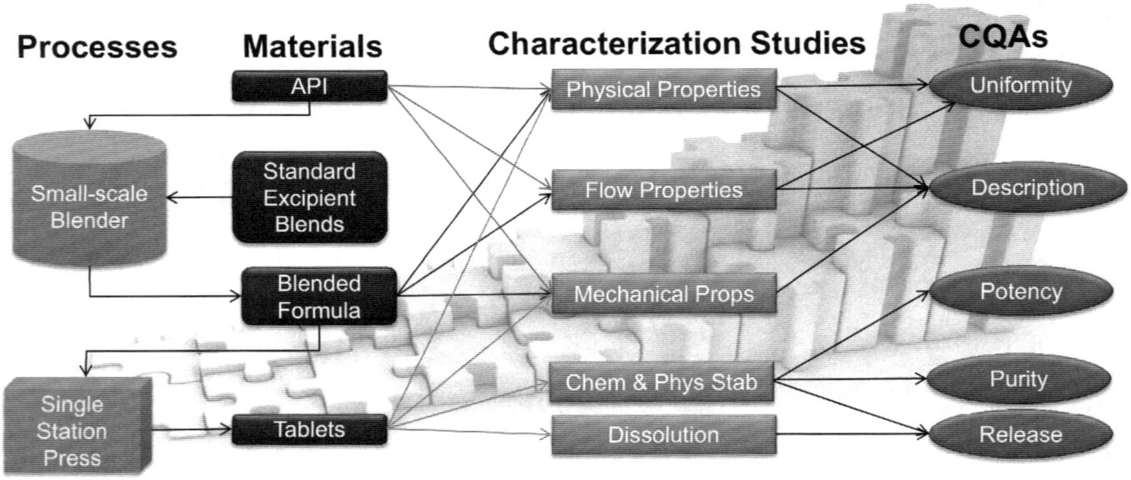

4

Small-Scale Development Blending

DS Lot	X90 (um)	Hausner Ratio	FFC (flow) values
A	192.4	1.41	4.10
B	179.8	1.46	9.18
C	129.7	1.41	4.86
D	135.7	1.34	4.84
E	72.2	1.56	2.85

Five drug substance (DS) lots of varying particle size distributions (PSD), each 50 gram quantity

Each DS lot blended into a common formulation for comparison, each blend targeting 50 gram quantity

Material Name	Common Formulation (%)
Drug Substance	28
Lactose Monohydrate	16
Microcrystalline Cellulose	48.5
Superdisintegrant	5
Lubricant	2.5

5

Small-Scale Development Tablets

- Standardized tablets:
  - 6 mm round, flat, 100 mg target weight

A preliminary blend at 42% drug load (DL). Observation of picking/sticking.

Reduced drug load

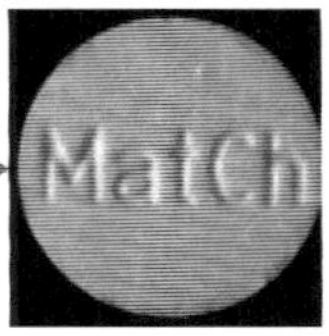

Comparison of five differing PSD DS lots conducted in blends at 28% DL. No visual picking/sticking.

- Single station instrumented press
- Compression profile and stability testing

6

Initial Tablet Results - Common Blend

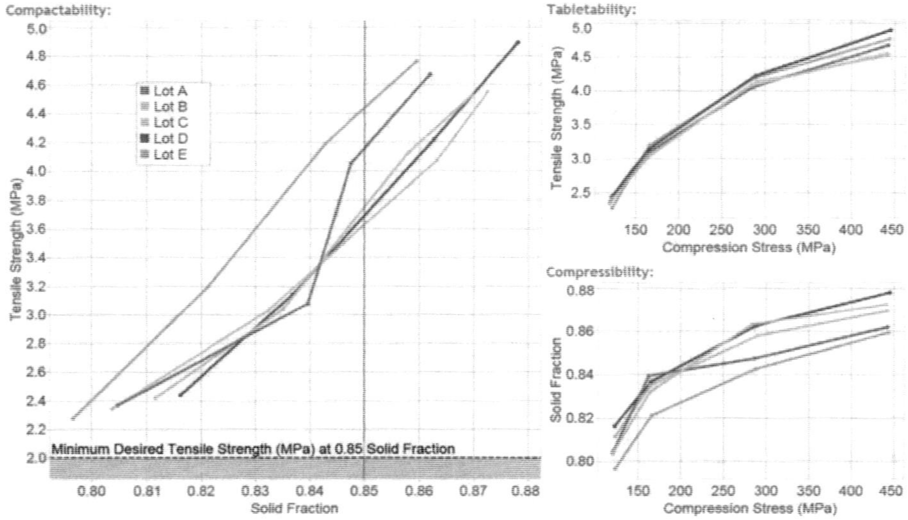

7

Adjustments and confirmation

- Initial risk assessment:
  - Smallest API, poor flow
  - No other red flags
- Formulation development:
  - Reduce drug load
  - Eliminate lactose

8

CM Direct Compression Flow Diagram

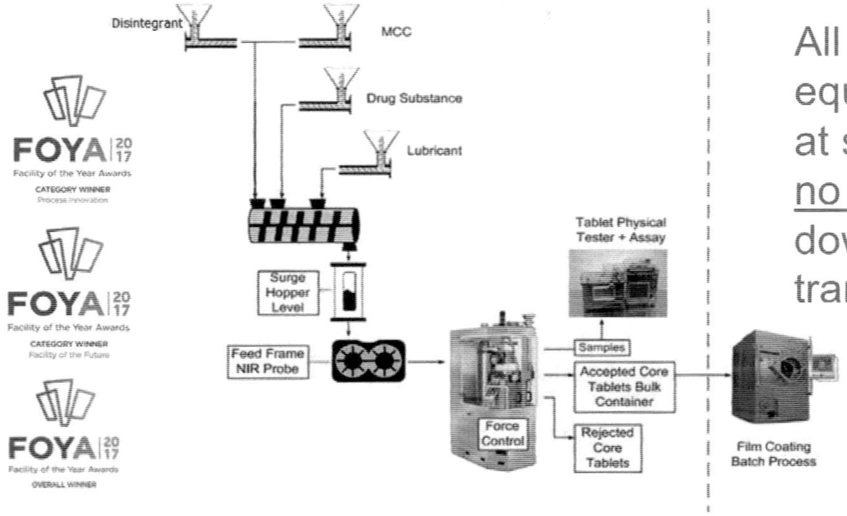

All CM equipment trains at same scale; no scaling up or down with tech transfer

9

CM Direct Compression Animation

10

Leading into Demonstration Batches

- Off-line **feedability** (LIW feeder) for DS:
  - For twin-screw tooling selection
- Spectroscopy: **NIR** signal range of DS and DS blends
  - For tablet press feed frame NIR on-line monitoring
- Select **tablet tooling** shapes and sizes appropriate for CT
- Identify **coating** system and color appropriate for CT

5 mg strength
6 mm round
Pink coat

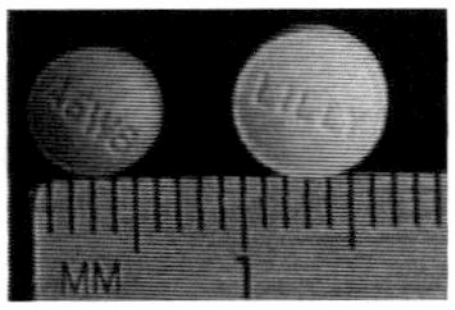

25 mg strength
7 mm round
Yellow coat

11

Development Demonstration Batches

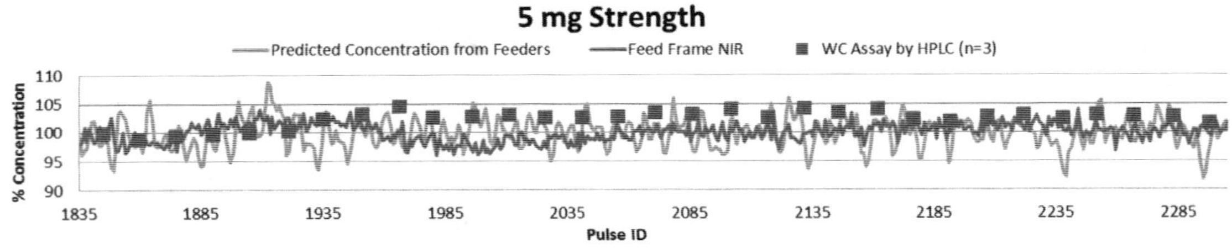

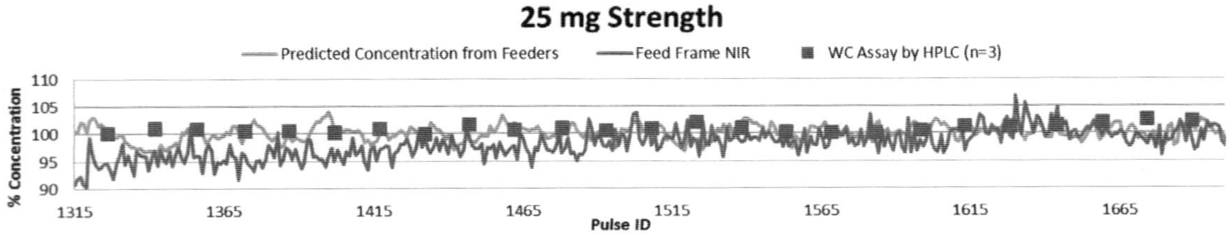

Less than 2 kg API required in total for both batches

12

Phase I Clinical Supply Decision

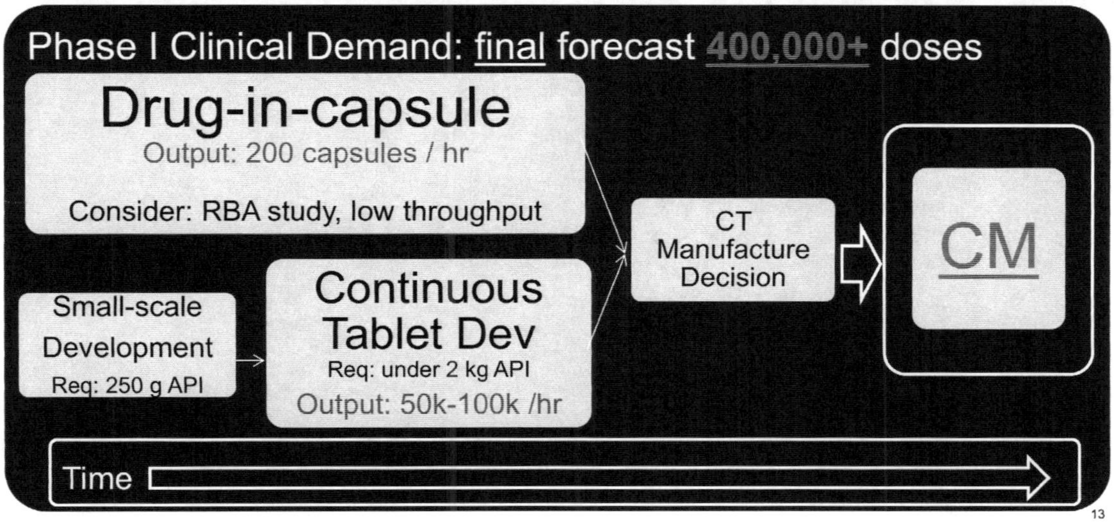

13

GMP Commercial Site Manufacture

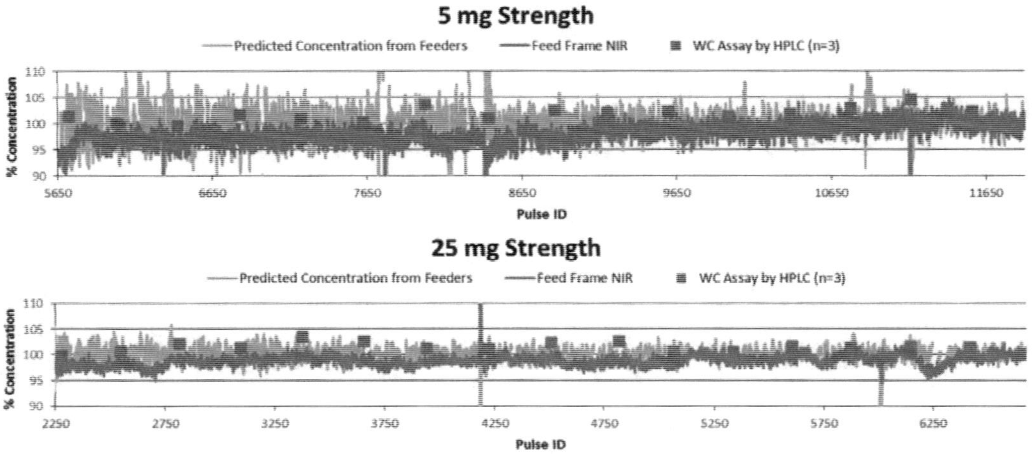

Both batches readily meet USP <905> and ASTM 2810 criteria

14

Timeline

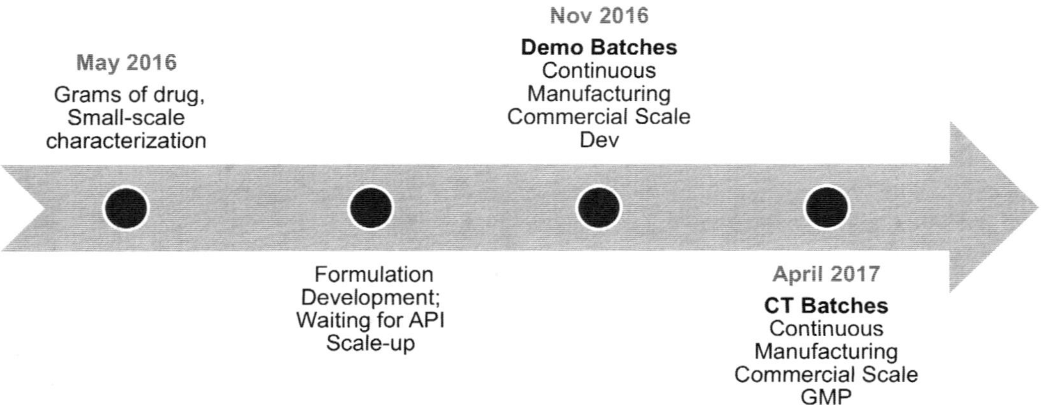

Grams to GMP commercial-scale in less than a year!

15

Conclusions

- Small-scale characterization and continuous manufacturing enables **rapid development** of a commercial-scale formulated tablet
- CM platform allows for **flexibility** in delayed decisions on supply requirement
- Early development work assures feasibility of CM platform to **buy down risk** with little API

16

Acknowledgements

Mark Polizzi
Dave Myers
Mike Miller
Pete Shi
Adam Butterbaugh
Josh Hanson
Melissa Keeney
Joe Martinelli
Adam McFarland
Moussa Boukerche
Dale Greenwood
Kirk VanderKamp
Tim Kramer
Sarah Schmitt

17

Questions?

18

487680 A Continuous and Controlled Pharmaceutical Freeze-Drying Technology for Unit Doses

Abstract:

Driven by growing needs in the biopharmaceutical market and regulatory pressure, a continuous and controlled freeze-drying technology for unit doses to preserve biopharmaceuticals has been developed. Such continuous process allows a more efficient, cheaper, greener and controllable manufacturing method compared to traditional batch production systems, offering competitive advantages and business opportunities.

Pharmaceutical freeze-drying (lyophilization) is a low-temperature drying process in which aqueous solutions of heat-labile biopharmaceuticals are converted into solids with sufficient stability for distribution and storage. Similar to all manufacturing processes of drug products (solids, semi-solids and liquids), conventional pharmaceutical freeze-drying is generally accomplished using batch processing that is considered time-consuming, costly, non-flexible and lacking robust quality control and real-time release.

Four major industrial drivers are demanding a more efficient and better controllable pharmaceutical freeze-drying technology for unit doses: cost-cutting, regulatory pressure, a fast growing biopharmaceutical market and an ageing population requiring more personalized medicines.

The continuous and controlled freeze-drying technology, developed following the principle of model based design, offers clear advantages over current batch production such as cost reduction (up to 50%), track-and-trace product quality control, and a significant reduction of processing time (> 40 times faster, e.g. 1 hour instead of 5 days at a vial level), reduced need for clean room and a substantial sustainability gain.

489155 Engineering Considerations for Dissolved Oxygen Sensor Response Across Silicone Membranes in 3-L Single Use Bioreactors

Abstract:

Single use bioreactors are becoming standard in biopharmaceutical development and manufacturing because they reduce capital expenditure, eliminate cleaning complications, and decrease down time between batches. Several successful bioreactor designs are available, but sensor integration poses different challenges between scales. For example, large-scale cultivations require a high degree of sensor fidelity, typically using costly disposable adaptors to sterilize and connect traditional reusable probes. In contrast, high throughput bioprocessing values cheaper sensor technology, often trading off robustness. With bench-scale bioreactors, a gap exists because their use in process characterization and scale-down studies still demands accuracy, precision, and longevity, but at a broad experimental scope that doesnâ€™t allow for high unit cost per run.

If industry leaders want the advantages of single-use vessels as well as the proven reliability of reusable sensors for 3L bioreactors, operators need to sterilize probes separately for aseptic insertion in a hood. To avoid the need for manipulation of dissolved oxygen (DO) sensors, we use a thin silicone membrane as a gas-permeable sterile-barrier. In doing so, we retain the measurement integrity in addition to the economic advantages of reusable probes.

This presentation discusses engineering considerations for DO sensor response when impeded by a secondary membrane. We applied first principles to separate the additional delay from overall probe response, quantified maximum lag times, and demonstrated a robust design window for acceptable performance of mammalian cell culture. We are demonstrating the long-term functionality of this solution by implementing across our routine process development cultivations.

496074 Process Development Tool to Rapidly Predict the Stability of Biopharmaceuticals

Abstract:

Process Development Tool to Rapidly Predict the Stability of Biopharmaceuticals

S. Hedberg, Y. Hutabarat, J. Heng, J. Haigh, D. Williams

ABSTRACT

Protein aggregation is a critical problem for the safety of biopharmaceuticals as they are linked to adverse immunologically related responses in patients. Much effort has been made to gain a better understanding of aggregation, however, the mechanisms leading to protein aggregation are still not fully understood. Protein-protein molecular interactions in solution are known to be involved in protein solution aggregation behaviour and are a common issue for the manufacturing of biopharmaceuticals such as monoclonal antibodies, mAbs.

Therefore, a major industrial and academic challenge is the development of fast and reliable methods, either theoretically or experimentally driven, which allow formulation scientists to identify optimal biopharmaceutical species and solution conditions which deliver stable solution products, thus avoiding aggregation. The osmotic second virial coefficient (B_{22}) is a fundamental physiochemical property that describes the thermodynamics of the protein-protein interactions in solution, which can be a useful tool for predicting the aggregation propensity of proteins.

One way of predicting aggregation propensity is self-interaction chromatography (SIC) that can be used for determining the osmotic second virial coefficient (B_{22}). This technique has recently demonstrated to be a tool that can successfully predict phase behaviour of proteins. However, due to the need to immobilise every protein in order to make B_{22} measurements makes this technique less ideal for use in process development. Another technique, cross-interaction chromatography (CIC), has shown to be an even more high-throughput technique than its predecessor with similar capabilities. In CIC only one protein needs to be immobilised in order to screen a range of other proteins.

In a case study a number of different monoclonal antibodies from both IgG1 and IgG4 subclasses, and a polyclonal antibody were immobilised and screened on each of the columns for a wide range of process development conditions. The aim was to see if the osmotic second virial cross coefficient (B_{23}) between two proteins can be used to predict B_{22}.

The results of the case study were very successful and proved to be a good prediction of the B_{22} value for all the proteins in all conditions. This will enable a rapid process development tool with only mg quantities of biopharmaceutical species that can be injected on a generic column e.g. a polyclonal IgG column.

Previously, the accuracy of the B_{22} values of several of the mAbs has been directly validated by the use of traditional gel filtration chromatography methods and dynamic light scattering, to investigate the size and quantity of aggregates over time. This has shown that the determined B_{22} held powerful information about the long term stability of biopharmaceuticals.

In summary, this work has developed a rapid and high-throughput tool, which manages to estimate the B_{22} values for different biopharmaceutical products using minimal amounts of protein. This will help scientists in the biopharmaceutical industry finding suitable conditions for process development as well as picking drug candidates with high long term stability.

NATIONAL INSTITUTE OF CHEMISTRY

Paper number ID# 489514

Model-Based Process Optimization for Upstream Bioreactor Production–Downstream Chromatographic Separation of CHO Cell Monoclonal Antibodies (mAb)

Uroš Novak, Miha Kastelic, Drejc Kopač, Miša Mojca Cajnko, Andrej Pohar, Matic Grom and Blaž Likozar

uros.novak@ki.si, blaz.likozar@ki.si

Minneapolis, 1.11.2017

next BioPharm DSP

BioPharm.Si

REPUBLIKA SLOVENIJA
MINISTRSTVO ZA IZOBRAŽEVANJE, ZNANOST IN ŠPORT

EVROPSKA UNIJA

Research areas of the department of Catalysis and Chemical Reaction Engineering

ADREM

ADREM (Spire, Horizon 2020)
Adaptable Reactors for Resource- and Energy-Efficient Methane Valorisation

Carbon dioxide activation

MefCO2 (SPIRE, Horizon 2020)
Methanol synthesis from captured CO2 with the use of surplus electricity from renewable resources

FReSMe

FReSMe (Horizon 2020)
From residual steel gases to methanol

Methane activation & conversion

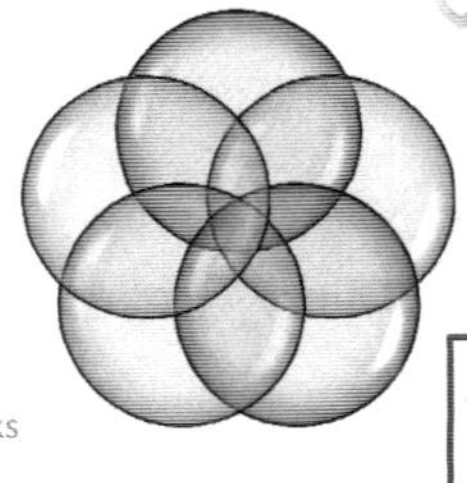

Hydrogen & fuel cells

NATO OTAN

NATO (SPS Programme)
Enhanced portable energetically self-sustained devices for military purposes

MarineBiotech

Biomass-derived building blocks

Interreg ITALIA-SLOVENIJA

CELKROG

Pharmaceutical process engineering

Biopharm.si :
New generation of biological drugs

nextBioPharmDSP (LEIT, Horizon 2020)
Next-generation Biopharmaceutical Downstream Process

https://www.ki.si/en/odseki/d13-department-of-catalysis-and-chemical-reaction-engineering/

Outline

- ❑ Introduction and aim of the work
- ❑ Upstream process:
  - mechanistical model describing the metabolism of CHO cells
  - metabolic flux analysis (MFA)
  - modeling of N-glycosylation
- ❑ Downstream process:
  - affinity chromatography model development
  - parametric sensitivity study
  - different options of operation (loading strategy)
- ❑ Conclusions

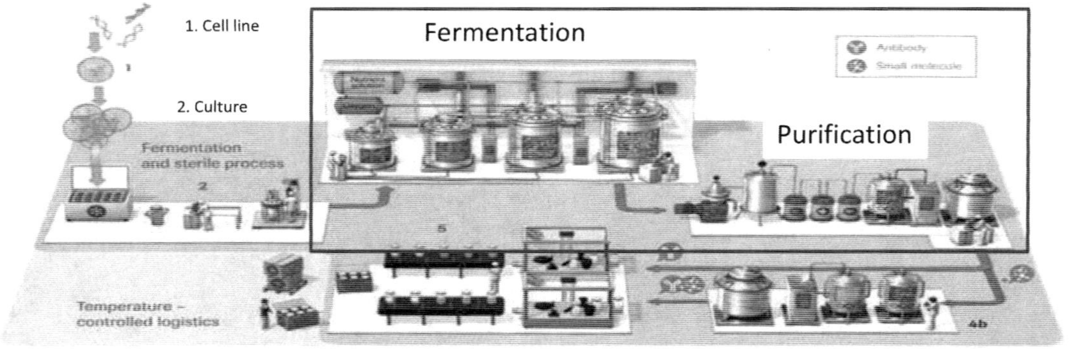

http://www.roche.com/dam/jcr:045db96e-6a0d-48ea-b37c-8ed42e5a59e7/en/supply_resp_2.jpg

Introduction

"The process is the product."

BioPharm.Si 2016-2020

The main goal of BioPharm.Si programme, involving high tech and biotech Slovenian companies and excellent research institutions, is development of continuous manufacturing processes and innovative devices and services that will facilitate transition from batch to fully continuous manufacturing.

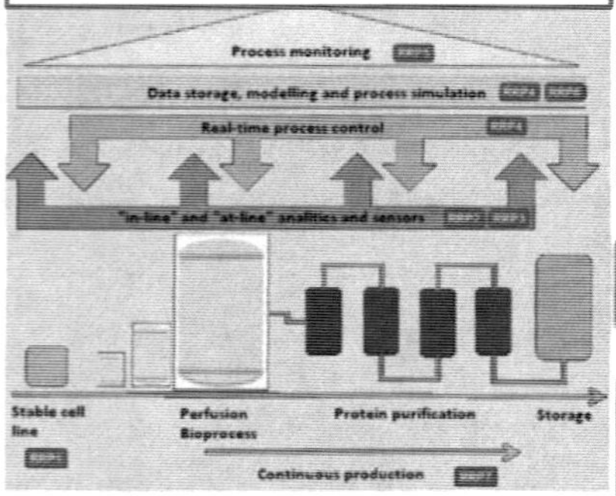

More info at www.biopharm.si/

Aim

Mechanistical mathematical modelling as a tool in biopharmaceutical development

Motivation

Systematic statistical evaluation of the data → Enhancing product quality control

Smart planning of experiments (analytics & formulation)

New process design and optimization → Reduce costs and increase productivity

nextBioPharmDSP 2015-2019

Implementation of a fully integrated manufacturing platform for biosimilar mAb based on continuous chromatography, in combination with single-use disposable technology for all unit operations of DSP on pilot/small production scale together with incorporation of advanced analytical tools.

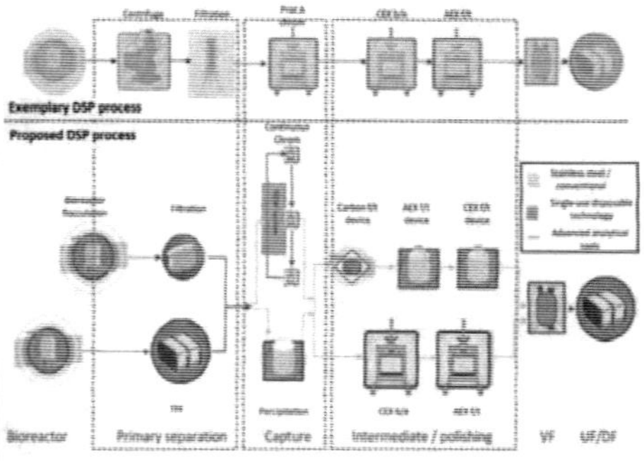

More info at http://nextbiopharmdsp.eu/

Upstream process - mechanistical model describing the metabolism of CHO cells

- Mathematical description of a series of bioreactions that have an impact on the dynamics of cell growth, metabolism and the production of intracellular products.
- Building a global model with a multi-model approach that takes into account the growth stages of the cell culture and could successfully predict the development of experimental data.
- Optimizing bioprocess *in-silico* through qualitative and quantitative insight into mechanisms through which conditions in bioprocess affect the glycosylation of therapeutic proteins.

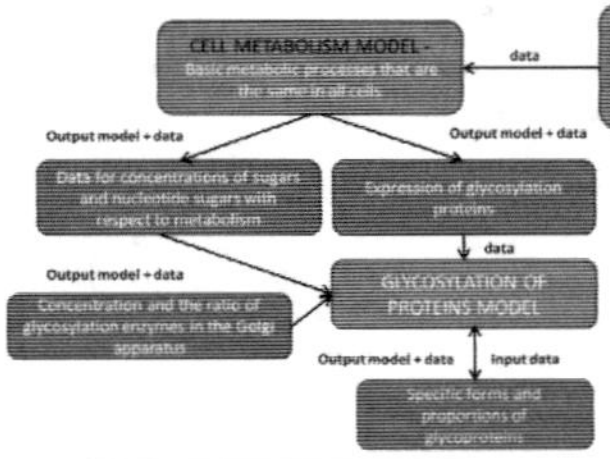

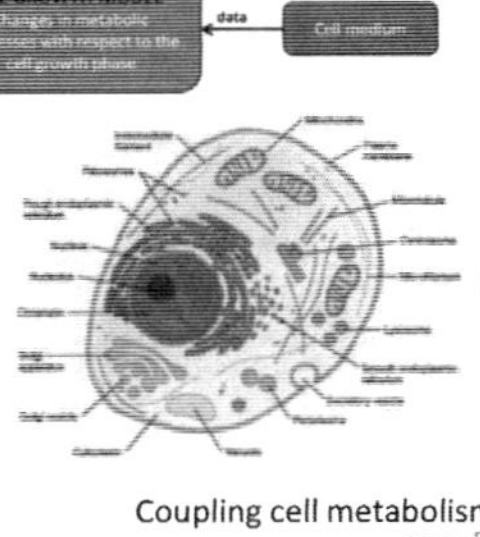

MFA- Methodology for quantification of metabolic pathways from limited number of experimental data

Scheme of metabolic flows in the CHO cell

Complex

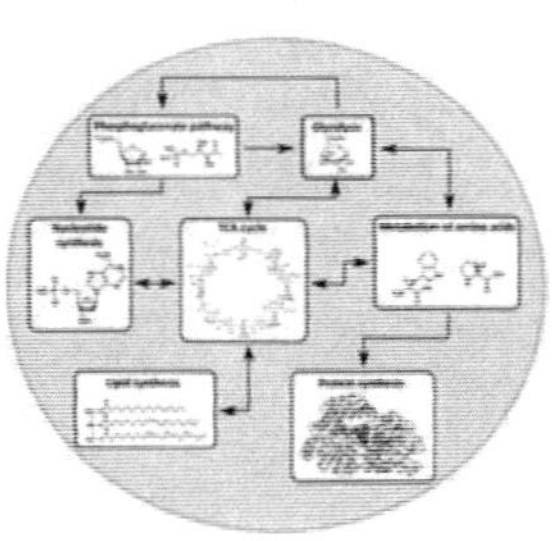

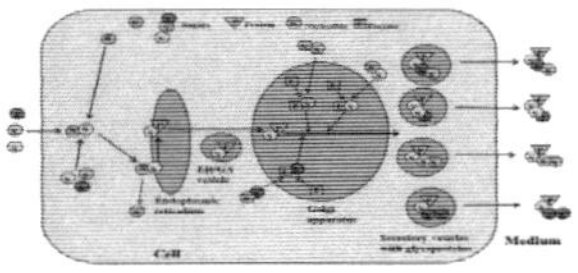

Coupling cell metabolism with N-glycosylation

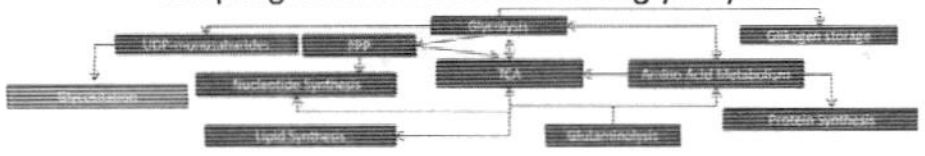

Metabolic flux anaylsis (MFA)

- The MFA allows you to calculate the value metabolic fluxes (v) for the selected metabolic scheme.
- Measured values of selected fluxes (typically: consumption and release of extracellular metabolites).
- The MFA allows you to shrink metabolic schemes to the several macro reactions.
- Easier modeling of reactions kinetics .

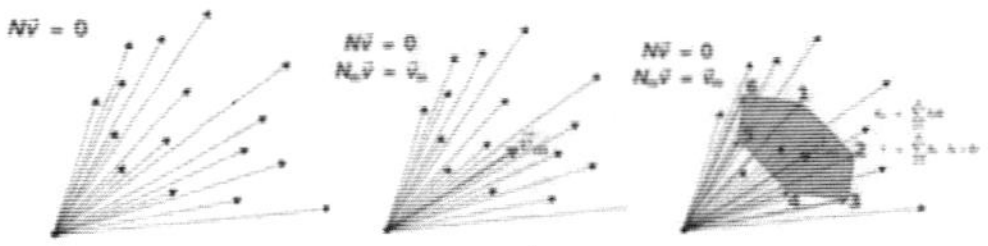

106 reactions → 18 Macroreactions

- Possibilities of addition of substrates and removal of products
- Dynamic profiles are available for each selected compound in metabolism

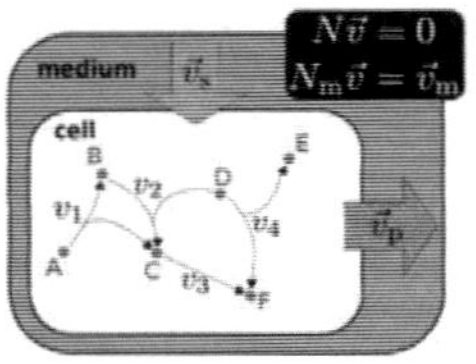

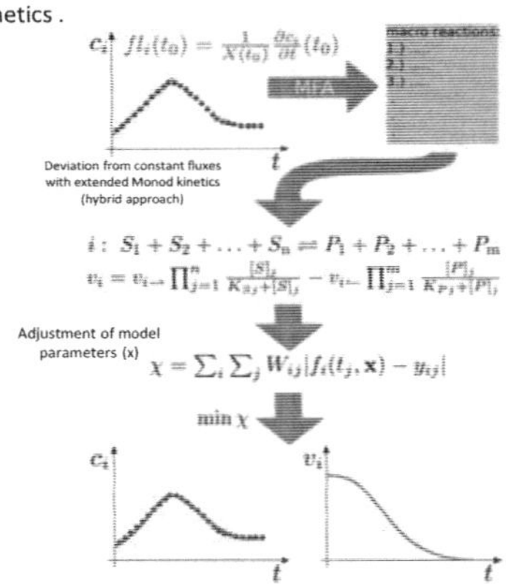

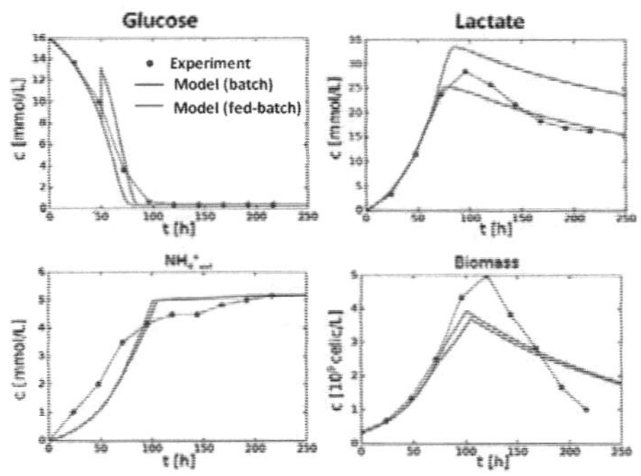

$\vec{N}$ - stoichiometric matrix

$\vec{N_m}$ - flux vector for reactions

$\vec{v}$ - matrix of measured metabolites

$\vec{v_m}$ - vector of measured fluxes

$\vec{v_{p.s}}$ - fluxes for membrane transport

Experimental data; Zamorano F., Wouwer A.V., Jungers R.M., Bastin G., J. Biotechnol., 164, 409-422, 2013

Modeling of N-glycosylation

- Glycan binding rules determine the decision trees or N-glycosylation pathway
- calculate the rates (rj) at each step of the tree, and choose the most likely path in a balanced way
- model one transition of the initial glycan m9 through Golgi apparatus, repeat 100 times (statistics)

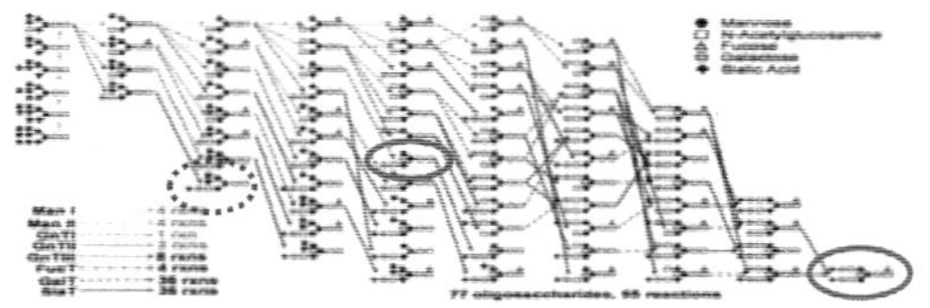

Jimenez del Val, et al., Biotechnol Prog., **27(6)**, 1730, 2011

I. Glycosylation depending on the position in Golgi, we take into account the enzyme concentrations vary according to the location in Golgi

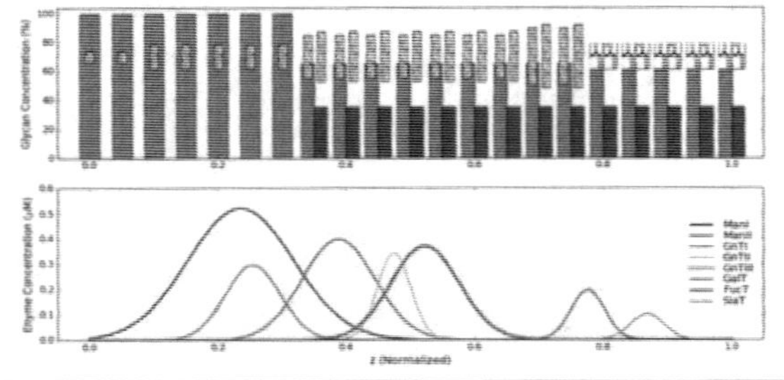

II. Finding parameters according to the final distributions of individual glycan

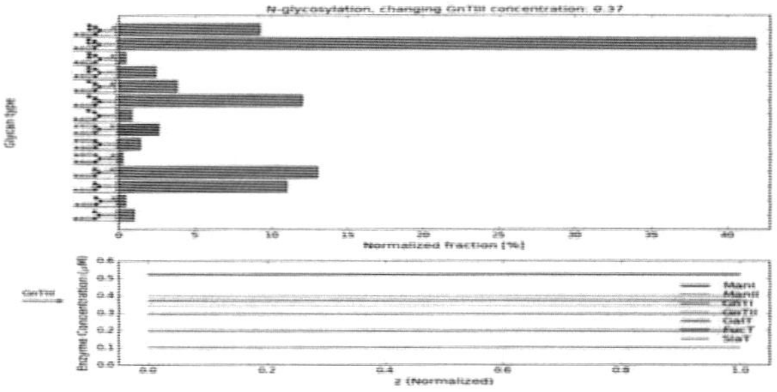

Coupling of cell metabolism with N-glycosilation

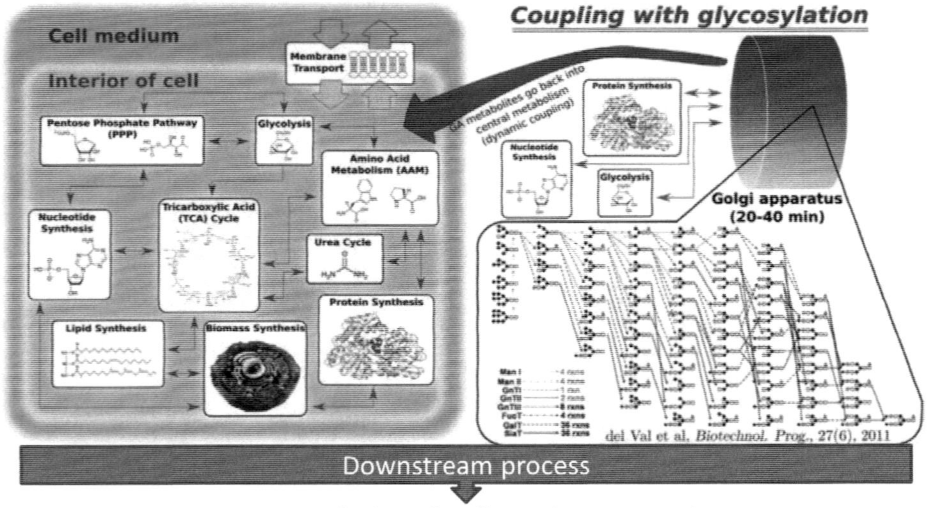

Capturing of gylcans by affinity chromatography

Downstream process – affinity chromatography

Batch affinity chromatography operates in cyclic mode.

1. loading of mAb on equilibrated affinity sorbent contained in chromatography column,
2. washing of column to remove impurities,
3. elution to desorb mAb bound on sorbent and
4. equilibration/regeneration step to prepare the column with affinity sorbent for next cycle.

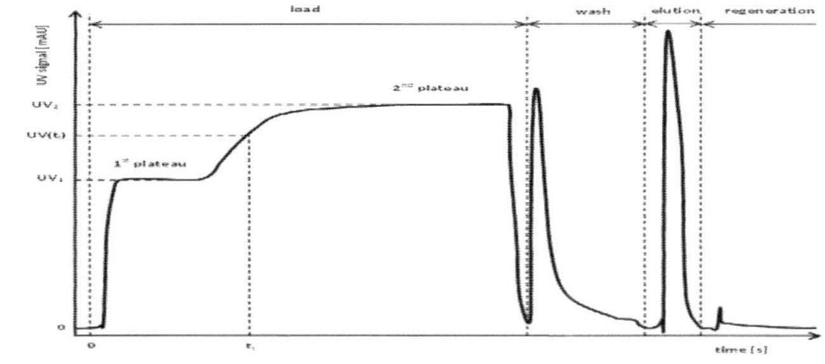

E. von Liers, J. Andresson, Computers and Chemical Engineering 34 (2010) 1180-1191

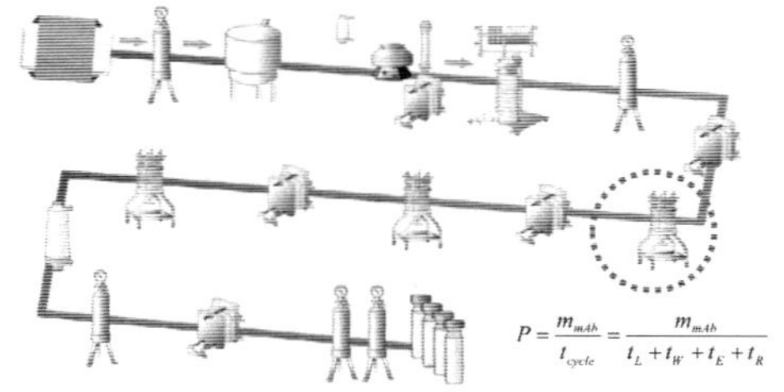

$$P = \frac{m_{mAb}}{t_{cycle}} = \frac{m_{mAb}}{t_L + t_W + t_E + t_R}$$

Affinity chromatography is an effective method to capture and purify mAb

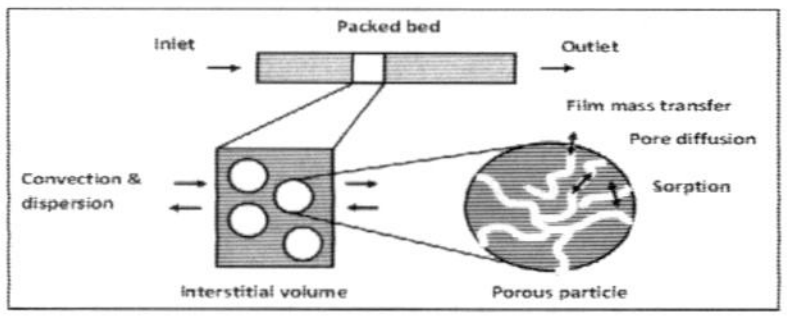

Mathematical model formation

Mass balance states that antibody accumulated in some (infinitesimally) point of bulk liquid is equal to sums of convection and exchange with affinity resin particles and liquid.

$$\varepsilon_c \frac{\partial c}{\partial t} = -u\varepsilon_c \frac{\partial c}{\partial z} + \varepsilon_c D_{ax} \frac{\partial^2 c}{\partial z^2} - (1-\varepsilon_c)\frac{3}{r_p} k(c - c_{p,s})$$

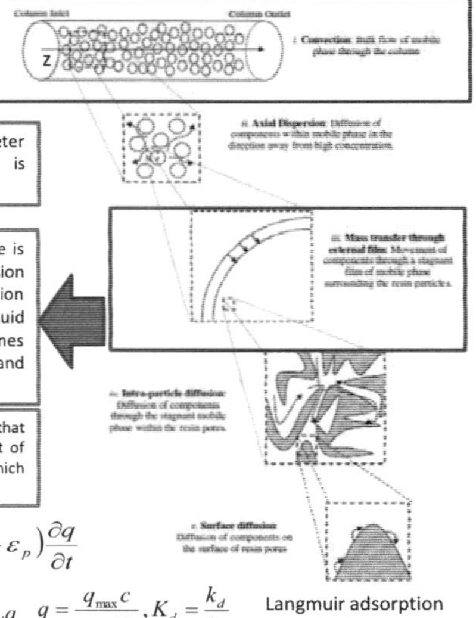

$$Sh = \frac{kd_p}{D_m} \quad Re = \frac{ud_p\rho}{\eta} \quad Sc = \frac{\eta}{D_m\rho}$$

$$Sh = \begin{cases} 1{,}85\left(\frac{1-\varepsilon_c}{\varepsilon_c}\right)^{0,33} Re^{0.33}\, Sc^{0.33}, Re < 40 \\ \frac{1{,}09}{\varepsilon_c} Re^{0.33}\, Sc^{0.33}, 40 \le Re < 55 \\ \frac{0{,}25}{\varepsilon_c} Re^{0.69}\, Sc^{0.33}, 55 \le Re < 1050 \end{cases}$$

If ratio of column length to particle diameter (L/dp) is above 350 axial dispersion is negligible.

When external film mass transfer resistance is negligible compared to internal diffusion resistance, particle surface concentration becomes equal to concentration in bulk liquid and diffusion at surface of particles determines mass transfer between bulk liquid and particles.

$$\varepsilon_c \frac{\partial c}{\partial t} = -u\varepsilon_c \frac{\partial c}{\partial z} + \varepsilon_c D_{ax} \frac{\partial^2 c}{\partial z^2} - (1-\varepsilon_c)\frac{3}{r_p} D_{eff} \left.\frac{\partial c_p}{\partial r}\right|_{r=r_p}$$

$$D_{ax} = \frac{d_p u \varepsilon_c}{0{,}339 + 0{,}033\, Re^{0,48}}$$

Mass balance for mAb in pores states that accumulation is equal to difference in transport of antibody by pore diffusion and reaction in which antibody binds to ligand within pores.

$$\varepsilon_p \frac{\partial c_p}{\partial t} = D_{eff}\left(\frac{\partial^2 c_p}{\partial r^2} + \frac{2}{r}\frac{\partial c_p}{\partial r}\right) - (1-\varepsilon_p)\frac{\partial q}{\partial t}$$

Boundary and initial conditions

$$c_p = c = c_{p,s}, r = r_p \quad \frac{\partial c_p}{\partial r} = 0, r = 0$$

$$c = c_0, 0 \le z \le L \quad c_p = c_{p0}, 0 \le r \le r_p, 0 \le z \le L$$

$$q = q_0, 0 \le r \le r_p, 0 \le z \le L$$

$$D_{eff} = K_p \frac{\varepsilon_p}{\tau} D_m \quad D_m = 2{,}74 * 10^{-9} M^{-1/3} \quad \frac{\partial q}{\partial t} = k_a(q_{max} - q)c_p - k_d q \quad q = \frac{q_{max} c}{c + K_d}, K_d = \frac{k_d}{k_a}$$

Langmuir adsorption isotherm

Parametric sensitivity study

Column and resin parameters (Mab Select SuRe and Mab Select SuRe LX) were taken from the literature (Hahn et al. 2005):

ε_c...column void fraction 0.36 [/]
ε_p...particle porosity 0.88 [/]
q_{max}...maximum binding capacity of affinity resin 5.47 [mol/m^3]
K_d...equilibrium desorption constant 7.7x10^{-4} [mol/m^3]
K_{fit}...parameter, taking resin particle pore diameter, antibody molecule diameter and pore tortuosity into account 0.083 [/]
c_{in}...concentration at piping system inlet 3.33 [mg/ml]
F... flow rate [m^3/s]
d_p...diameter of particles 85 [µm]
L...length of column 100 [mm]
D...column diameter 8 [mm]
D_{ax}...axial dispersion coefficient [m^2/s]
V_{PFR}...volume of plug flow elements [m^3]
V_{CSTR}...volume of perfectly mixed elements [m^3]

*R. Hahn, P. Bauerhansl, K. Shimara, C. Wizniewski, A. Tscheliessnig, A. Jungbauer, *J. Chromatogra. A*, 1093 (2005) 98-110

Changes at 80% of the initial value (%)		
Parameter	Mab Select SuRe LX	Mab Select SuRe
ε_c	14	3.3
ε_p	-13	-9.4
q_{max}	-19	-19
K_d	0.9	2.0
K_{fit}	-12	-8.7
d_p	24	12
L	-13	-9.6
D	-26	-21
k	/	3.9
D_{ax}	0.4	0.0
C_{in}	-1.5	-2.2
F	13	7.6
V_{CSTR}	0.0	-0.2
V_{PFR}	0.0	0.0

Dynamic binding capacity (DBC)
The amount of antibody that can bind to a chromatography media under flow conditions.
DBC_{10}...dynamic binding capacity at 10 % breakthrough [mg/mL]

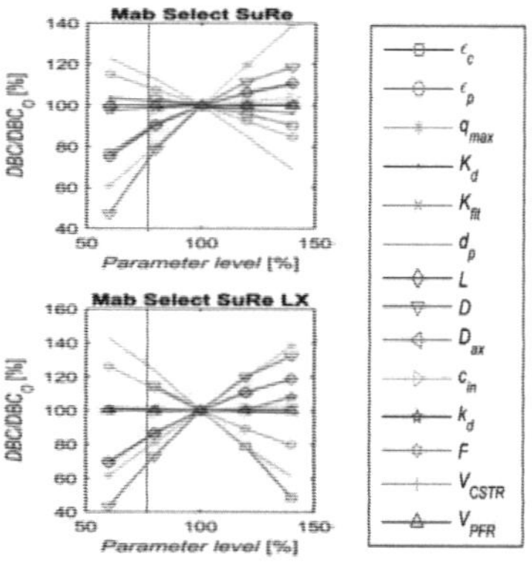

Different options of operation

Loading strategy	Constant flow rate	Flow ramp	Quadratic flow function	Reversed flow
Number	1	2	3	4
Description	Column is loaded with constant flow rate F until BT=0.6 % at which loading step ends.	Loading flow rate starts at value F_1 decreasing linearly to value of F_2 between times t_1 and t_2 until allowed BT=0.6 %.	Flow changes with quadratic function of time $F(t)=a_1+a_2t+a_3t^2$	Column is loaded with flow rate F_1 until time t_1 or BT=0.6 %. Then it is loaded from other side of column with flow rate F_2 until BT=0.6 %
Optimization parameters	Flow rate F	Flow rates F_1, F_2 and times t_1, t_2,	Quadratic function coefficients a_1, a_2, a_3	Flow rates F_1, F_2, time t_1
Reasoning	Optimum flow rate should correspond to compromise between amount of antibody captured per cycle (high amounts are favored by low flow rates) and time of load (short times are favored by high flow rates).	Column is first loaded quickly but less effectively with high flow rate, then with decreasing flow effectiveness loading increases while time needed also increases.	Similar as flow ramp only different flow function was tested.	When allowed breakthrough is reached, concentration of bound antibody falls from column inlet to column outlet. By reversing flow, fresh harvest first encounters less saturated part of column.
Productivity increase (%)	/	**13**	10	12

AC Column
Width 8 mm; Length 100 mm

$$P=\frac{m_{mAb}}{t_{cycle}}=\frac{m_{mAb}}{t_L+t_W+t_E+t_R}=\frac{m_{mAb}}{t_L+t_{WER}}$$

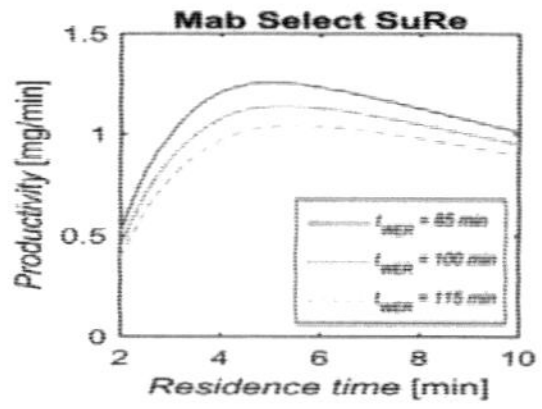

Affinity sorbent	Mab Select SuRe		
Parameter	RT	t_L	P
Unit	min	min	mg/min
t_{WER}=85 min	5.0	55	1.26
t_{WER}=100 min	5.3	60	1.14
t_{WER}=115 min	5.5	65	1.04

Optimum parameters of flow ramp loading strategy

Parameter	Unit	Mab Select SuRe
Start flow rate F_1	mL/min	3.18
Start of ramp t_1	min	0
End flow rate F_2	mL/min	0.77
End of ramp t_2	min	19
Loading time t_L	min	49
Productivity P	mg/min	1.42

Conclusions - Upstream

- An unique theoretical description that **continuously describes** all cell growth phases was developed.
- The **reversibility** of macro reactions is a key element that is needed for the correct description of the concentration profiles of measured components in the medium or cell.
- After determining model parameters any modifications in the medium could be simulated further enabling **smart design** of the experiments.
- The developed methodology can be **coupled** with the protein N-glycosylation and affinity chromatography (further work).

Conclusion - Downstream

- Developed model enables characterization of affinity resin properties with **minimum** experimental work requirements.
- Internal diffusion is **rate limiting step** of affinity chromatography process.
- Axial dispersion and film mass transfer resistance has **low influence**.
- Results of parametric sensitivity study showed, that volumes of external chromatography system have **minor effect**.
- Transport inside monoclonal affinity chromatography column is **controlled by pore diffusion**.
- Amongst loading strategies tested, **flow ramp loading** showed the best results compared to loading with constant flow rate.
- This strategy is the **simplest** to implement and required **no mechanical changes** to the existing equipment.

Acknowledgments

Assist. Prof. dr. Blaž Likozar
Dr. Miha Kastelic
Matic Grom
Dr. Andrej Pohar
Dr. Drejc Kopač
Dr. Miša Mojca Cajnko

BioPharm.Si

REPUBLIKA SLOVENIJA
MINISTRSTVO ZA IZOBRAŽEVANJE, ZNANOST IN ŠPORT

EVROPSKA UNIJA

next BioPharm DSP

Thank you for your attention

16

494003 Exploring the Multi-Minima Behavior of Organic Crystal Polymorphs at Finite-Temperature

Abstract:

Polymorph prediction is a critical aspect of pharmaceutical solid form design to ensure that the crystal with the highest stability and performance properties is selected for development. Computational crystal structure prediction (CSP) studies currently predict the experimental crystal forms by identifying minima on the lattice energy surface. One of the most pressing challenges in current CSP studies is that this minimized lattice energy approach typically suffers from a significant overprediction of stable crystal forms. Most polymorphic systems have on the order of ten stable solid forms. However, CSP studies can yield hundreds or thousands of unique lattice minima.

A common hypothesis put forward for why some lattice minima do not correspond with a unique experimental structure is that the thermal motions under working conditions will cause many lattice minima to rapidly restructure into a more stable form or cause multiple minima to coalesce into a single indistinguishable ensemble. Under this hypothesis, some potential energy basins explored at high temperature should be `rough' landscapes filled with multiple local energy minima rather than the classic picture of a smooth potential energy basin with a single global minimum-energy structure.

Here we demonstrate the existence of these rough basins with multiple lattice minima in a number of different polymorphic systems. The experimental crystal structures of twelve polymorphic systems were heated to a range of temperatures up to ambient conditions using molecular dynamics simulations. A set of configurations from each trajectory was then quenched to zero Kelvin using crystal energy minimization. We find that the high-temperature configurations of rigid-molecule crystals tend to all collapse to the same initial lattice minima after quenching. However, the configurations from more flexible molecules tend to minimize into a myriad of different lattice minima, suggesting that flexibility leads to multi-minima behavior. We also find that the number of minima found in the crystal basins increases with temperature, and the presense of multiple lattice minima within the free energy basin is polymorph-dependent in addition to being molecule-dependent. The existence of these multi-minima basins at ambient conditions lends credence to the theory that incorporating thermal motion can reduce the number of spuriously predicted structures in future CSP studies.

502617 Iterative Synthesis of Heteropolymers Using Organic Solvent Nanofiltration

Abstract:

Iterative Synthesis of Heteropolymers Using Organic Solvent Nanofiltration

Ruiyi Liu, Piers Gaffney, Patrizia Marchetti, Marc Schaepertoens, Ruijiao Dong, Andrew Livingston

Barrer Centre, Department of Chemical Engineering, Imperial College London

Sequence-defined heteropolymers underpin much of biology in the forms of proteins/ peptides and oligonucleotides (DNA and RNA). The pharmaceutical industry is investing heavily in the potential of these species to push through the bottleneck in the drug discovery pipeline. Such polymers feature a repeating backbone motif, but each monomer possesses a variable side-chain. It is the order and chemical nature of these side-chain residues that confers the function of the final heteropolymer. Beyond such familiar biomolecules, heteropolymers present opportunities in chemical bar-coding and security, and even for long-term information storage. Looking still further ahead, and taking a cue from Nature, engineered heteropolymers of non-biological monomers might provide uniquely tailored and functional engineering materials, if only they could be flexibly prepared at scale.

Oligonucleotides (oligos) and peptides are the dual test-beds upon which heteropolymer synthesis is founded. They are prepared by iterative synthesis, a process in which one monomer at a time is coupled to the terminus of a growing chain - like differently colored beads threaded onto a string. To ensure error-free sequence synthesis, other reactive groups are masked by protecting groups; permanent protecting groups block the side-chains throughout the synthesis, and a temporary protecting group is removed in the final stage of each chain extension cycle to expose a new reactive chain terminus. To simplify purification after chain extension, and to make the process easily generalizable, both oligos and peptides are widely synthesized on solid phase supports, where excess reagents are simply flushed from a packed bed or column of support beads.

Solid phase synthesis protocols are intrinsically difficult to up-scale. Furthermore, for the successful iterative synthesis of long sequences to succeed, the coupling efficiency must be as near to quantitative as possible, but mass transfer from the bulk liquid to the solid support is the rate-determining step for all but the very slowest coupling reactions. We overcome both these roadblocks to scalable iterative synthesis by performing coupling reactions in the bulk liquid phase and separating reaction debris by organic solvent nanofiltration (OSN).

OSN membranes separate solutes principally on the basis of size selectivity. During liquid phase iterative synthesis the growing heteropolymer is the largest solute present, whilst the monomer debris is the next largest. Therefore, the ability of the membrane to retain the growing heteropolymer, but to permeate the monomer debris, is the critical separation criterion. The molecular weight of the monomer debris from oligo or peptide synthesis falls in the range 400-700 Da. Indeed, it is hard to conceive of functional heteropolymers, either natural or man-made, with smaller monomer debris. Thus the development of OSN membranes with permeation pathways open enough to filter such large species, and yet maintain their physical and chemical integrity during repeated synthesis cycles, was an essential enabling technology for liquid phase heteropolymer synthesis.

We have previously shown that short peptides and oligos could be prepared by means of liquid phase synthesis and purification by single-stage OSN. We have now achieved pharmaceutical length oligo synthesis (20-mer) using a two-stage membrane purification apparatus that takes the cycle yield to near-quantitative recovery of the growing polymer. Despite harsh solvents (DMF and acetonitrile), and both acidic and basic treatments, the OSN membranes maintained constant performance over multiple cycles (> 19).

The same principles have been applied to the synthesis of poly-ethers. Uniform (or monodisperse) poly-ethylene glycols up to 60-mer were constructed by the stepwise concatenation of short units, purifying the growing chain by OSN after each cycle. We are now extending these principles to the synthesis of engineered heteropolymers with a chemically robust poly-ether backbone and functionalised chiral side-chains.

496577 Challenges in Process Development of Antibody Drug Conjugates (ADCs)

Abstract:

Antibody drug conjugates (ADCs) are a class of biopharmaceutical drugs in which a cytotoxic agent is chemically linked to an antibody that recognizes a tumor-associated antigen. The basic strategy is to combine target selectivity of antibodies with the potent cytotoxic molecules, with a goal of generating drugs that are highly efficacious and also safe.

Pfizer Biotherapeutics Pharmaceutical Sciences is responsible for developing scalable manufacturing processes for Pfizerâ€™s portfolio of ADCs. Process development group is engaged in initially receiving the technology from discovery groups/external collaborators and developing the technology further for preparation of these ADCs from small scale (several grams) to large scale (hundreds of grams to kilograms). Pfizerâ€™s early stage ADCs utilize several conjugation chemistries via cysteines, lysines or glutamines on the antibody as well as several different linkers and payloads. Each ADC molecule, even of the same type of conjugation chemistry, poses unique challenges for process and analytical development groups in the development and implementation of preparative and analytical technologies.

This presentation highlights conjugation and purification development of ADCs to illustrate Pfizerâ€™s development paradigm. In the process development arena the approach to maintain acceptable yields while maintaining quality attributes will be discussed. Our emphasis on thorough characterization of the final product will be discussed as well as consistency across multiple batches.

497845 Improved Manufacturing Process for Semi-Synthetic Calicheamicin Linker-Payload *En Route* to Antibody-Drug Conjugates

Abstract:

Calicheamicin is produced by fermentation of *Micromonospora echinospora* ssp. calichensis and isolated from the broth. The natural product is then synthetically modified to attach a linker, and the â€œlinker-payloadâ€ product is reacted with antibodies to produce antibody-dug conjugates (ADCs) used in the treatment of cancer.Â We have developed an improved manufacturing process for the production of calicheamicin linker-payload and gained a better understanding of the chemistry at each step. Process improvements include fewer synthetic steps, fewer unit operations, reduced solvent usage, improved yield, and increased robustness. The improved process is now approved for commercial production of calicheamicin linker-payload as a drug substance intermediate for ADCs. Highlights of the process, learnings, development strategy, and benefits of implementation will be presented, along with selected aspects of the conjugation process to produce calicheamicin ADCs.

495978 Technoeconomic Evaluation of Cyclosporine Crystallisation Intensification Using a Cascade of M.S.M.P.R. Crystallisers

Abstract:

Continuous Pharmaceutical Manufacturing (CPM) has been recognised as a promising new production paradigm over the currently implemented batch methods, which imply high volumes of waste and inefficiencies [1]. Whilst the literature demonstrates many continuous flow syntheses of active pharmaceutical ingredients [2], technological innovation is required to realise the benefits of integrated end-to-end CPM processes, including purification and separation unit operations. Process systems engineering (PSE) methodologies allow the screening of candidate CPM processes for their feasibility and viability prior to committing to significant time and financial investments in experimentation and scaleup [3]. Recent efforts in modelling and optimisation of the upstream CPM of APIs have demonstrated the importance of conducting PSE studies in establishing economically viable separation processes [4], realising issues in modelling of particulate processes for solids handling in continuous operation [5] and establishing limitations attainable process improvements in pharmaceutical manufacturing [6].

Cyclosporine is an immunosuppressant drug for the prevention of organ transplant rejection. The continuous cooling crystallisation of cyclosporine has recently been investigated using continuous mixed suspension, mixed product removal (MSMPR) crystallisers, due to their broad applicability and ease of integration into existing pharmaceutical processes [7]. The effect of process configuration [8], mother liquor and solids recycle [9], number of stages [10] on crystallisation yields, purities and product size distributions have been studied to demonstrate the feasibility of continuous crystallisation of the API. Technoeconomic comparisons between batch and continuous crystallisation configurations for cyclosporine to validate the viability of continuous crystallisations are yet to be implemented.

The present paper work develops a steady-state process model for the continuous cooling crystallisation of 100 kg per annum of cyclosporine from acetone including a systematic technoeconomic comparison of various continuous process configurations with the batch method, with a focus on computationally exploring and demonstrating process intensification. The effect of the number of MSMPR stages and varying recycle considerations (mother liquor versus solids recycle) on attainable yields, process material efficiencies and costs are compared. Mother liquor recycle requires vacuum evaporation of solvent, whilst solids recycle uses gravity driven separation to allow recycle of concentrated slurry to MSMPR stages. The developed model simultaneously solves MSMPR mass balances, crystal population balances and crystallisation kinetics to calculated stage and overall crystallisation yields. Prediction of temperature-dependent cyclosporine solubilities in acetone uses a regressed surrogate polynomial established from published experimental data [11]. Temperature- and concentration-dependent growth and nucleation rates are computed on the basis of published kinetic parameters [10] for accurate calculation of attainable yields. Material efficiencies are quantified via popular green chemistry metrics, namely the process mass intensity (PMI) and the mass productivity (MP), which offer quantitative insight into process intensification potential, for each configuration considered. Unit volumes and material throughputs required to meet the plant API capacity are scaled to account for stage inefficiencies. Economic analyses are conducted via an established methodology for pharmaceutical manufacturing processes [12].

The continuous MSMPR crystallisation configurations investigated show potential for application in the CPM of cyclosporine from the calculated attainable material efficiencies and total costs. The sensitivity of process performance to various operating parameters has been explored as well, indicating the significance of rigorous design space investigation of CPM processes. Our work encourages subsequent technoeconomic studies and life cycle assessments (LCA) of the CPM of cyclosporine for further investigation of process feasibility and viability.

The present study exhibits the importance of conducting PSE studies of candidate continuous separation processes for CPM prior to further process development. We demonstrate the economic viability of CPM to facilitate the transition from batch methods, and quantify the potential intensification benefits of improved material efficiency of continuous separation processes in comparison to their batch counterparts. This study expedites further process modelling for the CPM of cyclosporine and other candidates, as well as optimisation of continuous separation processes to realise the full benefits of continuous operation in the pharmaceutical industry.

LITERATURE REFERENCES

[1] Poechlauer, P., Colberg, J., Fisher, E., Jansen, M., Johnson, M. D., Koenig, S. G., Lawler, M., Laporte, T., Manley, J., Martin, B., O'Kearney-McMullan, A., 2013. Pharmaceutical roundtable study demonstrates the value of continuous manufacturing in the design of greener processes, *Org. Process Res. Dev.* **17**(12): 1472–1478.

[2] Baumann, M., Baxendale, I. R., 2015. The synthesis of active pharmaceutical ingredients (APIs) using continuous flow chemistry, *Beilstein J. Org. Chem.* **11**: 1194–1219.

[3] Teoh, S. K., Rathi, C., Sharratt, P., 2015. Practical assessment methodology for converting fine chemicals processes from batch to continuous, *Org. Process Res. Dev.* **20**(2): 414–431.

[4] Diab, S., Gerogiorgis, D. I., 2017. Process modelling, simulation and technoeconomic evaluation of separation solvents for the Continuous Pharmaceutical Manufacturing (CPM) of diphenhydramine, *Org. Process Res. Dev.* DOI: 10.1021/acs.oprd.6b00386, in press.

[5] Rogers, A., Ierapetritou, M., 2014. Challenges and opportunities in pharmaceutical manufacturing modelling and optimization, *Comput. Aided Chem. Eng.* **34**: 144–149.

[6] Benyahia, B., Lakerveld, R., Barton, P. I., 2012. A plant-wide dynamic model of a continuous pharmaceutical process, *Ind. Eng. Chem. Res.* **51**(47): 15393–15412.

[7] Su, Q., Nagy, Z. K., Rielly, C. D., 2015. Pharmaceutical crystallisation processes from batch to continuous operation using MSMPR stages: Modelling, design, and control, *Chem. Eng. Process. Process Intensif.* **89**: 41–53.

[8] Alvarez, A. J., Singh, A., Myerson, A. S., 2011. Crystallization of cyclosporine in a multistage continuous MSMPR crystallizer, *Cryst. Growth Des.* **11**(10): 4392–4400.

[9] Li, J., Trout, B. L., Myerson, A. S., 2016. Multistage continuous mixed-suspension, mixed-product removal (MSMPR) crystallization with solids recycle, *Org. Process Res. Dev.* **20**(2): 510–516.

[10] Li, J., Lai, T. C., Trout, B. L., Myerson, A. S., 2017. Continuous crystallization of cyclosporine: the effect of operating conditions on yield and purity, *Cryst. Growth Des.* **17**(3): 1000–1007.

[11] Wong, S. Y., Tatusko, A. P., Trout, B. L., Myerson, A. S., 2012. Development of continuous crystallization processes using a single-stage mixed-suspension, mixed-product removal crystallizer with recycle, *Cryst. Growth Des.* **12**(11): 5701–5707.

[12] Schaber, S. D., Gerogiorgis, D. I., Ramachandran, R., Evans, J. M. B., Barton, P. I., Trout, B. L., 2011. Economic analysis of integrated continuous and batch pharmaceutical manufacturing: A case study, *Ind. Eng. Chem. Res.* **50**(17): 10083–10092.

489029 Application of Advanced Process Control Tools in Continuous Processing

Abstract:

Continuous manufacturing is now widely discussed in the pharmaceutical industry. The availability of turn-key lines offered by various suppliers allows the transfer of established and new products to the continuous manufacturing approach. At the moment the focus is mainly on solid oral dosage forms including the common routes via wet and dry granulation or direct compaction. From a processing point-of-view the shift to continuous mode is supported by the fact that most common unit operations are continuous machinery by design, such as feeding, tableting or granulation. One challenge is the integration of all units into a single line, which needs implementation of interfaces, transport units and buffers. The remaining challenge is the establishment of control concepts that enable a fully automated, yet still flexible process, capable of handling start-up and shut down, as well as of out-of-spec events. Tools for in-line monitoring of critical quality attributes and real-time acquisition of process data have been widely demonstrated. Process models, fast enough for design of controllers, have been developed recently. However, most continuous processes filed by pharmaceutical industry still remain on a level 2 control strategy based on in-process controls without control loops.

In this work different tools for advanced process control were developed and investigated and their implementation in processes or unit operations will be shown. This includes novel process analytical technologies enabling real-time data acquisition of statistical representative numbers or even each single dosage form, concepts for out of spec material handling based on modelling of residence time distribution or model predictive control. The latter approach was developed for a feeding and blending unit. The rotation speed of the blender is manipulated to control the mass hold-up and the outlet mass flow rate in order to mitigate fluctuations of homogeneity caused by feeder events. Another example deals with the discharge of out-of-spec material, in combination with an operating concept for a tablet press based on different pre-defined set points of the turret speed. Recent developments in the field of PAT will complete the presentation.

491286 Nonlinear First-Principle Model-Based Control of a Continuous Pharmaceutical Manufacturing Process

Abstract:

Model-based process control is one of the pillars of continuous pharmaceutical manufacturing process by the QbD approach. The first-principle approach, which comes from in-depth understanding of the governing transport phenomena and performance of unit operations, yields developing of predictive models with which the continuous processes can be manipulated, controlled, and optimized. Despite empirical modeling, where limited operation conditions and process status are being mapped within a DoE, other unseen nonlinearities in the process can be captured in a nonlinear first-principle model to optimally control the process.

In this work, a nonlinear first-principle distributed dynamic model was developed to control a drying process for continuous manufacturing of orally dissolving films loaded with a pharmaceutical compound. The drying model is the core of active control for the process, in which the process parameters (such as air temperature, or production rate), feed compositions (like solid content and concentration), or final productsâ€™ properties (for instance the required final moisture content) can be the manipulating variables. Simulation of transport phenomena, diffusion controlled mass transfer (mutual solvent-polymer vs. moisture sorption isotherm), the spatial distribution of solvent concentration in the film and the moving boundary on the interface due to the thin film shrinkage during drying are some of the complexities of the model development. The final quality attribute of the product is set as the control set point and other process variables were manipulated in the nonlinear model control (MIMO) to converge the process in the design space of variable inputs. The model funds the steps for future phases of PAT implementation, RTRt, and process optimization.

497129 A Fault-Tolerant Control Design for Real-Time Release in Continuous Manufacturing of Solid Dose Using Direct Compaction

Abstract:

Â A Fault-tolerant Control Design for Real-time Release in Continuous Manufacturing of Solid Dose using Direct Compaction

Qinglin Su, Mariana Moreno, Jianfeng Liu, Sudarshan Ganesh, Yasasvi Bommiready, Marcial Gonzalez, Gintaras V. Reklaitis, Zoltan K. Nagy

School of Chemical Engineering, Purdue University,
West Lafayette, IN 47907

Thomas Oâ€™Connor, Tian Geng

CDER, Office of Pharmaceutical Quality, U.S. Food and Drug Administration,
Silver Spring, MD 20993

Abstract

Advantages of continuous manufacturing in the pharmaceutical industry have been demonstrated through extensive research studies in the last decade [1], but mostly at the conceptual or theoretical level. Only few continuous manufacturing processes have been approved by the US Food and Drug Administration thus far, e.g., the Vertex Orkambi and Janssen Darunavir [2]. To move from a conceptual continuous design to a practically operating pilot plant or manufacturing process, the efforts of regulators, academics, and industry are now focused more on quality control strategies [3]. Specifically, the active process control design for real-time release and its risk management in continuous manufacturing [4, 5] have been considered as critical for successful implementation of continuous manufacturing processes.

In this study, a fault-tolerant control design for real-time release is implemented, according to our previous systematic framework for control design and analysis [5], to a pilot plant at Purdue University for continuous manufacturing of solid dosage pharmaceuticals using direct compaction. First, beside the fundamental control loops provided by the equipment vendors for unit operations of feeding, blending, tableting, etc., to manipulate the critical processing parameters (CPPs), supervisory active feedback and feedforward process control strategies are designed to maintain the critical quality attributes (CQAs) based on process analytical technology (PAT) tools and final product quality testing, e.g., Sotax Auto Test 4. In the active process control design, the major system dynamics are identified with state-space models using historical operating data in the pilot plant, and a series of rational control design and analysis metrics are employed to evaluate the control performance and screen the potential design risks [5]. For example, a Relative Gain Array (RGA) analysis is used to evaluate the system interactions and select the best-performing control pairings in the feeding-blending units, where the diagonal pairing is preferred under low system frequency domain, and is risky under higher frequency as is the case when the feeder nozzle is accidently attached to the blender hopper frame.

In the proposed fault-tolerant control design, a risk mapping, assessment and planning (Risk MAP) strategy is also considered to address the potential risks associated with the above active process control design. The active process control system is designed to mitigate or avoid quality risks. However, implementation of risk reduction measures can introduce new risks into the system or increase the significance of other existing risks (ICH Q9). In this study accepted residual risks and new potential risks to product quality after the implementation of the active process control system for the continuous manufacturing process are examined. . For example, when faults occurred in the feeding-blending units, the process performance indices, e.g., process capabilities index, are below their minimum targets. Hence, for real-time release purposes, the proposed fault-tolerant control design is evaluated experimentally and in simulation mode under these risk scenarios, e.g., process disturbance, probe fouling, instrument calibration errors, etc. It is as expected that the fault-tolerant control design with advanced process systems engineering tools, such as process monitoring, data reconciliation, state estimation, etc., can mitigate those potential risks that are common causes of variations in an otherwise stable process.

Steady-state operation of the pilot-plant is also executed over longer durations to validate the proposed fault-tolerant control design. Concluding remark will also be offered, suggesting potential guidelines for achieving real-time release in continuous solid dosage manufacturing.

References

[1] M. Ierapetritou, F. Muzzio and G. Reklaitis, "Perspectives on the continuous manufacturing of powder-based pharmaceutical processes," *AIChE Journal,* vol. 62, no. 6, pp. 1846-1862, 2016.

[2] L. Yu, "Continuous Manufacturing Has a Strong Impact on Drug Quality," US Food and Drug Administration, 12 April 2016. [Online]. Available: https://blogs.fda.gov/fdavoice/index.php/2016/04/continuous-manufacturing-has-a-strong-impact-on-drug-quality/. [Accessed 11 April 2017].

[3] S. L. Lee, T. F. O'Connor, X. Yang, C. N. Cruz, S. Chatterjee, R. D. Madurawe, C. M. V. Moore, L. X. Yu and J. Woodcock, "Modernizing pharmaceutical manufacturing: from batch to continuous production," *Journal of Pharmaceutical Innovation,* vol. 10, no. 3, pp. 191-199, 2015.

[4] R. Singh, M. Ierapetritou and R. Ramachandran, "An engineering study on the enhanced control and operation of continuous manufacturing of pharmaceutical tablets via roller compaction," *International Journal of Pharmaceutics,* vol. 438, pp. 307-326, 2012.

[5] Q. Su, M. Moreno, A. Giridhar, G. V. Reklaitis and Z. K. Nagy, "A systematic framework for process control design and risk analysis in continuous pharmaceutical solid-dosage manufacturing," *Journal of Pharmaceutical Innovation,* p. in review, 2017.

496819 Intensified Crystallization of Sulfathiazole and Sulfanilamide By Direct Co-Crystal Assembly

Abstract:

Most of the studies mainly focus on the co-crystallization step rather than on the synthesis and co-crystallization together. Based on the concept of direct co-crystal assembly, we have successfully produced 1:1 co-crystals of sulfathiazole-theophylline (STZ-THE) and 1:1 co-crystals of sulfathiazole-sulfanilamide (STZ-SNM) by integrating the chemical synthesis and co-crystallization. Interestingly, the yields of the integration method were higher than the ones of normal reactive crystallization methods, and the amount of purities were also reduced. For the chemical reaction that suffered from the low yields or had the impurity which was hard to be separated, the formation of co-crystals might be an alternative strategy to improve the efficiency of purification and the value of drugs as well. The modification of loading stoichiometry of reactants to co-formers could lead to the difference of product compositions. Both STZ-THE and STZ-SNM exhibited different pH solubility profiles. In other words, co-crystallization could change the pH effect active pharmaceutical ingredient.

497068 Direct Co-Crystal Assembly from Synthesis to Co-Crystallization

Abstract:

Pharmaceutical co-crystals were assembled directly via chemical synthesis by cooling under three new strategies. The screening method for direct co-crystal assembly by a chemical reaction was developed based on liquid-assisted grinding. The ternary phase diagram including a chemical reaction and co-crystallization was established for a 2 : 1 co-crystal of benzoic acidâ€“sodium benzoate.

502452 Continuous Preparation of 1:1 Haloperidol-Maleic Acid Salt By a Novel Solvent-Free Method Using a Twin Screw Melt Extruder

Abstract:

Salts are generally prepared by acid-base reaction in relatively large volumes of organic solvents, followed by crystallization. In this study, the potential for preparing a pharmaceutical salt between haloperidol and maleic acid by a novel solvent-free method using a twin-screw melt extruder was investigated. The pH-solubility relationship between haloperidol and maleic acid in aqueous medium was first determined, which demonstrated that 1:1 salt formation between them was feasible (pH_{max} 4.8; salt solubility 4.7 mg/mL). Extrusion of a 1:1 mixture of haloperidol and maleic acid at the extruder barrel temperature of 60 °C resulted in the formation of a highly crystalline salt. The effects of operating temperature and screw configuration on salt formation were also investigated, and those two were identified as key processing parameters. Salts were also prepared by solution crystallization from ethyl acetate, liquid-assisted grinding, and heat-assisted grinding and compared with those obtained by melt extrusion by using DSC, PXRD, TGA, and optical microscopy. While similar salts were obtained by all methods, both melt extrusion and solution crystallization yielded highly crystalline materials with identical enthalpies of melting. During the pH-solubility study, a salt hydrate form was also identified, which, upon heating, converted to anhydrate similar to that obtained by other methods. There were previous reports of the formation of cocrystals, but not salts, by melt extrusion. ^{1}H NMR and single-crystal X-ray diffraction confirmed that a salt was indeed formed in the present study. The haloperidol-maleic acid salt obtained was nonhygroscopic in the moisture sorption study and converted to the hydrate form only upon mixing with water. Thus, we are reporting for the first time a relatively simple and solvent-free twin-screw melt extrusion method for the preparation of a pharmaceutical salt that provides material comparable to that obtained by solution crystallization and is amenable to continuous manufacturing and easy scale up.

500888 Prediction of Tablet Dissolution By Process Parameters in Continuous Manufacturing

Abstract:

Tablet dissolution is an attribute that is critical to the product quality and should be well controlled in order to produce an acceptable pharmaceutical product. Tablet dissolution testing is a lengthy procedure that is time-consuming, expensive, and requires media preparation. In addition, it is a destructive test, and the sample will no longer be available for further characterization. Hence, there has been a growing interest in the development of non-destructive predictive tools for tablet dissolution to enable the real time release testing (RTRt) strategy. Application of such a strategy would not only lower laboratory and personnel costs but also increase product quality. A statistical model for predicting tablet dissolution profiles will enable rapid techniques in continuous assessment of tablets ensuring the desired quality attributes.

In this work, we developed a statistical model to predict tablet dissolution profile in a continuous direct compaction manufacturing line. Tablets consisted of model API, Acetaminophen (APAP), lactose monohydrate as the filler, and magnesium stearate as lubricant. Multivariate effects were analyzed using Quality-by-Design methods. The experiment was based on a fractional factorial design with 30 experiments, including 3 repeated center points, to study the influence of four process variables – drug concentration (%), compaction force (KN), blending speed (rpm), and feed frame speed (rpm) – on the final product. Dissolution profiles of 180 tablets (six tablets for each experimental run) were obtained and fit into the Weibull model (including parameters of α and β), and the values of α and β were calculated. Multi variate analysis of variance (MANOVA) was used to determine the statistical differences between the dissolution profiles. The results indicated that compaction force and API concentration were the main parameters affecting tablet dissolution in this process. Therefore, a predictive model was developed to correlate the dissolution model parameters α and β to the process parameters. The model was verified using an independently manufactured validation set. The values of α and β of the validation set were estimated by the model. The dissolution profiles of these tablets were predicted using the estimated values of α and β. The predicted dissolution profiles obtained from the model were in agreement with the experimental results using the similarity factor f_2 value greater than 50. This work establishes an approach to the real time release testing (RTRt) of pharmaceutical tablets and the possibility of assessing the quality characteristics of tablets non-destructively.

501282 Integrating Sensors for Monitoring Blend Content in a Pharmaceutical Continuous Manufacturing Plant

Abstract:

In a pharmaceutical manufacturing process, Critical Quality Attributes (CQAs), as defined by the FDA, need to be monitored not only for the final product but also for intermediates. Blend uniformity of powders is one such attribute that needs to be measured to ensure the quality of the final product. In-line sensors, to monitor the blend content of the powders, were implemented within a Direct Compaction (DC) continuous tablet manufacturing line. In most cases, since the primary ingredient of interest is the active pharmaceutical ingredient (API), the amount of the API was monitored/predicted over the course of manufacturing. For the calibration model building process, a unique setup involving dynamic powder spectral acquisition method was used. This setup was aimed at mimicking the powder flow characteristics within the manufacturing line. A Raman probe and a portable NIR were used as sensors and they were placed at the exit of the blending process before the tableting stage. The performance of the two sensors and their respective models were evaluated in terms of accuracy, precision, operating range, measurement frequency, placement, reliability, robustness. Additionally, their performances were validated by off-line traditional analytical measurements.

501455 Scientific and Regulatory Considerations for Developing the Control Strategy of a Continuous Manufacturing Process

Abstract:

Advances in manufacturing technology over the last decade have provided the pharmaceutical industry opportunities for adopting innovative ways to develop flexible, efficient and robust processes. In recent years, continuous manufacturing has emerged as a new technology for the manufacture of pharmaceuticals. Compared to traditional batch process, continuous manufacturing has various advantages, such as, shorter processing times, smaller facility footprint, etc. On the other hand, it has unique elements that lead to uncertainty on how to implement a continuous process within the current regulatory framework. The Food and Drug Administration supports the adoption of innovative technology and is communicating with pharmaceutical companies and academia to facilitate the development of continuous processes. This presentation will discuss several scientific, current good manufacturing practice (CGMP) and regulatory considerations for developing the control strategy of a continuous manufacturing process.

500629 Uncertainty Quantification and Global Sensitivity Analysis of Drug Delivery Primary Containers to Understand Process Capability and Key Risk Factors

Abstract:

A simple high-fidelity model was created to predict the plunger position and air gap in a pre-filled syringe primary container to be used as a component in a combination product. The product-container system has six sources of uncertainty and two uncertain variables related to operating conditions. An uncertainty analysis was carried out to compute the joint distribution of plunger position and air gap performance parameters with the goal to understand the operating ranges for these CQA (critical quality attributes). In order to relate these to operational practices, process capability indices (Cpk) were calculated via a naïve approach where the marginal plunger-depth distribution was fit to a normal distribution in order to enable the index (Cpk) calculations. This simplified approach was compared with a full propagation of factor uncertainties via the rigorous nonlinear model, which resulted in a remarkable difference in the CpK values. This indicates that the Gaussian approximation was unsatisfactory and more refined techniques are required. As a final step, a global sensitivity analysis based on the Sobolâ€™ indices technique was carried out in order to identify the key factors affecting process capability and inform continuous process improvement initiatives. Overall, four of the eight factors were found to be unessential (non-key).

This work demonstrates the use of rigorous mathematical process models and advanced uncertainty/sensitivity techniques to shed light on design and operation performance characteristics (QbD and QbC). The model and techniques were implemented in a web-based deployment platform in order to leverage the rigorous models/techniques and share these simple-to-use applications with all users across the organization, regardless of their background.

500701 Nucleation Behavior of Eszopiclone - Butyl Acetate Solutions from Metastable Zone Widths

Abstract:

Experimental data on the critical supercooling ΔT_{max}, a measure of metastable zone width (MSZWs), as a function of saturated temperature T_0 and cooling rate R were analyzed to describe the nucleation behavior of unseeded eszopiclone-butylacetate solutions. The Nývlt's semiempirical model and Sangwal's theory were employed, respectively. And moreover, we propose a modified model based on the Sangwal's theory in which the nucleation parameters are not dependent on saturation temperature T_0 and nucleation temperature T_1 which can fit the experimental data better than Sangwal's theory. Furthermore, the pre-exponential A and solid – liquid interfacial energy γ can be obtained by employing this modified model, thus, sheds light on the relationship between nucleation parameters and the cooling rate R. Meanwhile, based on the Classical Nucleation Theory, with the estimated interfacial energies from the slope in the modified model, the critical nucleus size and critical Gibbs free energy were calculated and discussed. Then based on the Sangwal's theory, the nucleation activation barrier is found to be dependent on ln R with an exponential relationship, hence, the relationship between nucleation barrier and pre-exponential A can be constructed. At last, the effects of agitation rate and working volume on MSZWs were investigated and discussed.

Development, Implementation, and Use of an On-line Laser Diffraction Particle Size Method in a Spray Drying Manufacturing Process

Pharmaceutical Discovery, Development and Manufacturing Forum
2017 AIChE Annual Meeting

Martin Warman, Martin Warman Consultancy Ltd, and Alon Vaisman, Malvern Instruments

Malvern Panalytical

Why is this important - background

- The Biopharmaceutics Classification System (BCS)
  - Class I - High Permeability, High Solubility
  - Class II - High Permeability, Low Solubility
  - Class III - Low Permeability, High Solubility
  - Class IV - Low Permeability, Low Solubility
- In general, product performance for Class I + III compounds is driven by API concentration and tablet disintegration
- However product performance for Class II + IV compounds is driven by many more factors

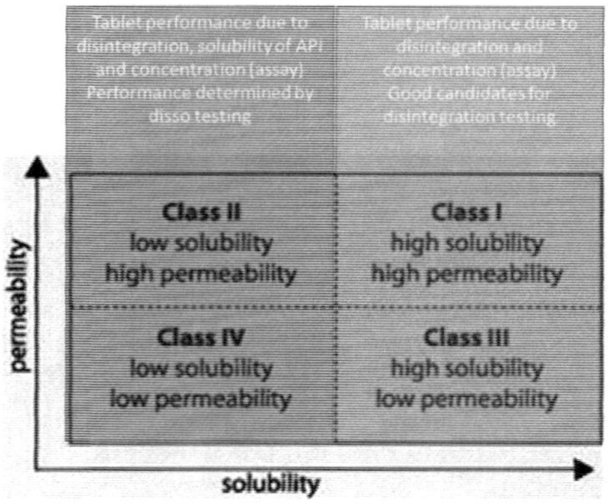

Malvern Panalytical

What is disintegration and weight are not enough

- For BCS II and IV API can't use simple indicators of tablet performance
  - Need to establish 'rate of release' data
  - Disso testing
- However the disso performance is a combination of numerous chemical and physical properties of the matrix
  - Concentration
  - Solubility
  - Hardness
  - Density
  - Porosity
  - Particle size
  - Inter-particulate forces
  - Etc
- Many of these are impacted by the physical API properties but also the intermediate particle properties
  - Spray Dried Dispersion
    - % crystallinity
  - Granulation

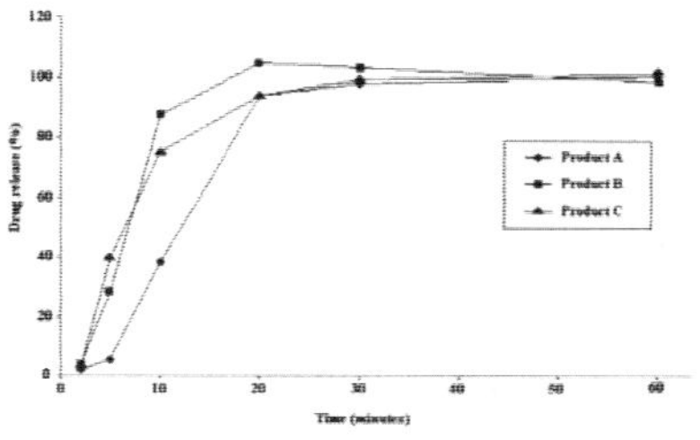

Malvern Panalytical

Case Study - High-level description of Process

- Crystalline API production
  - Reaction steps, crystallization, drying
- Amorphous API spray dried dispersion
  - Spray solution prep, spray drying, secondary drying
- Formulation
  - Pre lube blend, post lube blend
- Compression
  - Compression, film coat
- Described as full QbD, with elements of real-time release

Malvern Panalytical

Identifying steps critical to tablet performance

- For this product we could show the product performance was a result of
  - Tablet compression
  - SDD bulk density and particle size
  - SDD particle size could be controlled byt the spary drying process

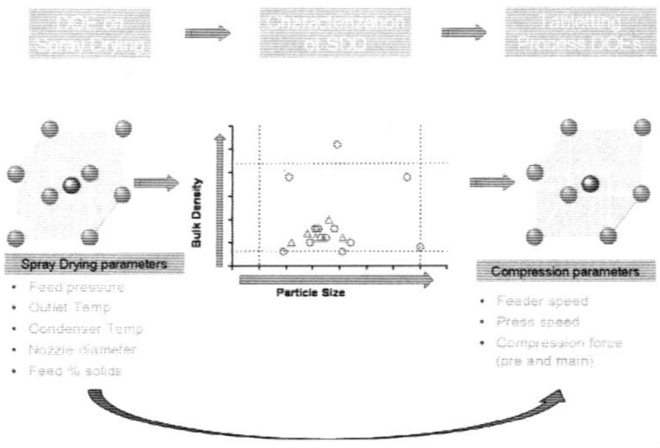

Malvern Panalytical

Identifying control models

- Regression analysis showed
  - The particle size of the wet SDD was determined by nozzle diameter and feed pressure
  - Whereas bulk density related to speed of drying, with droplets that dried fast having symmetrical shape (often with entrained gas bubbles) and droplets that dried slowly having high surface area/higher bulk density
- The key message is these could be controlled independently

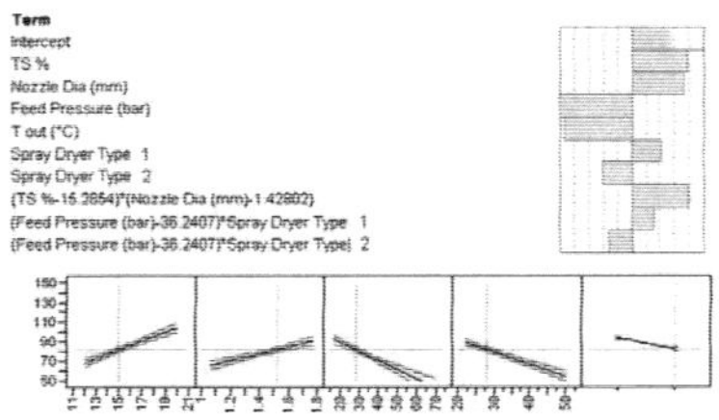

Particle size

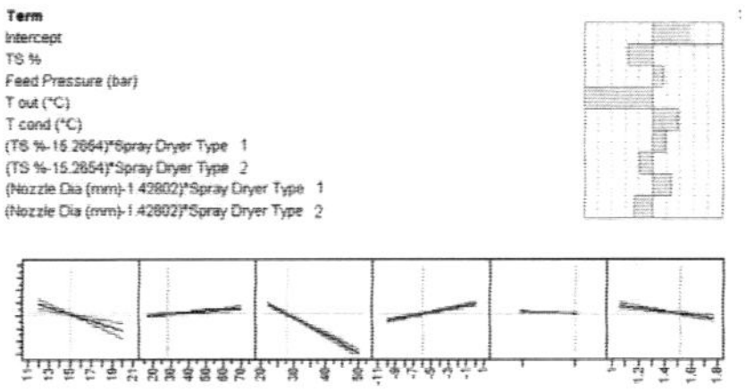

Bulk density

Malvern Panalytical

Graphical Illustration of Design Space

- More so we could demonstrate a three dimensional Design Space for product performance (disso)
- Control strategy complexity pyramid
  - First published by FDA ETT in the AAPS Journal, Vol 16, Number 4, July 2014
  - Included in the paper, Modernizing Pharmaceutical Manufacturing: From Batch to Continuous Production, Journal of Pharmaceutical Innovation, Vol 10, Issue 3, Sept 2015,
- The product filling pre-dates the complexity pyramid by 4 years but in hindsight the product was filled with a level 2 CS, but with the process knowledge to allow level 1 control

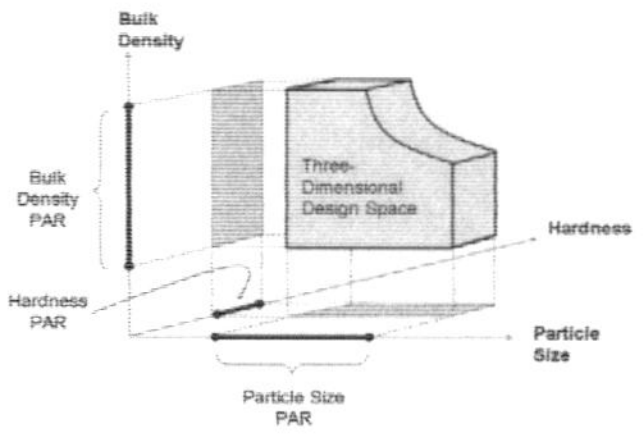

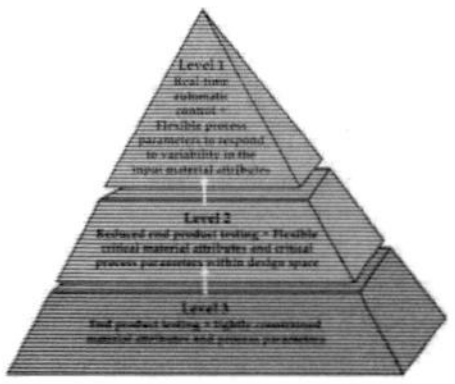

Malvern Panalytical

Control Strategy Implementation

- Dashboard sits on plant control system showing
- Predicted disso ranges shown to SDD process operator
  - If particle size and bulk density stay in the green range, the tablet compression set points are wide
  - If particle size and bulk density are in the yellow range, the tablet compression set point becomes critical
  - If the particle size goes red, there are no approved compression set points that will give acceptable product performance
- In reality the process runs in a NOR where we are always making good product
- Key message is we have real-time indicator of CQA predictions from PAT and parametric data

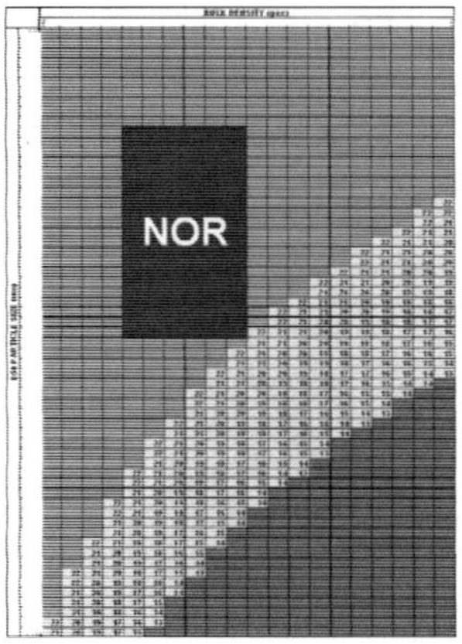

Malvern Panalytical

Key Question
How do we get that data?

Malvern Panalytical

Particle Size Application
Ensuring Representative Measurement

- Appropriate selection of sampling location
- Powder flow is linear (post vortex breaker)
  - However this means it is the 'shadow' of the vortex breaker
- Tried all commercial sampling systems
  - J-type, single funnel, etc
- Poor correlation with off-line data
- Implemented sampling arrangement is the result of empirical testing of various restrictions and funnel configurations
- Has been proven to generate the equivalent results
- Covers 25% of the cross sectional area, i.e. 25% of production runs through the analyser!

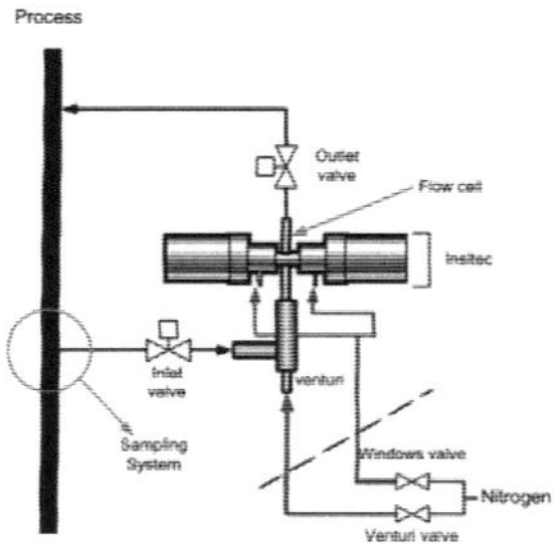

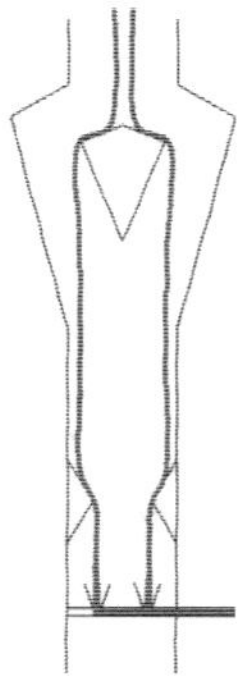

Malvern Panalytical

Prediction models

- First thing to realize is these the regression analysis is to SDD particles which had already been through the secondary dryer
  - Output of spray drying still has some residual solvent
  - Need to predict 'dry' values from 'wet' data
  - Models using process data predict dry particle size and dry bulk density from achieved set-points
- Secondly off-line methods are not the same as PAT methods
  - PAT methods need to predict off-line values as off-line values have been used in the control models
- In terms of particle size
  - Off-line method uses same fundamental measurement technique (laser, scattering, Mastersizer)
    - BUT Mastersizer uses the Fraunhofer approximation, and the Insitec used Mie scattering theory
    - And the off-line method uses a pressure titration to distribute the particles, whereas the PAT method uses flow titration (velocity)
  - PAT method needs to predict off-line values

Malvern Panalytical

Initial correlation between PAT and off-line

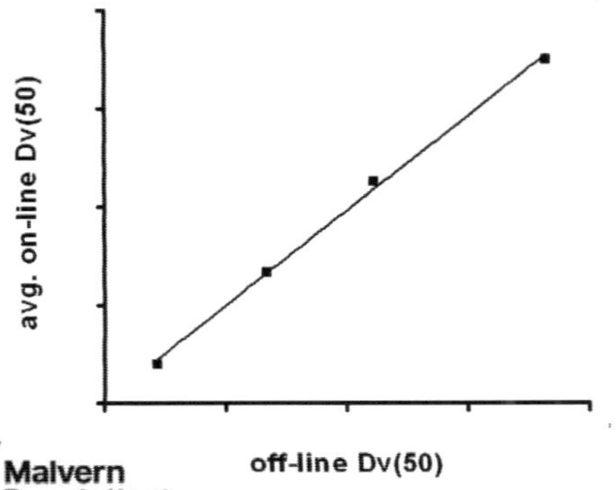

	Parameter	Value
Hardware	Online Laser Diffraction System	Insitec X
	Lens	200 mm (1-400 micron focal length)
	Venturi	90 degree
	Flow rate	9.5 Nm^3/hr
Material Settings	Model	Mie Scattering
	Particulate Refractive Index	1.5 + 0.1i
	Media Refractive Index	1
	Particle Density	1
Data Analysis	Min Particle Diameter (microns)	0.1
	Max Particle Diameter (microns)	600
	Scattering Threshold	8
	Multiple Scattering	Yes
	First Scattering Ring	5
System Suitability	Averaging	8-minute
	Transmission	≥ 80 %
	Counts at Sc(15)	≥ 500

Malvern Panalytical

How do we know this works?

- Example of a process change
- Correlation of process parameters to intermediate attribute means we can change process parameters to cause desired change in attribute
- Make a change expected to move the Dv50 by 3um
  - Dv50 responds immediately and stabilizes in 8mins (one block of data)
- More importantly we can set the press parameters to give the disso performance we require.

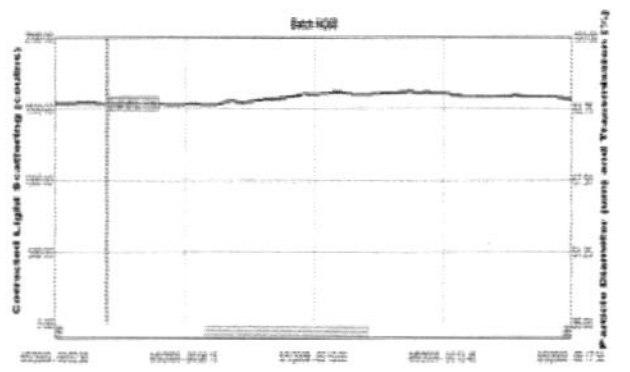

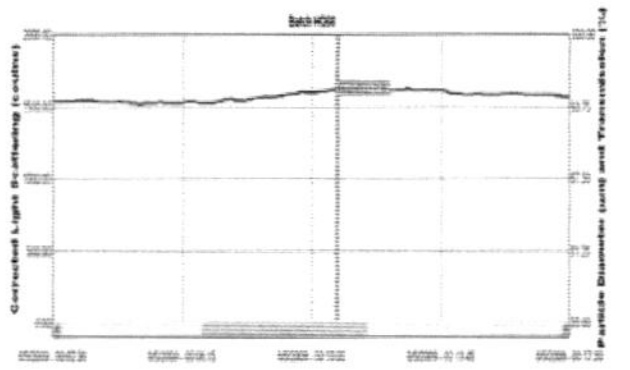

Malvern Panalytical

What does this all look like together?

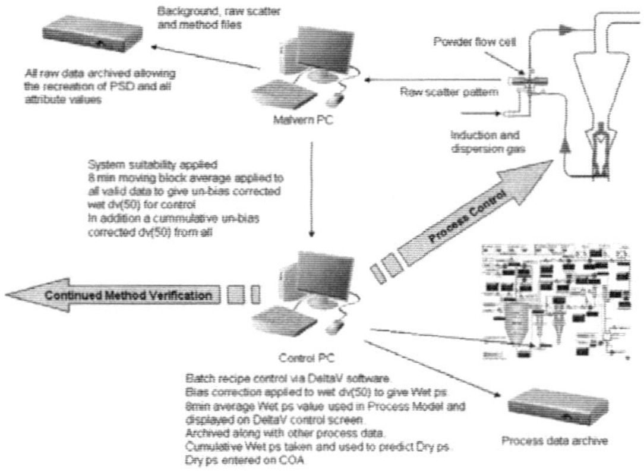

Malvern Panalytical

PAT Method Lifecycle

- Trend analysis of key performance indicators
  - Difference between predicted values (from achieved set-points)
    - 'parallel testing' of soft-sensor methods
    - Method updates
  - Difference between PAT measured values to off-line
  - Causes of PAT method system suitability (transmission at Sc15)
  - Predicting how process variance would impact the above
    - Proactive triggers for method updates

Malvern Panalytical

Differences between PAT and off-line methods

- Early PAT prediction methods of dry, off-line data where simple linear regressions (bias adjustments)
- However over time, and by using expanded ranges, we could show that the slope of all three PSD attributes (d(50), d(10) and d(90)) were not 45 degrees
- They are close enough to predict in the operational ranges
  - Fixed bias changed to slope adjustment
  - Fixed was OK, slope was better

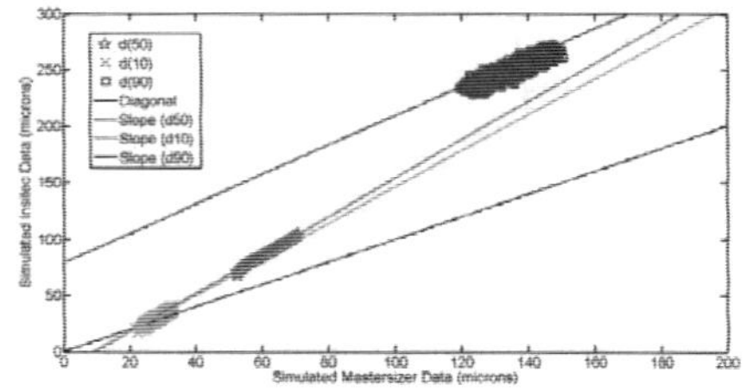

Malvern Panalytical

Characterizing Sc15 and transmission 'failures'

- PAT method uses transmission and counts on Sc15 as a system suitability
- Initially these were simple acceptance values
  - Found to be more sophisticated
- When transmission passes the PSD is mono-modal and SC15 has highest count
- As actually the system suitability should be on signal 'shape'

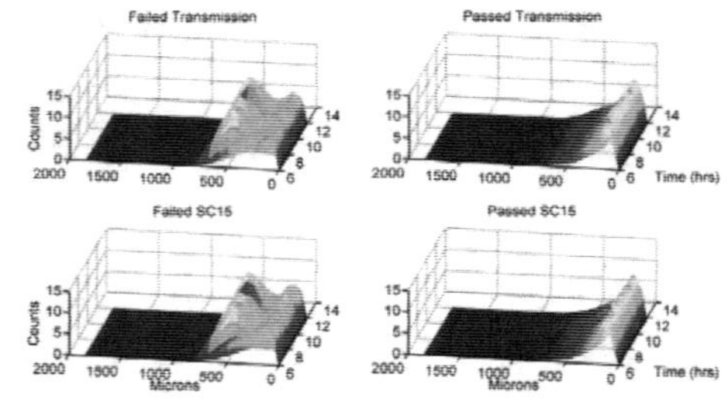

Malvern Panalytical

Impact of viscosity of d(50)

- During development an acceptance range for API/polymer solution was defined
- During production it was found that the viscosity of API/polymer solution varied (within range) and that if it varied the d(50) that would be generated changed
  - Suggested from regression analysis (positive correlation with %TS)
  - Fine if we stay in the narrow operating range, but if the range was expanded then a set-point correction should be applied to correct for cP
- However during the life cycle of the product various factors were identified which impacted viscosity
  - New acceptance tests for raw materials used in spray solution

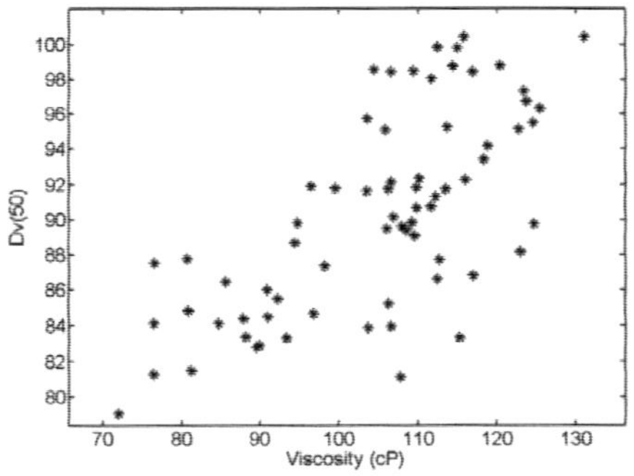

Malvern Panalytical

Continued application development

- Product underwent tech transfer to a bigger spray dryer, in a different facility
- Insitec installation moved upstream
  - Directly below drying tower
  - Before cyclone
- Measurement across entire discharge pipe
- Even then the installation required fine tuning
  - Lens were originally parallel, found to be reflection (multiple bounces) impacting scatter pattern
  - Original installation was across the center of the pipe, Insitec now rotated to look 'up-down' rather than 'left-right'
    - To account for stratified flow in the pipe
  - Method updated to include setting the 'zero' when process is spraying solvent
    - Changes refractive index

Malvern Panalytical

Conclusions

- In general, understand the inputs to tablet performance
  - Fishbone/Ishikawa diagrams are very useful
- Collect the data (either directly or via modelling) to have representative inputs into those diagrams
  - Use high frequency data collection rates to for control models and design space time limits, averaged data to determine product quality attributes
  - Justify data collection and decision frequency
- Make sure the data is weighted appropriately
  - For example in tablets with high drug load, weight variance is more important than low drug load, where variance in concentration has a bigger impact
- If you do, then full real-time release is achievable and has been achieved.

Malvern Panalytical

Acknowledgements

- Vertex
  - Joseph Medendorp, John Bric, Kelly Tolton, Greg Connelly
- Hovione
  - Filipe Gaspar, Nuno Matos, Filipe Neves
- Malvern Panalytical
  - Sensors and automation solutions team

Thank you for your attention.....

Malvern
Panalytical

490592 Impurity Rejection in Crystallization â€"Quantification and Mechanisms

Abstract:

Other than isolating the product as a solid, the perhaps most important objective in crystallization of pharmaceuticals is purification. Nevertheless, most of the engineering research in academia and industry has been focused on particle size and polymorph control. Notwithstanding their importance, crystallization as a unit operation is relied on not just in the API step but across the entire synthetic scheme to reject a number of different impurities. These can be reagents, reaction by-products, process impurities, degradants, chiral impurities, inorganic impurities and even genotoxic impurities. Purification via crystallizations is thus critical in producing a drug substance that is suitable for pre-clinical and clinical trials and eventually for market demands. A stringent regulatory oversight is also in place to ensure that the API meets all purity and quality attributes prior to administration to humans.

In Pharma industry, most of the impurity control in crystallization has traditionally been carried out by synthetic chemists who mainly rely on a before and after approach that is highly empirical in nature. Presented herein is the use of engineering principles to accurately quantify rejection of impurities throughout the crystallization. Various mechanisms of impurity rejection are identified and linked to thermodynamic and/or kinetic effects. Case studies from pharmaceutical development are presented that provide a clear understanding on the impurity purge mechanisms and how they are affected by process conditions. This work thus enables simple control measures to be introduced in the crystallization process that are capable of producing materials with targeted purity profile, whether that is for toxicity studies or for clinical trials.

492027 Bi-Directional Control of Glycosylation of Antibodies

Abstract:

Monitoring and control of critical quality attributes (CQAs) is important for biologics manufacturing. CQAs need to be controlled within desirable ranges in order to manufacture a clinically effective product. These ranges can be particularly tight for biosimilars, because similarity to the innovator molecule must be demonstrated. To address these challenges, new flexible methods of monitoring and modulating CQAs are required. Multivariate model predictive control (MPC) can be used to target CQAs within desirable ranges to ensure the delivery of high quality biologic products. In this study, MPC was applied to control the CQA of interest (CQA1) of a recombinant antibody by using two media additives and a cell culture parameter as modulation levers. A partial factorial experiment (n=25) was first conducted in bench-top bioreactors to understand the impact of these levers on CQA1. Both media additives were able to modulate CQA1 but in opposite directions. The cell culture parameter alone was able to modulate CQA1 in both directions. After these levers were identified and characterized, an empirical multivariate model was created to predict and control CQA1 during the cell culture process. Next, the robustness of this control strategy was tested in 3 L bench-top and 100 L pilot-scale bioreactors. For the test, CQA1 was purposefully deviated in either direction (high or low) away from the target using the cell culture parameter as a lever. Our results indicate that the model effectively predicted the amount of media additive required to bring CQA1 back to the target. The target CQA1 range was met for both scales at the end of the cell culture. Thus, the use of media additives to achieve bi-directional model predictive control of CQA1 has been achieved. This study demonstrates that MPC is a robust tool to modulate CQAs within desired targets for cell culture processes.

492107 Bioprocess Optimization for Production of Recombinant Proteins in Chinese Hamster Ovary Cells in Consideration of Metabolic Shifts By Shear Stress

Abstract:

Biopharmaceuticals such as monoclonal antibody drugs have attracted considerable interest in recent years for currently untreatable diseases. A series of large production plants have been built in different parts of the world. Chinese hamster ovary (CHO) cells are widely used as a host cell line and cultured in fed-batch cultivation in a large stirred bioreactor for commercial production of biopharmaceuticals. When a bioreactor is designed for commercial-scale production, it is indispensable to evaluate the impact of shear stress on cell growth and cell viability due to their mechanical fragility. In this study we focused on the impacts of shear stress on cellular metabolism to optimize fed-batch cultivation for production of tissue plasminogen activator (tPA) in CHO cells.

We went through the two steps to optimize a fed-batch cultivation process. The first step is the quantitative evaluation of the effects of shear stress. In the case of a stirred bioreactor, a quantitative evaluation is difficult to be performed due to the distribution of shear stress in a tank. Therefore we introduced a flow chamber which can impart a steady laminar shear stress on the cells for the evaluation of the effects of shear stress and found that metabolism shift occur with increased shear stress. We determined the appropriate range of shear stress for the production from the results. The second step is the optimization of an impeller configuration and feed medium in consideration of the metabolism shift. We designed 3L bioreactors which sustain appropriate shear stress in a tank using computational fluid dynamics (CFD) and feed medium in consideration of metabolism shift. CHO cells producing tPA protein were cultured with our optimized bioprocess. As a result, the production of tPA increased by 1.5 times compared to the conventional fed-batch process.

We found that metabolism shift occur by increasing shear stress and the optimization of a bioprocess taken into account the metabolism shift contributes to increasing productivity.

Using machine learning tools in bioprocess scale up when the number of batches is small

Viktor Konakovsky
Minneapolis, MN, US
Nov 2nd 2017

FUJIFILM Diosynth biotechnologies · Bio RAPID · Newcastle University · DTU Danmarks Tekniske Universitet · BIOPRO World Talent Campus

Topics

- What was happening in my project BioRapid so far
- Today I will talk about using PCA in combination with TOST time series
- The case study will focus on quantifying equivalence between 2L and 200L

FUJIFILM Diosynth biotechnologies · Bio RAPID · Newcastle University · DTU Danmarks Tekniske Universitet · BIOPRO World Talent Campus

2

My project BioRapid

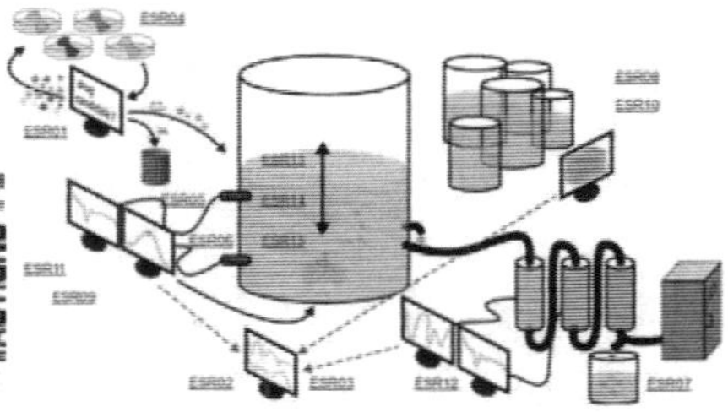

Meet my Team

http://www.bio-rapid.eu/
Financial support from the EU-Horizon 2020 Marie Skłodowska-Curie actions (MSCA) ITN project BIORAPID (no. 643056).

FUJIFILM Diosynth biotechnologies | Bio RAPID | Newcastle University | DTU Danmarks Tekniske Universitet | BIOPRO World Talent Campus

Motivation

- This work was initiated while writing a report...

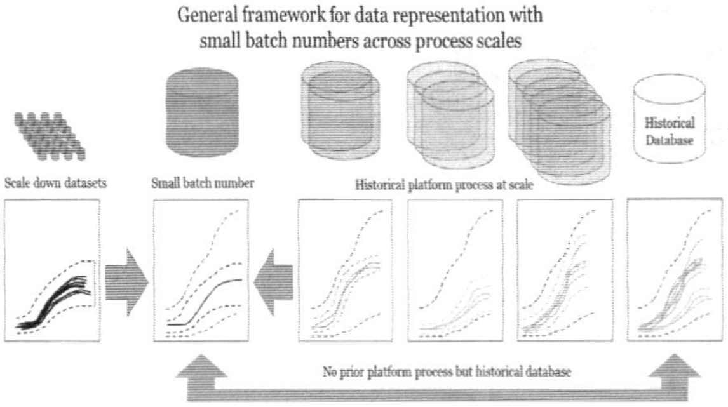

Bio RAPID

"The Rapid Bioprocess Development ITN, effective training framework in rapid development of novel bioactive molecules from the very early stages of potency and efficacy testing to the bio-manufacturing process characterization and monitoring"

EU-Horizon 2020 - Marie Curie ITN project BIORAPID (Grant Agreement n° 643056)

Deliverable D2.5: General framework for data representation with small batch numbers, across process scales

- ...and we are writing up to publish the results soon in Biotechnology Journal

FUJIFILM Diosynth biotechnologies | Bio RAPID | Newcastle University | DTU Danmarks Tekniske Universitet | BIOPRO World Talent Campus

Scientific output

HT process characterization across scales

Which scale down model heuristics should we use in ambr bioreactors?

Data automation & Augmentation

Application of machine learning on biotechnology datasets in R

Scale down model characterization

An industrial perspective on using ambr as scale down models in mammalian cell culture

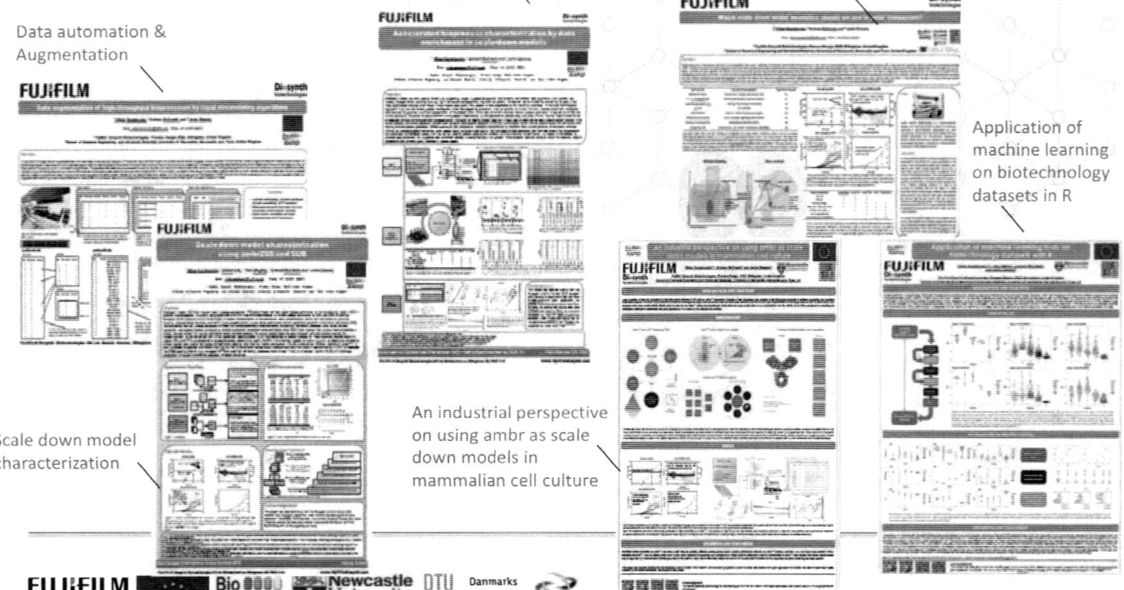

FUJIFILM Diosynth biotechnologies | Bio RAPID | Newcastle University | DTU Danmarks Tekniske Universitet | BIOPRO World Talent Campus

Introduction

My part is to get value out of databases, and to develop predictive modeling and machine learning tools on dedicated industrial projects. Those projects were mostly focused on high throughput upstream data sources in mammalian cell culture but also extend to CLD and to microbial manufacturing

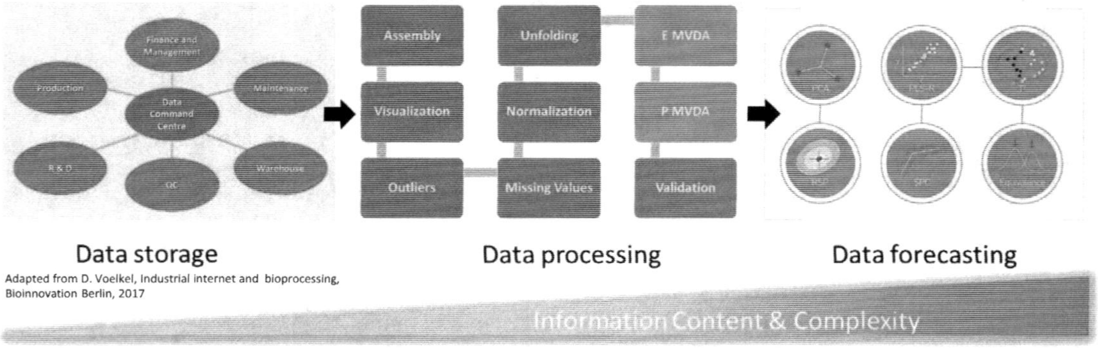

Adapted from D. Voelkel, Industrial internet and bioprocessing, Bioinnovation Berlin, 2017

FUJIFILM Diosynth biotechnologies | Bio RAPID | Newcastle University | DTU Danmarks Tekniske Universitet | BIOPRO World Talent Campus

The case

- In HT (high throughput) we have a lot of data, but as we scale up this is not the case any more
- Often at this stage we want to know whether batches are equivalent, or if not, what is different?
- This dataset is a CHO cell culture fed-batch

Scale	***Regime***	***n***
2L	C	6
2L	D	9
10L	E	2
200L	D	1
sum		**18**

The TOST

- Question: Is data from group A equivalent to group B?
- Typically would use TOST (two one sided t-test) to quantify a significant difference
- Statisticians in bioprocessing often use 2 standard deviations (SD) of a reference group to calculate delta

If a data distribution is approximately normal then **about 68 percent** of the data values are within one standard deviation of the mean (mathematically, μ ± σ, where μ is the arithmetic mean), **about 95 percent** are within two standard deviations (μ ± 2σ), and **about 99.7 percent** lie within three standard deviations (μ ± 3σ ...

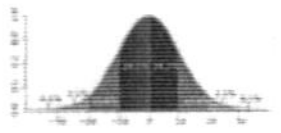

Standard deviation - Wikipedia
https://en.wikipedia.org/wiki/Standard_deviation

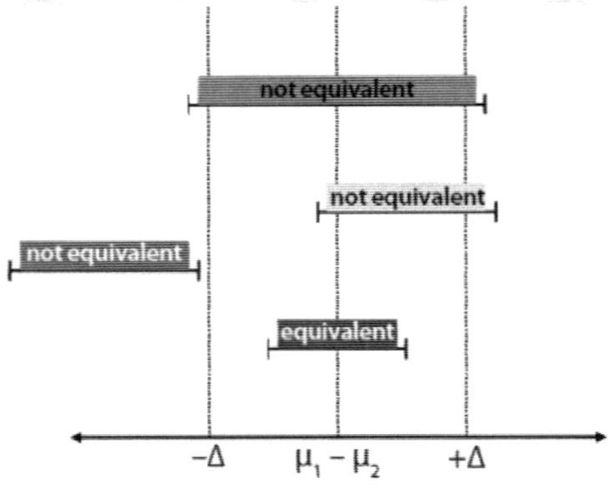

Ref: C. DiCesare et al.: Development, Qualification, and Application of a Bioreactor Scale-Down Process: Modeling Large-Scale Microcarrier Perfusion Cell Culture." *BioProcess International, 2016*

The challenge

- What if the number in one group is just 1? -> Time series TOST testing
- Definition of how many SD are used to form the equivalence interval
- Time series could be problematic, rules for passing/failing equivalence are required
- What if one variable is equivalent, another is not, the next is, the nth variable again is not...

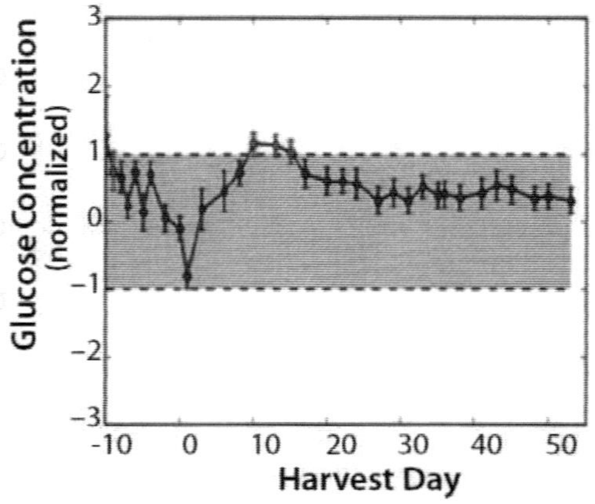

DiCesare used a rule of ‚three in a row' as maximum acceptable equivalence outliers

Ref: C. DiCesare et al.: Development, Qualification, and Application of a Bioreactor Scale-Down Process: Modeling Large-Scale Microcarrier Perfusion Cell Culture." *BioProcess International, 2016*

FUJIFILM Diosynth biotechnologies Bio RAPID Newcastle University DTU Danmarks Tekniske Universitet BIOPRO World Talent Campus

The challenge

- Especially relevant for CQAs or CPPs (critical quality attributes or critical process parameters)
- If it was critical to be equivalent, it becomes relevant how close equivalence was missed
- Sometimes those differences can be explained, but often this is means lengthy investigations have to be done

Variable	***Equivalent?***	***Was it a CQA or CPP?***	***P value debate***
V1	Pass	Yes	OK
V2	Pass	Yes	OK
V3	Pass	No	OK
V4	Pass	Yes	OK
V5	Fail	Yes	Clear fail
V6	Pass	Yes	OK
V7	Fail	Yes	Borderline fail
V8	Pass	Yes	OK
V10	Pass	Yes	OK
...			

FUJIFILM Diosynth biotechnologies Bio RAPID Newcastle University DTU Danmarks Tekniske Universitet BIOPRO World Talent Campus

The opportunity

- Are there any tools that allow us to capture most of the variance in the process?
- Can we get a quantitative measure of how equivalent the runs are in %, including which variables are equivalent and which are not?
- Today I present how PCA can be combined with a TOST test on time series to do exactly that using the TOSTER package in R

Ref: D Lakens, Two One-Sided Tests (TOST) Equivalence Testing, Package 'TOSTER', 2017

Goal

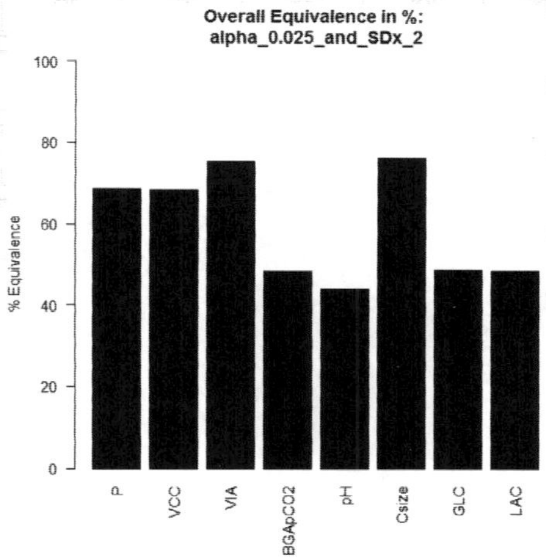

Total Equivalence (TEQ): 59.78%

Data

- Prepare dataset by filtering the data into subsets of all batches per day
- Data was mean centered and divided by the SD before extracting the scores with PCA

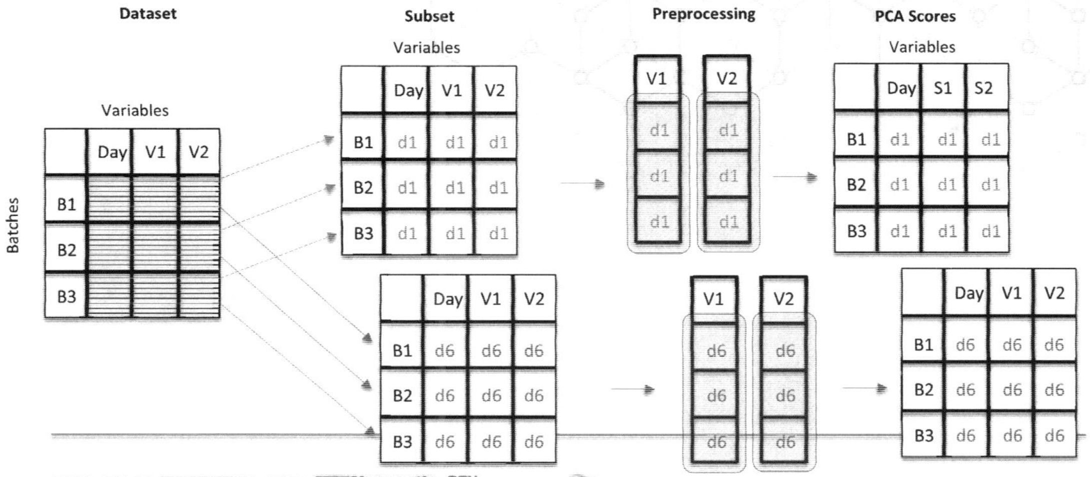

Workflow

- Take the scores and run a TOST test on it using the parameters alpha and SD multiplier to define whether these PCs on that day are equivalent or not

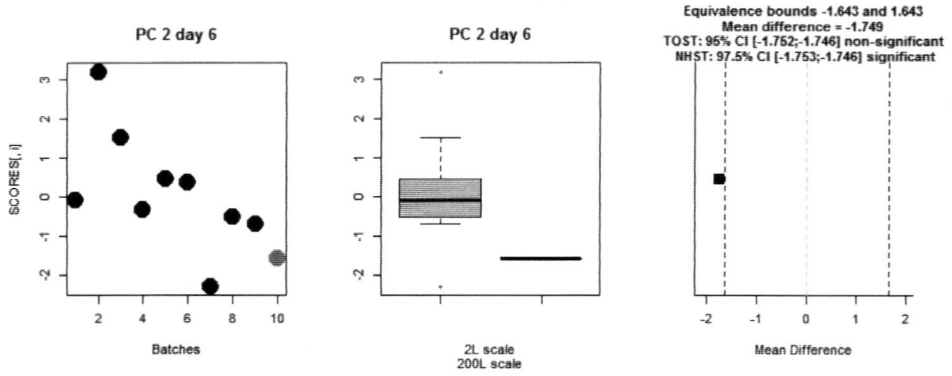

- alpha_0.025_and_SDx_1

FUJIFILM Diosynth biotechnologies · BioRAPID · Newcastle University · DTU Danmarks Tekniske Universitet · BIOPRO World Talent Campus

13

Workflow

- Take the scores and run a TOST test on it using the parameters alpha and SD multiplier to define whether these PCs on that day are equivalent or not

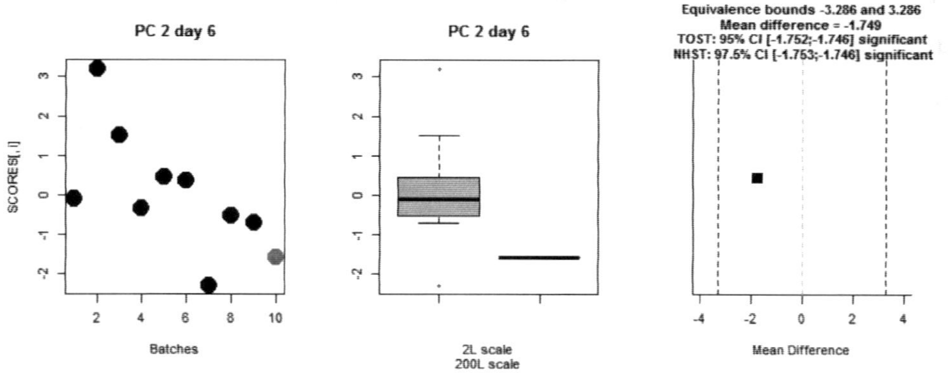

- alpha_0.025_and_SDx_2

FUJIFILM Diosynth biotechnologies · BioRAPID · Newcastle University · DTU Danmarks Tekniske Universitet · BIOPRO World Talent Campus

14

Workflow

- Take the scores and run a TOST test on it using the parameters alpha and SD multiplier to define whether these PCs on that day are equivalent or not

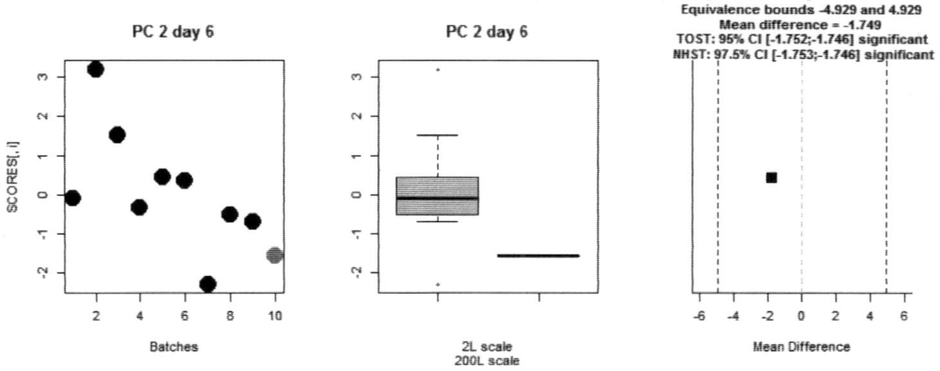

- alpha_0.025_and_SDx_3

FUJIFILM Diosynth biotechnologies | Bio RAPID | Newcastle University | DTU Danmarks Tekniske Universitet | BIOPRO World Talent Campus

Workflow

- The parameter alpha is not relevant in this example as there is only one observation in one of the groups. If would determine if the data point lies within or outside of equivalence.

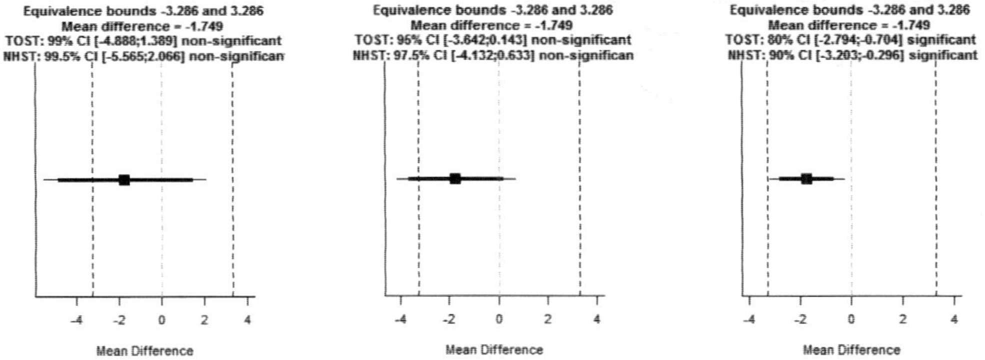

- alpha_0.005_and_SDx_2 alpha_0.025_and_SDx_2 alpha_0.1_and_SDx_2

FUJIFILM Diosynth biotechnologies | Bio RAPID | Newcastle University | DTU Danmarks Tekniske Universitet | BIOPRO World Talent Campus

Workflow

- Different combinations of conditions, alpha and SDmultiplier, are evaluated in a grid search because they have a strong impact on TOST

alpha = 0.005, 0.010, 0.025, 0.050, 0.100
SDmultiplier = 1.0, 1.5, 2.0, 2.5, 3.0

- Create a weight matrix: which PCs were equivalent or not

Name	PC1	PC2	PC3	PC4	PC5	PC6	PC7	PC8
alpha_0.005_and_SDx_1	1	0	0	1	0	0	0	0
alpha_0.01_and_SDx_1	1	0	0	1	0	0	0	0
alpha_0.025_and_SDx_1	1	0	0	1	0	0	0	0
alpha_0.05_and_SDx_1	1	0	0	1	0	0	0	0
alpha_0.1_and_SDx_1	1	0	0	1	0	0	0	0
alpha_0.005_and_SDx_1.5	1	1	0	1	0	0	1	0
alpha_0.01_and_SDx_1.5	1	1	0	1	0	0	1	0
alpha_0.025_and_SDx_1.5	1	1	0	1	0	0	1	0
alpha_0.05_and_SDx_1.5	1	1	0	1	0	0	1	0
alpha_0.1_and_SDx_1.5	1	1	0	1	0	0	1	0
alpha_0.005_and_SDx_2	1	0	0	1	0	1	1	0
alpha_0.01_and_SDx_2	1	1	0	1	0	1	1	0
alpha_0.025_and_SDx_2	1	0	0	1	0	1	1	0
alpha_0.05_and_SDx_2	1	1	0	1	0	1	1	0
alpha_0.1_and_SDx_2	1	1	0	1	0	1	1	0
alpha_0.005_and_SDx_2.5	1	1	0	1	0	1	1	0
alpha_0.01_and_SDx_2.5	1	1	0	1	0	1	1	0
alpha_0.025_and_SDx_2.5	1	1	0	1	0	1	1	0
alpha_0.05_and_SDx_2.5	1	1	0	1	0	1	1	0
alpha_0.1_and_SDx_2.5	1	1	0	1	0	1	1	0
alpha_0.005_and_SDx_3	1	1	0	1	0	1	1	0
alpha_0.01_and_SDx_3	1	1	0	1	0	1	1	0
alpha_0.025_and_SDx_3	1	1	0	1	0	1	1	0
alpha_0.05_and_SDx_3	1	1	0	1	0	1	1	0
alpha_0.1_and_SDx_3	1	1	0	1	0	1	1	0

Workflow

- Let us have a look what those PCs represent - weighting of all PCs, on day 6

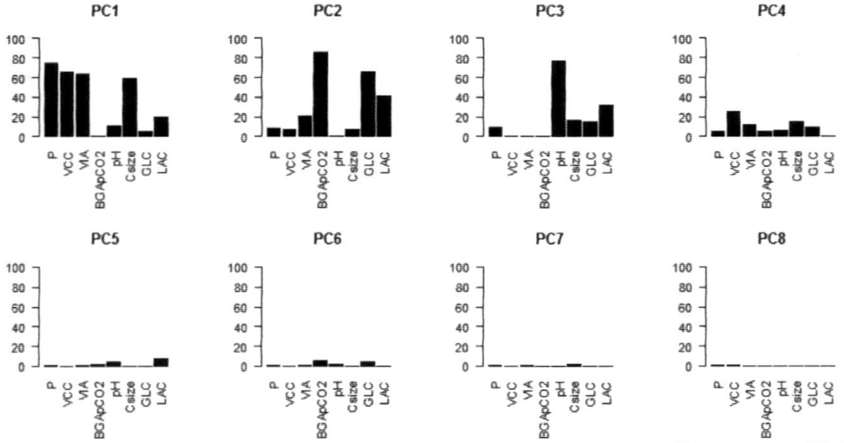

Name	PC1	PC2	PC3	PC4	PC5	PC6	PC7	PC8
SD	1.72	1.54	1.23	0.88	0.41	0.38	0.2	0.13
Var%	37	30	19	10	2	2	1	0
CumVar	37	67	86	95	98	99	100	100

FUJiFILM Diosynth biotechnologies
Bio RAPID
Newcastle University
DTU Danmarks Tekniske Universitet
BIOPRO World Talent Campus

Results

- Sum up PCs passing the TOST test
- For every condition
- For every variable
- For every day
- The variance covered in the passing PCs gives a measure of overall equivalence in %

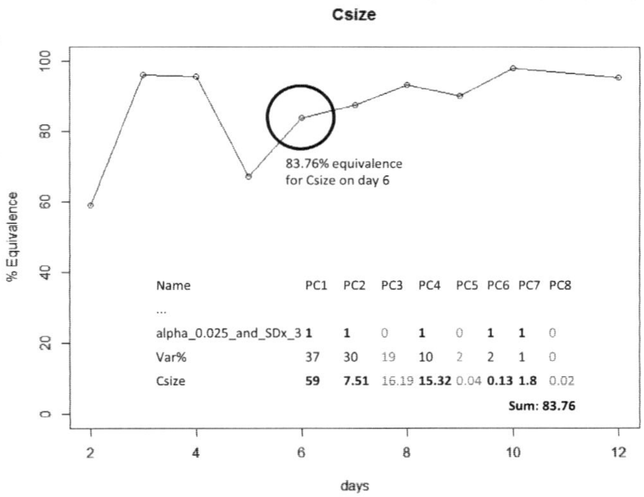

Results

- Visualize explained equivalence per variable as time profiles, for all days and all variables

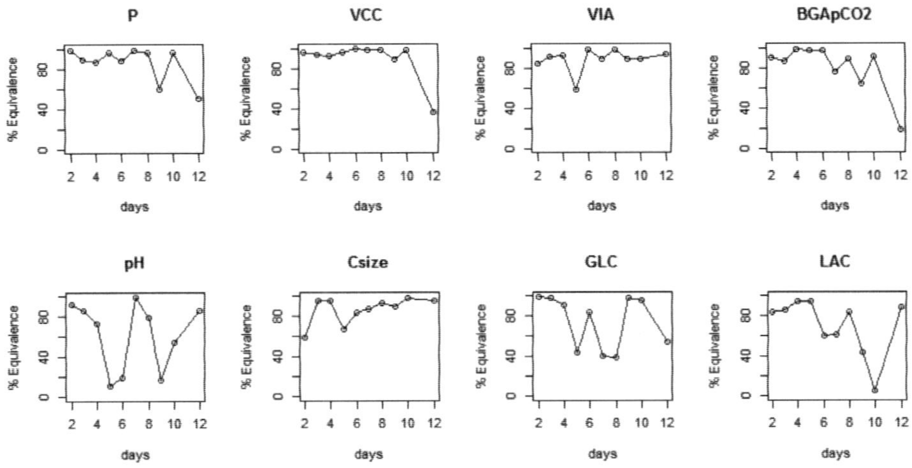

Results

- Get an overall metric for equivalence with a certain condition of alpha and SDmultiplier

Mean of all equivalent variables over the whole process time:

Total Equivalence (TEQ): 59.78%

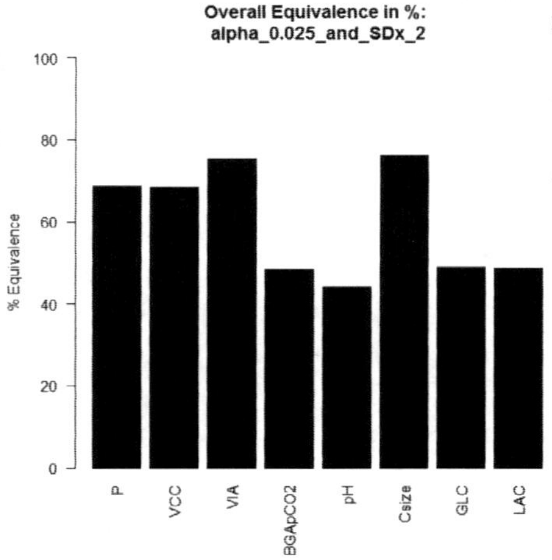

Results

- Visualize main differences per variable as time profiles

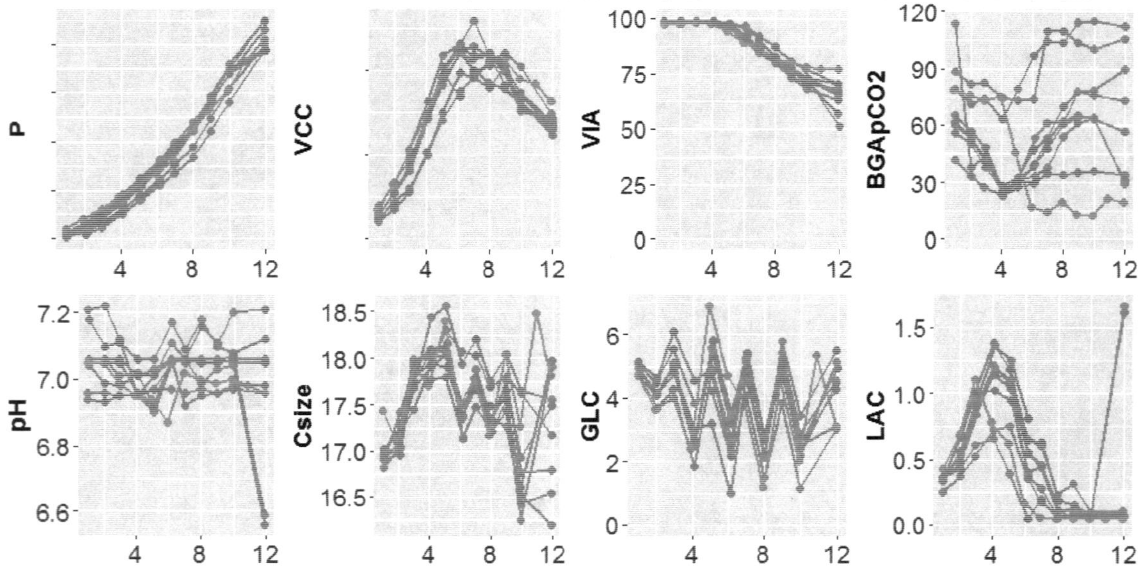

Results

- Get an overall metric for conditions tested at different levels of alpha and SD multipliers

TEQ: 28.17% **TEQ: 59.78%** **TEQ: 79.80%**

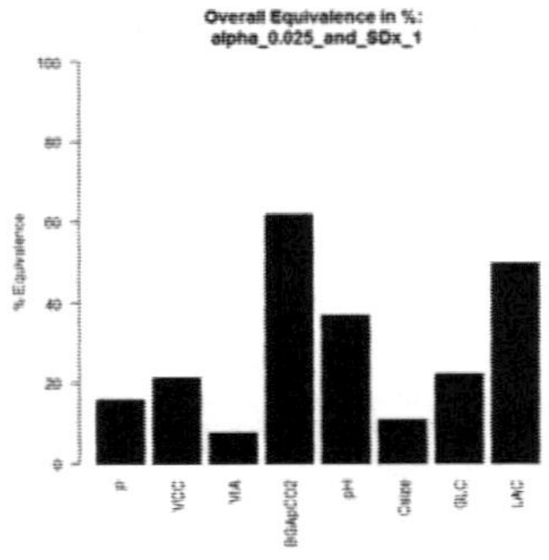

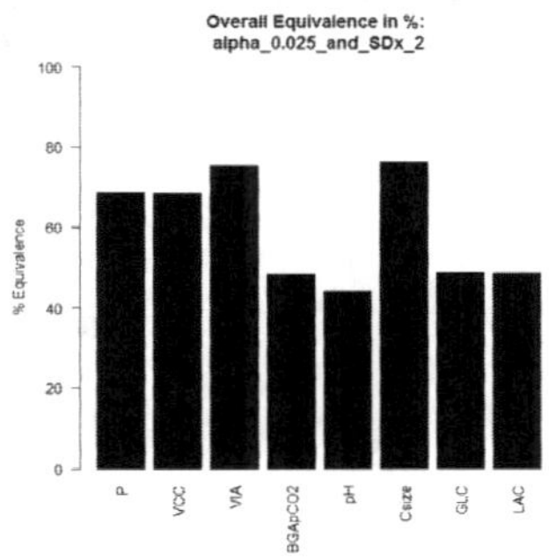

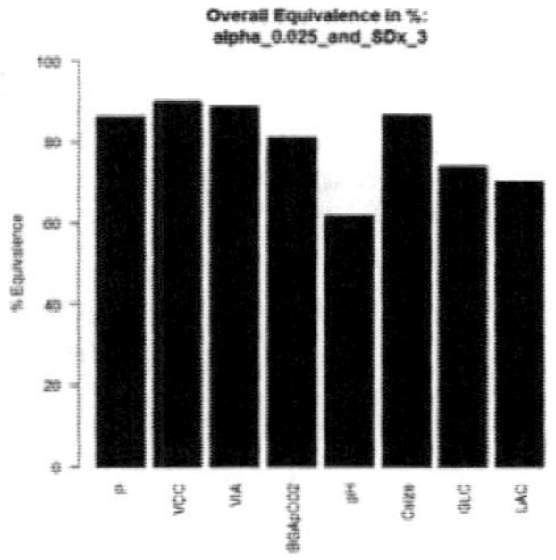

- alpha_0.025_and_SDx_1 alpha_0.025_and_SDx_2 alpha_0.025_and_SDx_3

FUJIFILM Diosynth biotechnologies · Bio RAPID · Newcastle University · DTU Danmarks Tekniske Universitet · BIOPRO World Talent Campus

23

Conclusions

- A method was presented that enables us to quantify equivalence with a novel metric TEQ
- The metric is based on PCs, which inherently explain the most variance in a dataset, and could be better suited than regular variables to compare groups holisticly using time series TOST
- Two major factors influencing the TOST (alpha and SDmultiplier) were evaluated for the new metric to allow for a better understanding of the TOST results
- This method summarizes which variables were most and least equivalent, which may help to find the root cause of undesireable process behaviour faster
- It is designed for HT datasets, but also worked for small batch numbers of n = 1, when data was scarce during scale up

FUJIFILM Diosynth biotechnologies · Bio RAPID · Newcastle University · DTU Danmarks Tekniske Universitet · BIOPRO World Talent Campus

24

Acknowledgements

Thanks to my Mentors
Graham McCreath, Head of Process Design, Fujifilm Diosynth Biotechnologies
Jarka Glassey, Professor, University of Newcastle

Thanks for continued support at Fujifilm Diosynth Biotechnologies!

Thanks to my colleagues and friends especially at NCL and DTU!

This project has received funding from the European Union's Horizon 2020 research and innovation programme under the Marie Curie Skłodowska-Curie grant agreement No 643056.

FUJIFILM Diosynth biotechnologies · Bio RAPID · Newcastle University · DTU Danmarks Tekniske Universitet · BIOPRO World Talent Campus

25

Results

- A different (non equivalent) case with more data: Regime C vs Regime D

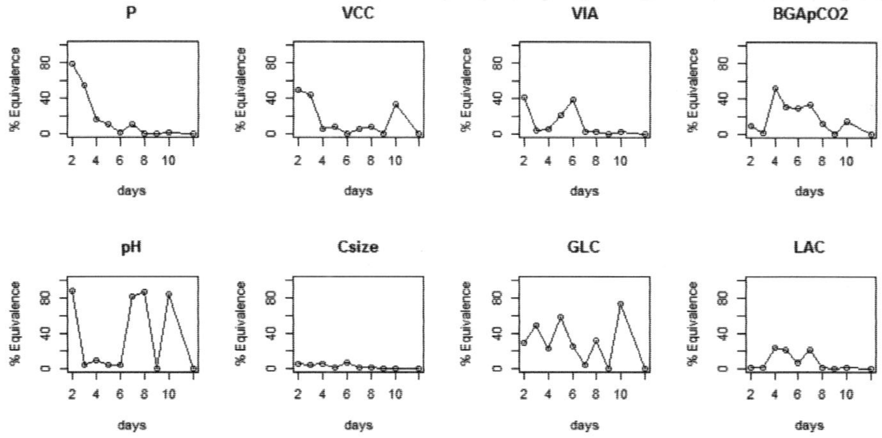

FUJIFILM Diosynth biotechnologies
Bio RAPID
Newcastle University
DTU Danmarks Tekniske Universitet
BIOPRO World Talent Campus

Results

- A different (non equivalent) case with more data: Regime C vs Regime D

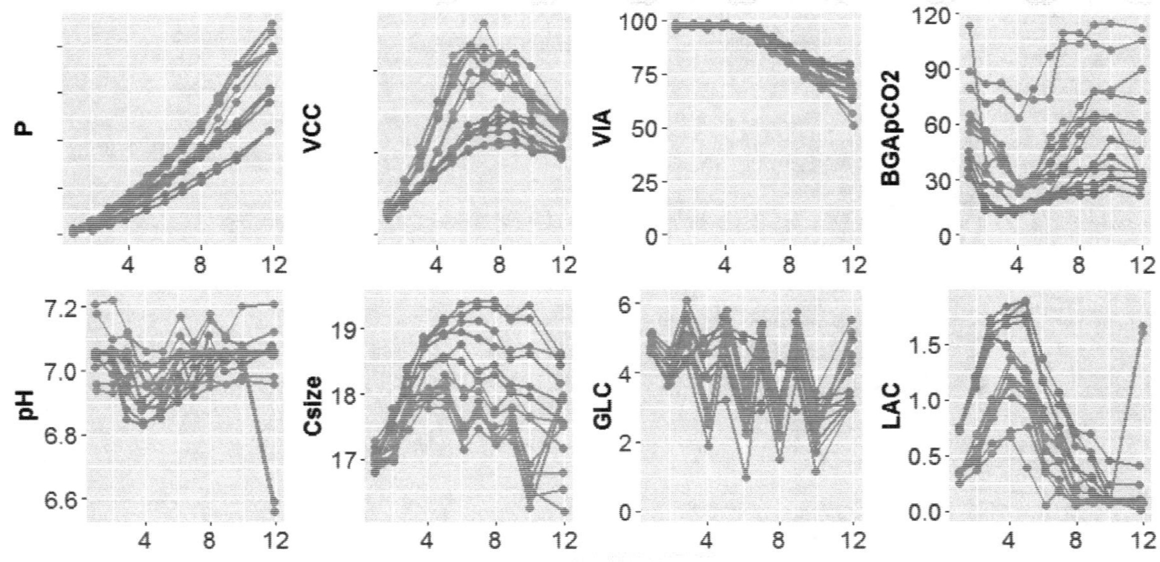

Results

- A different (non equivalent) case with more data: Regime C vs Regime D

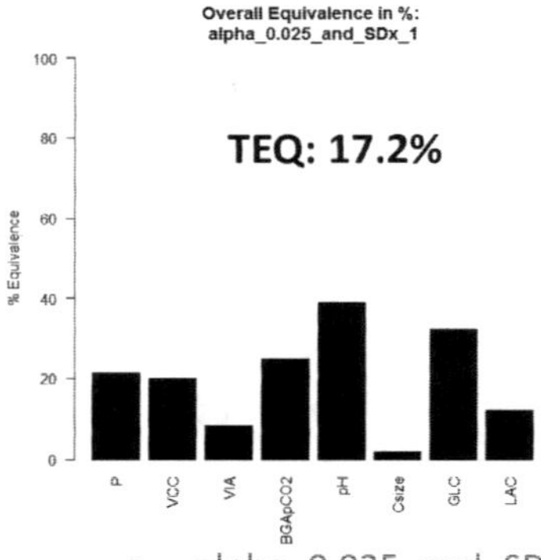

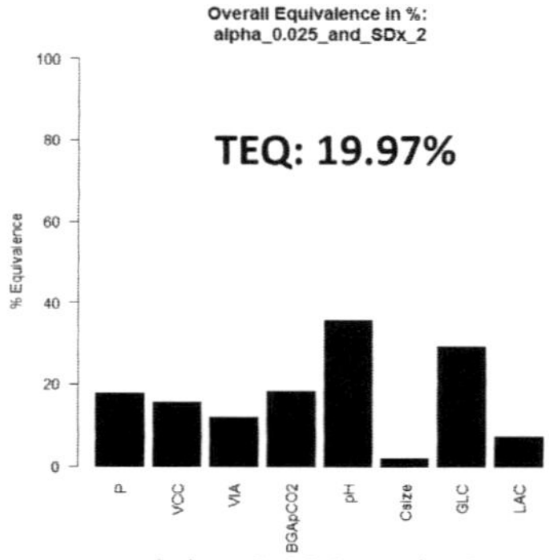

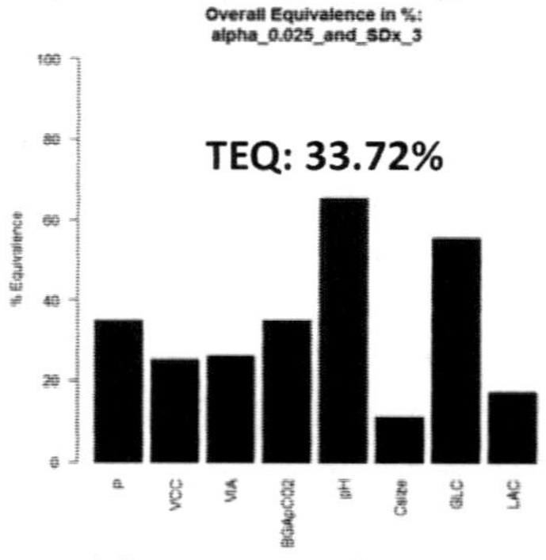

- alpha_0.025_and_SDx_1 alpha_0.025_and_SDx_2 alpha_0.025_and_SDx_3

FUJIFILM Diosynth biotechnologies · BioRAPID · Newcastle University · DTU Danmarks Tekniske Universitet · BIOPRO World Talent Campus

Workflow

- Combine weighted equivalence matrix with variance covered by individual variables.
- Not all PCs will be used, only those that passed the TOST test.

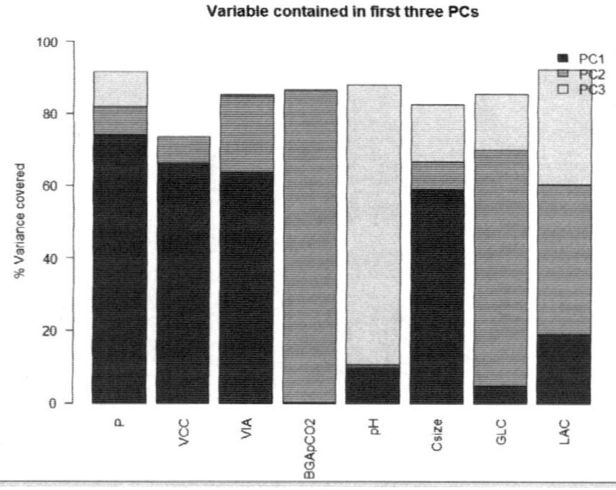

Name	PC1	PC2	PC3	PC4	PC5	PC6	PC7	PC8
...								
alpha_0.025_and_SDx_3	1	1	0	1	0	1	1	0
SD	1.72	1.54	1.23	0.88	0.41	0.38	0.2	0.13
Var%	37	30	19	10	2	2	1	0
CumVar	37	67	86	95	98	99	100	100
P	74.16	7.83	9.67	5.32	1.36	0.57	0.57	0.52
VCC	66.2	7.23	0.07	25.7	0.01	0.02	0.18	0.59
VIA	63.69	21.1	0.45	11.11	1.17	1.07	1.27	0.15
BGApCO2	0.18	86.35	0.12	4.95	2.57	5.81	0	0.01
pH	9.98	0.7	77.25	6.34	4.05	1.66	0.01	0.01
Csize	59	7.51	16.19	15.32	0.04	0.13	1.8	0.02
GLC	4.73	65.23	15.54	9.14	0.22	4.84	0.14	0.16
LAC	19.06	41.18	32.02	0.24	7.27	0.01	0.08	0.15

FUJIFILM Diosynth biotechnologies · BioRAPID · Newcastle University · DTU Danmarks Tekniske Universitet · BIOPRO World Talent Campus

A workflow-based framework for managing product analytical data and statistical results for lot release

G. Joglekar[1], Q. Cai[2], P. DeLaurentis[1], L. Mockus[1], K. Morris[2], G.V. Reklaitis[1]
[1]Davidson School of Chemical Engineering, Purdue University
[2]Lachman Institute for Pharmaceutical Analysis, Long Island Univ

AIChE Annual Meeting
Paper 665(e)
November 2, 2017

Outline

- Open data requirements
- KProMS system
- The collaborative research project
- KProMS benefits
- Conclusions

Objective

- Demonstrate the use of a workflow-based knowledge management system to support a collaborative research project
  - ✓ Data gathering
  - ✓ Data curation
  - ✓ Data analysis
  - ✓ Multiple groups
  - ✓ Open data

Open Data

- Directive to funding agencies to ensure that public has access to the results of government-funded research (Office of Sc and Tech Policy, 2013)
- The directive requires that data are
  - ✓ Richly described with machine-actionable metadata
  - ✓ Complete, self-explanatory, and accurate
  - ✓ Accessible and preserved beyond the life of a grant
  - ✓ Protected for confidentiality and privacy
- Private funders require that data be released with some degree of openness
- Companies would have similar requirements for internal use of confidential data

Data Sharing

- Even though more data are becoming available, they are not used widely because
  - ✓ Full context is not available
  - ✓ Meta-data is not available
- Data sharing among academic researchers very sporadic
- Tech transfers often time consuming
- The need for well curated data is expected to grow in the future
- There are no guidelines for good practices, or standards for publishing data
  - ✓ Putting it on the cloud is not sufficient

KProMS: Knowledge Provenance Management System

- KProMS framework produces shareable and reusable data by
  - ✓ Capturing the context of data
  - ✓ Streamlining data submission into a repository
  - ✓ Organizing data for machine and manual access over the web

KProMS Architecture

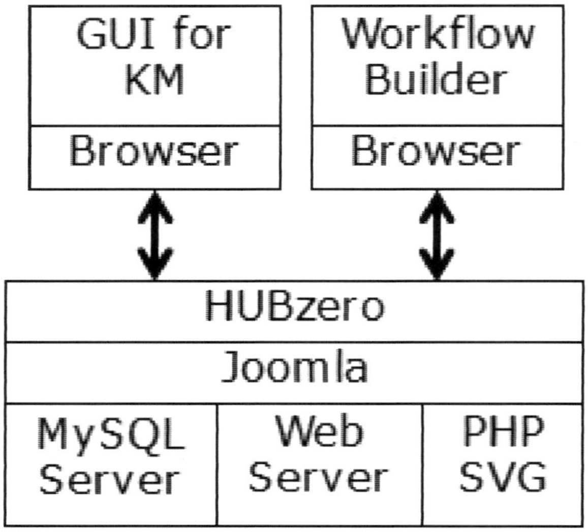

Joglekar, G., Giridhar, A., Reklaitis, G.V. "A Workflow Modeling System for capturing data provenance", Computers & Chemical Engineering (DOI: 10.1016/j.compchemeng.2014.04.006)
Joglekar, G., Giridhar, A., Reklaitis, G.V. "A Novel Knowledge Management System Based on Workflows", Computer Aided Chemical Engineering Series, Vol 33, Elsevier.

Collaborative Project

- Perform analytical tests associated with lot release on drugs purchased in open market
  - ✓ Dissolution
  - ✓ Content uniformity
  - ✓ Physical properties: Weight, Dimension, Mechanical strength
- Perform Bayesian analysis for hypothesis testing
- Funded by the FDA
  - ✓ Long Island University, Purdue University

Project Execution

- Model each experimental procedure as a workflow
- Procure drug to be studied
- Record each experiment as an instance of the associated workflow model
  - ✓ Parameter values
  - ✓ Upload instrument data
- Perform appropriate data checks
- Perform data analysis

Interactions of Working Groups

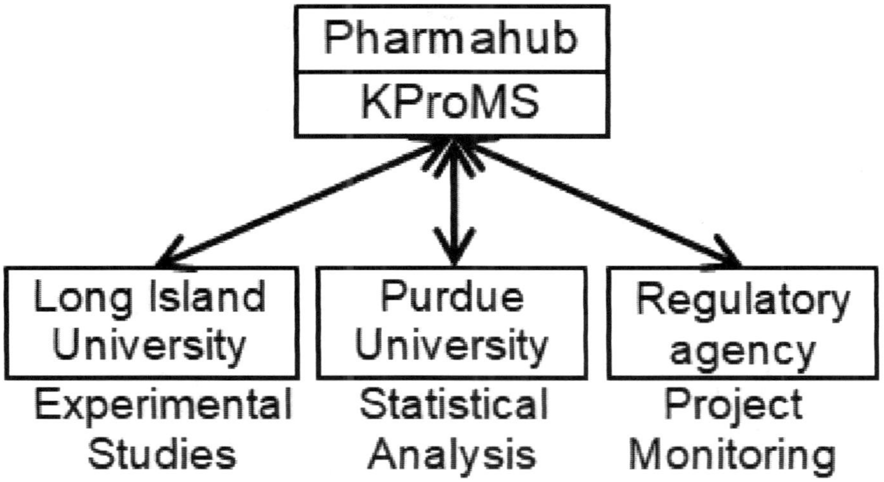

Dissolution Workflow

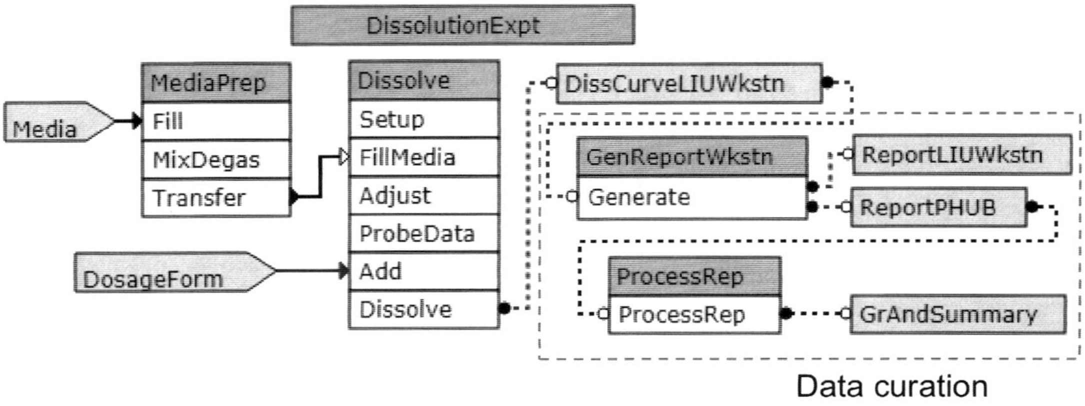

Parameters

		Parameter
Material:Media		Amount, liter
Material:DosageForm		Material lot, ID
Task:MediaPrep	*Subtask*:Fill	pH, number
	Subtask:MixDegas	Degas, Yes/No
Task:Dissolve		Number of vessels, integer
	Subtask:Setup	Bath temperature, C Accessory, Paddle/Basket
	Subtask:FillMedia	Amount per vessel, ml
	Subtask:ProbeData	Exposure Autoset, Yes/No Standard A blank, Yes/No Standard B blank, Yes/No Sample blank, Yes/No Standard A, g/l Standard B, g/l Response factor, Number Wavelength max, nm Path length, mm Calibration curve, Yes/No Baseline correction, Yes/No
	Subtask:Add	Number of units per vessel, number
Task:GenReportWkstn	*Subtask*:Generate	Program name, OPT-DISS
Task:ProcessRep	*Subtask*:ProcessRep	Library program, DigData Input file type, Dissolution report Output data, Pct diss and graph/Zip data

GrAndSummary Data Node

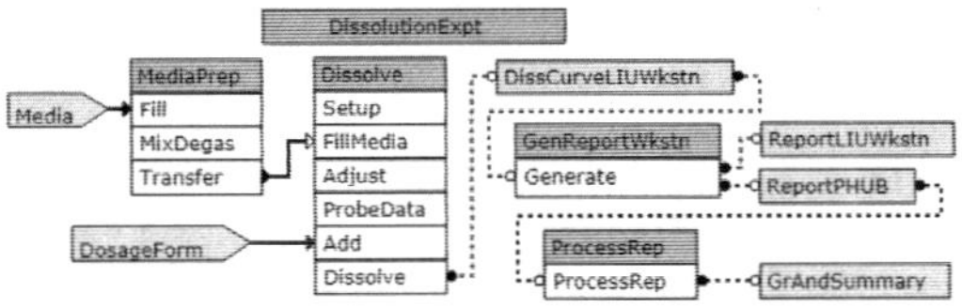

Number of tablets	6					
Dissolution end time	00:30:00					
Pct dissolved at end time	Ch1	Ch2	Ch3	Ch4	Ch5	Ch6
	104.95	100.99	99.90	102.41	102.74	97.26
Av Pct dissolved	101.3750					
Std dev	2.6447					

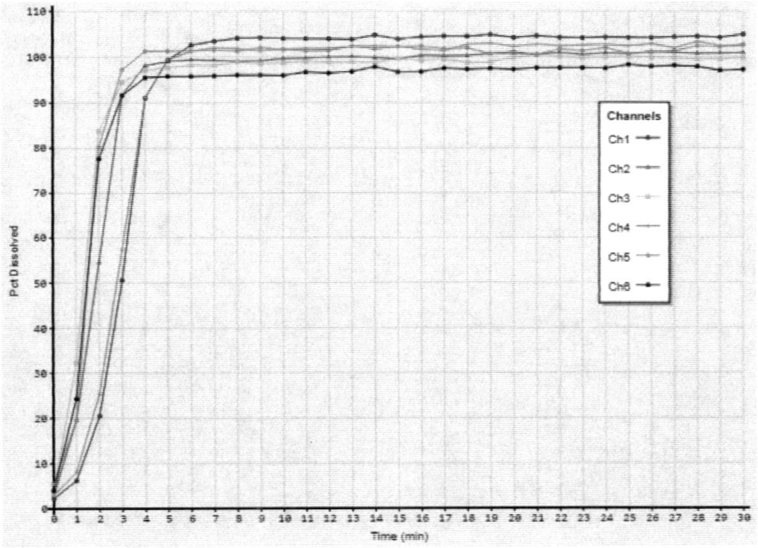

Workflows

Dissolution

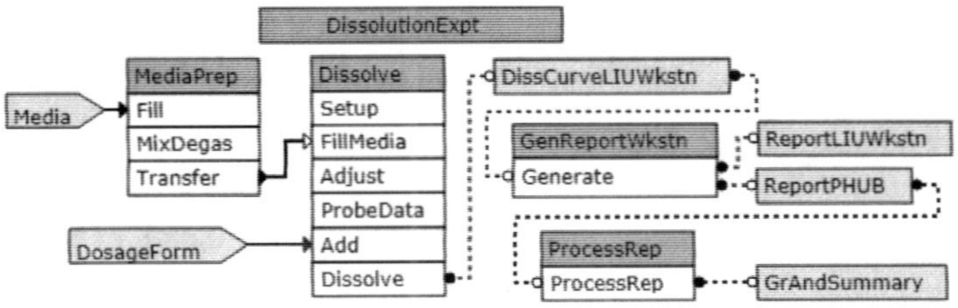

Content uniformity

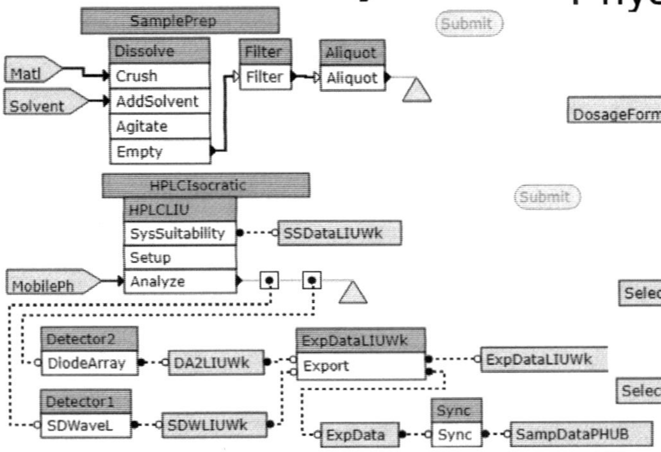

Physical Properties

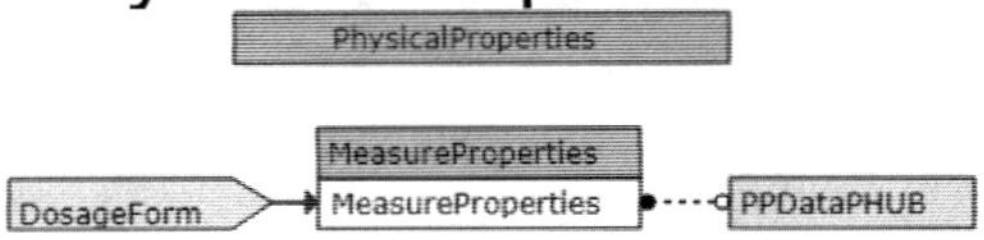

Bayesian Analysis

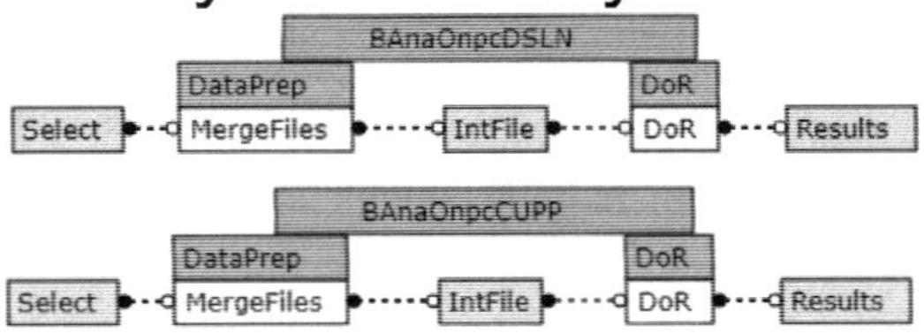

Bayesian Analysis

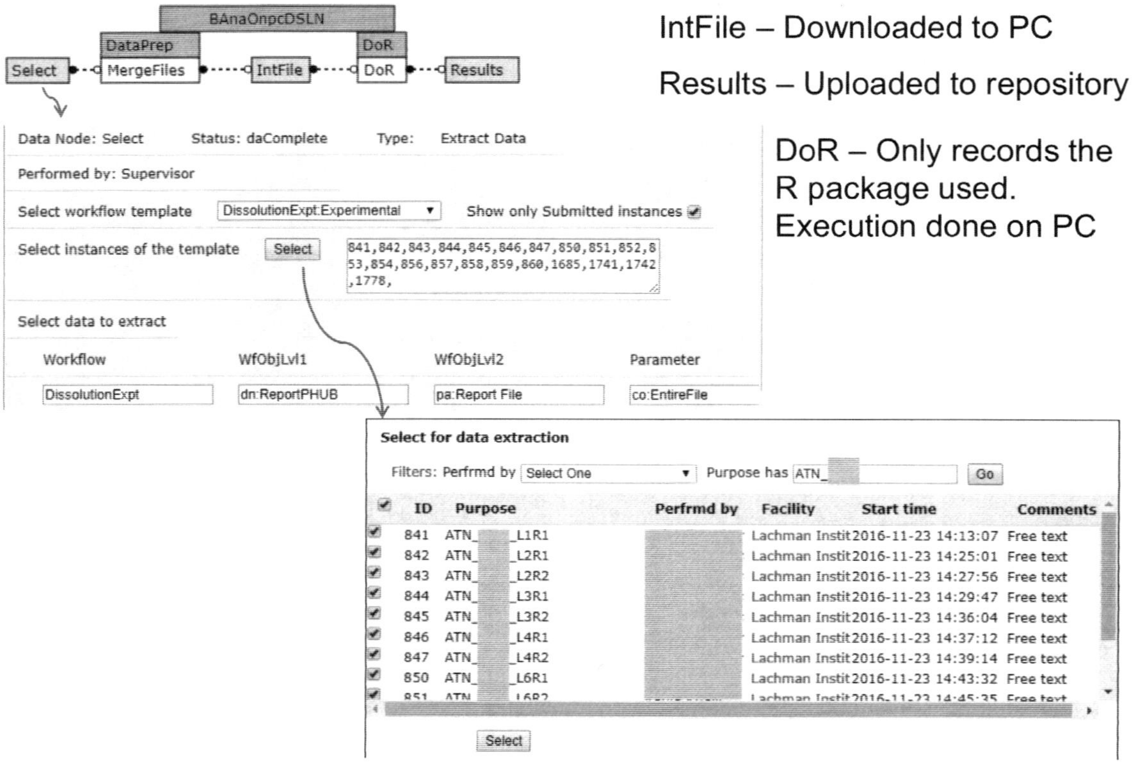

IntFile – Downloaded to PC

Results – Uploaded to repository

DoR – Only records the R package used. Execution done on PC

Benefits of Using KProMS

- Workflows tailored to the underlying procedure
- All steps and parameters explicitly defined
- Workflow developed based on input from all collaborators
- Improved quality of data
- Data available on the web as soon as it is submitted
- Facilitates discovery
- Uploaded data is protected

Summary

- KProMS framework provides complete context to the underlying data
- KProMS satisfies the requirement of Open Data archival system

Questions?

495398 Evaluation of Recovery Pathways for Improved Bioavailability of Clofazimine Nanoparticles

Abstract:

Clofazimine is an anti-mycobacterial agent used as part of a multidrug treatment for leprosy for over four decades. Recently clofazimine has shown promising activity against multidrug resistant tuberculosis. Due to its extremely poor aqueous solubility and high lipophilicity, the formulation development has focused on the solubility enhancement. Nanoparticles enhance bioavailability by two mechanisms. First, relative to larger drug crystals, nanoparticles have higher surface-to-volume ratios, which enhances dissolution rates. Second, rapid nanoparticle precipitation can trap drug molecules in higher energy, amorphous states, which have higher solubility. Here we use Flash Nanoprecipication (FNP) to form clofazimine-loaded nanoparticles with hypromellose acetate succinate (HPMC-AS) as polymer stabilizers, and investigate how to recover the clofazimine-loaded nanoparticles from the aqueous dispersions. Dry nanoparticle powders that can be readily reconstituted would enable delivery of drug dispersions to children and infants. Three approaches were used to produce dry powders: direct salt precipitation, spray drying and lyophilization. The yield efficiency, particle morphology, drug loading, dissolution rate and the amorphous/crystalline state of clofazimine in the dry powders are determined and compared among the processes. Based on these results, we evaluate the advantages and disadvantages of each process regarding the maintenance of the bioavailability after drying.

497815 Assessment of the Impact of Interfacial and Shear Stress on Biologics Drug Product Via Mini-Piloting Tools

Abstract:

Assessment of the Impact of Interfacial and Shear Stress on Biologics Drug Product via Mini-piloting Tools

During biologics development, processes such as membrane filtration, filling, and mixing are critical in the drug product manufacturing. However, these processes, as well as turbulent flow conditions, and multiple passes through pumps, valves and â€œpinchâ€ points; impose various stresses on the protein. These stresses can lead to protein instability. Shear, cavitation, and interfacial stress are often attributed as the main causes of protein unfolding and aggregation.

Proper risk analysis needs to be in place to understand the susceptibility of the protein to unfold and aggregate in presence of interfacial and shear stress. There are certain techniques, such as agitation/shaking studies that have been traditionally used to understand the impact of these stresses on the protein physical stability. However, the stresses applied in these systems are convoluted, making it difficult to define the control strategy, (i.e. adjustment in process parameters to reduce foaming/bubble formation, etc.). In this study, we have developed two mini-piloting tools that allow for the isolation of interfacial and shear stress respectively. The interfacial stress is applied to the sample by generating uniform bubbles when recirculating air from the headspace of the vial in a closed loop. The bubbles generated are passed through the solution for finite periods of time. For the shear stress, an estimation of the shear rates present during processing is made by fundamental transport relations and CFD modeling. The small scale shear stress tool simulates the shear experienced by the protein during processing, using a small-scale high pressure pump connected to a fine stainless steel tubing.

These systems help simulate the normal operating ranges as well as proven acceptable ranges for unit operation such as TFF, mixing and filling. Different protein formulations were subjected to controlled interfacial and shear stresses. The quality attributes evaluated after the stress were focused on size variant (particulate matter and HMW species), which is most likely the mechanism associated with interfacial and shear stress. It was possible to identify the susceptibility of the protein to aggregate depending of the stress that was applied. In addition, these tools contributed in formulation screening by evaluating which formulation would better mitigate the impact of the imposed stress.

During development, it is important to identify the protein formulations that have a tendency to aggregate in order to implement appropriate risk mitigation strategies. And also, to have scale down models to validate the design space for the process parameters of key unit operations.

499509 Suitable Raw Material Attributes for Continuous Manufacturing of a Drug Product Via Direct Compression

Abstract:

Continuous manufacturing for small molecule drug products is a rapidly maturing technology for innovator pharmaceutical companies and contract manufacturing organizations, supported by intensive study at academic research institutions. Of particular interest are the raw material properties which predict behavior in the novel unit operations of feeding and continuous blending, as well as their impact on residence time distributions. This study presents a thoughtful approach to the selection and justification of raw material property measurements based on process and product risks, predictive power, laboratory resource utilization, accuracy, robustness, and long-term sustainability. A suitable risk-based raw material monitoring plan for a continuous direct compression product is proposed and discussed.

501486 Wettability of Pharmaceutical Powders of Different Particle Size By Droplet Penetration Technique

Abstract:

We have recently developed a droplet penetration method to characterize the wettability of pharmaceutical powders. This method relies on two key observations during the penetration of a drop deposited on a powder bed: 1) the capillary pressure inside the porous powder is the dominant driving force for the imbibition process and 2) the contact area between the penetrating drop and the powder bed remains constant during most of the process. Under these conditions, we have shown that the penetration process in non-dimensional variables is universal. Hence, using a *reference* liquid it is possible to decouple the properties of the powder bed from the assessment of the contact angle of the *test* liquid under consideration. In addition, measuring the penetration times is sufficient to measure the contact angle, without having to solve the penetration problem.

In this work, we use the droplet penetration method to investigate the effect of particle size distribution on the wetting behavior of pharmaceutical powders. We consider two powder materials, an excipient and an active pharmaceutical ingredient (API). The excipient is lactose monohydrate, and was sieved to obtain particles in the following size ranges: 38-45, 45-53, 53-63, 63-75, 75-90, 90-106 mm. The API used is caffeine anhydrous with particle size distribution characterized by d50 = 15, 19 and 30 mm, obtained by jet milling. In all cases, the test liquid is deionized water and the reference liquid used is silicone oil. Lactose powders show a relatively constant contact angle, except for the largest particles. In the case of caffeine, we observe that larger particles yield smaller contact angles. The observed differences could be attributed to differences between milling and sieving processes. Other possible mechanisms leading to the observed dependence of the contact angle on particle size will be discussed.

492184 Investigation of Microsphere Strength for Spray Drying Applications By Means of Acoustic Levitation

Abstract:

Investigation of Microsphere Strength for Spray Drying Applications by Means of Acoustic Levitation

M. Kreimer[1], I. Aigner[1], S. Sacher[1], M. Krumme[2], T. Mannschott[2], P. van der Wel[3], A. Kaptein[3], H. Schröttner[4,5], G. Brenn[6], J.G. Khinast[1,7]

[1] Research Center Pharmaceutical Engineering (RCPE) GmbH, Inffeldgasse 13, 8010 Graz, Austria

[2] Novartis Pharma AG, Novartis Campus, 4056 Basel, Switzerland

[3] Hosokawa Micron B.V., Gildenstraat 26, 7005 BL Doetinchem, Netherlands

[4] Graz University of Technology, Institute for Electron Microscopy and Nanoanalysis, Steyrergasse 17, 8010 Graz, Austria

[5] Austrian Centre for Electron Microscopy and Nanoanalysis (FELMI-ZFE), Steyrergasse 17, 8010 Graz, Austria

[6] Graz University of Technology, Institute of Fluid Mechanics and Heat Transfer, Inffeldgasse 25 F, 8010 Graz, Austria

[7] Graz University of Technology, Institute for Process and Particle Engineering, Inffeldgasse 13, 8010 Graz, Austria

Growing demands on flexibility and agility in the production of pharmaceutical compounds have led to a paradigm shift, away from batch processing towards advanced continuous manufacturing. Continuous production of solid oral dosage forms starts with API synthesis, continuing with crystallization, purification, filtration, drying, blending, granulation and typically ends via tableting/capsule filling.

While tableting is a continuous process, several other unit operations currently exist mainly in batch mode. One example is the drying of crystals. However, spray drying is a commonly used technology to transform a fluid feed stream into dry powders. An advantage of spray drying is that it can be applied even to heat-sensitive products, since drying occurs close to the wet bulb temperature. Moreover, solutions and suspensions can be dried by evaporation of the liquid phase in a gaseous drying medium.

In the case of spray drying of suspensions, the objective is not to change material properties, such as particle size distribution. However, after crystallization, washing and filtration, different amounts of dissolved solids are present in the suspension slurry. During spray drying of the slurry, the dissolved material will solidify and form solid bridges between the suspension particles. Clearly, the amount of the dissolved material has a crucial impact on the final particle properties. The more material is dissolved in the slurry, the larger and stronger the bridges between primary particles will be. This has significant impact on the final particle properties and the associated processability.

This work highlights the influence of suspension composition on the material properties after drying. Drying of a model compound in a binary-liquid mixture was studied via acoustic levitation. This method enabled observations of the drying behavior of single droplets on a microscopic scale. Changes in the suspension composition had a strong impact on the mechanical properties and morphology of the final particles (microspheres). The morphology varied from very loose agglomerates to densely packed microspheres with crust formation. Hardness of single microspheres was measured via a compression strength test and correlated to the suspension composition.

494080 Nanostructural Interfaces of Additives on the API Surfaces Influencing Process Performance

Abstract:

The objective of this research is to detect and quantify the % extent of nanosmearing distribution of hydrophobic functional groups on multiple pharmaceutical particle surfaces during high shear mixing conditions of pharmaceutical formulations affecting the critical blend flow properties. Two sets of pharmaceutical blends were prepared by mixing 90% fast flo lactose as excipient, 9% micronized acetaminophen (Acetyl Para Amino Phenol - APAP) as Active Pharmaceutical Ingredient (API) and 1% hydrophobic Magnesium Stearate (MgSt) as lubricant at two shear rates of 100 rpm and 500 rpm in a v-blender. In the first stage, APAP particles from the two blend samples were studied under Raman Spectroscopy to detect and confirm the presence of hydrophobic functional groups of MgSt nanosmears and then to map the nanosmeared area distribution on the surface of multiple APAP particles. In the second stage, using Image-J histogram analysis, the Raman images of MgSt mapping on APAP particles were analyzed and the % nanosmeared area covered by the hydrophobic MgSt nanosmears on multiple APAP particles were quantified. For the first time in the field of particle technology, Raman Spectroscopy Mapping method was employed on the organic particulate matter to develop a novel methodology to quantify the presence and % extent of nanosmeared area distribution of hydrophobic MgSt additive on multiple particles. The results indicated that the applied process shear has a significant effect on the percentage area covered by MgSt nanosmears. The % nanosmeared area of MgSt on multiple APAP particles increased from 22% to 41% as the applied process shear during mixing increased from 100 rpm to 500 rpm. The increased nanosmeared area improved the powder blend flow properties. The nanosmears on APAP particles were found to be non-uniform and partially coated. A direct correlation between % nanosmearing area with applied process shear and powder cohesion was found.

501114 Rational Strategies for Physical Characterization of Amorphous Solid Dispersions during Drug Product Development and Manufacture

Abstract:

As amorphous solid dispersions become more widely exploited in pharmaceutical products to enhance active ingredient solubility, strategies to more efficiently confirm the absence of undesired crystallinity during manufacture and shelf life become necessary. In recent years, rapid improvement in the application of solid state characterization techniques, such as calorimetry, spectroscopy, XRPD and ssNMR, has permitted order-of-magnitude increases in method sensitivity and development of reliable quantitative limit tests to justify control strategy selection. Verification of these strategies by transfer of solid state techniques to traditional quality control laboratories remains a challenge. Companies seeking to avoid extensive use of solid state techniques in routine manufacturing release testing may instead choose to leverage such techniques in well-designed development experiments based upon fundamentals.

Using several recent cases of drug products (prepared with hot melt extrudate or spray dried intermediates), we outline efficient approaches for identification of credible phase transition risks and development of methods to explore the propensity for such risks to occur. We find methods can often achieve limits of detection of ~5% amorphous to crystalline conversion on an active ingredient basis (particularly when the drug loading exceeds 10wt%). Experience from both marketed and late stage development programs are used to illustrate a rational, risk-based strategy can be demonstrated during process validation. In contrast to extensive, multi-dimensional stability studies and measurements aiming to anticipate the infinite set of excursion scenarios that might be anticipated during the product lifecycle, we instead advocate a more fundamental exploration of physical stability boundaries. These are often associated with solid dispersion glass transition temperatures. We emphasize the use of well-designed product characterization studies to understand the failure modes of solid dispersion intermediates and drug products, and compare a more limited set of experiential measurements to predictions from nucleation and growth theories. Specialist solid state methodologies can then be confined mainly to development laboratories, yet manufacturing facilities are still provided with robust amorphous control strategies and the knowledge to react to manufacturing excursions comprehensively and efficiently.

494868 Investigation of Compression Induced Amorphization of Crystalline API

Abstract:

Active Pharmaceutical Ingredient (API) often exists in several solid forms (crystalline or amorphous). The different solid forms are known to influence the physical, chemical and mechanical properties of the API. Despite of significant API form screening effort, the selected and thoroughly studied API form may still undergo phase transformations during typical pharmaceutical processing. Among the various pharmaceutical process induced phase transforms, crystalline to amorphous phase transformation and its underlying mechanism has been rarely reported. This study will discuss the recent findings on compression induced crystalline to amorphous phase transformation. The effect of axial and shear stresses as well as excipient and formulation will be discussed.

485877 Routine Process Monitoring for Automated Clean-in-PlaceÂ Systems

Abstract:

While manufacturing process monitoring is of widespread use in the pharmaceutical industry, there are few examples where these principles are applied to cleaning processes, despite their criticality in preserving product quality and ensuring the reliability of product supply. The level of automation in modern manufacturing facilities, supported by automated cleaning- in- place (CIP) systems, provide for the continuous assessment of the performance of the cleaning process. This work describes a routine process monitoring (RPM) approach applicable for a fully automated CIP processes. A risk based approach for the screening and identification of key cleaning parameters for the establishment of a RPM Program will be discussed. RPM enables the transition from a reactive to a proactive approach for the monitoring of cleaning processes. Examples where RPM tools provide for minimizing equipment down time and strengthening of a cleaning validation lifecycle approach will be presented.

499094 Scientific Considerations and Regulatory Challenges for Low Dose Drug Manufacturing Process Monitoring and Control

Abstract:

For solid dosage form manufacturing, assay and content uniformity (CU) are two critical quality attributes (CQAs) of drug product that are impacted by powder blending, blend transferring, and tableting or capsule filling. Clinically, assay, CU, and dissolution together ensure that patients receive the strength claimed in the label and thus impact clinical efficacy and safety. The criticality of the powder blending operation and its outcome (e.g., blend uniformity (BU) and blend assay) increases for low dose drugs where the amount of active is very small in comparison to the amount of excipients. Therefore, low dose drug manufacturing process monitoring and control are uniquely challenging.

In this presentation, we will discuss scientific case studies excerpted from research studies and drug applications, challenges and opportunities, along with considerations for low dose drug manufacturing process monitoring and control. Specific attention will be paid to the easily overlooked or under-reported areas such as: 1) risk-based PAT monitoring interface location considerations for both batch processing and continuous manufacturing; 2) evaluation of engineering process scale-up criteria as applied to powder blending process; 3) sampling and analytical methods. Our discussion will illustrate a holistic approach supported by solid scientific understanding for low dose drug manufacture.

496636 Frontiers of Industrial Crystallization Science and Technology

Abstract:

Industrial crystallization is an environmentally friendly unit operation mainly used to separate or purify a variety of materials. Compared with crystallography, industrial crystallization is a large-scale process that focuses on improving physical properties of powders and on the effects of operating conditions and process control on nucleation and crystal growth. Over the past decades, great progress has been made in many aspects of industrial crystallization, including polymorphism, morphology, online control technology, computer simulations of crystallization and crystallizer design. The modern industrial crystallization technology is the key common technologies for preparation of high-end products in pharmaceutical industry, materials industry, marine chemicals, energy chemical, fine chemical industry, food industry and etc.

Crystallization has been quite routinely practiced for low molecular weight compounds in the laboratory and in industrial manufacturing. Using crystallization, especially in industrial manufacturing processes, has many advantages such as reducing costs, greater stability of the crystals and increased functional characteristics such as product formulation, transport and storage.Furthermore, the crystallization of macromolecular compounds such as proteins and nano materials such as Quantum dots are generally practiced at the laboratory-scale and used as a research tool. However, there is very few rule in protein and Quantum dots crystallization as compared to low molecular weight compounds, especially for industrial scale protein crystallization.

Being targets for over 50% of all modern medicinal drugs, membrane proteins have essential physiological functions. Unfortunately, compared to soluble proteins, very less has been known about membrane proteins. By the end of 2015, there are about 110,000 structures available in protein data bank, but only 1% is membrane proteins.

The major aim of our work in membrane protein crystallization was to structurally and functionally characterize MOP proteins. Five type of MOP proteins, 1 from PST family, and 4 from MATE family were from different organisms and successfully overproduced in E. coli. Then we focus on the MATE transporter from Pseudomonas stutzeri, on which extensive functional characterization and crystallization trials have been performed.

The green engineering technology of crystallization involving coupling crystallization processes; process intensification; simulation and optimization of process process analytical technologyas well as feedback control of processes (Technology + Equipment + CAD + +CAC+CAM etc). Considering all of these criteria, we selected the following objects for our international cooperation research project:

1. Biological macromolecular medicine drug delivery systemsï¼›

Mechanisms of crystallization for biological macromolecular medicine;

2. Crystal engineering of membrane protein;

3. Crystallization process intensification (PI) and micro-crystallizer.

Over the past decades, the significant advance have been made in our understanding of some mechanism of industrial crystallization processes even though the crystallization still situated in a semi-science and semi-art stage. We expect to enhance the international information exchange and the academic cooperation about crystallization process development for speeding up the advance in chemical industry over the world.

497105 Models for Comparative Characterization of Complex Mixture and Biological Drug Products

Abstract:

Comparative characterization of drug products via different analytical techniques is often needed to support product development. This comparison can be used for the purposes of biosimilar or generic drug product development or to evaluate changes during the drug product lifecycle. These may include manufacturing process changes, and especially those initiated post-approval. Such comparisons may also help ensure therapeutic consistency while advancing through the phases of a clinical trial. For homogeneous small molecule drugs these comparisons can be relatively straight forward. However, for heterogeneous drugs such as biotechnology products or complex mixtures, this process can prove much more challenging. For these cases, mathematical models may aid in determining the degree of sufficiency of *in vitro* characterization needed for such comparisons. This presentation will examine such models developed via FDA collaborations to: (i) characterize multiple batches of heterogeneous drug products, (ii) examine structure, physicochemical stability, and biological activities informed by existing knowledge using a wide range of analytical technologies, and (iii) develop data visualization tools and novel mathematical algorithms to integrate the chemical, physical, and biological data to assess critical quality attributes (CQAs).

495313 Self-Aggregation of Gabapentin Under Confinement of Primary Particle By Smpt

Abstract:

1. **Introduction** Solution-mediated phase transformation (SMPT) is one of the most common manufacturing methods used in pharmaceutical industry to control of the microstructure and the properties of the APIs [1-4]. In the traditional SMPT process, the crystal size and shape of stable form are uncontrolled by that of metastable form and is affected by the crystallization dynamics in the whole process. But now we can obtain the well-defined shape and size of stable form particles. The particles have a self-aggregation process under confinement of primary methastable particles during the phase transformation.

Gabapentin (GBP) is known to exist as a zwitterion in three solvent-free forms: II, III and IV, with form II being the most stable and form IV the least stable [5]. In practical production process, preparing of form I is the first step and then recrystallize to obtain form II. Form I can transform to form II through SMPT in organic solvent. The object is research the self- aggregation of GBP under confinement of primary particles. **2. Materials and methods**

Form I and form II powder of GBP, with mass purity of 99%, were supplied by Zhejiang Chiral Medicine Chemicals Co., Ltd., China. Methanol, ethanol, propanol, acetone, acetonitrile and ethyl acetate were purchased from Tianjin Kewei Chemical Technology Co., Ltd., China.

The phase transformation was investigated in the form II saturated solution. It can also reduce the impact of dissolving process of form I. Excess amount of form II powder were dissolved in 150 mL of methanol, ethanol, propanol, acetone, acetonitrile and ethyl acetate, respectively. For each experiment, 110 mL saturated solution was filtered and added to 150 mL scale crystallizer, then 4.4 g form I was added to the solution. A Raman spectrometer and Fourier transformed infrared spectroscopy were used to monitor the process.

3. Results

3.1 Morphology characterization of the product

The SEM of the solid products reveals that the crystal habit of form II is diversity in different solvent (Figure 1). The particle obtained from methanol, ethanol or propanol has a good dispersion, and the particles have slightly aggregate in methanol. The form II nucleated by homogeneous primary nucleation from saturated solution. The form II particles grew significantly at the expense of disappearing of form I.

The result in acetone, acetonitrile and ethyl acetate is totally different from that in alcohols. The products are all aggregates and the morphology of the aggregates is plate-like shape, which is same with form I particles. Some breakage is due to the stirring or the sample preparation for test. It is apparently, the morphology and size of the aggregates are startlingly close to that of form I particle. We speculate that the predecessor of one aggregate is one

particle of form I (Figure 1).

3.3 Non-directional self-aggregation under confinement of primary particle and the mechanism of the SMPT. Take the case in acetone as an example. The results of intermittent sampling analysis in acetone indicate that there was no obvious dissolution after the form I particles added into the solution (Figure 2). we analyzed the water content of the suspended solid at different time. When the form I particles was suspended in acetone, there is an obvious dehydration process, especially in the first 5 hours. This causes the self-cleaving in the particle and produces many small lamellae (Fig. 3B). The rough surface and defects provide many nucleation sites for stable form. Few hours later, a large number of nucleuses appear on these small lamellae (Fig. 2C). Owing to the very low solubility, the particles do not dissolve and there are almost no molecules of GBP spread into the solution. These nucleuses only can grow along the lamellae, and the molecules of GBP transfer from the lamellae to the nucleuses. Due to the various locations of the nucleuses and the stable form crystals have a variety of growth direction. The whole process advances slowly. After it is completely over, the product is an aggregate composed by a large number of need-like crystals .

The individual crystallites in the aggregate have no specific directional and they can aggregate only in the limited space â€“ the space of the original particle. We call this process is non-directional self-aggregation under confinement of original particle during phase transformation. We have proposed a solvent penetration mediated phase transformation mechanism[6].

Â **4. Conclusion**

The phase transformation of gabapentin from form I to the stable form II in methanol, ethanol, propanol, acetone, acetonitrile and ethyl acetate were studied. Form II nucleation is most likely homogeneous primary nucleation from saturated solution in methanol, ethanol and propanol. Due to the very low solubility of GBP in acetone, acetonitrile and ethyl acetate, there is no dissolution process of metastable form. A self-cleaving phenomenon appears on the original particle because of the extremely low water activity in these organic solvents. The hydrate is not stable and the water molecules need escape from the structure. The dehydration causes many defects on the particle and contributes to heterogeneous nucleation of stable form. The growth of these new nucleuses is confined by the original particles. Non-oriented self-aggregation phenomenon is observed during this phase transformation process. So the concept of non-directional self-aggregation under confinement in desolvation mediated phase transformation is proposed. The products of the stable form are aggregates, but the shape and size of these aggregates coincide exactly with that of the metastable particles. And the morphology of the individual crystallite can be needle-like, plate-like or even bulk shape.

495757 Feasibility of NIR Spectroscopy for Monitoring Enzymic Digestion Process of Proinsulin Producing Insulin Glargine

Abstract:

Enzymic digestion is an important step in insulin glargine (IG) manufacturing. Due to the fact that proinsulin has multiple tryptic digestion sites, excessive digestion leads to lower yields and more undesirable by-products. However, analytes of interest have not been monitored in real time during digestion process and end-point of the process could not be determined accurately. In this study, feasibility of using near-infrared (NIR) spectroscopy to monitor four analytes of interest was investigated. Partial least squares (PLS) calibration models were developed using reference measurements by high performance liquid chromatography (HPLC) and *in situ* NIR spectra acquired continuously by an immersion probe. The achieved root mean square errors of prediction (RMSEP) were 0.06 mg/mL for intermediate product (IP), 0.06 mg/mL for IG, 0.03 mg/mL for mono-Arg impurity (mAI) and 0.05 mg/mL for isomer of IG (iso-IG). Results indicated that change trends of the four analytes were defined by the developed models, but probably with poor robustness. More variations might be involved in model calibration so that NIR spectroscopy can be utilized for digestion process monitoring and end-point determination, which contributes to increased yields and decreased by-products in this step.

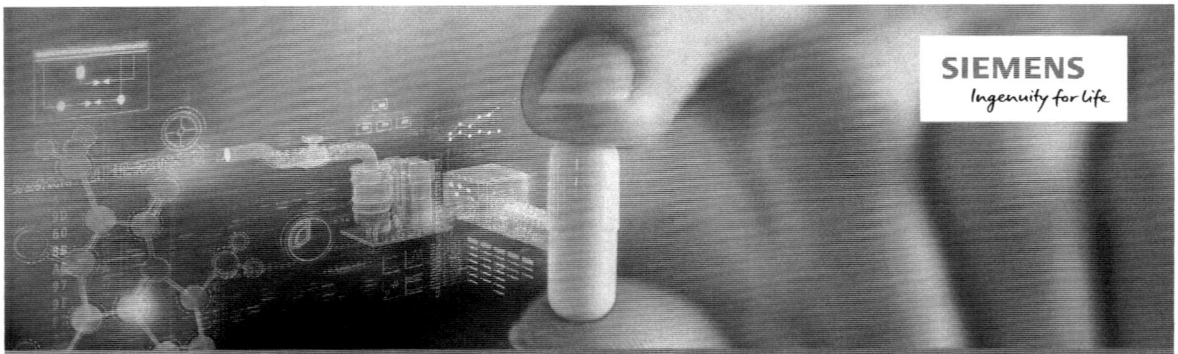

The Benefit of Advanced Process Control (APC) in OSD Continuous Manufacturing

Pamela Docherty
Siemens

What is PAT Data Management?

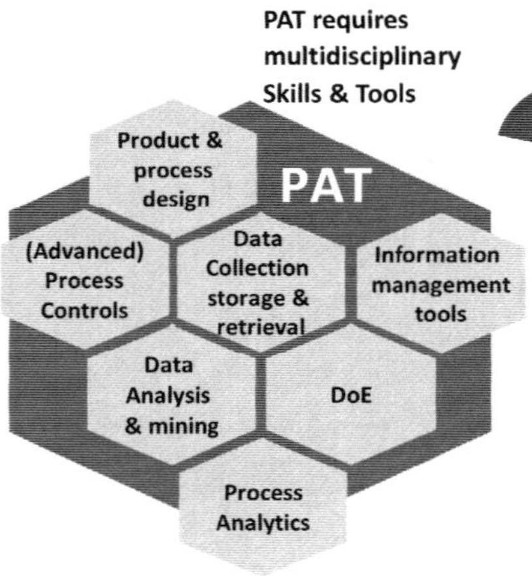

All PAT Tools are included in a single PAT system architecture, easy to integrate and scalable

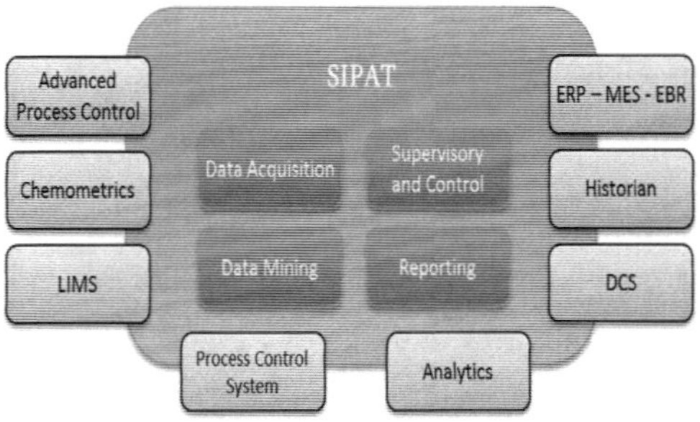

What is a Process Control System (PCS 7)?

SIEMENS
Ingenuity for life

What is a Process Control System (PCS 7)?

SIEMENS
Ingenuity for life

SIEMENS
Ingenuity for life

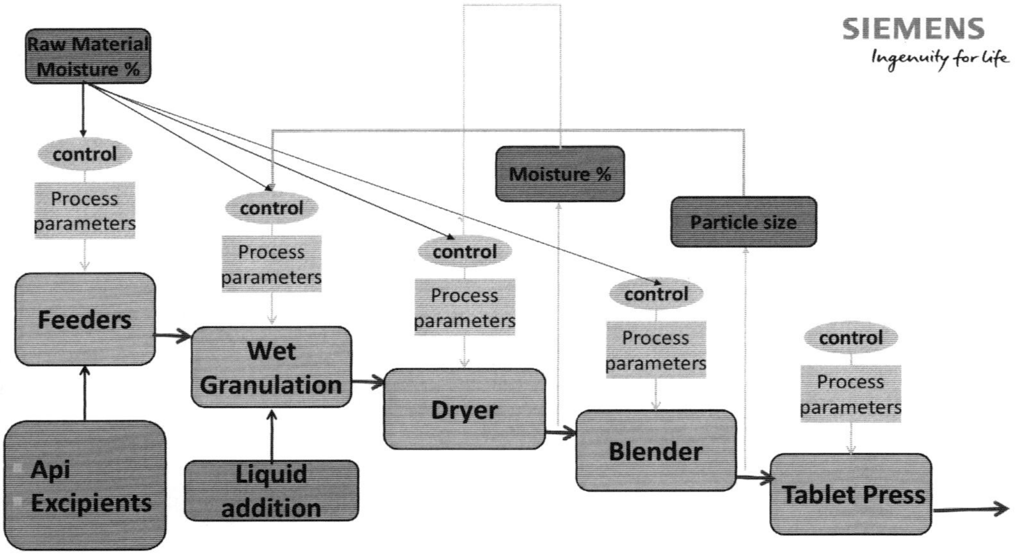

Page 5 Nov 02, 2017 Pamela Docherty/ Siemens

SIEMENS
Ingenuity for life

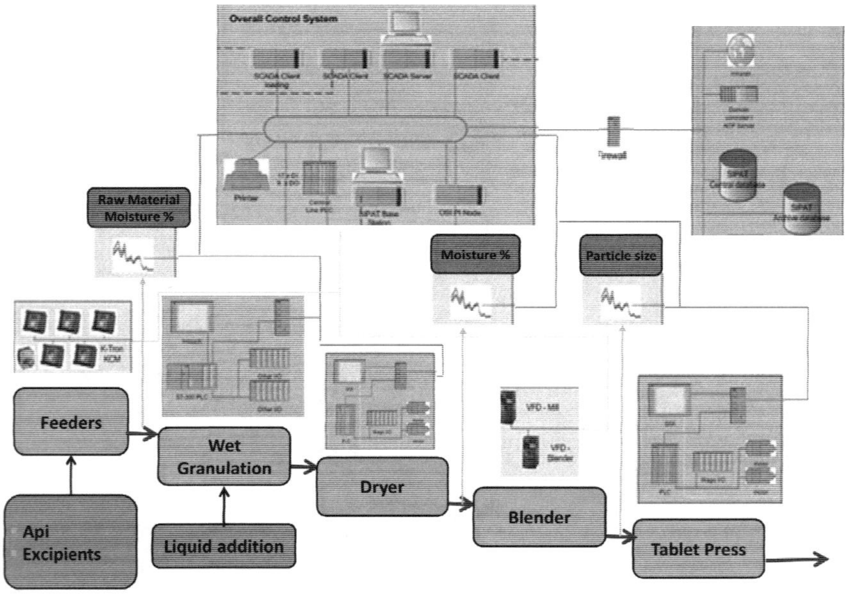

Page 6 Nov 02, 2017 Pamela Docherty/ Siemens

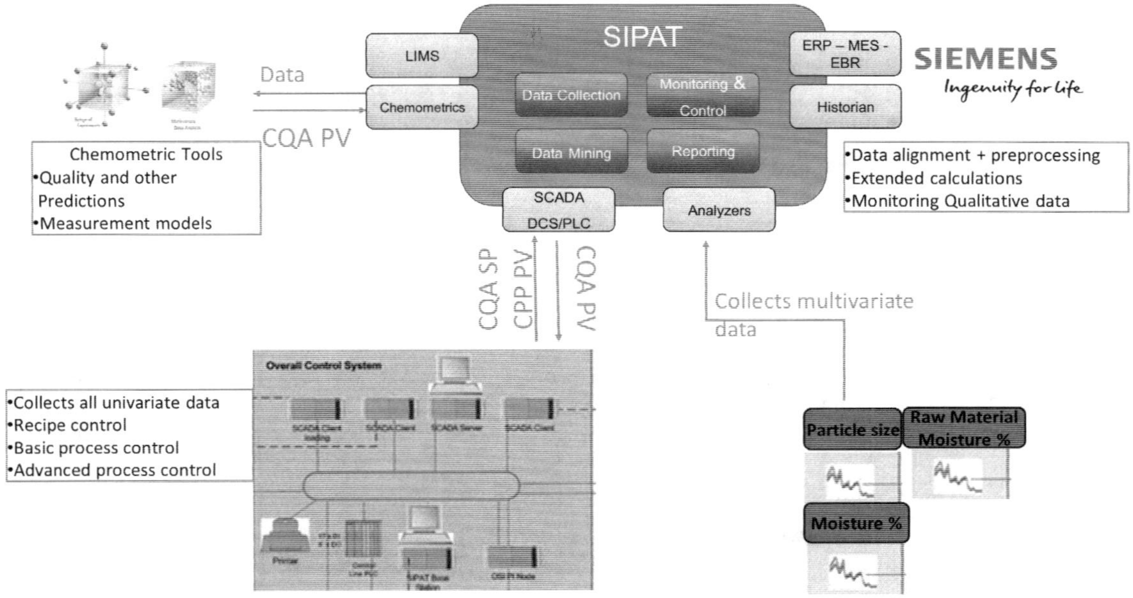

Page 7 Nov 02, 2017 Pamela Docherty/ Siemens

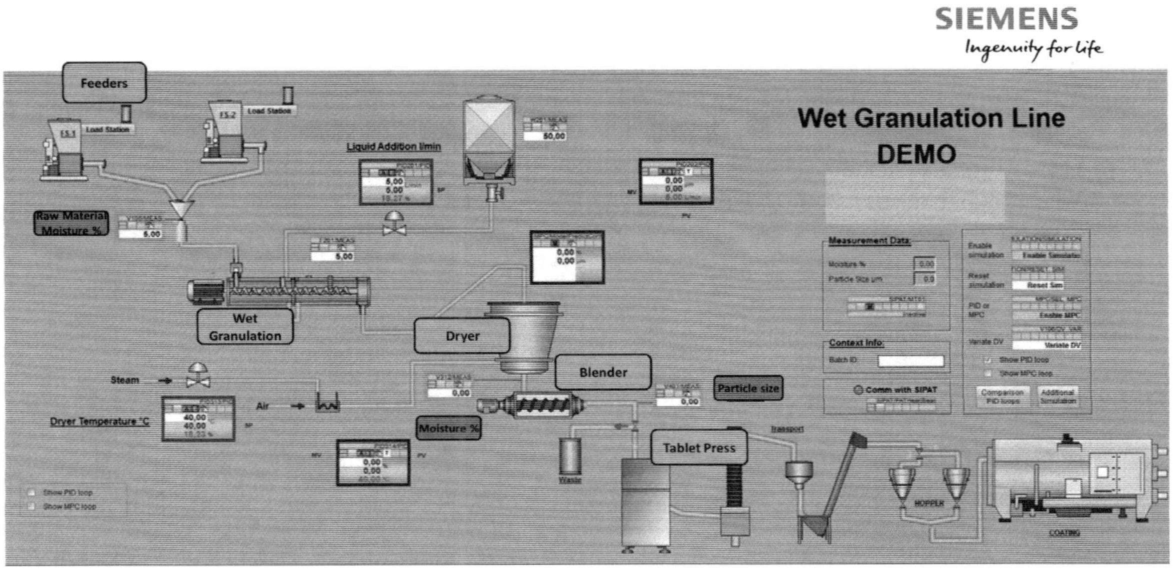

Page 8 Nov 02, 2017 Pamela Docherty/ Siemens

SIEMENS
Ingenuity for life

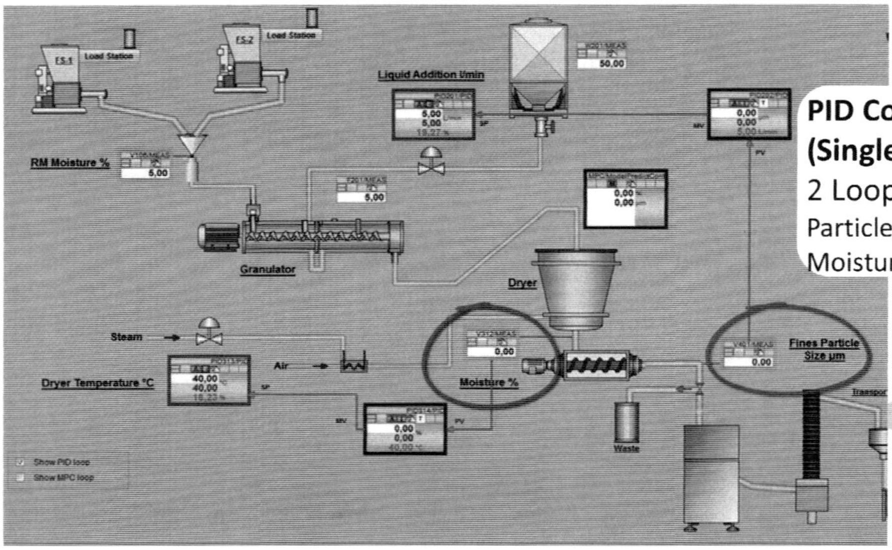

PID Control (Single In – Single Out)

2 Loops

Particle Size → Added Liquid

Moisture → Dryer Temp

(large dead time)

Liquid ↑	Particle Size ↑
Liquid ↑	Moister ↑
Moister ↑	Dryer Temp ↓
Dryer Temp ↓	Particle size ↓

SIEMENS
Ingenuity for life

A proportional–integral–derivative controller (PID controller) is a control loop feedback mechanism (controller) commonly used in industrial control systems. A PID controller continuously calculates an error value as the difference between a desired setpoint and a measured process variable and applies a correction based on proportional, integral, and derivative terms (sometimes denoted P, I, and D respectively)

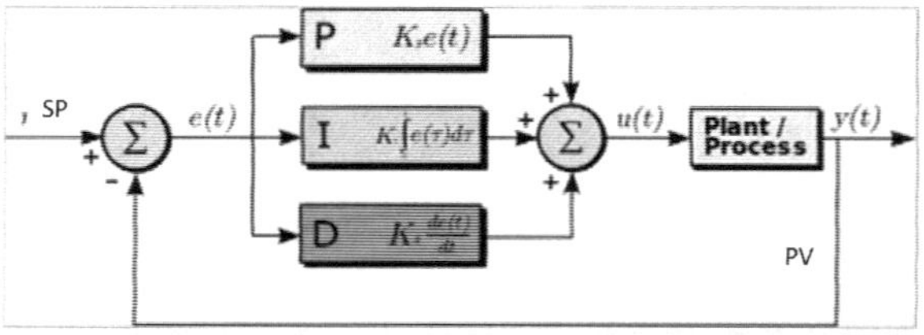

A block diagram of a PID controller in a feedback loop. r(t) is the desired process value or "set point", and y(t) is the measured process value.

https://en.wikipedia.org/wiki/PID_controller

SIEMENS
Ingenuity for life

PID314: Moisture to Dryer temperature

- Kp = -5
- Ti: 65s
- Td: 0s

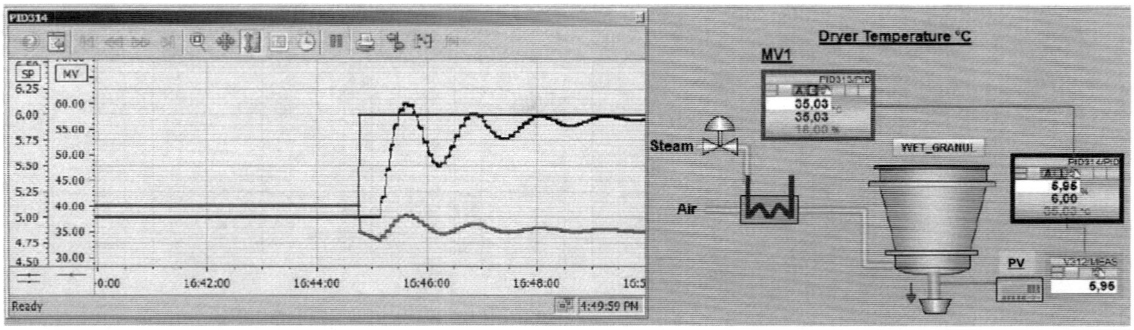

Advanced Process Control (APC)

SIEMENS
Ingenuity for life

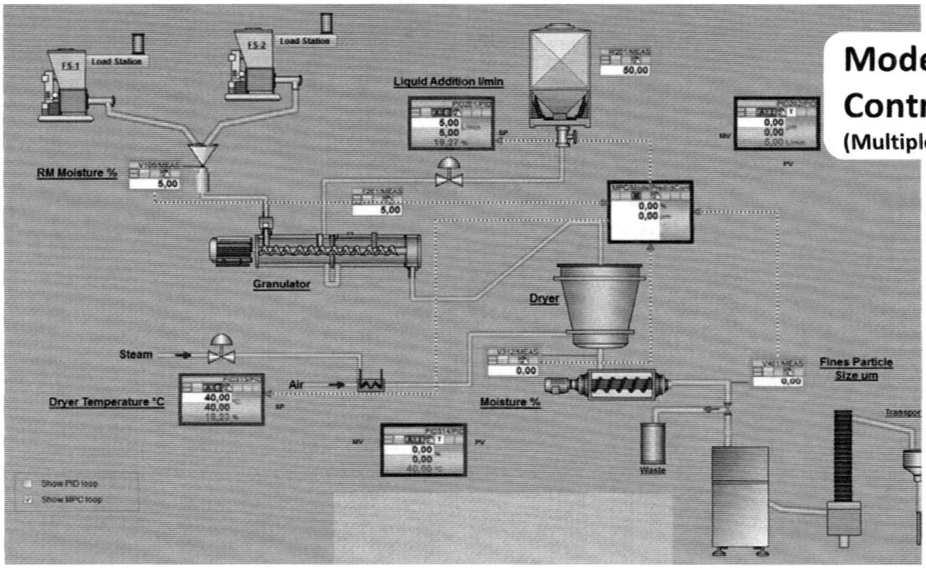

Model Predictive Control (MPC)

(Multiple In – Multiple Out)

3 Inputs

RM Moisture - DV

Moisture

Particle Size

2 Outputs

Liquid Addition

Dryer Temp

SIEMENS
Ingenuity for Life

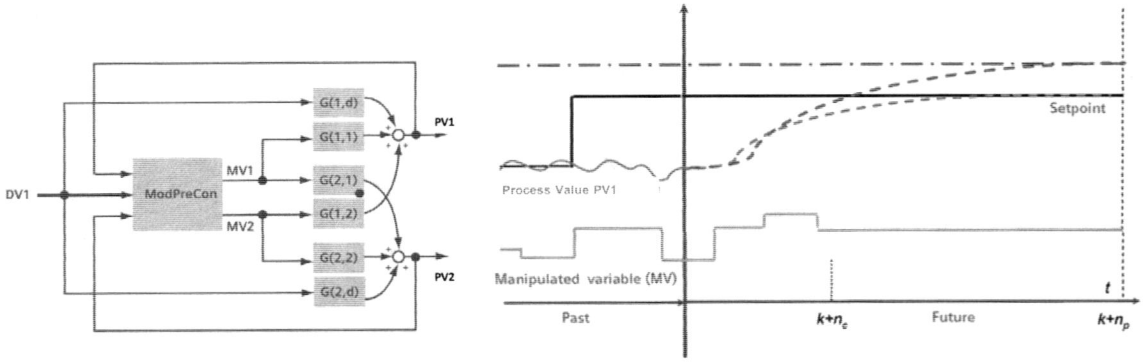

http://w3.siemens.com/mcms/process-control-systems/en/distributed-control-system-simatic-pcs-7/simatic-pcs-7-technologies/advanced-process-control/pages/apctutorial.aspx

Page 13 Nov 02, 2017 Pamela Docherty/ Siemens

Change in setpoint of moisture from 5 to 6 %:

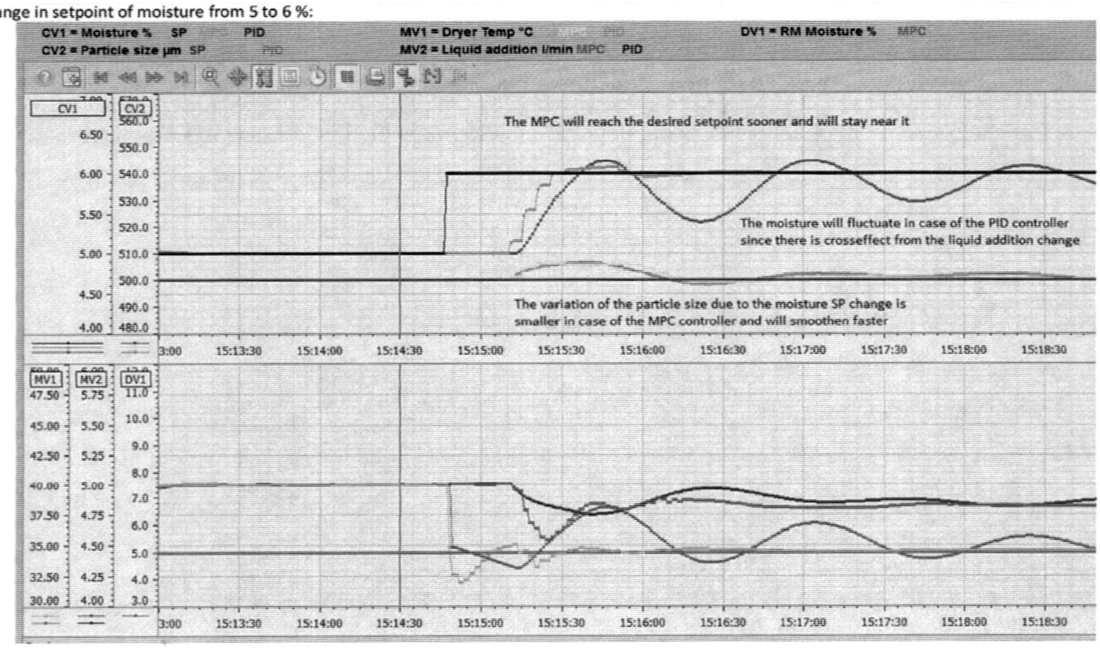

Change in setpoint of particle size from 500 to 550µm :

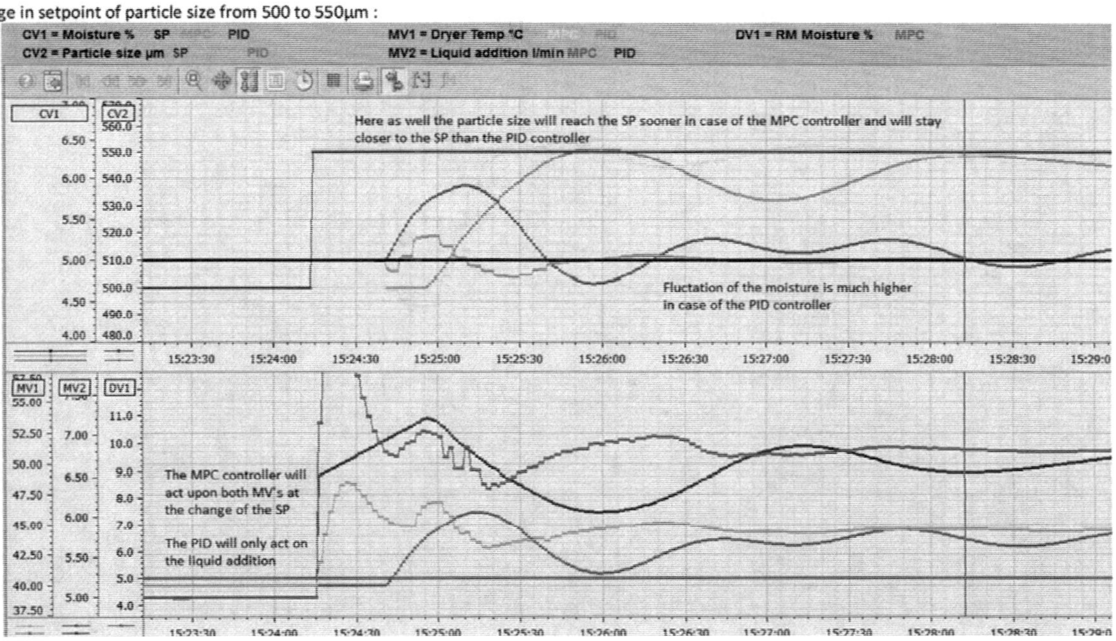

Variation of disturbance variable (raw material moisture):

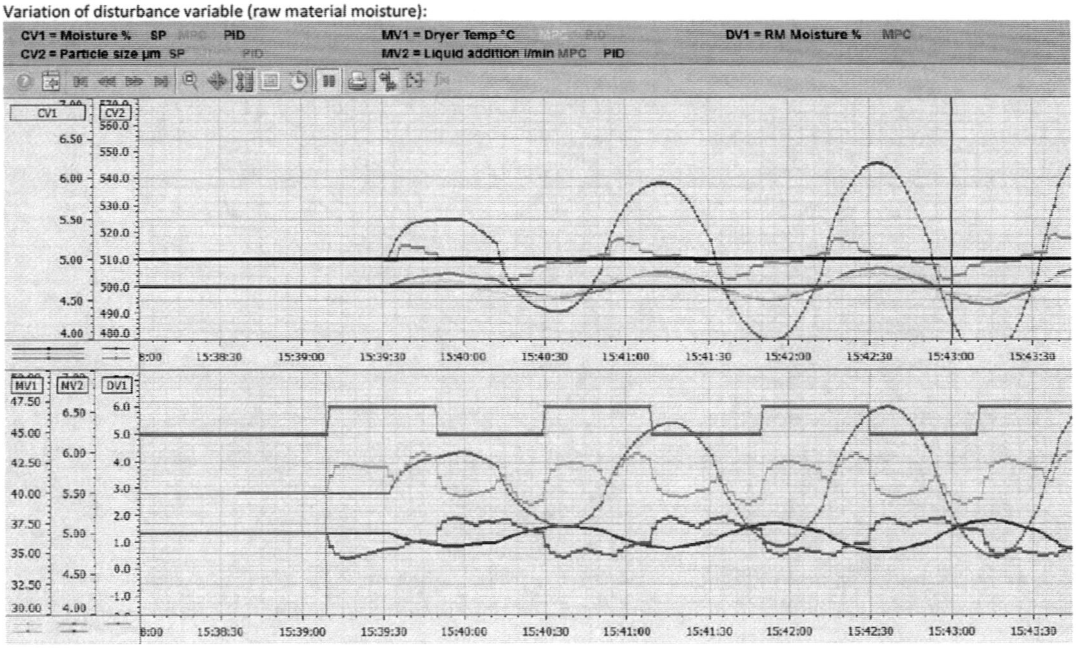

Conclusion

PAT data management is critical
An overall control system is mandatory

- Complex processes benefit from Advanced Process Control

Thank you

SIEMENS
Ingenuity for life

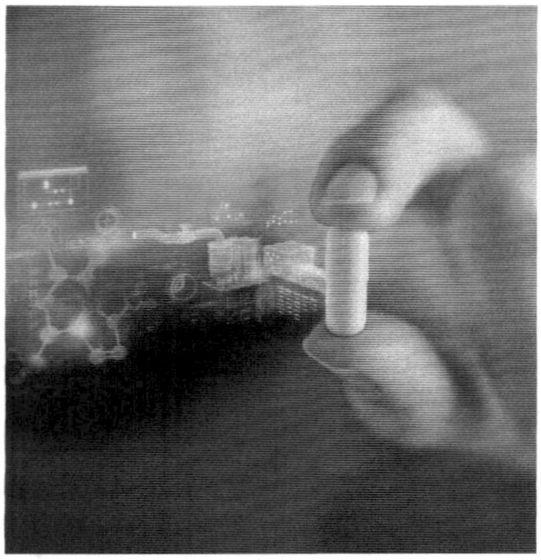

Pamela Docherty
Life Sciences Industry Manager- USA

1201 Sumneytown Pike
Spring House, Pa

Mobile: (267) 207-4931
E-Mail: pamela.docherty@siemens.com

USA.Siemens.com/pharma

499423 Oral Formulations: Perspectives from Additive Manufacturing

Abstract:

Selection of oral formulation strategy plays a significant role in development of new drug candidates throughout the development and commercial product stages; requirement for oral bioavailability must be balanced against additional complexity in the manufacturing process[1,2]. Formulation strategies may present manufacturing challenges, particularly for poorly water-soluble drugs, resulting from the necessity of particle size reduction and multiple active excipients to achieve bioavailability targets[3]. As an alternative to established manufacturing strategies, much recent work has focused on additive manufacturing, partially due to its capacity for accurate and precise delivery of drug amounts in the micro- to milli-gram range[4]. The technology also exhibits potential as a flexible platform capable of producing doses in a number of forms from solvent- and melt-based forumulations[5,6], which may include colloidal particles[7].

The manufacture of multiple capsule-based formulations using non-colloidal particulate suspensions is presented in this study. Particles selected from micronized and non-micronized powder grades are characterized in terms of particle size and bulk powder properties, which, in concert with rheology experiments are used to guide printing fluid formulations. Quantitative measures of product physical properties and respective production rates are presented for four active substances in 1-100 milligram amounts, delivered by hydro-philic/phobic carrier fluid platforms. Based on these results, perspectives are given on the ability of the platforms to complement current formulation selection strategies for BCS Class I and Class II drugs and on the potential for the technology serve at both the personalized medicine and industrial manufacturing scales.

References

1. M. Kuentz, R. Holm, and D. P. Elder, 2016, Methodology of oral formulation selection in the pharmaceutical industry, European Journal of Pharmaceutical Sciences, 87, 136-163.
2. M. Ierapetritou, F. Muzzio, and G. Reklaitis, 2016, Perspectives on the Continuous Manufacturing of Powder-Based Pharmaceutical Processes, AIChE Journal, 62, 1846-1862.
3. H.D. Williams, N.L. Trevaskis, S.A. Charman, R.M. Shanker, W.N. Charman, C.W. Pouton, and C.J.H. Porter, 2013, Strategies to Address Low Drug Solubility in Discovery and Development, Pharmacoligical Reviews, 65, 315-499.
4. R. Daly, T.S. Harrington, G.D. Martin, and I.M. Hutchings, 2015, Inkjet printing for pharmaceutics â€" A review of research and manufacturing, International Journal of Pharmaceutics, 494, 554-567.
5. L. Hirshfield, A. Giridhar, L.S. Taylor, M.T. Harris, and G.V. Reklaitis, 2014, Dropwise Additive Manufacturing of Pharmaceutical Products for Solvent-Based Dosage Forms, Journal of Pharmaceutical Sciences, 103, 496-506.
6. E. IÃ§ten, A. Giridhar, L.S. Taylor, Z.K. Nagy, and G.V. Reklaitis, 2015, Dropwise Additive Manufacturing of Pharmaceutical Products for Melt-Based Dosage Forms, Journal of Pharmaceutical Sciences, 104, 1641-1649.
7. J. Pardeike, D.M. Strohmeier, N. Schrödl, C. Voura, M. Gruber, J.G. Khinast, and A. Zimmer, Nanosuspensions as advanced printing ink for accurate dosing of poorly soluble drugs in personalized medicines, 2011,

   International Journal of Pharmaceutics, 420, 93-100.

487351 Predicting the Free Energy of Solvation: A Hybrid QSPR Model for Organic Solute/Solvent Pairs

Abstract:

The free energy of solvation is a useful quantity in the journey of a pharmaceutical molecule from discovery through to commercial process development and product design. In this paper, we review some of the roles played by this property across the moleculeâ€™s life cycle, including in the early assessment of the quantities such as the octanol/water partition coefficient or on the assessment of solvent effects on reaction energetics for screening purposes. Initial efforts to propose quantitative structure property relationships (QSPR) and predict the Gibbs free energy of solvation for a wide range of solute/solvent pairs are presented here. They include the use of partial least square (PLS) and multivariate linear regression (MLR) methods. A dataset consisting of 303 solutes in 215 solvents with total number of 1800 Gibbs free energy data points has been built. The proposed models combine experimental descriptors with quantum mechanical (QM) descriptors. For the solvents, experimental descriptors are proposed, as these provide a good description of the bulk properties of solvents and the range of solvents considered is usually relatively narrow; twelve of these descriptors are used in the proposed model (surface tension, dipole moment, octanol-water partition coefficient, dielectric constant, refractive index, boiling point, molecular weight, enthalpy of vaporization, molar volume, critical temperature, critical pressure, and critical volume). For the solutes, on the other hand, QM descriptors are used, making it possible to model solutes that have not been synthesized or that are short-lived/unstable; nine QM solute descriptors are proposed (polarizability, energy of the highest occupied molecular orbital, energy of the lowest unoccupied molecular orbital, dipole moment, most positive partial charge on hydrogen atom in molecule, most negative atomic partial charge in molecule, molecular van der Waals volume, total energy at 0 K, and entropy at 298 K). The accuracy of these initial models is found to be acceptable for fast calculation of the Gibbs free energy of solvation of a wide range of solutes in different solvents and to compare favourably with existing methods. Building on these encouraging results further work to explore whether this level of accuracy is suitable for a range of activities, including screening and process design will also be of interest.

492658 *In silico* Prediction of Absolute Rates of Crystal Growth from Solution

Abstract:

Predicting rates of crystal growth from solution *in silico* remains extremely challenging, but presents a lucrative engineering opportunity to more efficiently design crystals for pharmaceutical, catalytic, and other purposes. Mechanistic crystal growth models [1-3] capture key thermodynamic and kinetic phenomena, and they have successfully predicted steady-state crystal shapes. [4,5] These models can facilitate *in silico* crystal growth rate predictions if rates of attachment to and detachment from kink sites are known; kinks are favorable surface sites for attachment of growth units (atoms, molecules, etc.). Complex computational techniques are currently the only approach to calculate these rates. Moreover, these methods are necessary to gain mechanistic insight to aid future development of a model to predict attachment/detachment rates.

In this work, we demonstrate that *in silico* predictions of absolute crystal growth rates can yield favorable agreement with experimentally measured growth rates. We examine a proof-of-concept system â€" NaCl in water â€" which provides valuable insights that will aid future investigations of organic molecules and organic salts. We compute the rates of Na^+ and Cl^- attachment and detachment at kink sites and also elucidate the mechanistic role of water molecules during these events. [6] These rates are implemented in the mechanistic spiral growth model [7,8] to make *in silico* NaCl crystal growth rate predictions, which are in approximate agreement with experimental measurements. [9-11] These results promote the investigation of additional systems and, ultimately, the development of a model to predict attachment and detachment rates. It will be interesting to explore the role of rotational degrees of freedom and functional groups during kink attachment/detachment of organic molecules.

[1]Â J. Li, C.J. Tilbury, S.H. Kim, M.F. Doherty, *Prog. Mater. Sci.* **2016**, *82*, 1-38.

[2]Â M.A. Lovette, A.R. Browning, D.W. Griffin, J.P. Sizemore, R.C. Snyder, M.F. Doherty, *Ind. Eng. Chem. Res.* **2008**, *47*, 9812-9833.

[3] P. Dandekar, Z.B. Kuvadia, M.F. Doherty, *Annu. Rev. Mater. Res.* **2013**, *43*, 359-386.

[4] S.H. Kim, P. Dandekar, M.A. Lovette, M.F. Doherty, *Cryst. Growth Des.* **2014**, *14*, 2460-2467.

[5] C.J. Tilbury, D.A. Green, W.J. Marshall, M.F. Doherty, *Cryst. Growth Des.* **2016**, *16*, 2590-2604.

[6] M.N. Joswiak, M.F. Doherty, B. Peters, *in prep.*

[7] W.K. Burton, N. Cabrera, F.C. Frank, *Phil. Trans. R. Soc. Lond. A* **1951**, *243*, 299-358.

[8] R. Snyder, M.F. Doherty, *P. Roy. Soc. A-Math. Phy.* **2009**, *465*, 1145-1171.

[9] H. Offerman, G. von Brachel, A. Al-Sabbagh, F. Farelo, *Cryst. Res. Technol.* **1995**, *30*, 651-658.

[10] S.B. Zhang, J.J. Yuan, H.A. Mohameed, J. Ulrich, *Cryst. Res. Technol.* **1996**, *31*, 19-25.

[11] S. Al-Jibbouri, J. Ulrich, *J. Cryst. Growth* **2002**, *234*, 237-246.

501156 Mathematical Modeling and Simulation of Magnetophoresis. Application to Drug Delivery Using Magnetic-Field

Abstract:

The effects of magnetic field on paramagnetic species such as ions, biological molecules and solid particles have been investigated theoretically[1-4] and experimentally.[5,6] These paramagnetic species in the presence of an applied magnetic field experiences three kinds of body forces such as paramagnetic gradient force due to concentration gradient, field gradient force due to magnetic field gradient, and Lorentz force due to interaction of moving charged species with the magnetic field.[3,7] The flux of a population of paramagnetic species due to these magnetic body forces have been previously modeled using four different approaches- (A) the magnetic forces are included as the external body forces in the Navier-Stokes equation and the resulting velocity is added to the convective term of the flux expression,[2,3,8] (B) the velocity of paramagnetic particles is approximated as a product of mobility and a net magnetic force, which is then included in the convective term of the flux expression,[4,9,10] (C) the mobility in the migration term of the Nernst-Planck equation is modeled as a matrix containing Hall and Transversal mobilities, which are the functions of magnetic fields,[1] and (D) the diffusion coefficients are modified to include the effects from Lorentz force and the velocity in the convective term is modeled as a product of mobility and field gradient force.[11] The approaches A-D have a few common limitations that these models are not applicable to the concentrated solutions of paramagnetic particles, and they do not account for all magnetic forces including magnetic dipole interactions between particles. Here we propose a novel model for the flux of paramagnetic species that overcomes all the limitations of the preceding approaches. Moreover, the proposed model is based on the first principles that accounts for fluxes due all three magnetic forces with inter-particle interactions.

References:

(1) Aaboubi, O.; Chopart, J.; Douglade, J.; Olivier, A.; Gabrielli, C.; Tribollet, B. Journal of the Electrochemical Society 1990, 137, 1796.

(2) Waskaas, M.; Kharkats, Y. I. The Journal of Physical Chemistry B 1999, 103, 4876.

(3) Hinds, G.; Coey, J.; Lyons, M. Electrochemistry communications 2001, 3, 215.

(4) Furlani, E. Journal of Applied Physics 2006, 99, 024912.

(5) Lueshen, E.; Venugopal, I.; Soni, T.; Alaraj, A.; Linninger, A. Journal of biomedical nanotechnology 2015, 11, 253.

(6) Venugopal, I.; Pernal, S.; Fusinatto, T.; Ashkenaz, D.; Linninger, A. Nano Biomedicine & Engineering 2016, 8.

(7) Weston, M. C.; Gerner, M. D.; Fritsch, I. Analytical chemistry 2010, 82, 3411.

(8) Waskaas, M.; Kharkats, Y. I. Journal of Electroanalytical Chemistry 2001, 502, 51.

(9) Furlani, E.; Ng, K. Physical review E 2006, 73, 061919.

(10) Furlani, E. P.; Xue, X. Microfluidics and nanofluidics 2012, 13, 589.

(11) Kinouchi, Y.; Tanimoto, S.; Ushita, T.; Sato, K.; Yamaguchi, H.; Miyamoto, H. Bioelectromagnetics 1988, 9, 159.

486626 Polyalkoxylated Alcohols As Excipients for Poorly Soluble Drugs

Abstract:

Polyalkoxylated Alcohols as Excipients for Poorly Soluble Drugs

Thomas H. Kalantar[1], Jin Zhao[2], Mladen Ladika[1], Keith Harris[1], Chris Tucker[1], Michael Tulchinsky[1], TC Kuo[1], Joe Kiefer[1], Xiaoyun Chen[1], Robert Krystosek[1]

(1) The Dow Chemical Company, Herbert D. Doan R&D Center, Midland, Michigan, 48674, USA

(2) Pharmaceutical Excipients R&D, Dow Food, Pharma & Medical Films, Midland, MI 48674

There has been a steady increase in the number of active pharmaceutical ingredients (APIs) that are characterized with low aqueous solubility. It has been estimated that 40 to 80% of new drug candidates suffer from poor aqueous solubility and/or slow dissolution, reducing their bioavailability and efficacy, and even preventing their general use [1]. This study explored a solubilization approach using low-T_g alkylated polyoxyalkylene copolymers to enhance drug solubility and dissolution rate[2], where the drug was combined with the polymeric excipient to form an amorphous solid dispersion at the molecular level. Such low-T_g excipients yield liquid or semi-solid drug/excipient formulations.

In this study, a novel class of alkylated polyols was synthesized using renewably sourced alkyl initiators, with a specific emphasis on terpene alcohol precursors. These alcohols were first alkoxylated with butylene oxide then further ethoxylated with ethylene oxide. The effect of solubilization by these copolymers was studied using four model drugs (ketoprofen, danazol, griseofulvin, and probucol). Raman spectroscopy suggested that several copolymers from this study were effective in stabilizing ketoprofen in its amorphous form. The aqueous solubility and dissolution rate of drugs in presence of these excipients were studied using nephelometry and optical microscopy. Several drug/excipient samples showed a dramatic reduction in turbidity relative to the turbidity of drugs alone in water. Optical microscopy of those drug/excipient samples in water also indicated an increase in drug dissolution. Drug solubility was also measured by an HPLC method in which a capsule filled with probucol/excipient mixture was placed in phosphate buffer solution for 30 minutes after capsules were disintegrated.

Medium to high enhancement of solubility were observed with blends of common commercial polymeric excipients with three drugs: phenytoin, griseofulvin, and probucol. Probucol solubility was significantly enhanced when mixed with excipient polymers at a molecular level. When mixtures of probucol and novel alkyl-BO-EO excipients were loaded into capsules and subjected to ambient conditions for one week and accelerated aging at 40 °C and 75% RH for 4 months, the observed increases in probucol solubility were significant (up to 8,000-fold) under both sets of storage condition.

Many polyoxyalkylene copolymers can be used as solubilization-excipients for poorly soluble drugs; however the

structure of alkyl-BO-EO copolymers can be optimized to specific drug structures to enable otpimization of solubilization.

References

Â

1. Benet, L. Z., The role of BCS (biopharmaceutics classiï¬cation system) and BDDCS (biopharmaceutics drug disposition classiï¬cation system) in drug development. *J Pharm Sci.* **2013,** *102* (1), 34-42.

2. (a) Tang, J.-L.; Sun, J.; He, Z.-G., *Current Drug Therapy* **2007,** *2*, 85-93; (b) Nielsen, F. S.; Petersen, K. B.; Mullertz, A., *Eur. J. Pharm. Biopharm.* **2008,** *69*, 553-562; (c) Kohli, K.; Chopra, S.; Dhar, D.; Arora, S.; Khar, R. K., *Drug Dicsovery Today* **2010,** *15*, 958-965; (d) Holm, R.; Jorgensen, E. B.; Harborg, M.; Larsen, R.; Holm, P.; Mullertz, A.; Jacobsen, J., *Eur. J. Pharm. Sci.* **2011,** *42*, 416-422.

503694 2D Printed Product Design of Patient Centric Dosage Forms for Adaptive Clinical Trials

Abstract:

Adaptive clinical trial design has been encourage by both FDA and EMA as a flexible clinical conduct which allows the sponsors to bring successful drugs to the market faster while saving money and resources. As a consequence, for exploratory clinical studies, the conventional, rigid structure for manufacture, warehousing and distribution of Investigational Medicine Products (IMPs) can be advantageously replaced by manufacturing technologies which can tailor the dosage strength with real time adjustment in due course alongside with the open clinical trial and minimize the waste of drug compounds.

In this work, we investigate the potential of industrial inkjet printing technology as a flexible and scalable approach of producing clinical supplies and also overcoming biopharmaceutics challenges of difficult-to-deliver drugs such as poor solubility. A systematic development of poorly soluble model drug was accomplished through (1) rational design for amorphous solid dispersions (ASDs) formulation, (2) inkjet printing process understanding, development and optimization for printing ASD producing ink solution and (3) product design for ASD-printed film that could be filled in hard capsules (PFC).

We will present the results of our ink development approach which selects the desired drug-polymer-solvent systems based on systematic solid-state screening, jetting/printing performance and product key attributes (e.g. dosage size and drug load) and substrate selection. PFC prototypes were produced using a R&D inkjet system and preliminary *in vitro* dissolution and short-term stability studies were carried out. Both the potential and remaining challenges of inkjet printing technology in the context of adaptive drug development will be discussed and summarized.

501018 How Crystal Specifications Influence Pharmaceutical Tablets?

Abstract:

Paracetamol (acetaminophen) is an active pharmaceutical ingredient (API) known for its poor tabletability, mainly because of low flow- and compression-abilities. Several approaches have been proposed to improve this, including modifying crystal shape and size, changing the maximum punch compaction pressure during tablet making, and utilizing formulations with higher percentage of excipient.

In this work, we are using the first approach by investigating the influence of crystal shape and size. To do that, we use three different crystals of paracetamol with different habit and sizes and will characterize the resulting tablets (pure as well as complete formulations) manufactured using direct compression both as single tablets and in continuous fashion.

For pure paracetamol tablets, we have observed that the tensile strength and compaction profile are strongly size and shape dependent. For formulation tablets, tensile strength, friability, dissolution and disintegration profiles are being investigated. We hypothesize that the crystal habit (shape) is the influencing factor in such settings. The present study provides a comprehensive data set that enables the prediction of the relation between crystal specification, direct compression setting, and quality of the final products.

500338 Using a Focused Beam Reflectance Measurement (FBRM) Probe to Characterize Tablet Disintegration Behavior As a Function of Drug Product Processing Conditions

Abstract:

Standard USP disintegration testing describes only the macro scale break-up of a tablet in a basket-rack assembly through a woven mesh. The method provides no insight into micro scale behavior, such as the particle size distribution of disintegrated tablet fragments, changes in this distribution which occur concurrently with dissolution of the active ingredient, or the dissolution rate at which of the active itself. Here we present the application of FBRM (Focused Beam Reflectance Measurement) to evaluate the evolution of particle size during the disintegration process.

Immersion of an FBRM probe in the media of a modified USP dissolution vessel enables the tracking of both particle count and size as disintegration progresses. We demonstrate how differences in upstream tablet processing steps are manifest in differences in micro scale disintegration behavior, and how FBRM-based analytics can provide insights absent from the traditional USP-based disintegration assessment. Specifically, in a roller compaction QbD assessment, granule density differences are traceable in FBRM particle size data on final tablet disintegration. Process ranges for roller compaction can be tied to specific physical properties such as granule sizes during disintegration, and help set operating ranges. Infrared and Raman techniques can be employed in parallel, monitoring active as it dissolves, and interrogating the physical form of residual solid material during the process. The resulting approach offers a more detailed analysis of initial phenomena tied to tablet disintegration, providing a deeper understanding of the relationships between material properties, processing conditions, and formulation performance.

501753 Finding a Surrogate Material to an Expensive API By Applying a Multivariate Analysis on a Material Database

Abstract:

Implementing a QbD approach in the pharmaceutical industry implies full understanding of processes effects into product performance. It also requires performing quantitative risk assessments. Risk relies on understanding the role played not only by processing parameters, but also by raw and intermediate material properties. From this, the importance and necessity of implementing techniques to characterize material properties and to relate them with product performance through processing parameters. Within this context is that determining critical material properties is an important step towards defining the design space. Usually, the number of characterization techniques is too many. Decreasing that number without significantly losing descriptive power is the goal of this presentation, for that we use Principal Component Analysis....

490023 Statistical Evaluation of Modeling Approaches of Drug Release Profiles for HPMC Matrix Tablets

Abstract:

Statistical Evaluation of Modeling Approaches of Drug Release Profiles for HPMC Matrix Tablets

Wenzhao Yang[1], Jin Zhao[2], Jamie Curtis-Fisk[2], Karen Balwinski[2], True Rogers[2], and Shrikant Khot[2]

(1) The Dow Chemical Company, Herbert D. Doan R&D Center, Midland, Michigan, 48674, USA

(2) Pharmaceutical Excipients R&D, Dow Food, Pharma & Medical, Midland, MI 48674

Matrix tablets containing hypromellose or hydroxypropyl methylcellulose (HPMC) are widely used in oral drug delivery modulate drug release over time. Drug release kinetics is primarily influenced by the kinetics and extent of polymer swelling and erosion, and drug (or active pharmaceutical ingredient (API)) dissolution, diffusion, and/or erosion through the polymer matrix. Mathematical modeling of drug release profiles has been developed in the past to enable pharmaceutical scientists to design and optimize matrix tablet formulations [1 2 3 4]. This research focuses on using statistical analyses to compare and assess the suitability of various models for METHOCEL™ Premium HPMC matrices.

Linear mixed model (LMM) without empirical models and two-stage model (TSM) with empirical models were used to develop predictive models. Multiple regression methods (Stepwise Regression, Partial Least Square and Generalized models) in JMP were applied to incorporate drug property factors for modeling drug release rate in TSM. The dataset analyzed included API with solubility ranging from 0.2 to 1000 mg/ml and API concentration from 12.5 to 50 wt%, HPMC grades of METHOCEL™ K100M, METHOCEL™ K4M, METHOCEL™ K100LV and HPMC concentration from 20 to 40 wt%. The two models were compared based on the split validation approach using the regression coefficient (r^2) and mean square error (MSE). The LMM has better overall prediction accuracy compared to the TSM. However, TSM outperformed LMM in predicting drug release rate during the first two hours following introduction to dissolution media. It has also been found that the fitting parameters of empirical models such as Korsmeyer-Peppas and Peppas-Sahlin models have strong correlations as shown in Figure 1. In addition to the drug characteristics, the effect of polymer viscosity grade and API concentration were found as key impacting factors in the model.

The predictive model developed here provided mechanistic insight into kinetic release profile of HPMC matrices. This can be used to aid pharmaceutical scientists with efficient and streamlined formulation design and *in vitro* drug release experimentation to deliver desired modified-release performance. A more comprehensive model will be further developed to predict the effect of key controllable variables such as type and concentrations of fillers, polymer concentrations, and compression force.

Figure 1. Correlation of fitting parameters of Korsmeyer-Peppas Model.

References

1 J Sipemann and N Peppas, "Mechanism of drug release from delivery systems based on HPMC," Adv Drug Deliv Rev **48**, 139 (2011).

2 N. A. Peppas and B. Narasimhan, J. Cont. Rel. **190**, 75-81 (2014).

3 LS Koester, GG Ortega, P Mayorga, and VL Bassani, "Mathematical evaluation of in vitro release profiles of hydroxyl propyl methylcellulose matrix tablets containing carbamazepine associated to cyclodextrin.," Eur J Pharm Biopharm **58**, 177–179 (2004).

4 MC Gohel and MK Panchal, "Novel mathematical method for quantitative expression of deviation from the Higuchi model.," AAPS Pharm Sci Tech **1** (1-6) (2000).

FACULTY OF
PHARMACEUTICAL SCIENCES

DEPARTMENT OF PHARMACEUTICAL ANALYSIS
LABORATORY OF PHARMACEUTICAL PROCESS ANALYTICAL TECHNOLOGY

MULTIVARIATE DATA ANALYSIS OF RAW MATERIAL PROPERTIES FROM PHARMACEUTICAL POWDERS

PREDICTING COMPACTION BEHAVIOR USING MVA AND FEA

AIChE – November 2nd, 2017
Jens Dhondt, Ashish Kumar, Bernd Van Snick, Valérie Vanhoorne, Johny Bertels, Didier Klingeleers, Chris Vervaet, Thomas De Beer

GHENT UNIVERSITY LP PAT janssen

OVERVIEW

- Introduction
- Raw Material Properties
- Blend Characterization
- Use of FEA Models
- Conclusion

GHENT UNIVERSITY LP PAT janssen

INTRODUCTION

GHENT UNIVERSITY LP PAT janssen

3

INTRODUCTION

- Development of continuous pharmaceutical processes still challenging
  - Insufficient material characterization and understanding
  - Inadequate fundamental process understanding
  - Formulation development mostly trial and error approach
  - Limited usage of sophisticated process simulation tools
  - Need for increased flexibility towards formulation changes

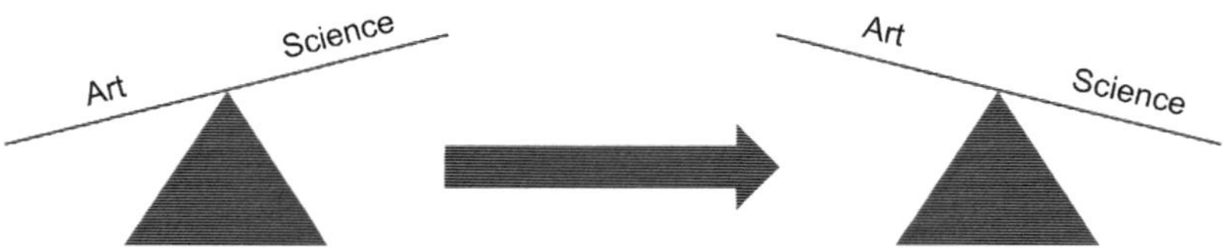

GHENT UNIVERSITY LP PAT janssen

4

INTRODUCTION

- Systematic approach towards state-of-the-art scientific formulation and process development
  - Database with material characteristics (raw materials, blends, tablets…)
  - Identification of critical material attributes at each unit operation
    - Time-saving towards future formulation and process development
    - Less consumption of material in early R&D phase
    - Identify subset of informative characterization techniques towards different processes
    - Define preferred API and excipient properties

GHENT UNIVERSITY LP PAT janssen

5

RAW MATERIAL PROPERTIES

GHENT UNIVERSITY LP PAT janssen

6

MATERIALS & METHODS

- 22 API's
- 25 Fillers (17 DC + 8 WG)
- 10 Binders
- 5 Disintegrants
- 2 Lubricants
- 2 Glidants

- Total of 67 powders included
  - Commonly used in DC and WG processes

GHENT UNIVERSITY LP PAT janssen

7

MATERIALS & METHODS

- Particle Size Distribution
  - Laser Diffraction
  - Static Image Analysis
- Flow Properties
  - Tangens Alfa (AoR)
  - GranuHeap (AoR)
  - FlowPro (Flow through an orifice)
  - FT4 Stability and Variable Flow Rate
  - GranuDrum (Avalanche Angle)
  - Schulze Ring Shear Tester
- Powder Density
  - Bulk & Tapped Density
  - He-pycnometry (True Density)
- Wall Friction
  - Schulze Ring Shear Tester
  - FT4 Shear Cell
- Compressibility
  - FT4 Vented Piston
- Charge Density
  - Dispersion Charging

GHENT UNIVERSITY LP PAT janssen

8

MATERIALS & METHODS

- Powder Fluidization
  - FT4 Aeration
  - FT4 Permeability
- Specific Surface Area
  - N_2-adsorption (BET)
- Particle Shape
  - Static Image Analysis
- Moisture Content
  - Dynamic Vapor Sorption
  - Loss on Drying
- 20 testing methods
  - Multiple responses per method
    115 material descriptors

GHENT UNIVERSITY LP PAT janssen

9

MATERIALS & METHODS

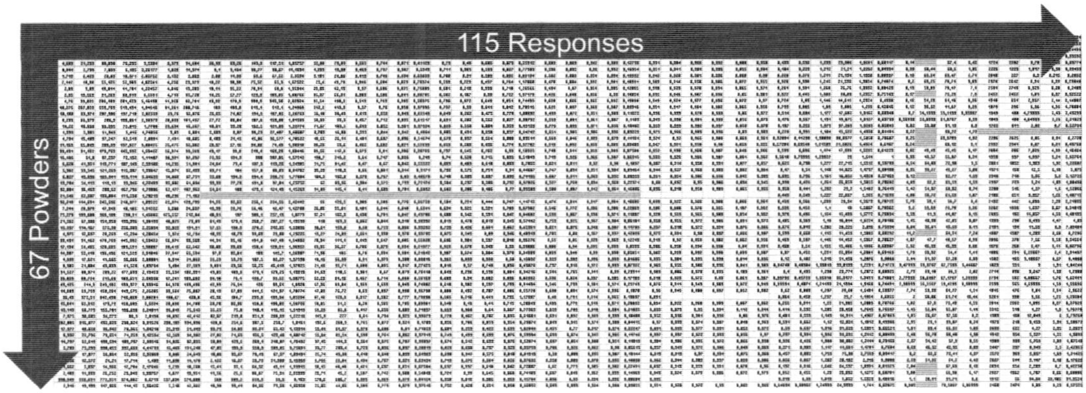

GHENT UNIVERSITY LP PAT janssen

10

USE OF PCA

- Big Data table: are there hidden secrets?
- MultiVariate Data Analysis (MVDA)

  = fast track from data table to graphical interface

  Software: SIMCA™ (Umetrics, Umea, Sweden)
  Phi_1.7 toolbox (Salvador Garçia-Muñoz), MATLAB™ (MathWorks, Natick, MA, USA)

  - Which variables / observations (un)correlate?
  - Unique patterns?
  - How much of the variability can be described and predicted?

GHENT UNIVERSITY LP PAT janssen

11

DATA TREATMENT

- Exclude outliner materials
- Absolute value of Charge Density
- Unit Variance scaling
- Log transformation (e.g. d_v50)
- Limit number of correlated properties
- Exclude responses with $Q^2 < 0$

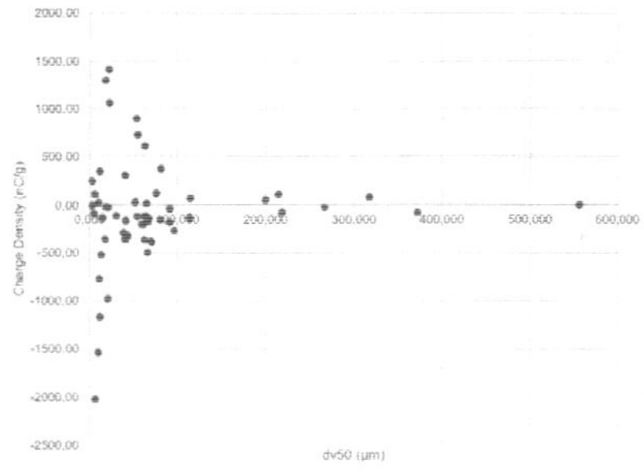

GHENT UNIVERSITY LP PAT janssen

12

RAW MATERIAL MODEL

– M2: R^2 70,1% | Q^2 51,3%

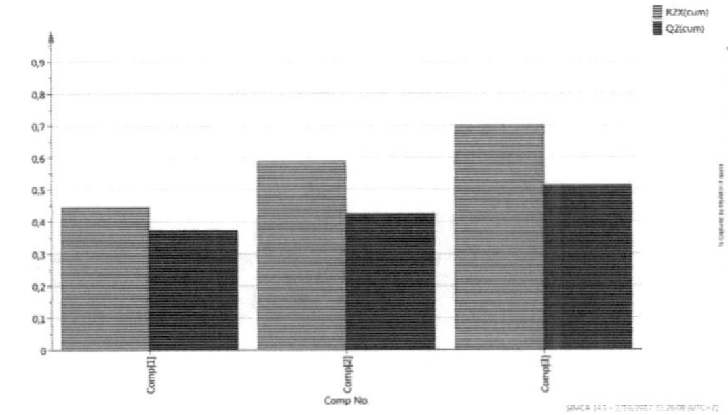

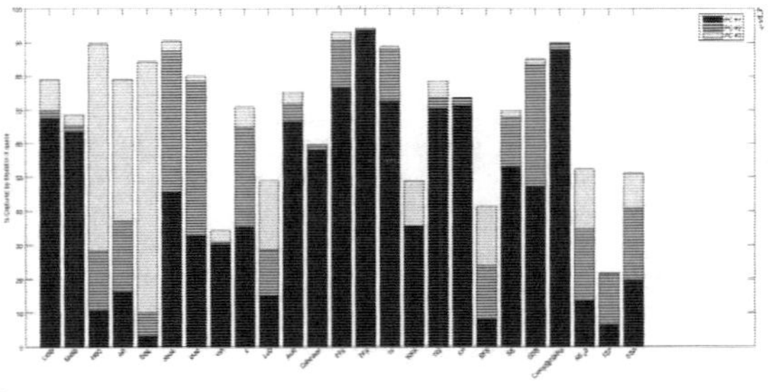

GHENT UNIVERSITY LP PAT janssen

RAW MATERIAL MODEL

– PC1 vs PC2

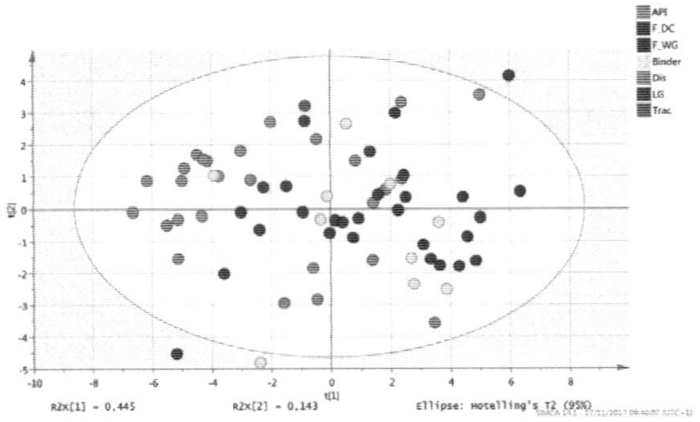

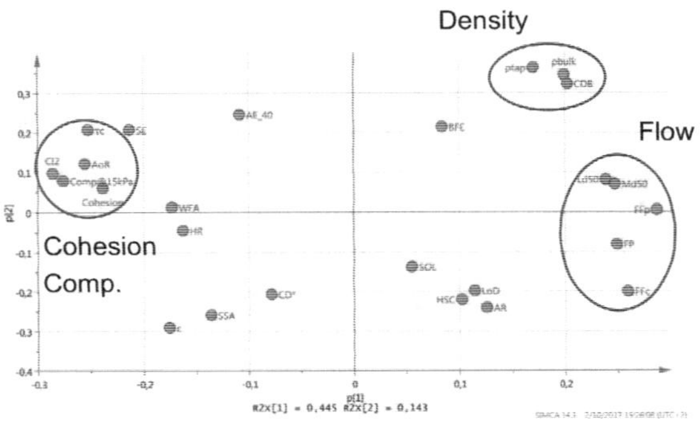

GHENT UNIVERSITY LP PAT janssen

RAW MATERIAL MODEL

– PC1 vs PC2

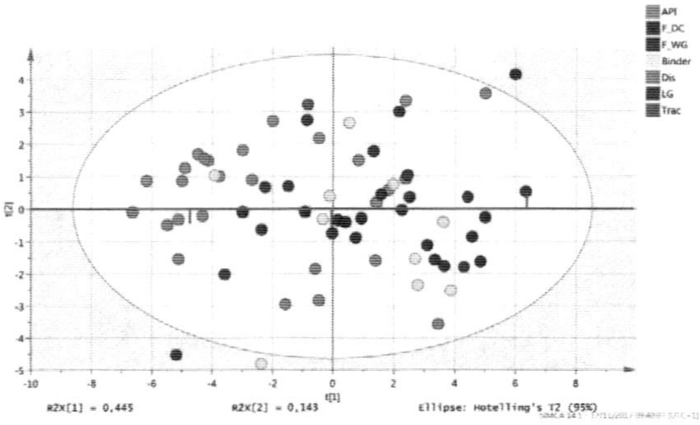

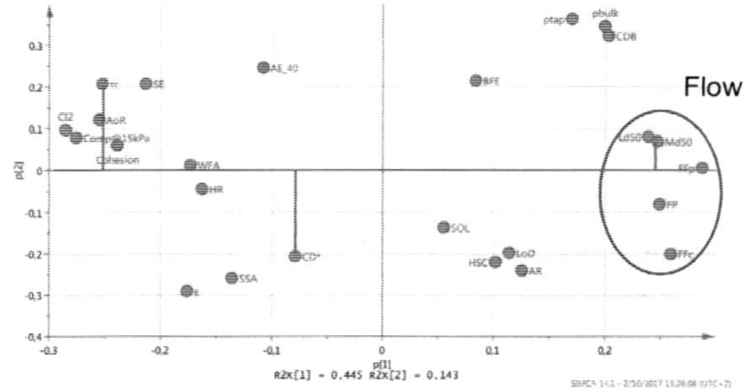

RAW MATERIAL MODEL

– PC1 vs PC3

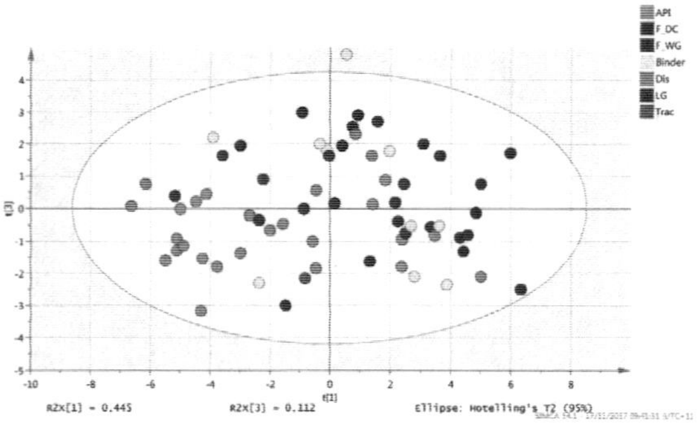

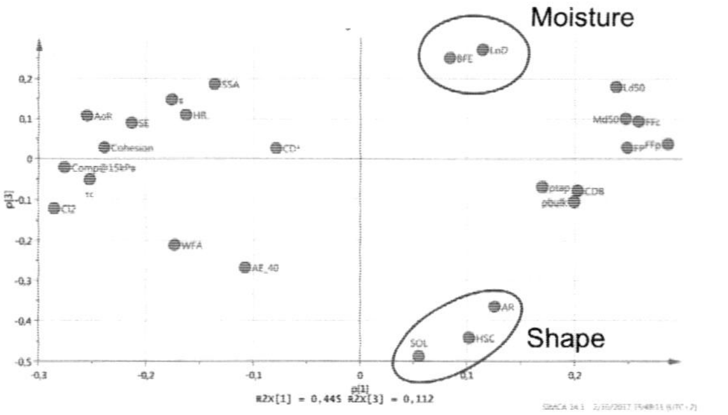

RAW MATERIAL MODEL

- Principal components cover up to 70% of variability in data set
  - PC1: Flow related properties
  - PC2: Density related properties
  - PC3: Shape/Moisture related properties
- Reduced characterization set based on PCA
  - Maximal positive and negative loading on PCs
  - PC1: Ring Shear Test / FT4: Compressibility
  - PC2: Bulk-Tapped-True Density / FT4: Variable Flow Rate
  - PC3: FT4: Aeration / Loss on Drying
- 20 initial tests → 7 tests
  - **66% reduction in workload**

GHENT UNIVERSITY LP PAT janssen

17

RAW MATERIAL MODEL

- Principal components cover up to 70% of variability in data set
  - PC1: Flow related properties
  - PC2: Density related properties
  - PC3: Shape/Moisture related properties
- Reduced characterization set based on PCA
  - Maximal positive and negative loading on PCs
  - PC1: Ring Shear Test / FT4: Compressibility
  - PC2: Bulk-Tapped-True Density / FT4: Variable Flow Rate
  - PC3: FT4: Aeration / Loss on Drying
- 20 initial tests → 7 tests
  - **66% reduction in workload**

GHENT UNIVERSITY LP PAT janssen

18

CROSS VALIDATION

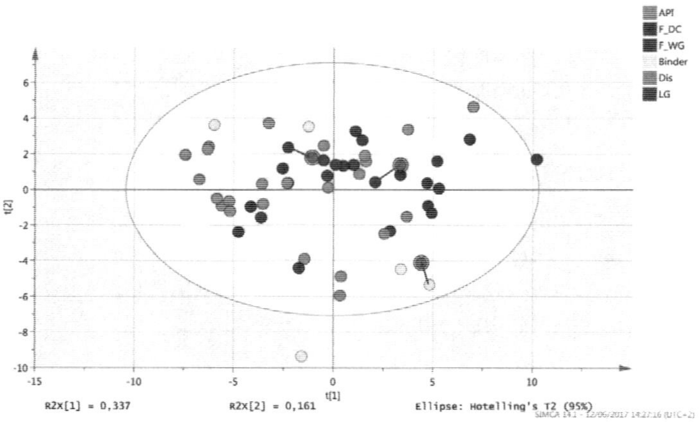

GHENT UNIVERSITY LP PAT janssen

19

BLEND CHARACTERIZATION

GHENT UNIVERSITY LP PAT janssen

20

BLEND SELECTION

- Subanalysis for API's and DC Fillers

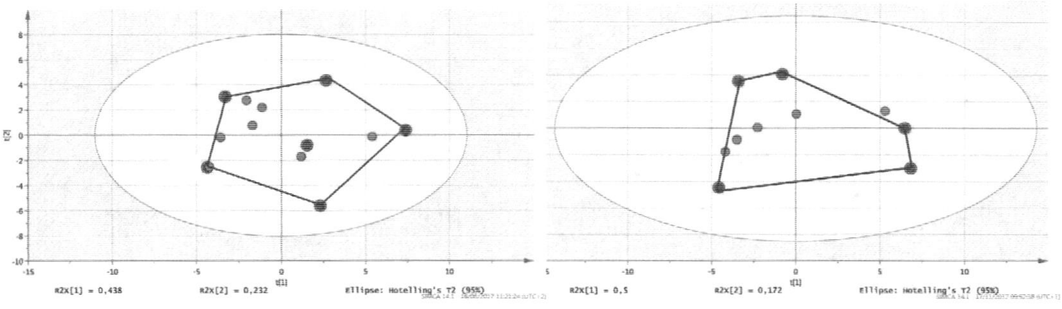

GHENT UNIVERSITY LP PAT janssen

21

BLEND CHARACTERIZATION

- Fixed blend ratio: 30/70 API/Filler
- Reduced characterization set based on PCA
- Principal components cover up to 70% of variability in data set
  - PC1: Ring Shear Test / FT4: Compressibility
  - PC2: Bulk-Tapped-True Density / FT4: Variable Flow Rate
  - PC3: FT4: Aeration / Loss on Drying

  - Blend uniformity (Raman spectroscopy)

GHENT UNIVERSITY LP PAT janssen

22

BLEND MODEL

– M3: R^2 85,0% | Q^2 69,1%

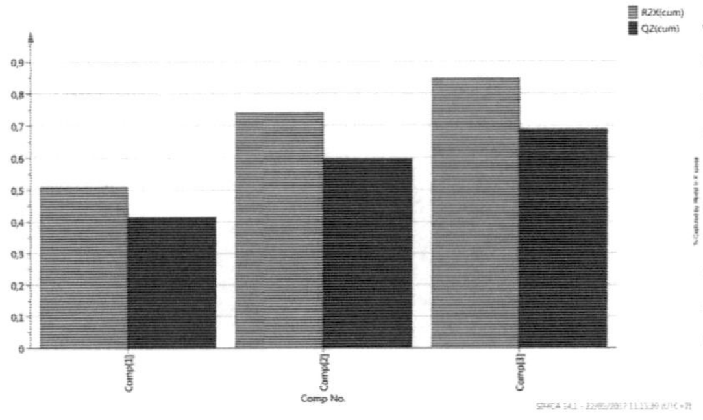

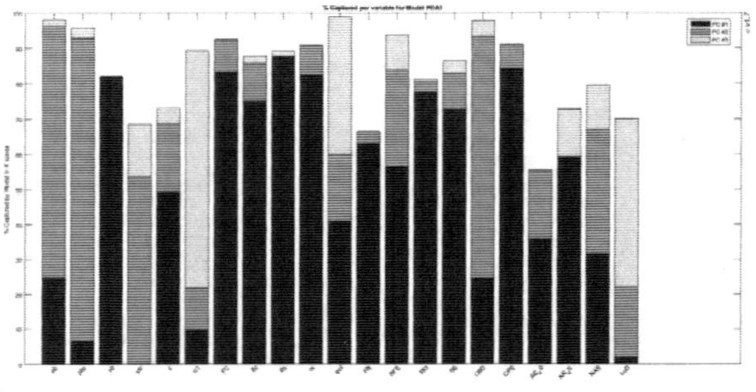

GHENT UNIVERSITY

LP PAT

janssen

23

BLEND MODEL

– PC1 vs PC2

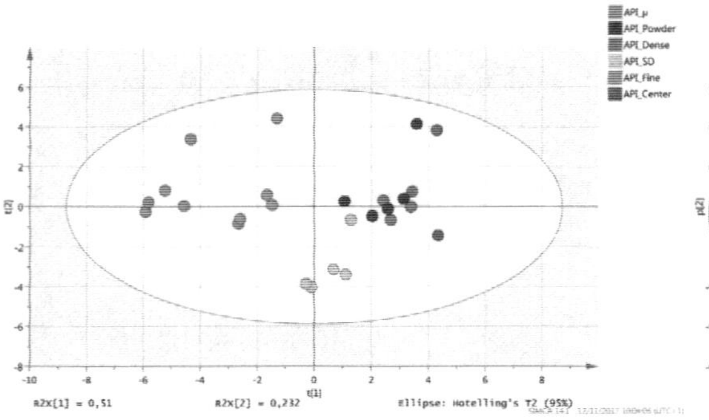

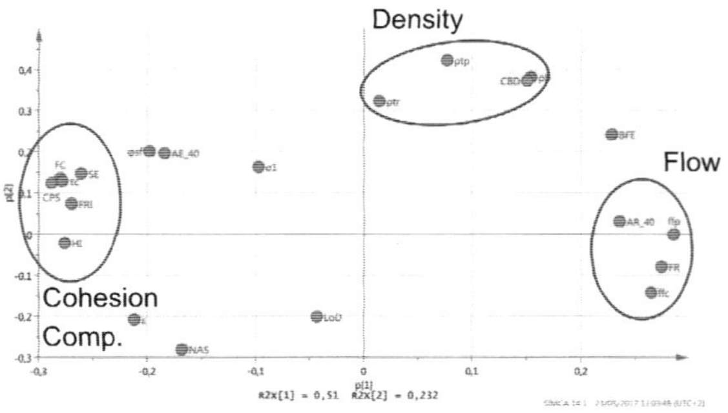

GHENT UNIVERSITY

LP PAT

janssen

24

BLEND MODEL

- PC1 vs PC3

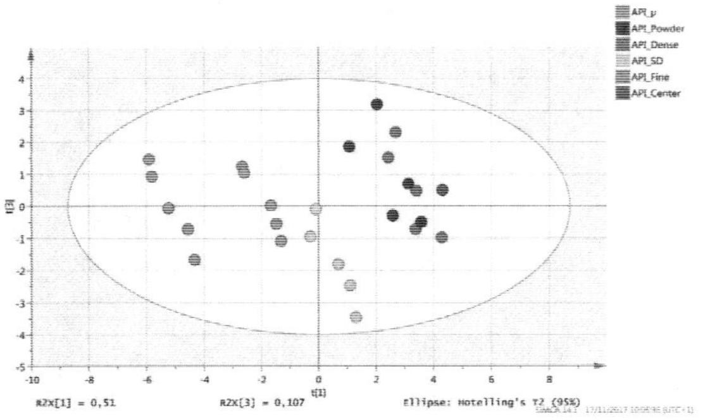

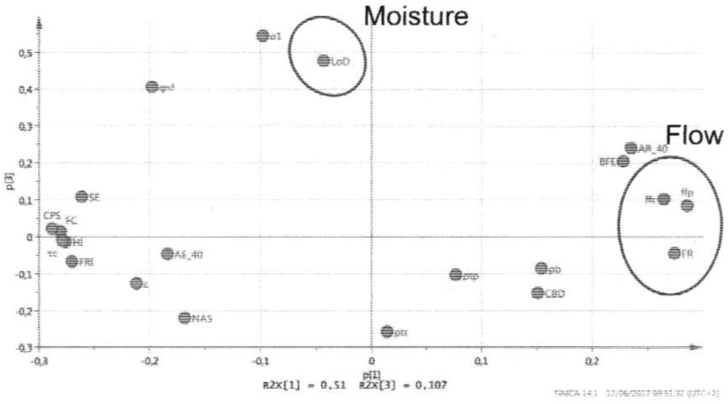

GHENT UNIVERSITY LP PAT janssen

25

BLEND MODEL

- Principal components cover up to 85% of variability in data set
  - PC1: Flow related properties
  - PC2: Density related properties
  - PC3: Moisture content
- Clustering along PCs based on API or filler depending on property
- Results in accordance with raw material PCA

GHENT UNIVERSITY LP PAT janssen

26

USE OF FEA MODELS

GHENT UNIVERSITY LP PAT janssen

27

CALIBRATION OF DPC MODEL

- Modified Drücker-Prager Cap model
  Han et al. (2008)

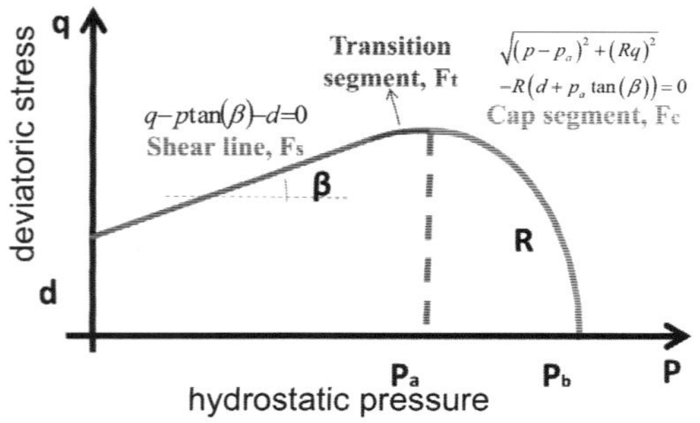

Elastic properties			
E	Young's modulus	Die compaction test, unloading	
V	Poisson's ratio		
Plastic properties			
d	Cohesion	Uniaxial compression and diametrical compression test	
β	Internal friction angle		
R	Cap eccentricity	Instrumented die compaction test, loading	
Pa	Evolution parameter		
Pb	Hydrostatic yield stress		

GHENT UNIVERSITY LP PAT janssen

28

CALIBRATION OF DPC MODEL

– Experimental Calibration using Huxley Bertram Compaction simulator

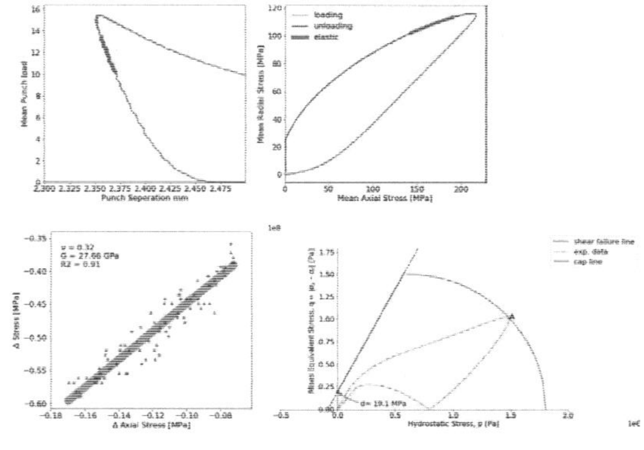

GHENT UNIVERSITY LP PAT janssen

FEA SIMULATIONS

Geometrically symmetric

Geometrically asymmetric

GHENT UNIVERSITY LP PAT janssen

FEA SIMULATIONS

Geometrically symmetric

Axisymmetric FE model-2D

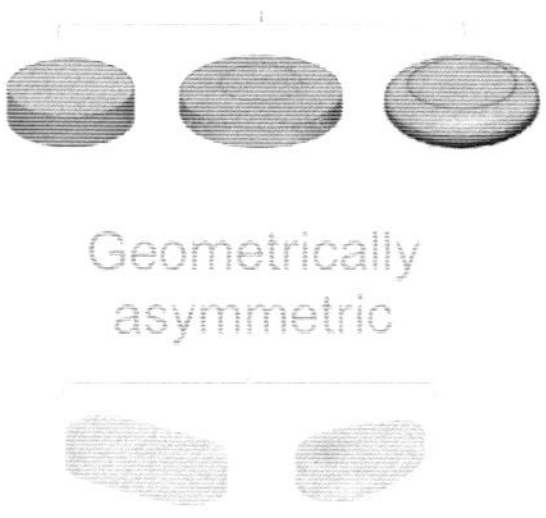

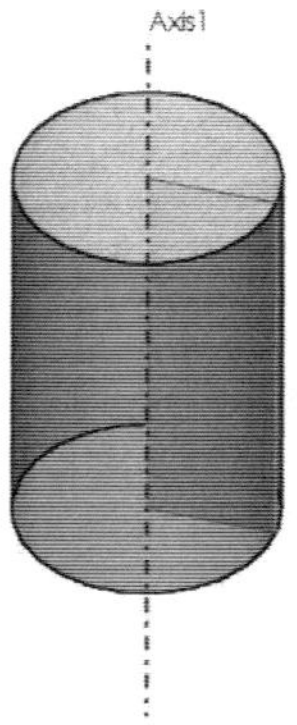

GHENT UNIVERSITY LP PAT janssen

FEA SIMULATIONS

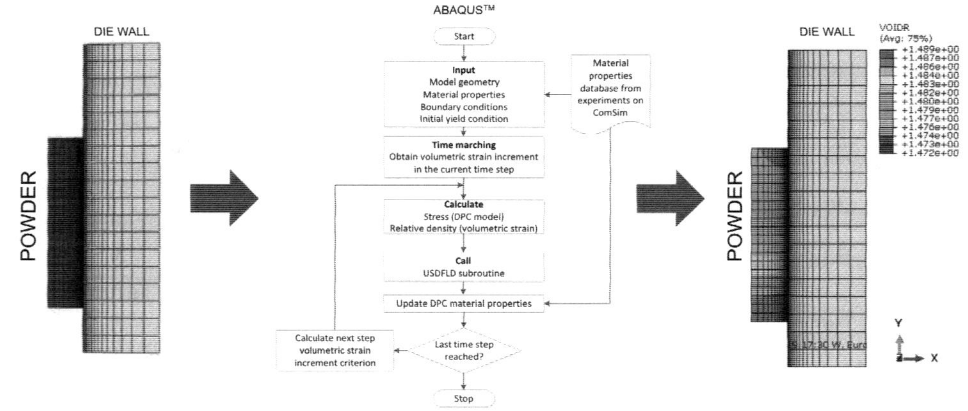

GHENT UNIVERSITY LP PAT janssen

FEA SIMULATIONS

– Visible difference in stress distribution in different powders when compacted similarly

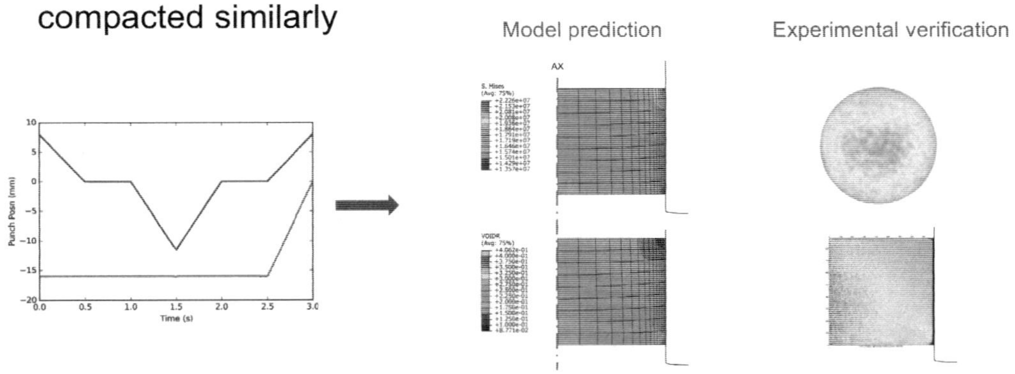

GHENT UNIVERSITY   LP PAT   janssen

33

CONCLUSION

GHENT UNIVERSITY   LP PAT   janssen

34

CONCLUSION

- 1 particle or bulk property does not provide enough information to differentiate between powders
- Use of MVDA:
  - Provides valuable insights in large databases
  - PCA to rationalize number of critical techniques for routine characterization
  - Use submodels to compare materials of similar nature
  - Minimize use of API during R&D by maximizing information/kg
  - Accelerate launch of new products on continous manufacturing systems

GHENT UNIVERSITY LP PAT janssen

35

CONCLUSION

- Database should be maintained as "self-learning" platform
  - Expand with new materials and powder properties (e.g. Compaction behaviour)
  - Add end-product quality attributes and process settings
  - Complete predictive platform using L-PLS and T-PLS moddeling
    Garçia-Muñoz et al. (2014)
  - Implementation towards formulation and process development
    - Direct Compression
    - Feeding Performance
    - Wet Granulation
    - ...

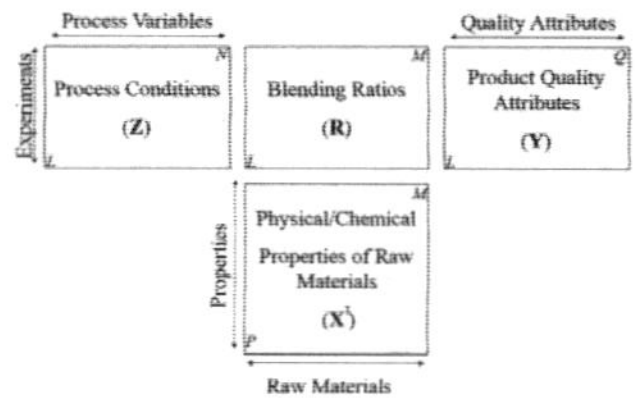

Fig. 1. "T" shape collection of matrices with process, materials, ratios of usage and quality information.

GHENT UNIVERSITY LP PAT janssen

36

IMPLEMENTATION

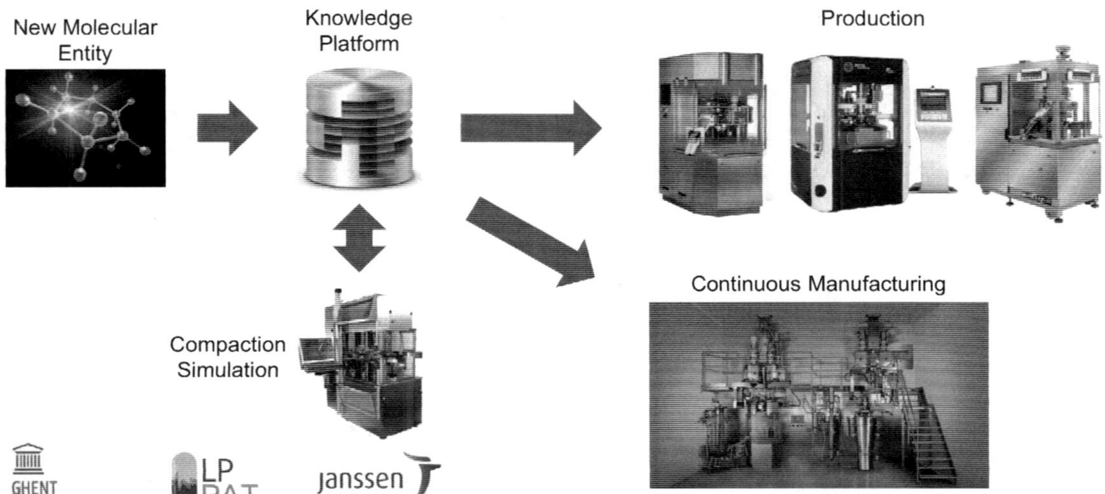

GHENT UNIVERSITY LP PAT janssen

37

ACKNOWLEDGEMENTS

- PreTaP Team
  - Dr. Ashish Kumar
  - Prof. Dr. Thomas De Beer
  - Johny Bertels
  - Dr. Didier Klingeleers
- ICONS Team
  - Bernd Van Snick
  - Prof. Dr. Chris Vervaet
  - Dr. Valérie Vanhoorne
  - Kenny Pandelaere
  - Dr. Giustino Di Pretoro
  - Dr. Ivo Van Assche
  - Prof. Dr. Matthieu Boone
  - Dr. Jan Aelterman
  - Laurens Meeus
  - Prof. Dr. Antonios Zavaliangos

AGENTSCHAP INNOVEREN & ONDERNEMEN Vlaanderen is ondernemen

janssen PHARMACEUTICAL COMPANIES OF Johnson-Johnson

GHENT UNIVERSITY LP PAT janssen

38

FACULTY OF
PHARMACEUTICAL SCIENCES

Jens Dhondt
PhD Student

LABORATORY OF PHARMACEUTICAL PROCESS ANALYTICAL TECHNOLOGY

E JensJ.Dhondt@UGent.be
T +32 9 264 81 18

http://www.ugent.be/fw/pharmaceutical-analysis/en/research/pat

GHENT
UNIVERSITY

LP
PAT

janssen

492735 Simulation of Particle Dissolution Using the Phase Field Approach

Abstract:

Simulation of Particle Dissolution using the Phase Field Approach

D. Sleziona[1], D. R. Ely[2], M. Thommes[1]

[1]Technical University Dortmund, Emil-Figge-Str. 68, 44227 Dortmund, Germany

[2]Ivy Tech Community College, 3101 S Creasy Ln, Lafayette IN 47905, USA

PURPOSE

The aim of this work is to describe the dissolution behavior of single crystals with a numerical model. Already existing analytical approaches for the description of crystal dissolution like Fick's law of diffusion[1], the Nernst-Brunner equation[2,3], the Hixson-Crowell cubic root law[4] and the Higuchi[5] equation are suitable. However they do not take into account the anisotropy or the particle shape. The motivation of this work is to model the anisotropic dissolution behaviour of single crystal particles by using a finite volume method. Dissolution is understood as the transformation of a solid (crystalline state) in a liquid environment in a solution. This process is divided into two parts. The phase transition and diffusion. The first part can be modelled numerically with the so called phase field simulation[6]. In this numerical tool a partial differential equation is used to replace boundary conditions at the interface[6]. This leads to an evolving ancillary field. In this field the state of the materials is described by a non-conserved order parameter, which is a function of position and time[6]. It models the phase as a continuous variable ranging from disordered (liquid) to perfectly ordered (solid)[6,7]. Also the second part, diffusion, can be calculated with a finite volume method. This approach offers the advantage that no diffusion layer has to be implemented. It is formed independently during the simulation.

RESULTS

An isotropic, circular, two-dimensional Xylitol particle dissolving in water (25 °C) was simulated as a model system. The time evolution of the phase (phase transition) of the binary system could be described by the Allen-Cahn-equation[6,8], as follows:

Equation 1 gives the rate change of the phase ($\partial F/\partial t$) in terms of the mobility of the phase (M_F), the free energy density (f) and the gradient energy coefficient of the phase (e_F). The corresponding concentration field (diffusion) was modelled using the approach of Cahn and Hilliard[8,9] given by the equation,

As seen in equation 2, the rate change of the concentration field ($\partial c/\partial t$) is expressed as a function of the concentration mobility (M_c), the free energy density (f) and the gradient energy coefficient of the concentration (e_c). Equations 1 and 2 form a system of two coupled partial differential equations linked via the free energy density $f(F,c,T)$, which is a function of phase, concentration, and temperature. The system was solved numerically to obtain the phase transition and diffusion (concentration field) as a function of time as shown in Figure 1.

Figure 1: dissolution of a 100 µm xylitol particle in water (25 °C)

A simultaneously evolving phase and concentration field could be observed. Because of the isotropic properties of the simulated, two dimensional particle a validation of the numerical results based on the previous mentioned Nernst-Brunner equation is possible. Due to the fact that the Nernst-Brunner law describes the dissolution behaviour of a three dimensional, isotropic spherical particle, a new validation equation for the circular two dimensional, isotropic, Xylitol crystal was derived.

This equation 3 represents the dissolution process of an infinite long, three dimensional cylinder which dissolves only over its lateral surface. There the particle radius (r) is a function of the starting radius (r_0), diffusion coefficient (D), saturation concentration (c_S), concentration (c), diffusion layer thickness (d), particle density (ρ) and time. Geometrically this reflects the two dimensional dissolution of the simulated circular particle. Figure 2 shows the course of the simulated remaining particle radius over the simulation time, which follows continuously the derived validation equation and ends up in a similar total dissolution time.

Figure 2: particle radius of a 100 µm, circular, isotropic, two dimensional xylitol particle in water (25 °C)

So a successful coupling of the Allen-Cahn and Cahn-Hilliard equation can be presented, which leads to a simulation where the change in phase transition and diffusion can be observed over time.

CONCLUSION

A finite volume simulation was developed for the investigation of the dissolution behaviour of crystal particles. Phase transition and diffusion could be coupled by linking the Allen-Cahn- and Cahn-Hilliard-equation over the free energy density function $f(F,c,T)$. The dissolution of a circular, isotropic, two dimensional Xylitol particle in water at 25°C was simulated and validated with a derived two dimensional Nernst-Brunner equation. The extension of this simulation into the third dimension and incorporation of anisotropy into this model is part of the ongoing work.

REFERENCES

1. Fick, A. (1855) Ueber Diffusion, Annalen der Physik 170, 59-86.
2. Nernst, W. (1904) Theory of reaction velocity in heterogenous systems, Zeit. physikal. Chem 47, 52-55.
3. Brunner, E. (1903) Reaktionsgeschwindigkeit in heterogenen Systemen, Georg-Augusts-Universitat, Gottingen.
4. Hixson, A. and Crowell, J. (1931) Dependence of Reaction Velocity upon surface and Agitation, Industrial &

Engineering Chemistry 23, 923-931.

5. Higuchi, T. (1961) Rate of release of medicaments from ointment bases containing drugs in suspension, Journal of Pharmaceutical Sciences 50, 874-875.

6. Boettinger, W. J., Warren, J. A., Beckermann, C.,Karma, A. (2002) Phase-field simulation of solidification, Annual review of materials research 32, 163-194.

7. Cui, Z., Gao, F., Cui, Z.,Qu, J. (2011) Developing a second nearest-neighbor modified embedded atom method interatomic potential for lithium, Modelling and simulation in materials science and engineering 20, 015014.

8. Allen, S. M. and Cahn, J. W. (1979) A microscopic theory for antiphase boundary motion and its application to antiphase domain coarsening, Acta Metallurgica 27, 1085-1095.

9. Cahn, J. W. and Hilliard, J. E. (1958) Free energy of a nonuniform system. I. Interfacial free energy, The Journal of chemical physics 28, 258-267.

We make tomorrow's drugs possible.

Optimization of a Low-Dose Dosator Capsule Filling Process for Dry Powder Inhalation (DPI) Applications using In-line PAT Approaches

Sandra Stranzinger[1,2], Eva Faulhammer[1], Otto Scheibelhofer[1], Stefano Biserni[3], Vittorio Calzolari[3], Amrit Paudel[1,2], Johannes G. Khinast[1,2]

(1) RCPE GmbH, Graz, Austria

(2) Graz University of Technology, Graz, Austria

(3) MG2, Bologna, Italy

K1 Competence Center - Initiated by the Federal Ministry of Transport, Innovation and Technology (BMVIT) and the Federal Ministry of Science, Research and Economy (BMWFW).
Funded by the Austrian Research Promotion Agency (FFG), Land Steiermark and the Styrian Business Promotion Agency (SFG).

We make tomorrow's drugs possible.

Motivation

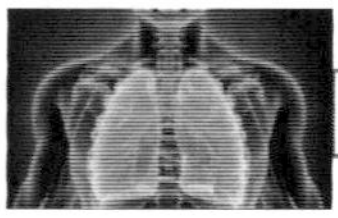

Lung Delivery

- The lung is a very attractive target for drug delivery
- Possible treatment of local diseases or systemic therapies
- **Advantages:**
  - No first pass effect
  - Fast onset of action
  - Less side effects

Dry Powder Inhalers

- Inhalable powder formulations
- Lung deposition of API: Aerodynamic size 1-5 µm
- Carrier based systems: Mixture of API particles with larger excipient

  → **Majority of DPI products on the market**

Capsule Filling

- Capsules used in DPIs are typically filled with less than 40 mg of powder

  → **Dose Uniformity!!!**

  → **Optimization of capsule filling process for DPI products**

Dosator Nozzle Capsule Filling – Process at a Glance

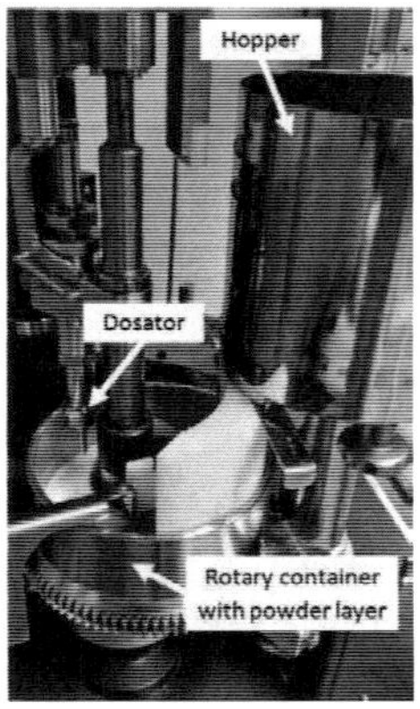

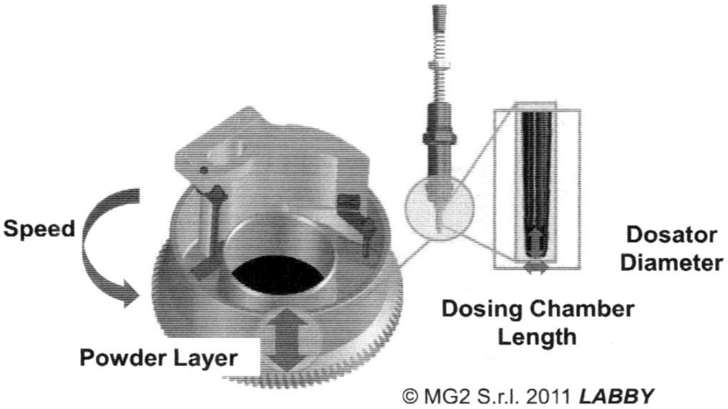

Dosator Nozzle Capsule Filling – Multiple Step Process

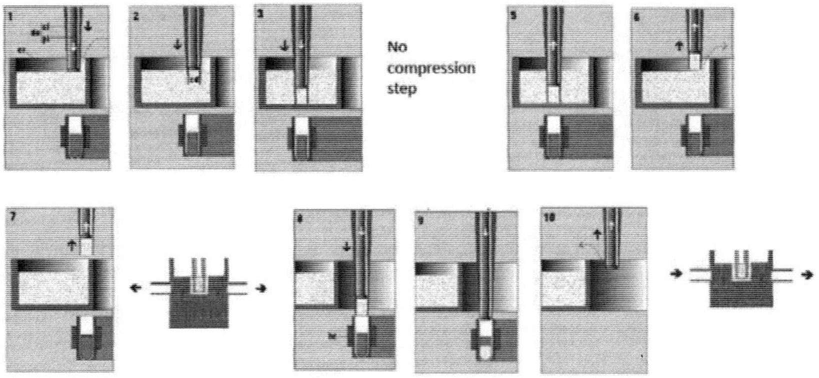

Gamberini, E., 1985. Method for filling containers with metered quantities of powdered materials.

Dosator Nozzle Capsule Filling – Challenges

- **Many factors influence the effectiveness and quality** of a capsule filling process with a dosator nozzle machine (powder and nozzle properties, size of the dosator, speed of capsule filling, etc.).
- For efficacy/safety **the dosed mass** is very critical (therapeutic effect, side effect).
- For volumetric dosing systems a **constant powder bed density** is essential to get a constant mass.
- Yet, powder layer uniformity (powder bed density) is affected by powder properties and flowability of powders. → ***Challenging inhalation powders***
- The powder bed from which the nozzle collects a specific volume of powder, is subjected to vibration as it rotates. It is expected that the vibrations in the bowl induce particle movement, and possibly, long-term segregation. → ***Different segregation tendency of various powders (blends)***

Objective of the Study – Optimization using Different Strategies

Dosator Nozzle Capsule Filling – Multiple Step Process

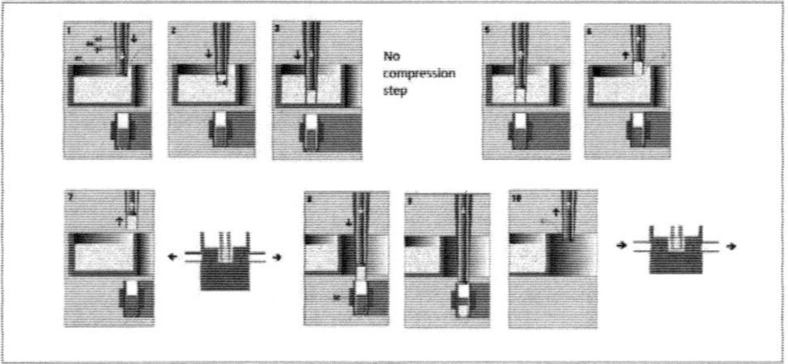

Powder Characterization/ Material Science

In-Process Monitoring – <u>Proof of Concept Study</u>

Mechanical Modifications

Simulation Studies

Decoupling of the Process

Improved Product Performance
- -Weight uniformity
- -Dosage
- -Content uniformity

<u>Final Goal:</u>

Low fill weight (down to 1 mg) with RSD < 5 %

research
center
pharmaceutical
engineering

We make tomorrow's drugs possible.

Proof of Concept Study

In-line Monitoring using RAMAN Spectroscopy

research
center
pharmaceutical
engineering

We make tomorrow's drugs possible.

Experimental Setup: Materials

API

Salbutamol Sulphate (SS)

Air jet milling

- Common method to generate inhalable sized particles
- Aerodynamic diameter: 1µm to 5µm

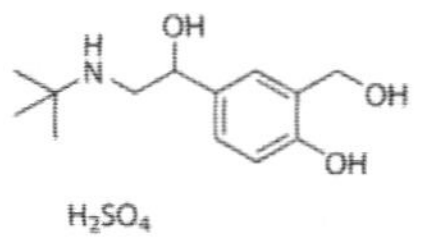

Carrier

α-Lactose Monohydrate	Particle size – x_{50} [µm]
Lactohale 100	155.24 +/-0.58

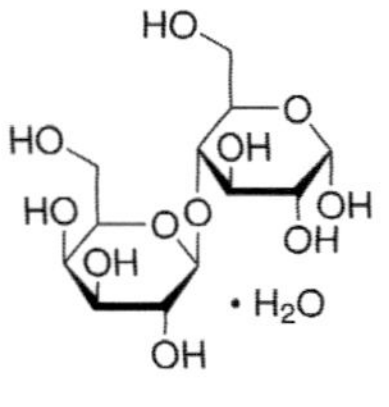

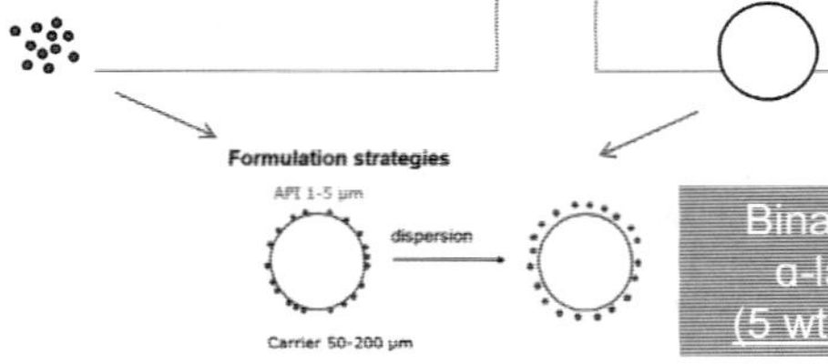

Binary mixtures of SS and α-lactose monohydrate (5 wt%, 10 wt% SS content)

Experimental Setup: Capsule Filling Settings

Speed:
3000 capsules/hour

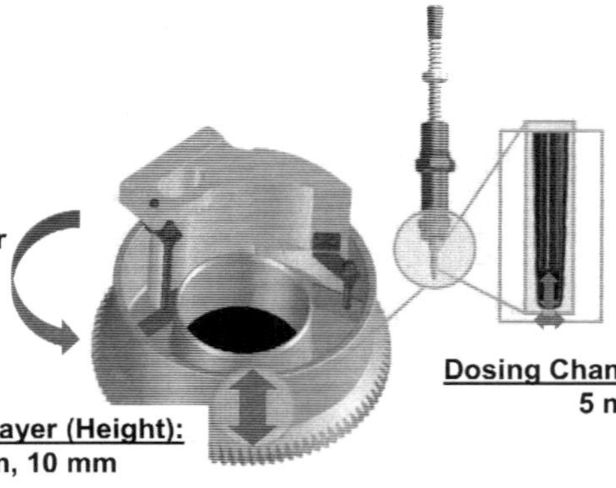

Dosator Diameter:
1.9 mm

Dosing Chamber Length:
5 mm

Powder Layer (Height):
5 mm, 10 mm

Experimental Setup: Mechanical Modifications

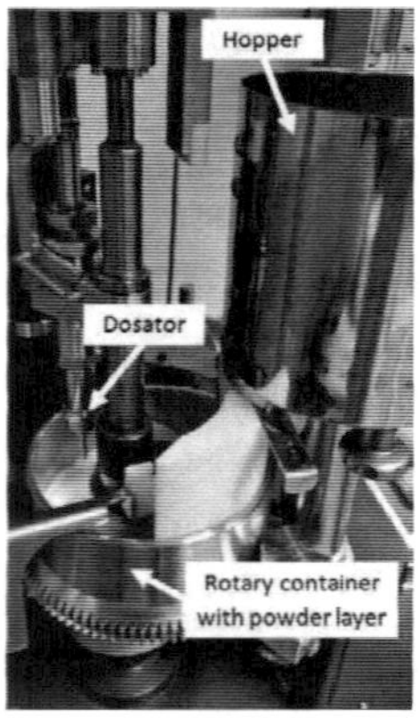

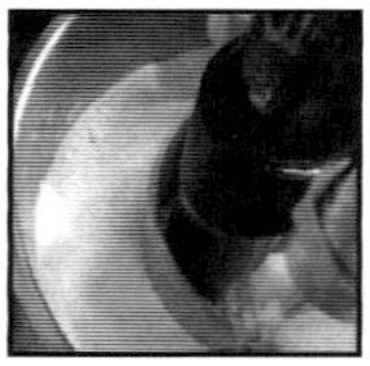

Filling performance without „Modified Scraper“

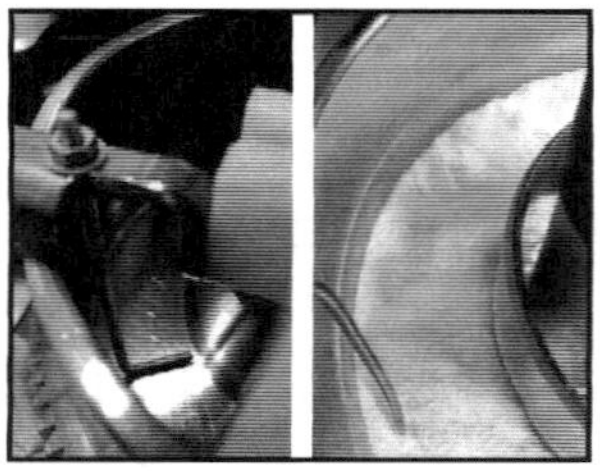

Filling performance with „Modified Scraper“

Experimental Setup: RAMAN Spectroscopy (In-line Monitoring)

- **In-line Sampling System:**
  - A filtered fiber-optic probe with uniform global illumination
    (PhAT probe, Kaiser Optical Systems, MI, USA)
- **RAMAN Settings:**
  - Spectra were collected every 7 seconds
  - 300-500 consecutive spectra per run (depending on run time)
- **→ Evaluation of Data/Spectra using MATLAB 2016a (MathWorks, USA), PLS and Statistic Toolbox**

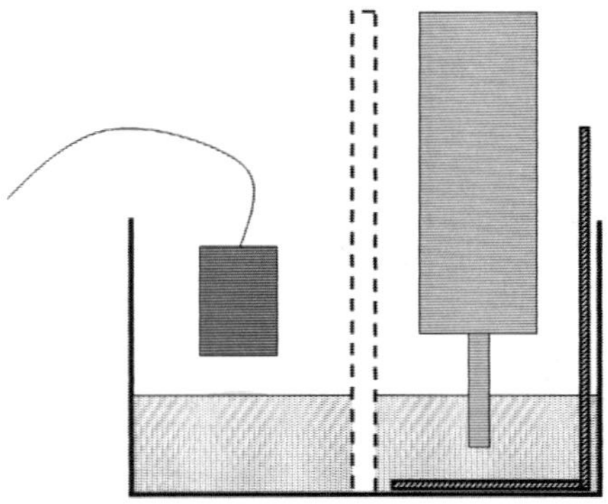

Development of a PLS Model: Principal Components

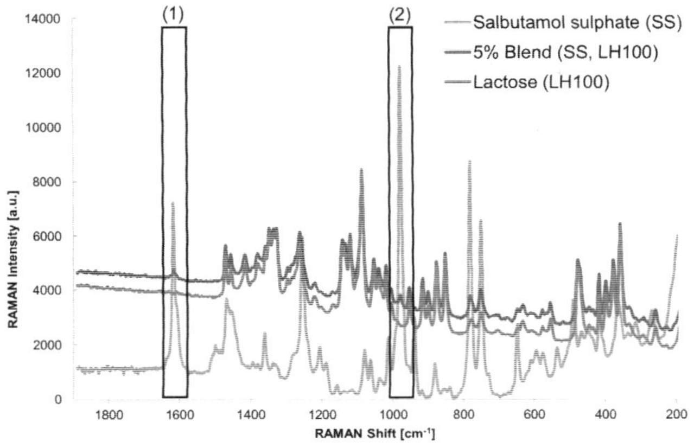

- Which spectral ranges were used for process monitoring?

Characteristic Spectral Ranges:

- Spectral range (1) from 1590 cm^{-1} to 1630 cm^{-1}
- Spectral range (2) from 970 cm^{-1} to 984 cm^{-1}

We make tomorrow's drugs possible.

Development of a PLS Model: Model Calibration

- **Calibration Set:**
  - Binary Mixtures of SS and Lactose
  - Mixtures ranging from 1 to 12 wt% SS content
  - Blending in Turbula® blender T2F (Willy A. Bachofen Maschinenfabrik, Muttenz, Switzerland) → 60 min at 68 rpm

Reference Measurements using HPLC

- Blend uniformity assessment of calibration samples

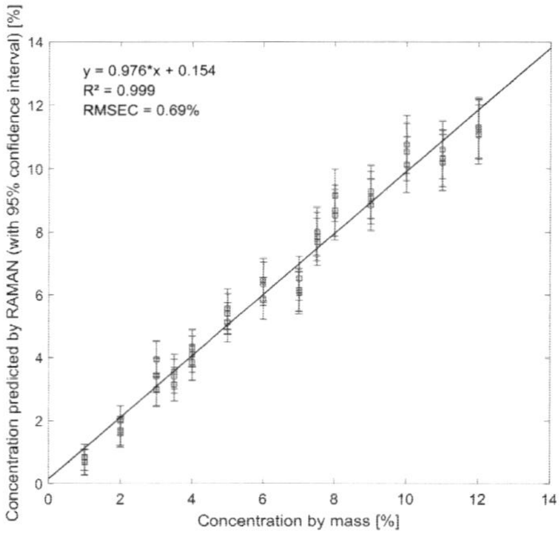

We make tomorrow's drugs possible.

Evaluation of Predictive Performance: Model Validation

- Can the off-line detected concentrations of Salbutamol be correlated with in-line measured concentrations?

→ Off-line Method Validation:

- Calibration samples in the powder bowl
- Manual rotating of powder bowl and collecting of RAMAN spectra (PhAT probe)
- Correlation off-line and in-line data

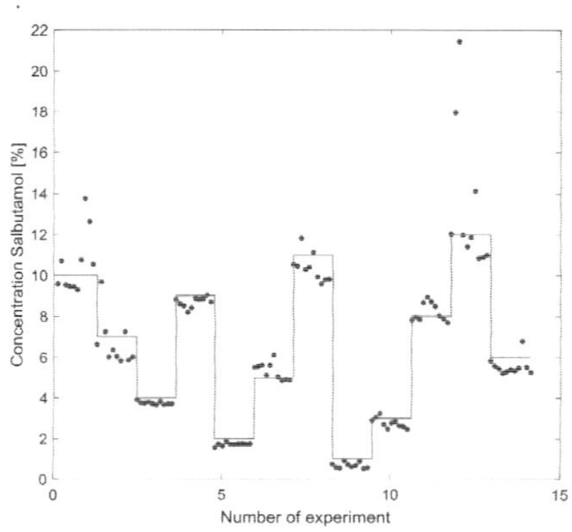

Application of the PLS Model: In-line Monitoring (Test)

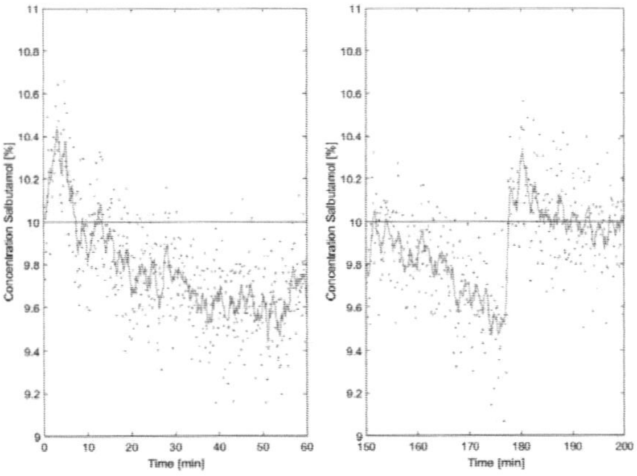

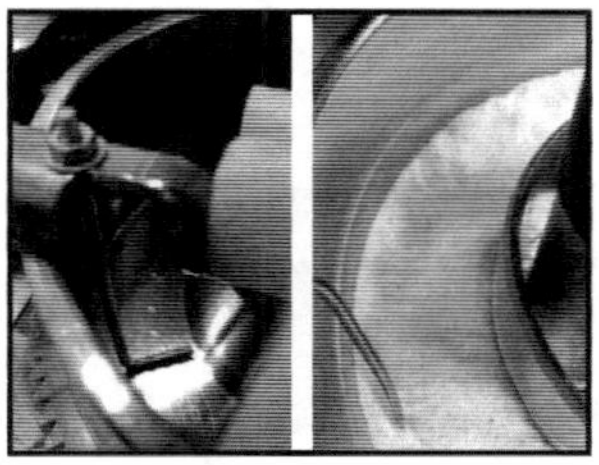

Findings:

- **"Modified Scraper" reduces segregation**
- **Decrease of API concentration of about 1 % without "Modified Scraper"**

Evaluation of Predictive Performance: Model Validation

- Is surface segregation representative for the entire powder bed?

→ Off-line Method Validation:

- Removal of powder after the process to get a sloping surface
- Manual rotating of powder bowl and collecting of RAMAN spectra
- Comparison of off-line data with in-line data

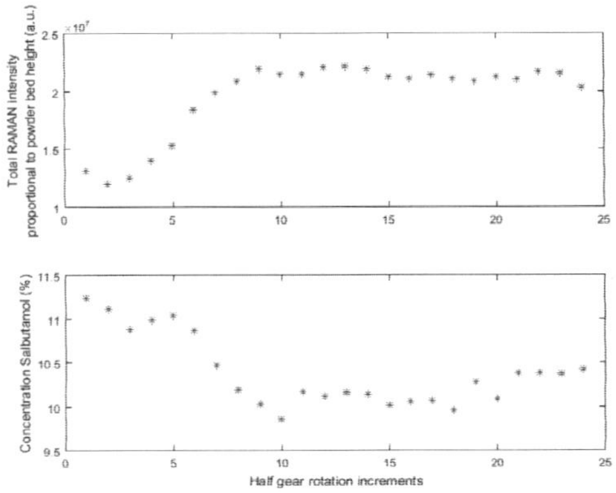

Conclusion: RAMAN Proof of Concept Study

- It was possible to use RAMAN spectroscopy to predict segregation tendency *in-line*
- The developed model can be used to test various mechanical modifications of a capsule filling process → **Improved filling performance**

- **Model will be further optimized for other binary mixtures**
- **Model will be tested for different capsule filling settings**
- **Model can also be used for evaluation of segregation tendency in the powder hopper**

Summary: Optimization using Different Strategies

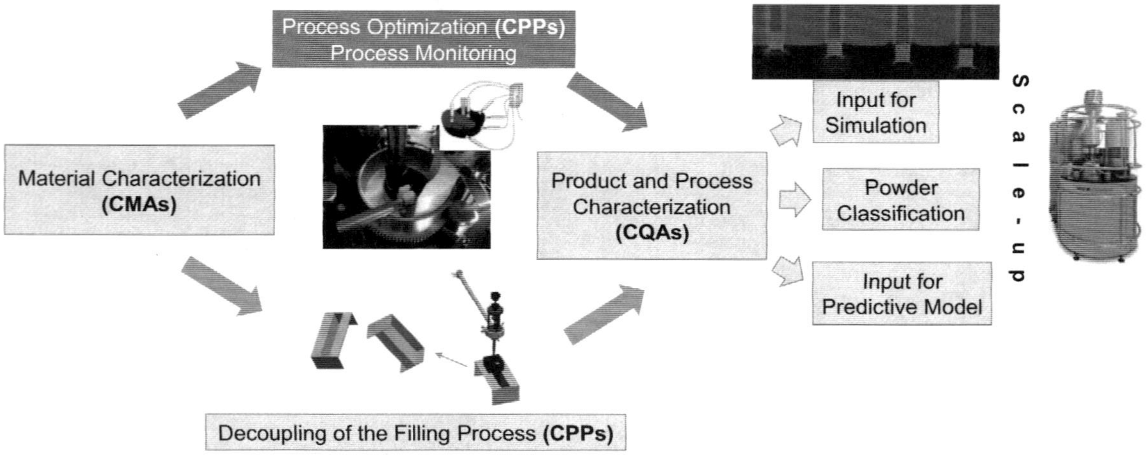

Publication: Mechanistic Modeling of a Capsule Filling Process

Dosing simulations for different bond numbers

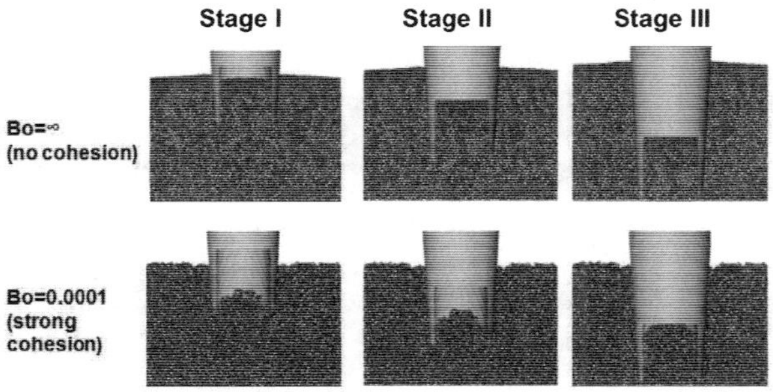

- Three stages of dosator filling (I, II, III) for non-cohesive powders
- Cohesive powders don't have the distinct stages → Influence of powder bed height on the fill weight
- For high cohesion it is difficult to fill the nozzle (problem of arching)

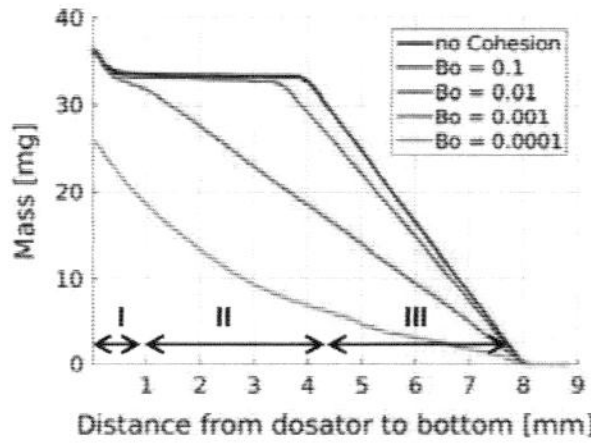

Loidolt, P., Madlmeir, S., Khinast, J.G., 2017. Mechanistic modeling of a capsule filling process. Int. J. Pharm. 532, 47–54.

Thanks to:

- The Austrian COMET Program under the auspices of the Austrian Federal Ministry of Transport, Innovation and Technology (BMVIT), the Austrian Federal Ministry of Economy, Family and Youth (BMWFJ) and by the State of Styria (Styrian Funding Agency SFG). COMET is managed by the Austrian Research Promotion Agency FFG.
- MG2 for their financial support.
- Eva Faulhammer and Otto Scheibelhofer for their help during the study.

Contact:
sandra.stranzinger@rcpe.at

493250 Application of Radiometric Sensor for Inline Mass Flow Rate Monitoring and Control of a Continuous Tablet Manufacturing Line

Abstract:

The accelerating research in continuous pharmaceutical manufacturing has already resulted in approval for two drug products produced via continuous manufacturing in 2016 by the US-FDA [1]. The two approvals and more filings for continuous manufacturing of drug product reflects the growing interest by pharmaceutical manufacturers and the regulatory body in the shift of manufacturing towards continuous operations. The implementation of an integrated tablet manufacturing process in continuous operation requires effective real time process management [2]. This involves integration of real time process data of the operation parameters and the quality attributes of intermediate blends, granules and tablets, from online and soft sensors into a model based and data driven framework.

The first step in the tableting process is to ensure blend uniformity and desired throughput from the continuous feeding and blending of excipients and API. Quality specifications at the outlet of the blender can be achieved by implementing ratio control of the feed streams [3]. This requires continuous monitoring of blend composition and mass flow rate at the outlet for implementation of the ratio control framework. Additionally, it is desirable to monitor the mass flow rate at the inlet of the tablet press, as the system could have material losses or bulk density variations by virtue of multiple processing units. This would enable implementation of robust feed forward and advanced control structures for the tableting operation. This work focusses on critical evaluation of a radiometric sensor for mass flow rate monitoring of particulate flow in continuous manufacturing of tablets using dry powder feeds.

The radiometric sensor, developed by Enurga Inc., (West Lafayette, IN) is based on x-ray attenuation and cross correlation velocimetry [4]. The monitoring technique is non-invasive and implemented for on-line monitoring of mass flow rate [5]. The sensor is evaluated for its capability to monitor mass flow rate in real time for the materials and operating conditions similar to that handled in the pilot plant facility at Purdue University. The mass flow rate is observed to be linearly proportional to the attenuation of x-rays by the material flowing through the sensor.

The performance of the sensor primarily depends on material handling and flow properties. The measurement precision for powder blends comprising of varying proportions of acetaminophen (APAP), microcrystalline cellulose (MCC) and lactose is evaluated. The sensor gives a precise measurement of the attenuation of the x-rays, with the same proportionality constant for 0 to 10% of APAP and MCC mixtures and 6 to 11 kg/h. The variance of measurement is higher at lower flow rates and for materials with poor flow. The sensor accuracy in real time is evaluated by monitoring flow rate with sampling at every second using MCC and a granulated sample of acetaminophen and MCC. The x-ray attenuation for a given material is calibrated over a known flow rate range. The sensor is used for monitoring a different flow rate in the range and compared with the measured flow using direct weighing. The flow rate monitored by the sensor depends on the calibration, measurement variance and material flow properties. The measured flow rate is observed to be within 5% of the actual flow rate [6].

After substantial offline testing, the sensor has been integrated into the tableting line and demonstration of real time monitoring and control of the feeder-blender unit will be reported. The added sensor provides needed redundancy in measurement of process variables in order to enable data reconciliation for the process line.

References:

1. https://blogs.fda.gov/fdavoice/index.php/2016/04/continuous-manufacturing-has-a-strong-impact-on-drug-quality/
2. Giridhar A, Gupta A, Louvier M, Joglekar G, Nagy ZK, Reklaitis G V. Intelligent Process Management for Continuous Operations in Pharmaceutical Manufacturing. *24th Eur Symp Comput Aided Process Eng*. 2014;33:391-396.
3. Moreno M, Reklaitis G V., Nagy ZK. Systematic Framework for the Design and Analysis of Control Systems for Continuous Pharmaceutical Manufacturing. In: *AIChE Annual Meeting*. Salt Lake City; 2015
4. Personal communication with Jongmook Lim, Enurga Inc., West Lafayette, IN
5. Barratt, I. R., Yan, Y., Byrne, B., & Bradley, M. S. A. (2000). Mass flow measurement of pneumatically conveyed solids using radiometric sensors, *11*, 223â€“235.
6. Ganesh S., Troscinski R., Schmall N., Nagy Z., Reklaitis G., In-line and non-invasive monitoring of mass flow rate using x-ray sensor in continuous tableting line, *In Preparation*

494699 Design Space Development for the Dropping Process of a Dripping Pill with Material Attributes Represented As the Scores from Principal Component Analyses of Spectra Data

Abstract:

Design space development for the dropping process of a dripping pill with material attributes represented as the scores from principal component analyses of spectra data

Xingchu Gong, Jichen Shen, Haibin Qu*

(Pharmaceutical Informatics Institute, College of Pharmaceutical Sciences,

Zhejiang University, Hangzhou 310058, China)

Abstract

Yinxingye Diwan is a dripping pill preparation treating angina pectoris and cerebral infarction. It is made from Ginkgo biloba extract and PEG 4000. The pass rate and productivity of dripping pills were considered as critical quality attributes (CQAs) of dropping process. Plackett-Burman designed experiments were performed to investigate the effects of materials and process parameters. The two materials of Ginkgo biloba extracts and PEG 4000 have significant effects on the process CQAs. The distance between dropping needle and condensate, and dropping speed were critical process parameters (CPPs). Split-plot designed experiments then were carried out to model the dropping process. Material properties were measured by near infrared spectroscopy (NIR) for Ginkgo biloba extracts and PEG 4000. The scores from principal component analyses (PCA) of NIR spectra were used in model development. Models with and without interaction terms between PCA scores and CPPs were compared. PCA scores with significant effects on the process CQAs were considered as critical material attributes (CMAs). After favorable model was selected, design space was developed based on the probability to attain process CQA limits. The probability was calculated using a Monte-Carlo simulation method. The design space then was verified using new materials. The quality levels of Ginkgo biloba extracts and PEG 4000 were also set according to PCA scores.

494086 Water Proton NMR for Noninvasive Chemical and Pharmaceutical Analysis

Abstract:

Water is by far the dominant component of a pharmaceutical solution. As the solvent, water has extensive interactions with the drug and the excipients. Hence abnormalities of a drug product could be reflected in the water proton NMR signal. This forms the basis of using the ^1H_2O signal for noninvasive analysis of pharmaceutical solutions with desktop NMR spectrometers. Examples of using the ^1H_2O signal to quantify protein concentration, protein aggregation, surfactant micellization, dendrimer folding and nanoparticle formulation will be presented. The unique advantage of a water NMR analytical technique is the ability to inspect a sealed drug product without any tampering. This opens the door for quantitative inspection of every vial of a pharmaceutical solution in its entire lifecycle, from the point-of-release to the point-of-care.

494295 Monitoring the Whole Blending End-Point of an-Gong-Niu-Huang Wan Using the QbD Concept

Abstract:

[**Abstract**] An-Gong-Niu-Huang Wan (AGNH), a famous Chinese Materia Medica (CMM) for brain diseases, contains herbal, animal and mineral medicines such as cinnabar, realgar, *Calculus Bovis, Cornu Bubali, Margarita, Borneolum Syntheticum, Rhizoma Coptidis, Radix Scutellariae*, and other components. Thus, the whole end-point determination of AGNH is very significant. Based on Quality by Design (QbD) concept, Laser induced breakdown spectroscopy (LIBS) was used to assess the cinnabar, realgar and pearl powder blending of AGNH in a pilot-scale experiment, especially the whole blending end-point determination. The blending variability of three mineral medicines including cinnabar, realgar and pearl powder, was measured by moving window relative standard deviation (MWRSD). The blending was separated four phases. The time profiles of cinnabar and realgar was in tune at the last phase of blending, though not always consistent in other phases. Meanwhile, pearl powder showed similar blending behavior. At the first phase, blending was effective and efficient. But at phase B, blending tended to stable. With further time, blending became violently at phase C. The steady state then appeared in phase D. Generally, the terminal blending stage of element arsenic, mercury and calcium was phase D. Thus, the whole proposal blending end- point of element arsenic, mercury and calcium was detemined rapidly. LIBS is a promising Process Analytical Technology (PAT). Unlike other elemental determination technology, LIBS does not need an elaborate digestion procedure, which is a promising and rapid technique to understand the blending process of CMM containing cinnabar, realgar and other mineral medicines. This study proposed a novel method for the research of large varieties of traditional Chinese Medicine.

[**Key words**] An-Gong-Niu-Huang Wan, whole blending end-point, Laser-induced breakdown spectroscopy, Quality by Design, Process Analytical Technology

492437 Steady-State Data Reconciliation of a Direct Compression Tableting Line

Abstract:

The assurance of critical quality attributes (CQA's) is a major concern in the pharmaceutical manufacturing industry. During traditional batch production, the CQA's of a batch are controlled by statistical sampling and batch rejection if sampling indicates deviations from target specifications, leading to significant waste and increased cost. Continuous manufacturing has been pursued by the pharmaceutical industry and encouraged by regulators to overcome these limitations [1]. Effective continuous manufacturing is generally implemented using a real-time process management strategy, which encompasses multiple components. The first foundational component consists of real-time measurement and monitoring of the system using Process Analytical Technology (PAT) tools [2]. The second involves a robust process control system. The third involves the detection, diagnosis and mitigation of exceptional events. In pharmaceutical manufacturing, a fourth component is also required, namely active procedures for tracking and isolating noncompliant product.

The objective of this work is to develop a model-based framework, which for a given sensor network configuration and measurement uncertainties, will predict the most likely state of the process. As a consequence, real time process decisions, whether in control or exceptional events management can be based on the most reliable estimate of the process state. The presentation focuses on real time measurement and monitoring of the direct compression continuous tableting line. Specifically, we report on the use of data reconciliation and gross error detection (GED) as tools for mitigating the effects of measurement errors and sensor malfunctions.

Data reconciliation (DR) has been widely used in other industries (i.e. oil and gas) to accomplish this goal. Data reconciliation is model based and thus consists of mechanistic (e.g., material balance) as well as empirical relationships between variables including properties [3]. Through DR, we can deal with random errors and have an estimation of unmeasured variables. However, in order to use steady-state data reconciliation (SSDR), redundancy with no gross error in the measurement is required. In order to detect gross errors in the system, we need to apply the global test and measurement test. These tests are similar to outlier detection [4]. DR also offers an opportunity to detect gross errors.

In this paper, we develop and demonstrate a DR framework for a direct compression (DC) line in real-time at steady-state. The DC line sensor network includes the on-line use of load cells for the weight of the feeders, the near-infrared (NIR) spectroscopy sensors for the powder blend composition, the X-ray sensor for the total powder flowrate and a soft-sensor model for the tablet hardness and weight. Steady-state data reconciliation (SSDR) is one of the simplest forms and offers short computational times. The results are compared to the results from the multivariate models (MVM), which are also use for the prediction of process states and error detection [5]. Since the DC line model at steady-state is linear, the results are similar [6]. The SSDR and MVM are implemented in MATLAB 2015b and the data is collected in the Emerson DeltaV server. The operator can see the real-time results in the process historian view and take a decision according to the result.

References

[1] S. L. Lee, T. F. O. Connor, X. Yang, C. N. Cruz, L. X. Yu, and J. Woodcock, "Modernizing Pharmaceutical Manufacturing : from Batch to Continuous Production," no. 1, pp. 1–25, 2015.

[2] M. Ierapetritou, F. Muzzio, and G. V. Reklaitis, "Perspectives on the Continuous Manufacturing of Powder-Based Pharmaceutical Processes," *AIChE J.*, vol. 62, no. 6, pp. 1846–1862, 2016.

[3] S. Narasimhan and C. Jordache, *Data reconciliation and Gross Error detection*. Houston, TX.: Gulf Publishing Company, 2000.

[4] C. Knopf, *Introduction to Data Reconciliation and Gross Error Detection*. .

[5] J. A. Romagnoli and S. M.C., *Data Processing and Reconciliation for Chemical Process Operations*. San Diego, CA, 200AD.

[6] O. Cencic and R. Fruhwirth, "A general framework for data reconciliation-Part I: Linear constraints," *Comput. Chem. Eng.*, vol. 75, pp. 196–208, 2015.

487554 Considerations for Control of Manufacturing Processes for Transdermal Systems â€“ FDA Perspective

Abstract:

FDA classifies transdermal systems (TDS) as complex drug-device combination products with unique critical quality attributes (CQAs). It is usually difficult to identify, characterize, and control TDS CQAs that maybe impacted by drug product manufacturing process. This presentation discusses considerations for developing a robust manufacturing process and control strategy for TDS and to ensure manufacture of consistent quality product throughout the lifecycle.

To start, we will classify transdermal manufacturing processes and discuss typical TDS CQAs. From there, we will highlight case studies demonstrating the identification, characterization, and adequate control of TDS CQAs. Discussion will focus on control strategy associated with typical unit operations employed for TDS manufacture e.g. mixing, coating, drying, die-cutting, pouching. For mixing, we will discuss control strategies for content uniformity and the microscopic arrangement of matter. For coating and drying, we will review control approaches for physical adhesive properties, permeation enhancer content, residual solvents, and residual adhesive impurities. We will also cover control strategies for complex die-cutting and pouching operations. Lastly, we will discuss control strategy considerations in the context of inspections and facility readiness. Case studies will be drawn from submissions reviewed by FDAâ€™s Office of Pharmaceutical Quality (OPQ).

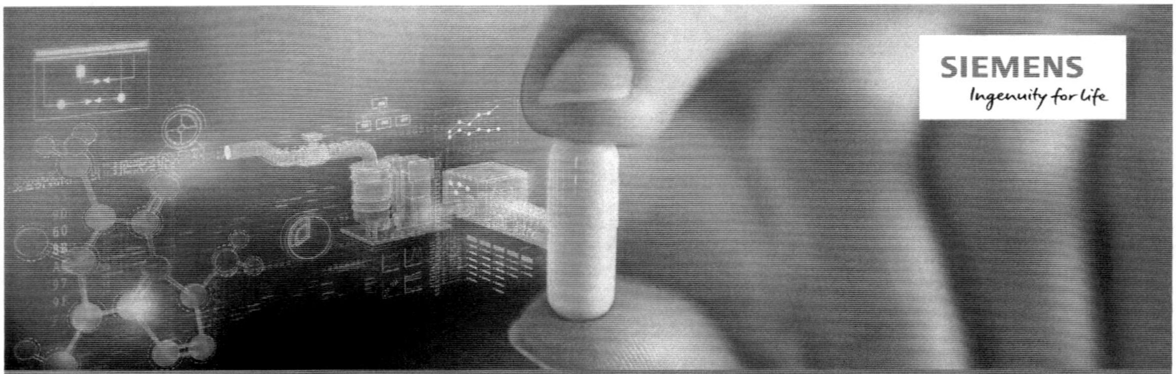

Real Time Release Testing Using a Quality Data Management Software - Sipat

Pamela Docherty
Siemens

Goal: Real Time Release Testing (RTRT)

SIEMENS
Ingenuity for Life

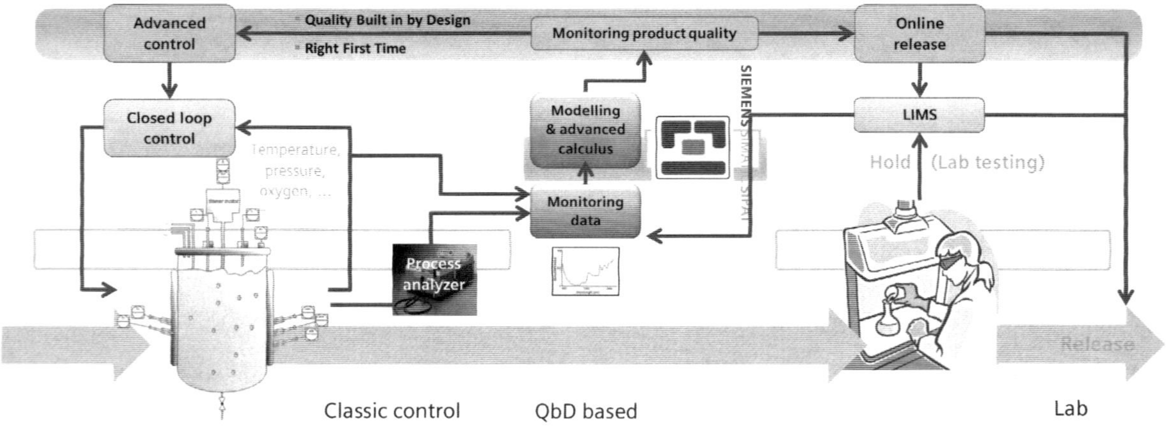

SIMATIC SIPAT
Our offering for integrated PAT Data Management

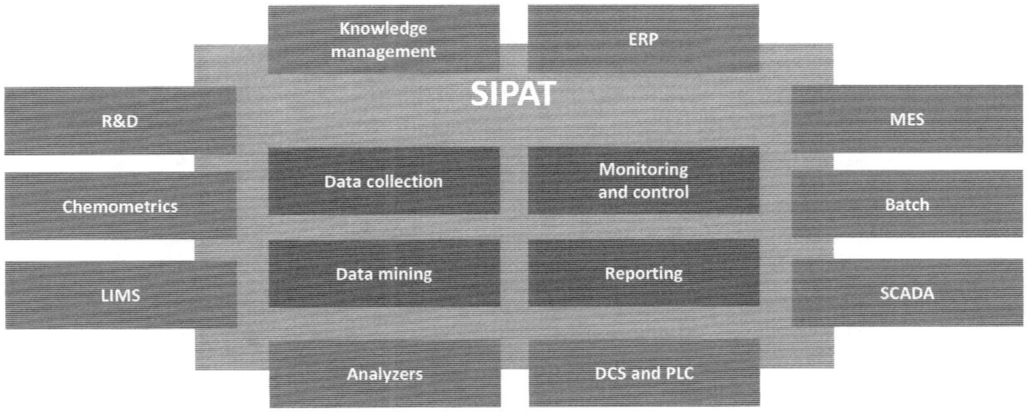

Page 3 XX.XX.20XX
Author / Department

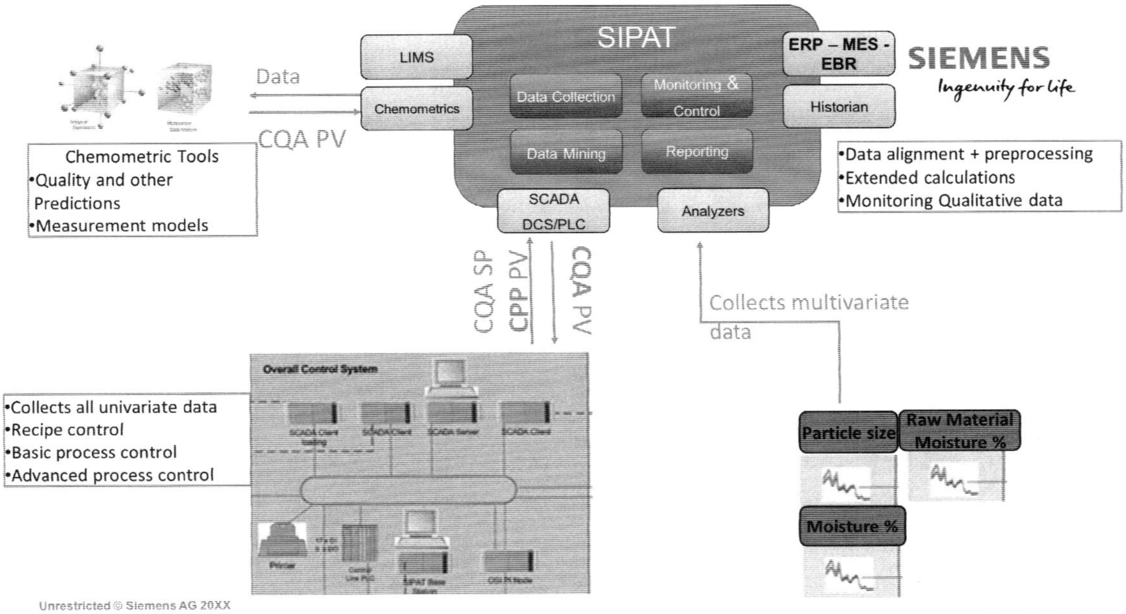

Page 4 XX.XX.20XX
Author / Department

SIEMENS
Ingenuity for life

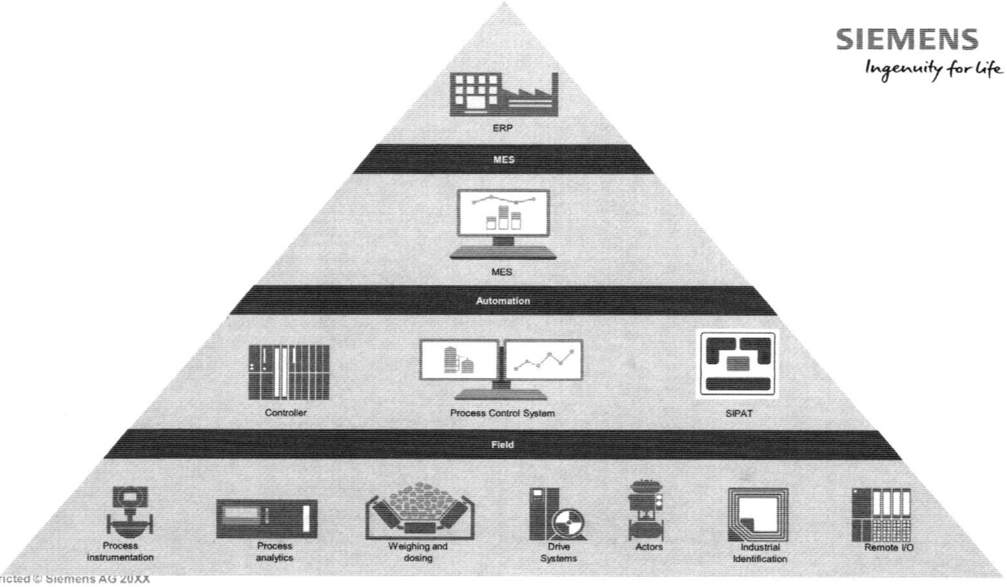

MES (Simatic IT eBR)
Quality Control & Real Time Release

SIEMENS
Ingenuity for life

Functionalities

Integrated Quality: at line, in line and off line testing

- At line quality management
  - In-process controls
  - Directed protocol execution
  - Reliable analysis result collection
  - Defect tracking and corrective actions
- In line quality management with **PAT**, SPC, Historian
- Off Line quality management (LIMS) with Unilab support integration

Central Repository of batch information enabling –

review by exception

- Deviation and Alert Management
- Detailed alert and execution reports
- Electronic Batch Records
- Electronic Signature

Realtime Time Release (RTR)

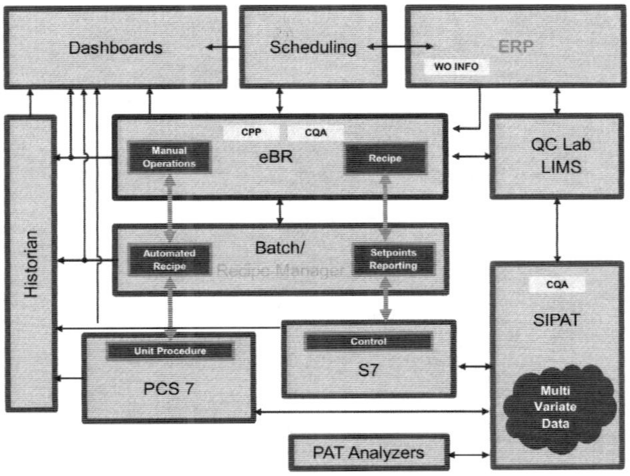

Goal: Real Time Release Testing (RTRT)

SIEMENS
Ingenuity for life

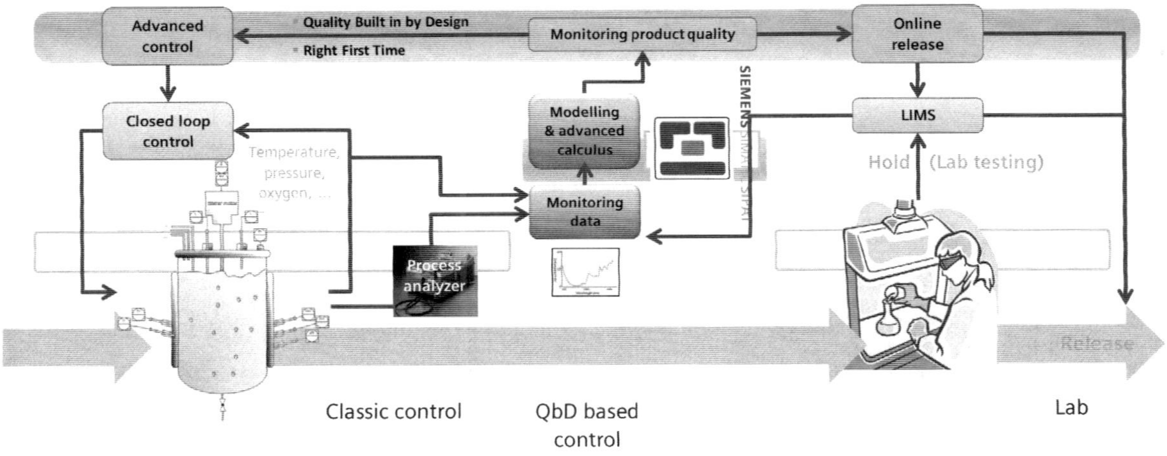

Goal: Real Time Release Testing (RTRT)

SIEMENS
Ingenuity for life

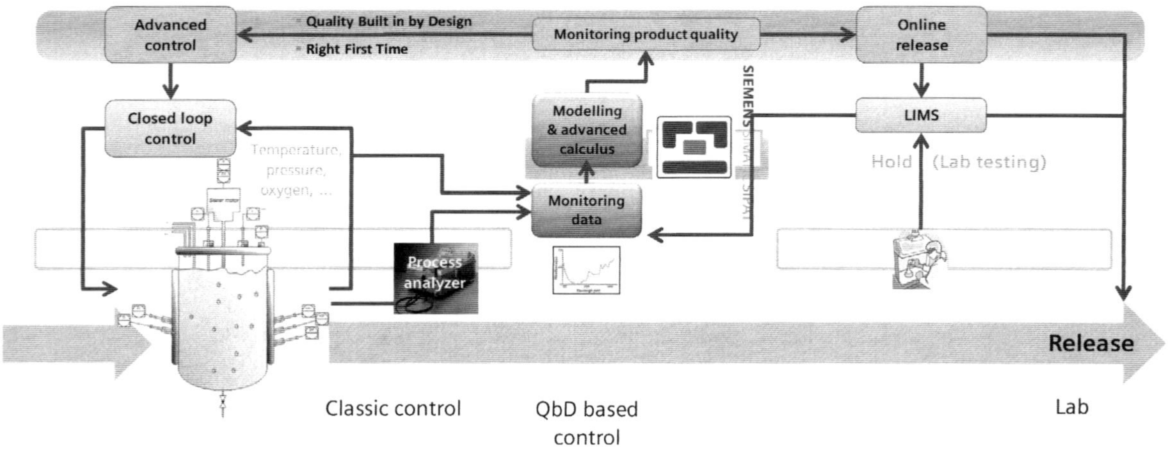

Thank you

SIEMENS
Ingenuity for life

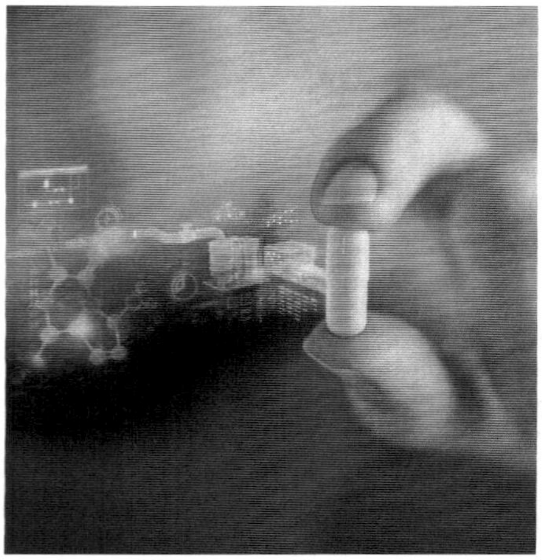

Pamela Docherty
Life Sciences Industry Manager - USA

1201 Sumneytown Pike
Spring House, Pa

Mobile: (267) 207-4931

E-Mail: Pamela.Docherty@Siemens.com

USA.Siemens.com/pharma

496472 Use of a Trickled Bed Reactor to Improve the Commercial Feasibility of the Hydrogenation of a Nitro-Compound

Abstract:

Use of a Trickled Bed Reactor to Improve the Commercial Feasibility of the Hydrogenation of a Nitro-Compound

Carla Luciani, Jonas Buser, Michael Laurila, Richard Cope, Kevin Cole, Bradley Campbell, Justin Burt, Martin Johnson, Joseph Martinelli, David Mitchell

Lilly Research Laboratories, Lilly Corporate Center, Indianapolis, IN-46285, USA

The use of continuous processing to manufacture intermediates and drug substances has grown significantly over the last few years. While the major reasons for the rapid adoption of continuous technologies in the pharmaceutical industry are typically linked to productivity and safety as well as quality benefits, there are cases where the use of continuous processing is required long term to ensure the process robustness required for commercial manufacturing.

This work investigates the Pd-catalyzed reduction of a nitro compound to produce an aniline during the second step of the proposed registered sequence. Major challenges resulting from the complexity of the reaction matrix were overcome by the use of advanced monitoring techniques (flow-NMR and on-line HPLC). Those monitoring techniques produced exceptionally detailed kinetic profiles that were used to fit kinetic models. The kinetic models were used to simulate the reaction profiles using both continuous and batch reactor types. In addition to the typical manipulated variables investigated for the batch option, such as temperature, pressure, catalyst loading, and catalyst poisons entering the process, several other process parameters were investigated to understand the trickle bed reactor performance (e.g., temperature and concentration profiles in both radial and axial direction, flowrates, wetted catalyst surface, catalyst deactivation mechanism, etc.). The model was used to compare the resulting feasible operating regions/design spaces for both batch and flow modes. Finally, the model was used to investigate re-packing strategies and optimal temperature profiles during operation.

500963 Use of Process Analytical Technology to Gain Mechanistic Insights into Spherical Agglomeration of Active Pharmaceutical Ingredients

Abstract:

Crystallization is a solid-liquid separation process widely used in the pharmaceutical industry to purify and establish physical properties of the drug substance. The performance of subsequent downstream unit operations often depends on the quality of crystals generated in the crystallization. Drug substance attributes (*e.g.,* crystal size, shape, etc.) directly impact the filtration and drying performance in addition to drug product manufacturing processes.

Although pharmaceutical manufacturing processes such as crystallization are well established, more intensified downstream unit operations are being pursued as alternative manufacturing strategies. Spherical agglomeration is a process intensification unit operation that combines crystallization, milling, and granulation into a single unit operation. Historically, growth dominant crystallization processes have been used to generate large drug substance crystals for improved filterability, impurity control, and process robustness. However, in some cases, the product crystals must then be milled to in order to achieve the desired bioavailability. The powder properties that result from the milling operation are often non-ideal for use in the drug product, since the powders can be poor flowing, cohesive, and electrostatic. A powder granulation step is often required to help mitigate these short comings. Spherical agglomeration might have the ability to address these short comings and combine three unit operations (crystallization, milling, granulation/blending) into a single step process, and deliver a powder with desired drug product manufacturing attributes.

In the present work, Process Analytical Technology[1] (PAT) is used to gain mechanistic insight into the effect of various process parameters on spherical agglomeration. Agglomeration of the drug substance has been carried out using a variant of spherical agglomeration, the Quasi Emulsion Solvent Diffusion (QESD) technique. In this case, a quasi-emulsion is formed and used to agglomerate drug substance crystals inside emulsion droplets. It has been shown in the open literature that changing key process parameters such as the turbulent energy dissipation rate, the concentration of solute in the initial solution, and the solvent to anti-solvent ratio changes the dynamics of quasi-emulsion that is formed and hence the quality of agglomerates that are produced. This contribution proposes a mechanistic understanding of QESD through the utilization of PAT, and provides insight into the fundamental rate processes that are required in order to design QESD processes with tailored powder properties.

References:

[1] Y. Yang *et al.*, â€œApplication of Ultra-Performance Liquid Chromatography as an Online Process Analytical Technology Tool in Pharmaceutical Crystallization.â€, *Cryst. Growth Des., 2016, 16, 7074â^'7082*

489889 Process Dynamics of Continuous Cooling Crystallization of Carbamazepine in a Mixed Suspension Mixed Product Removal (MSMPR) System

Abstract:

Current pharmaceutical manufacturing is mainly based on batch processing which has inherent drawbacks and disadvantages. To overcome these drawbacks, emerging technologies such as continuous manufacturing (CM) have been highlighted in the industry in the past few years [1, 2]. As the final purification and isolation step, crystallization has a significant impact on the final physicochemical properties of drug substance and is considered one of the critical process steps in the continuous manufacturing of drug substances. The ability to monitor process dynamics of a continuous process is critical for successful control of crystal quality attributes, such as size distribution. PAT tools such as Focused Beam Reflectance Measurement (FBRM) and spectroscopy based system (i.e. NIR, FTIR, and Raman) can be used to monitor the crystallization process. In this work, a systematic framework to evaluate and understand process dynamics using PAT tools is presented.

The continuous cooling crystallization of carbamazepine (CBZ) in ethanol in a multi-stage MSMPR system is used as the case study. FBRM and Raman spectroscopy are implemented for on-line monitoring. Various startup conditions are evaluated to determine the optimal scenario for the continuous system of interest. The impact of disturbances on the process dynamics and state of operation are evaluated by implementing pulse or step changes to the temperature, flowrate, feed solution concentration, and polymorphic form. Slurry samples for all experiments are obtained and characterized by off-line HPLC, laser diffraction and XRPD before and after the disturbances. The on-line and off-line data obtained are used to characterize the effect on crystal properties (i.e. size and form) and process parameters (i.e. solution concentration). Moreover, simulation studies for the crystallization process are conducted using population balance models to further understand the impact of process disturbances on quality attributes for the continuous cooling crystallization of CBZ.

1. Lee, S.L., et al., *Modernizing pharmaceutical manufacturing: from batch to continuous production.* Journal of Pharmaceutical Innovation, 2015. **10**(3): p. 191-199.

2. Poechlauer, P., et al., *Continuous processing in the manufacture of active pharmaceutical ingredients and finished dosage forms: an industry perspective.* Organic Process Research & Development, 2012. **16**(10): p. 1586-1590.

501155 Monte Carlo Stepwise Regression for More Accurate Selection of Critical Process Parameters during Process Characterization

Abstract:

Design of experiments (DOE) is commonly used during process characterization (PC) to define control strategies for the manufacture of drug substance. Although the use of sophisticated DOE techniques in establishing the conditions for PC is common, the analysis of the results is often done with a stepwise regression reliant on p-value thresholds. This technique can often produce artificially large models with multiple false positives, leading to unnecessary work to examine spurious variable relationships and potential implementation of overly complex layers of control in the manufacturing process. This work describes the use of Monte Carlo cross validation (MCCV) simulations to aid in the selection of critical process parameters (CPP). Once models were selected defining the relationship between CPPs and critical quality attributes (CQA), MCCV was used to assess the accuracy of those models. Monte Carlo simulations were then used to establish an appropriate control strategy with defined operating ranges (OR) predicted to meet quality standards.

Three recombinant monoclonal antibody cell culture processes were evaluated. The process parameters that were studied included dissolved oxygen, pH, temperature, inoculation density, and feeding strategy. The quality attributes that were evaluated included aggregate levels and charge variants. Each PC was carried out using a D-optimal design performed in 2L bioreactors. The resulting data was analyzed using MATLAB. Stepwise regression with backward elimination was performed to establish baseline models. MCCV stepwise regression with both forward and backward elimination was then performed to identify process parameters which may have been inappropriately included by the baseline method. When available, large-scale data was used to validate the accuracy of the models along with MCCV. ORs were established by Monte Carlo simulation by modifying CPP ranges until appropriate quality thresholds were met.

A comparison of stepwise regression with other model building techniques was also performed. The simplified interpretation of stepwise regression presented here has the advantage of clearly defining CPPs, despite the modest increases in predictive accuracy afforded by more advanced and complex methods, such as regularization (LASSO, ridge) and dimensionality reduction (PCA) . Applying these techniques to a PC program can lead to more robust interpretation of data, higher statistical confidence in CPP selection, and establishment of more reliable ORs.

Delivering a Design Space for a Continuous Drug Substance Manufacturing Process Using Simulation

Neil Hodnett
GlaxoSmithKline R&D

Agenda

1. Introduction
   - Overview of continuous manufacturing process
   - Motivation for this work
2. Approach
3. Results
   - Specific example of impurity control
4. Summary
   - Key learnings
   - Future opportunities

Introduction

Process overview

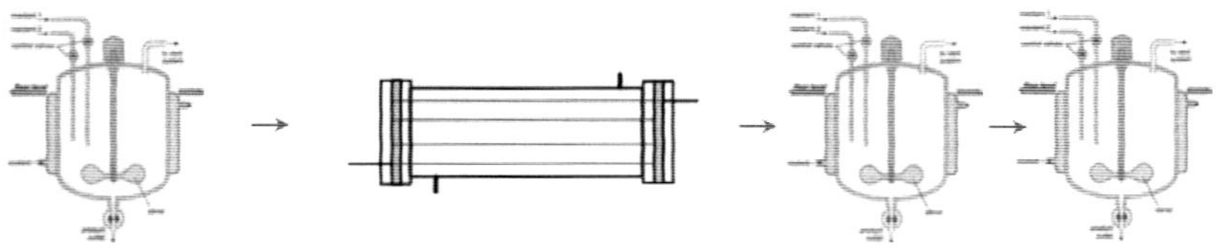

- 5 telescoped transformations
- 13 process parameters

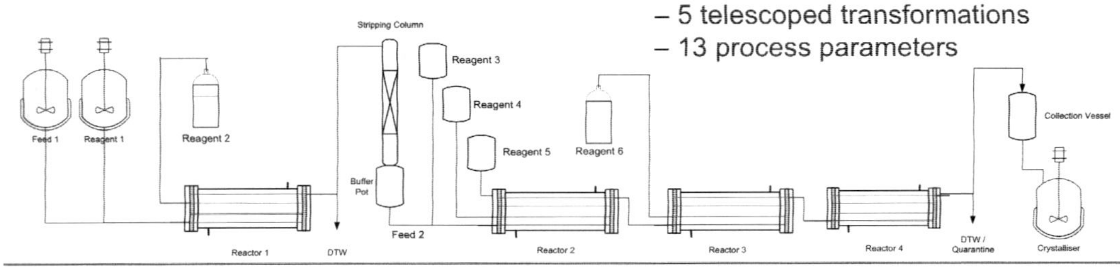

Motivation

Objective: Identify conditions that are most forcing wrt 9 specific process-derived impurities to run in R&D Pilot Facility (laboratory confirmation of Design Space)

- DoE approach unsuitable due to:
  - Challenge in translating lab scale batch/flow results to R&D Pilot and Commercial Manufacturing
  - High material and resource requirements for running DoE in R&D Pilot facility

- Alternative approach based on reaction kinetics and process simulation
  1. Develop a deep understanding of chemistry and capture in kinetic models
  2. Characterize equipment impact and couple with chemistry in process model
  3. Identify most forcing conditions and interactions for each impurity via simulation '*in silico* DoE'

Approach

Process Simulation Approach to Design Space

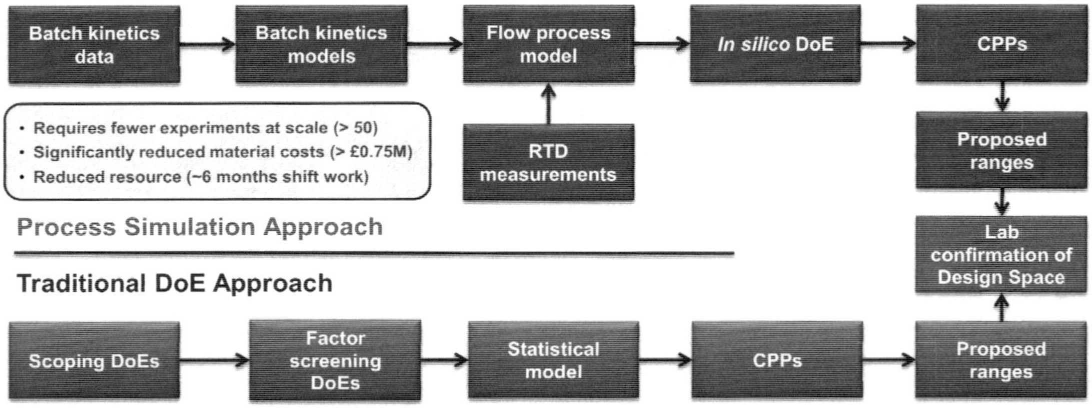

Generating the Batch kinetics data

- Structured sequence of batch experimentation for each chemistry transformation:

1. Profile at ~centre-point conditions
2. Repeat run (reproducibility)
3. Different stoichiometry (less reactant c.f. 1)
4. Different temperature (colder c.f. 1)

- Profiles fitted to proposed mechanisms (DynoChem)
- Outputs are rate constant (k) and activation energy (E_a) for each step

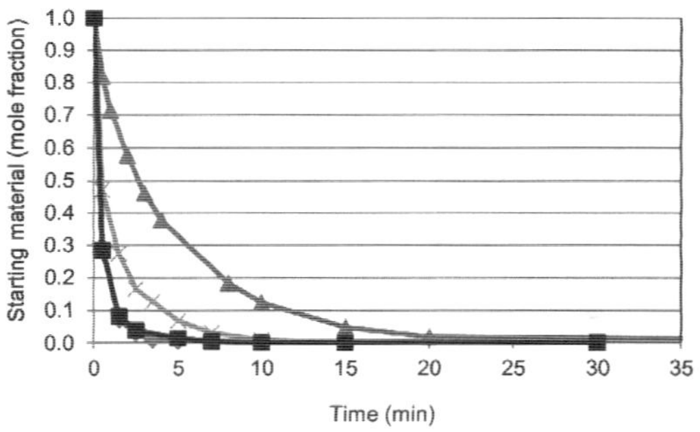

Developing the Batch kinetics models

- Chemistry captured as a network of coupled 1st and 2nd order equations
- Complex mixture at final stage (potentially >30 components)
  - Excess reagents
  - By-products
  - Impurities

Simplifying assumption: Protonation equilibria excluded from model due to buffering and minimal impact on rate expected within proposed ranges

Origin of process derived impurities:

Residual intermediates	Derived impurities	Alternative pathways	Over-reaction
4	3	2	0

Developing the Process model

- Characterization of continuous processing equipment allows calculation of true system volume
- Characterisation combined with kinetics parameters to model output concentrations of components
  - This model is used to simulate the continuous process (ideal plug flow)
  - Combines existing data (no additional data generated)

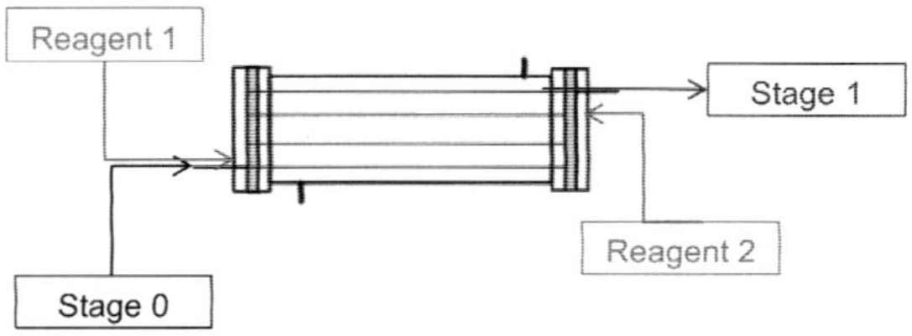

Developing the Process model

- Characterization of continuous processing equipment allows calculation of true system volume
- Characterisation combined with kinetics parameters to model output concentrations of components
  - This model is used to simulate the continuous process (ideal plug flow)
  - Combines existing data (no additional data generated)

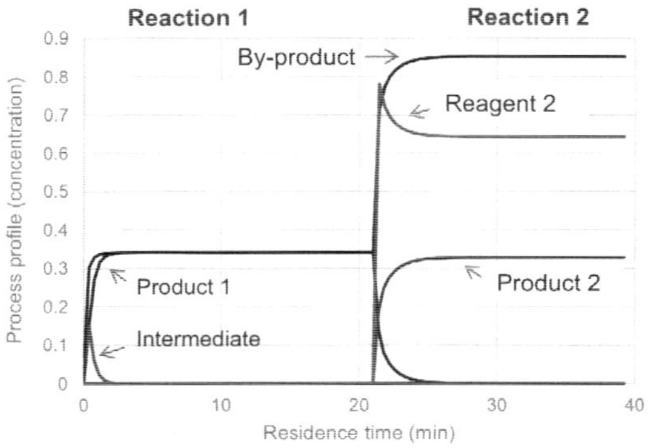

RTD Measurements

- RTD measurements were carried out on the R&D Pilot and commercial manufacturing equipment
- In both cases, all reactors can be approximated as ideal Plug Flow Reactors (PFRs)
- Data justifies the use of a process model where axial dispersion is ignored
  - Simulation demonstrates measured dispersion is unimportant to kinetics

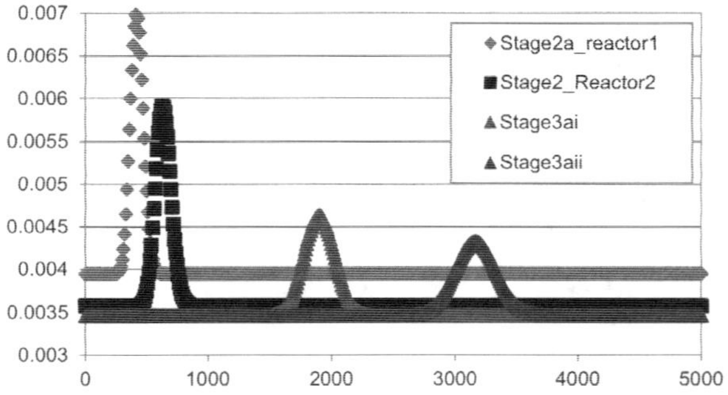

Evolution of the process model

- Process initially modelled as four discrete PFRs (ideal)
  - Gas-stripper excluded
  - Steady state simulations
  - Transcription intensive
- Later moved to single flowsheet model
  - Simpler to use
  - Impact of axial dispersion modeled
  - Dynamic simulation possible

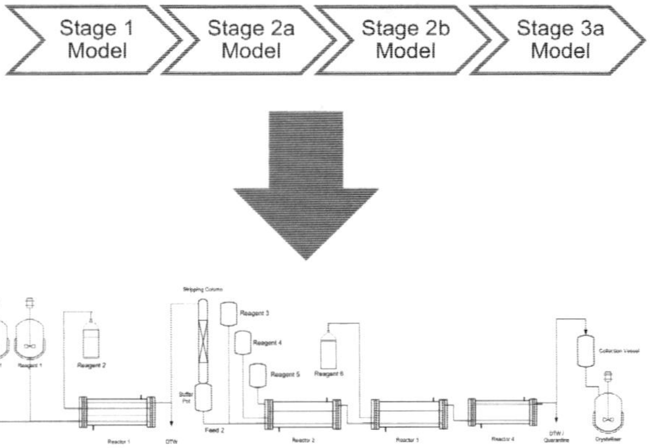

In silico DoE

- Impact of varying each process parameter on predicted levels of process impurities was investigated using a 2-level factorial design
- Enables visualisation of simulation results in a familiar format
- Most forcing conditions identified and ran in R&D Pilot Facility
  - Compared with optimisation results

Run	Parameter 1	Parameter 2	Parameter 3
1	+ 1	+ 1	+ 1
2	− 1	+ 1	+ 1
3	+ 1	− 1	+ 1
4	− 1	− 1	+ 1
5	+ 1	+ 1	− 1
6	− 1	+ 1	− 1
7	+ 1	− 1	− 1
8	− 1	− 1	− 1
9	0	0	0

2^{13} = 8192 possible combinations!

In silico DoE

- Impact of varying each process parameter on predicted levels of process impurities was investigated using a 2-level factorial design
- Enables visualisation of simulation results in a familiar format
- Most forcing conditions identified and ran in R&D Pilot Facility
  - Compared with optimisation results

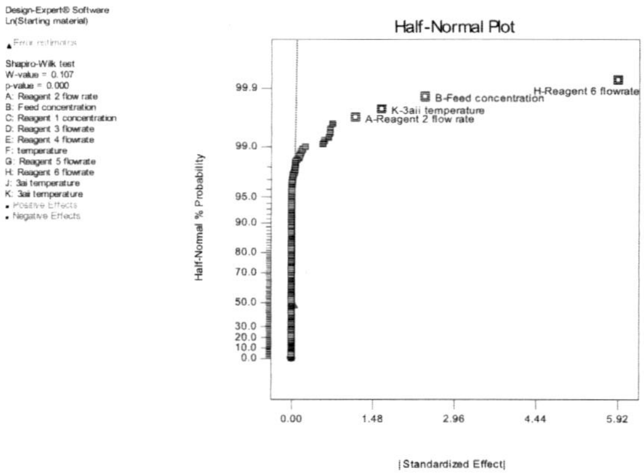

Results

Simulation output

- Output from process simulation is a matrix of most forcing conditions wrt each process derived impurity
- Conditions ran in R&D Pilot Facility

✓ Operating within Design Space delivers API that meets proposed specification

✓ Model-designed experiment successful in 8 / 10 cases

✓ Trends well predicted

  - Differences attributed to equipment performance and calibration errors

Process parameter	Impurity 1	Impurity 2
Concentration of Feed 1 solution	H	H
Concentration of Reagent 1 solution	L	L
Stage 1 residence time	L	L
Flow rate of Reagent 1 solution	L	L
Stage 1 temperature	L	L
Flow rate of Reagent 2	L	L
Flow rate of Reagent 3	H	L
Flow rate of Reagent 4	L	H
Stage 2 temperature	L	H

Impurity 1 = residual Stage 0, Impurity 2 = derived from Stage 0

Stage 3a

Control of process impurities

- Stage 3 input requires tight control in continuous process due to poor purging downstream
- Impurity formed by competing parallel pathway
- Options investigated via process simulation:
  - Increased residence time
  - Increased amount of Reagent 6
  - Increased Stage 3 temperature
- Simulation results highlighted need to tighten & shift process parameter ranges to manufacturing

$$\frac{d[\text{Stage 3a}]}{dt} = k_1[\text{Stage 2b}][\text{Reagent 6}]$$

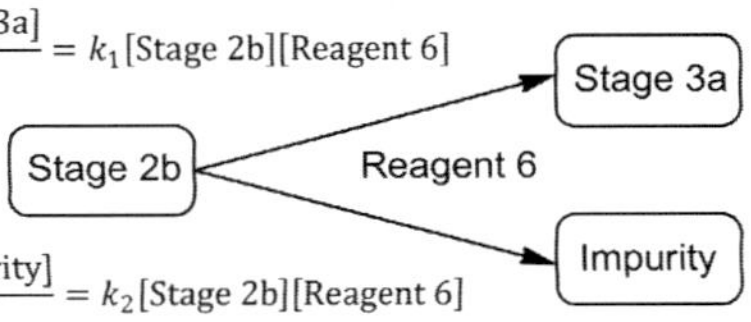

$$\frac{d[\text{Impurity}]}{dt} = k_2[\text{Stage 2b}][\text{Reagent 6}]$$

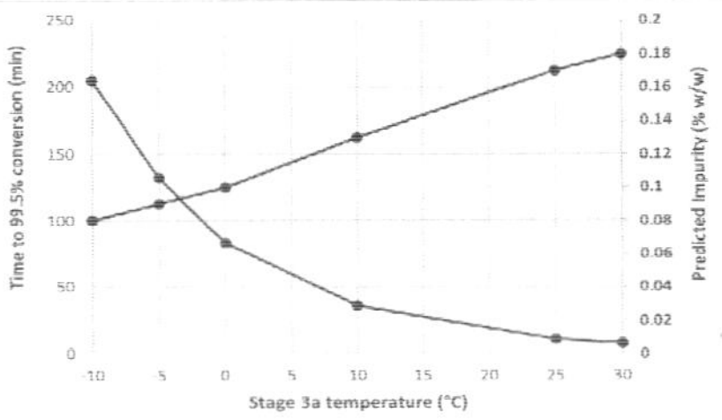

Stage 3a

Control of impurities

- **Two** reactors used: one cold, one warmer
- Compromise between selectivity and conversion in acceptable time frame

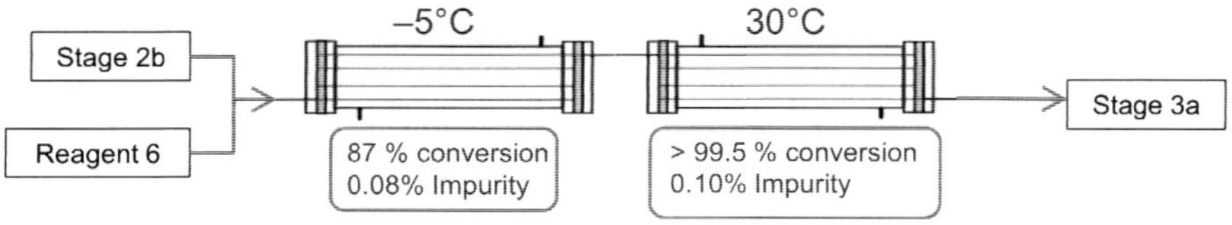

- Reactor temperature set points result from optimization using process model
- Optimized temperatures successfully demonstrated in R&D Pilot Facility

Input to Risk Assessment

Steady state risk assessment

- Process simulation results used to pre-populate risk assessment scoring
- Further simulations of interest were ran 'live'

Disturbance risk assessment

- Process simulation in dynamic mode
- Assess the impact of univariate process disturbances
- Output used to inform Divert to Waste strategy

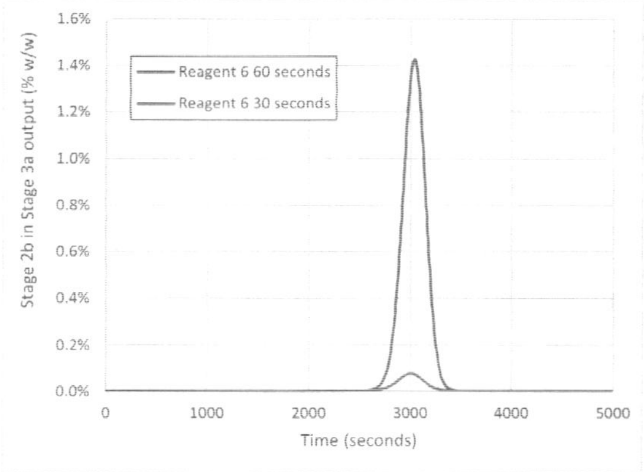

Summary

Summary

- A development strategy based on reaction kinetics and process simulation was used to identify the Design Space for a highly telescoped continuous API manufacturing process
- This approach resulted in significant savings in time, resource and materials
- Process model was a valuable tool for risk assessment - particularly for disturbance handling
- Visualizing the output from multiple simulations remains challenging

Acknowledgements

Malcolm Berry
Peter Shapland
Hannah Robinson
Peter Clements
Mark Hughes
Sabri Ukuser
Peter Hamilton
Irene Areri
Chris Clarke
Lesley Wong

Gary Breen
Augustine Ochen
Markus Andersson
Andrew Richards
Melanie Dumarey
Mohammad Yahyah

Andrew Bird
Joe Hannon

495685 Estimating NRTL-SAC Conceptual Segments of Molecules Using Sigma Profile

Abstract:

The Non-Random Two Liquid Segment Activity Coefficient (NRTL-SAC) model[1] as a solubility model is now routinely used in the industry[2-3]. In this model, the conceptual segments are defined to describe molecular surface characteristics, i.e., hydrophobicity, hydrophilicity, polarity. In practice, the conceptual segments of molecules must be identified from experimental phase equilibrium data with other solvents, and that limits the model applicability to molecules with sufficient phase equilibrium data. In this work, a methodology is presented to estimate the conceptual segments from the sigma profile, which is the molecular surface charge density distribution calculated from computational quantum chemistry, and is used in COSMO-based models[4-5]. We present the rationale and algorithm of segment numbers estimation from sigma profiles. Therefore, the applicability of NRTL-SAC model can be expanded to any components with known structures. The predicted phase equilibria, e.g., VLE, LLE, SLE, are compared with experimental data and predictions from other activity coefficient models to validate the model performance. The proposed approach allows the NRTL-SAC model to predict solubility in unconventional solvents, excipients, or polymers, where phase equilibrium data are insufficient to identify their segment numbers.

1. Chen, C.-C.; Song, Y., Solubility modeling with a nonrandom two-liquid segment activity coefficient model. *Industrial & Engineering Chemistry Research* **2004,** *43* (26), 8354-8362.

2. Gani, R.; JimÃ©nez-GonzÃ¡lez, C.; ten Kate, A.; Crafts, P. A.; Jones, M.; Powell, L.; Atherton, J. H.; Cordiner, J. L., A modern approach to solvent selection. *Chemical Engineering* **2006,** *113* (3), 30-43.

3. Tung, H. H.; Tabora, J.; Variankaval, N.; Bakken, D.; Chen, C. C., Prediction of pharmaceutical solubility Via NRTLâ€SAC and COSMOâ€SAC. *Journal of Pharmaceutical Sciences* **2008,** *97* (5), 1813-1820.

4. Klamt, A., Conductor-like screening model for real solvents: a new approach to the quantitative calculation of solvation phenomena. *The Journal of Physical Chemistry* **1995,** *99* (7), 2224-2235.

5. Lin, S.-T.; Sandler, S. I., A priori phase equilibrium prediction from a segment contribution solvation model. *Industrial & Engineering Chemistry Research* **2002,** *41* (5), 899-913.

498449 Discrete Element Modeling to Predict Pharmaceutical Powder Flow at Different Humidity Conditions

Abstract:

Pharmaceutical powder flow can alter significantly based on the exposed humidity conditions. Lack of predictive models to understand the flow variability at different humidity conditions may undermine its subsequent impediments. A Discrete Element Model (DEM) has been developed to predict the effects of humidity on powder flow, altering the cohesive forces based on granular bond numbers in simple hopper geometries. Experiments similar to the simulations are further performed for different pharmaceutical excipients at different humidity conditions. To predict cohesive properties for powders, those did not flow under gravity; vertical vibration has been introduced in both experiments and simulations. The equivalent DEM bond numbers are in good accordance with the experimental results. Pharmaceutical powders with lower moisture sorption show more variability in flow properties in comparison to cohesive powders having higher moisture contents. The equivalent DEM parameters suggest the same and are being implemented to predictpowder flow variability in different geometries and humidity conditions.

491765 Long-Term Coating Predictions for in a Wurster Fluidized-Bed Coater with Using a Combined CFD-DEM and Markov-Chain Monte Carlo Approach

Abstract:

Coating of particles, in particular, microspheres, using the Wurster fluid-bed coater is a common unit operation in the pharmaceutical industry. Such coatings are typically used to layer active drugs or other ingredients that control drug release, and also for taste/odor masking and enteric barrier coatings that prevent dissolution/disintegration. To ensure control over the drug-release rate and to achieve a consistent inter-particle drug loading (potency), a high degree of coating uniformity is essential. However, relating the coating uniformity of the beads to the process parameters (such as fluidization air-flow rate, atomizing air-flow rate, Wurster-gap height, and bead size) is not trivial.

Experimental methods to characterize bead-flow dynamics inside Wurster coaters are difficult for large systems; moreover, sampling techniques are insufficient to quantify the coating uniformity of millions of beads. In such situations, coupled CFD-DEM simulations (computational-fluid dynamics coupled with discrete-element method) can model the complex particle and airflow in a fluidized bed. However, due to the high computational costs (several days per process-second for an industrial system), simulating several hours of a coating process using CFD-DEM is infeasible. Therefore, a multi-scale model is required to predict the long-term coating uniformity, based on inputs from the CFD-DEM simulations. The Markov-chain Monte-Carlo-based compartment model is one such approach to develop a reduced-order model.

The compartment model subdivides the coater into non-overlapping regions, i.e., compartments. The exchange of particles between adjoining compartments is modeled as stochastic process described by a probability matrix *P*, where P_{ij} describes the probability of a particle moving from compartment *i* to compartment *j* during some small compartment-model time step Î"t_{CM}. This probability matrix *P* is calculated from the detailed CFD-DEM particle trajectories; therefore, the exact positions of the particles are not tracked in the compartment model. Changes in the process parameters affect the flow behavior of the beads which is reflected as changes in the probability-matrix entries P_{ij}.

Certain compartments near the spray nozzle are designated as the spray zone; every particle that visits these spray compartments gains some coating mass, based on the spray rate, particle size, and the compartment-model time stepÂ Î"t_{CM}. The probability matrix *P* is then used to update the positions of the particles in various compartments. By tracking the flow of a large, representative sample of beads, the evolution of the coating mass distribution can be tracked with time - these predictions can be generated in minutes for hours of real process time using the reduced-order Markov-chain Monte Carlo method.

Results from the compartment model for a given set of process conditions were successfully validated against

experimental results of particle diameters obtained using Sympatec QicPic and coating thickness from optical-coherence tomography (OCT). After validation, a virtual design of experiments (DoE) study with varying process conditions has been performed. The results show that a larger Wurster-gap height reduces the variation of the coating mass and thickness but, owing to the initial size distribution of the uncoated beads, the variation in the coating potency increases. Depending on the application, it may be more desirable to optimize the coating potency (to maintain a constant coating-mass fraction for coatings that contain active drug), or optimize the coating thickness (e.g., for protective coatings where a predefined barrier thickness is needed). The present Markov-chain Monte Carlo approach, in combination with CFD-DEM simulations, can be used to design the optimal process parameters for both applications.

496208 Microfluidization as an Enabling Technology for Solubility Enhancement

Abstract:

Microfluidization as an enabling technology for solubility enhancement

Iris Duarte, Clara Sa Couto, Tiago Porfirio, João Vicente and Marcio Temtem

R&D Drug Product Development, Hovione Farmaciência S.A, Sete Casas, 2674-506 Loures, Portugal

INTRODUCTION

Emerging drugs are essentially poorly water-soluble and such limited aqueous solubility has been one of the major hurdles in the development of oral-dosage forms, as poor solubility typically hinders oral bioavailability. This challenge has been addressed with different formulation strategies and enabling technologies, including: complexation with cyclodextrines (CD), production of amorphous solid dispersions (ASD) or particle-size reduction, such as the production of nanosuspensions and/or nanocrystals [1].

This work describes a technique which encompasses a microfluidization followed by a spray drying step, to produce nano- to microparticles of crystalline drug-alone, cyclodextrins complexes or amorphous solid dispersions. Three different case-studies will be detailed in this work. The manufacturing process is scalable to industrial scales and can operate either in batch or continuous manufacturing mode as benchmarked in the current work.

The microfluidization step is defined as a dynamic high-pressure process where two liquid streams, solution or suspension, pass through micro-channels toward an impingement area through which the fluids flow and interact [2]. The microfluidization step renders in either a liquid solution or liquid suspension that is typically followed by a spray drying step to remove the solvents and produce the dried particulate system. A schematic illustration of the process train is shown in Figure 1. This process was successfully demonstrated using fluticasone propionate as drug alone, itraconazole (ITZ) / Eudragit L-100 amorphous solid dispersions and ITZ / cyclodextrins inclusion complexes.

Case study I: production of micro / nanocrystals using microfluidization: Fluticasone propionate was dissolved in acetone. As anti-solvent, a mass of deionized water corresponding to 10 times that of the solvent was used. The two solutions were fed and mixed using PureNano™ in a single passage. The particle size of the isolated product was characterized through scanning electron microscopy, SEM (Figure 2). Roughly, particles with nano-scale were obtained [3].

Case study II: complexation of a poorly water soluble API using cyclodextrines: ITZ was dissolved in DCM and SBE-ß-CD dissolved in water. The two solutions were fed and mixed using PureNano™ in a single passage. Two different trials at different temperatures were done, a filtration step was done prior to the spray dryer to remove the

API not complexed. 3% was the maximum drug load obtained in the dried product [4].

Case study III: co-precipitation of nano-solid dispersions: ITZ and Eudragit L-100, in a mass proportion 1:9 and 40:60, were dissolved in a mixture of ethanol / acetone in a volume proportion of 1:1. Acidified cold water (pH=2) was used as anti-solvent with a ratio solvent:anti-solvent of 1:10 w/w. The results indicated that amorphization of ITZ was successful when using the present method (Figure 4) [5].

CONCLUSION

This work demonstrates that the technique herein described, consisting in a microfluidization step followed by an isolation step via spray drying is able to produce drug-alone nano-crystals, cyclodextrins complexes and amorphous solid dispersions. The process is scalable to industrial scales.

Figure 1 - Experimental set-up for the batch and continuous controlled precipitation process.

Figure 2 - SEM of spray dried particles of fluticasone propionate.

Figure 3 - i) X-ray diffractograms of ITZ solid dispersion at two different ratios and its crystalline form and ii) SEM of spray dried dispersion of ITZ.

[1] Perrie Y., Rades T. "Pharmaceutics - Drug Delivery and Targeting", Pharmaceutical Press, 2010;

[2] Rakesh P. Patel, Ashok H Baria, Nikunjana A Patel, An overwiew of size reduction technologies in the field of pharmaceutical manufacturing, Asian Journal of Pharmaceutics October-December (2008) 216-220.

[3] Fonseca, T., Duarte, I., Vicente, J., Temtem, M. "Continuous production of particles", Patent No. WO2016156841 A1, 2016.

[4] Lisboa, H., Santos, F., Vicente, J., Temtem, M. "Continuous complexation of active pharmaceutical ingredients", Portuguese Patent No. 109117, 2016.

[5] Temtem, M., Pereira, R., Vicente, J., Duarte, I., Gaspar, F., Duarte, I. "A method of preparing amorphous solids dispersions in submicron range by co-precipitation", Patent No. WO2016016665 A1, 2016.

491353 Droplet-Templated Anti-Solvent Spherical Crystallization of Hydrophilic and Hydrophobic Drugs with an *in situÂ* Formed Binder

Abstract:

Conventional pharmaceutical crystallization processes in batch reactors often lead to broad crystal size distributions with frequent batch-to-batch variations. These crystals subsequently undergo several secondary formulation processes for crystal and dosage uniformity. To consolidate these processes, spherical crystallization is a promising method, which combines crystallization, milling, and granulation into a single unit operation. Droplet-templated spherical crystallization methods have been developed recently to achieve highly monodisperse crystal microspheres with droplet microfluidics and evaporation techniques. We present here a novel droplet-templated anti-solvent spherical crystallization method applicable to both hydrophilic and hydrophobic drugs. In both cases, an alginate hydrogel binder forms in situ simultaneously, effectively encapsulating the drug crystals into highly monodisperse spheres. We study in detail the crystallization and gelation processes with mass transfer modelling, and the characterization of the obtained alginate/drug spheres. While glycine and carbamazepine are used as model hydrophilic and hydrophobic drugs, this method can be generalized to other drugs as well. In particular, this method offers additional benefits such as minimal thermal impact, fast crystallization rates, high drug-binder loading ratios, and high selectivity towards metastable polymorphs.

Tailoring CSD & Shape of Ibuprofen to Improve Flow Properties in Batch & Continuous Crystallization

Yunliang He, Brian Glennon, Yuan Gao
SSPC, School of chemical & bioprocess engineering, University College Dublin
Anne-Marie Healy, Zelalem Worku
SSPC, School of pharmacy & pharmaceutical sciences, Trinity College Dublin

❖ The significance of powder flow properties in pharmaceutical industry

In pharmaceutical industry, most API and excipients are manufactured or processed in the form of powders.
Flow properties are critical for handling, storing and formulation of powders
borderline flow properties for high speed tableting
flow factor: 6.74(±0.27) and bulk density and tap density is 0.353 and 0.476 g/ml respectively.

❖ The effect of size distribution & shape on flow properties

Common belief
Big particles show good flow properties, particles >300 microns are free-flowing
Plates and needles are always poor-flowing

❖ Tailor crystal size distribution and shape of ibuprofen

micronization
Granulation
Batch Crystallization
Continuous Crystallization

Is it possible to prepare ibuprofen for direct compression from crystallization?

Batch & continuous crystallization of ibuprofen

❖ Fast crystallization to prepare fine crystals

10~30g ibuprofen/100g 50%m Ethanol-Water or 20~30g ibuprofen/100ml hexane was heated above 50°C to dissolve and crash out by quick cooling to 5 °C(in 20 mins).

❖ Slow crystallization to prepare big crystals

10~30g ibuprofen/100g 50%m Ethanol-Water or 25g ibuprofen/100ml hexane was heated above 50C to dissolve and then cool to about 40 cool slowly (2~6 °C/h) from about 40C to about 5C

❖ Continuous crystallization to prepare narrow distribution big crystals

MSMPR(mixed suspension mixed product removal)
Feeding solution: 15g per 100g 50%m ethanol-water
Feeding rate:1~2ml/s
Working volume: 80ml
Residence time:40~80min
Generally, after 6~8 residence time, crystallization will reach steady state

Control of crystal size based on metastable zone

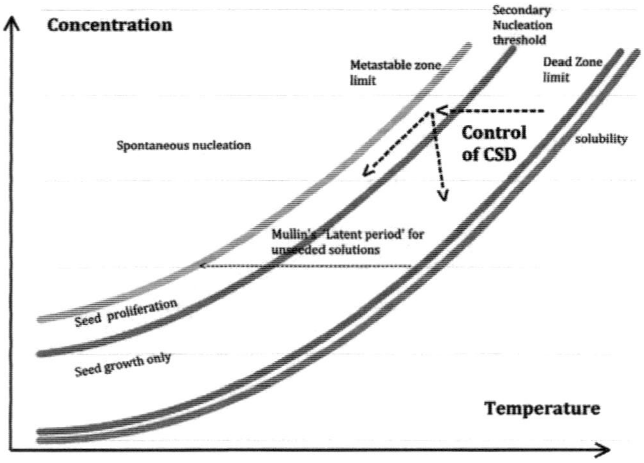

Primary and secondary metastable zone

Threlfall and Coles, Cryst. Eng. Comm. 2016, 18, 369

For batch crystallization

If crystallization was restricted in secondary metastable zone, there will be only crystal growth and no nucleation. The counts and average size R could be controlled by seed counts N.

$$R = \sqrt[3]{\frac{mass}{density * N * shape\ factor}}$$

For continuous crystallization

Crystallization must be operated in primary metastable zone. Continuous secondary nucleation will provide enough seeds to maintain steady state during continuous suspension removal. Crystal size is controlled by supersaturation and residence time.

Control of crystal size in batch crystallization

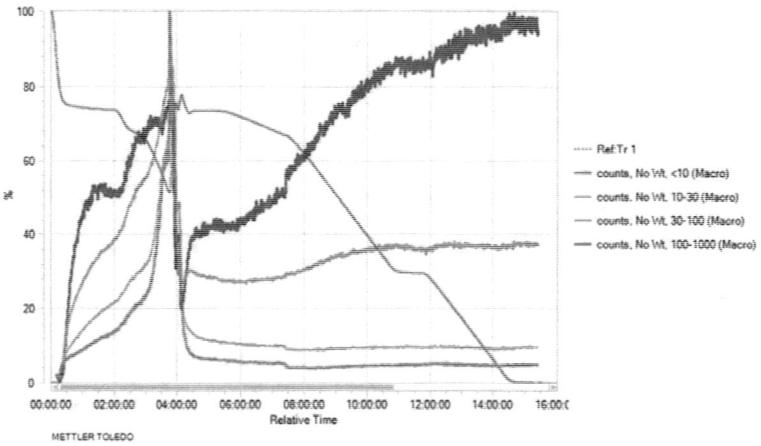

Failed Control of Crystal Size in Batch Crystallization

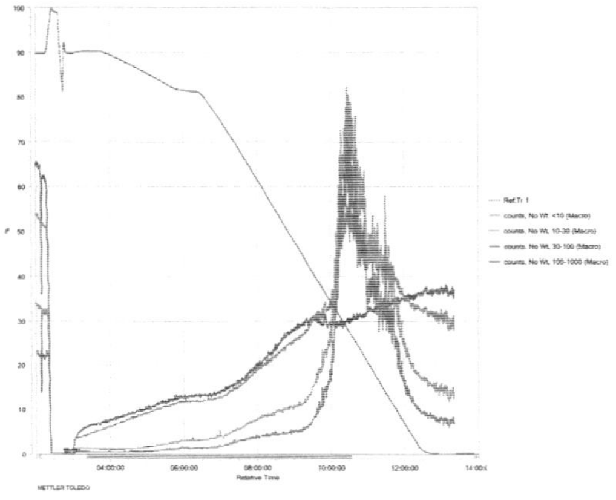

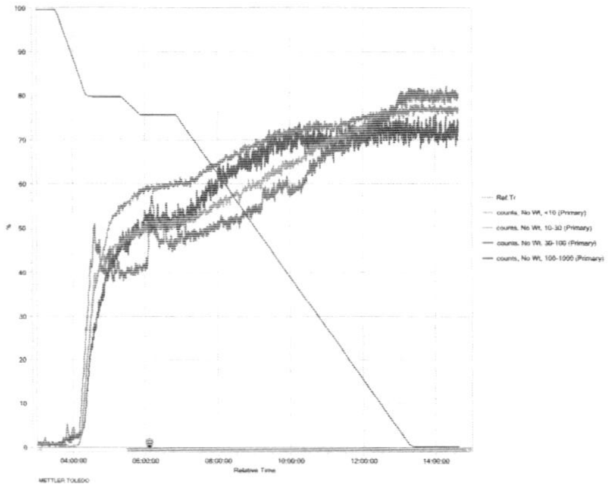

Control of crystal size in continuous crystallization

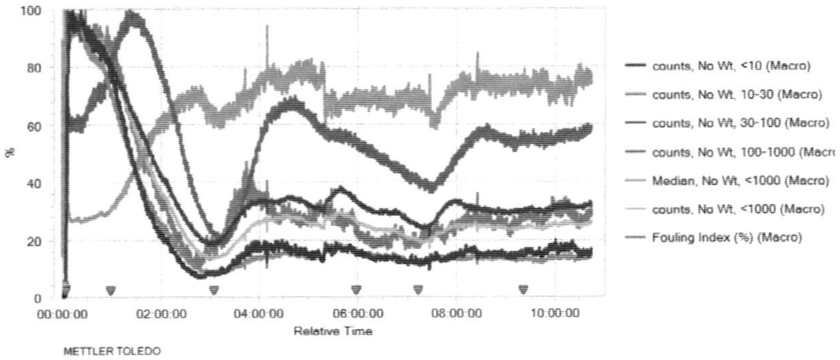

Crystal growth mode and control of shape

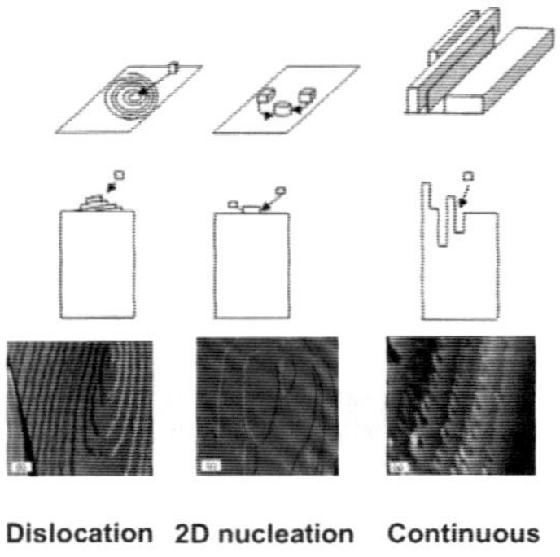

Y. Matsuzuki et al. J. Cryst. Growth 242(1-2):192

Factors on crystal shape

- ◆ Intrinsic crystal anisotropy
- ◆ Interfacial energy
- ◆ Supersaturation

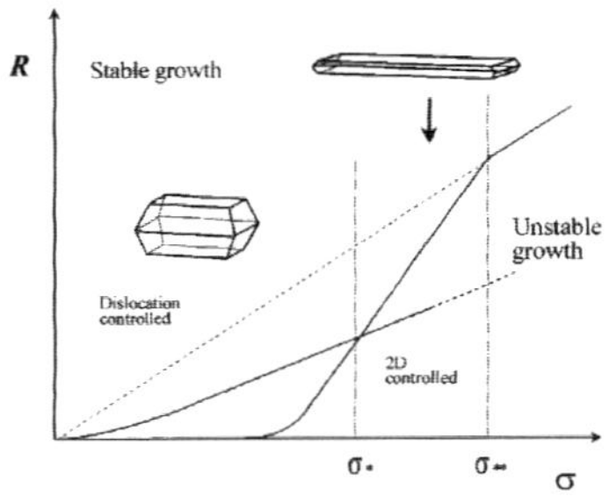

X. Liu and P. Bennema, Physical Review B 53(5):2314

Ibuprofen Crystal Shape From Controlled Crystallization

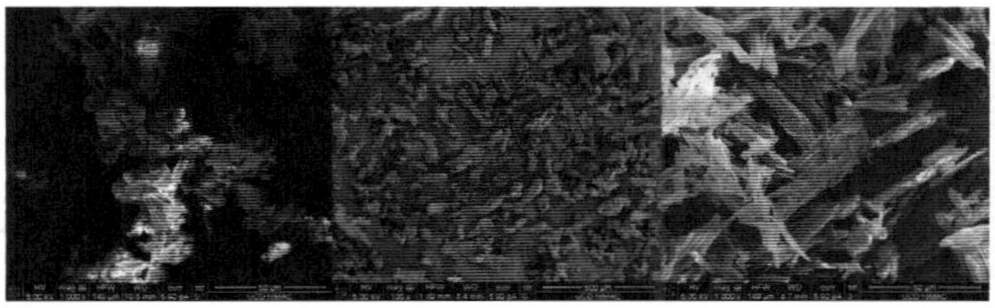

fast batch product from E-W(left & middle) and hexane(right)

continuous product from E-W

slow batch product from E-W

slow batch product from hexane

Comparison of CSD from batch & continuous crystallization

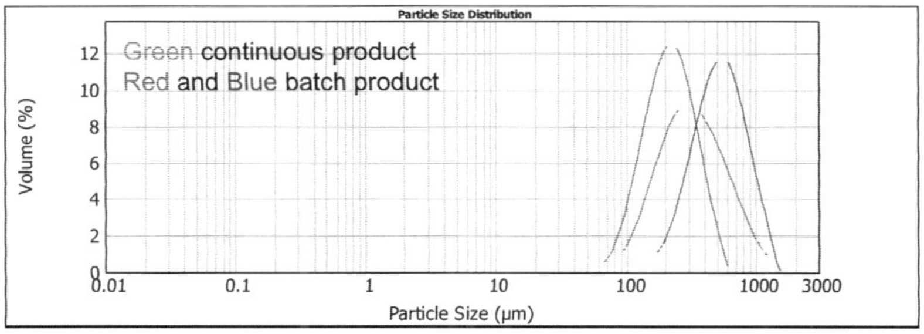

sample		Span	D [4, 3]	D [3, 2]	d (0.1)	d (0.5)	d (0.9)
continuous	Green	1.26	229.85	169.06	110.23	212.59	378.74
slow batch 1	red	1.88	364.50	163.45	123.72	303.34	693.37
slow batch 2	blue	1.40	542.40	277.64	218.59	506.72	929.39

ibuprofen crystal size distribution

Table 1 size distribution of batch and continuous crystallization product in ethanol-water

sample	Span	D [4, 3]	D [3, 2]	d (0.1)	d (0.5)	d (0.9)
Continuous1	1.42	373.07	259.46	156.62	341.15	639.69
Continuous2	1.44	275.20	194.24	119.19	247.18	475.34
Continuous3	1.32	200.34	143.47	90.89	184.74	335.40
Continuous4	1.39	252.26	166.16	108.05	231.41	430.38
Continuous5	1.18	215.49	164.80	109.80	200.61	346.29
Continuous6	1.26	229.85	169.06	110.23	212.59	378.74
Fast batch1	8.25	511.40	17.38	11.57	164.65	1370.03
Fast batch2	2.41	117.41	59.04	32.45	88.51	245.56
Fast batch3	1.57	79.73	49.34	31.46	69.95	141.52
Slow batch1	1.40	542.40	277.64	218.59	506.72	929.39
Slow batch2	1.88	364.50	163.45	123.72	303.34	693.37
Slow batch3	1.22	326.10	171.05	145.53	315.02	530.69
Slow batch4	1.18	160.11	114.33	77.85	140.03	243.60
Slow batch5	1.53	178.78	112.27	74.00	159.14	317.84

Table 2 Size distribution of product from n-hexane

sample	Span	D [4, 3]	D [3, 2]	d (0.1)	d (0.5)	d (0.9)
Fast batch	**3.29**	**44.07**	**14.40**	**7.19**	**18.31**	**67.43**
slow batch1	**3.57**	**354.63**	**105.74**	**67.86**	**219.16**	**849.73**
Slow batch2	**2.43**	**193.91**	**100.92**	**59.74**	**144.71**	**410.69**
Slow batch3	**2.06**	**288.17**	**134.67**	**85.25**	**236.76**	**573.11**

FT4 Powder Rheometer

FT4

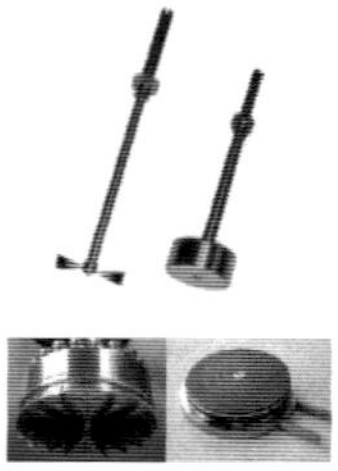

Accessories

Shear Cell Test

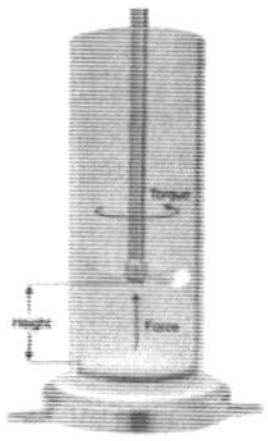

Stability &VFR

Shear cell test

FF vs D[3,2] and D[4,3]

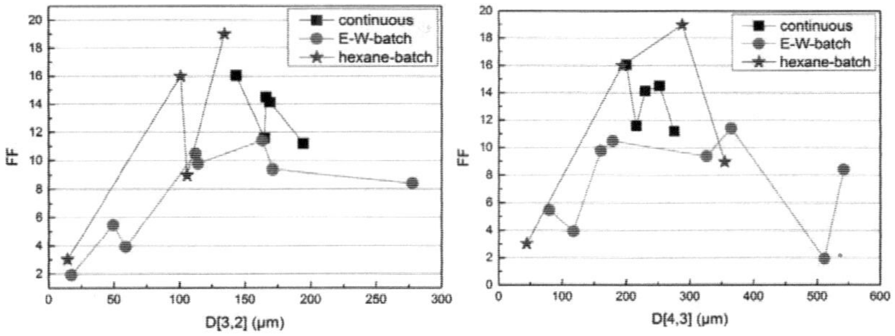

Type of flow	Flow Function Value
Free-flowing	10 < FF
Easy-flowing	4 < FF < 10
Cohesive	2 < FF < 4
Very cohesive and non-flowing	FF < 2

cohesion vs D[3,2] and D[4,3]

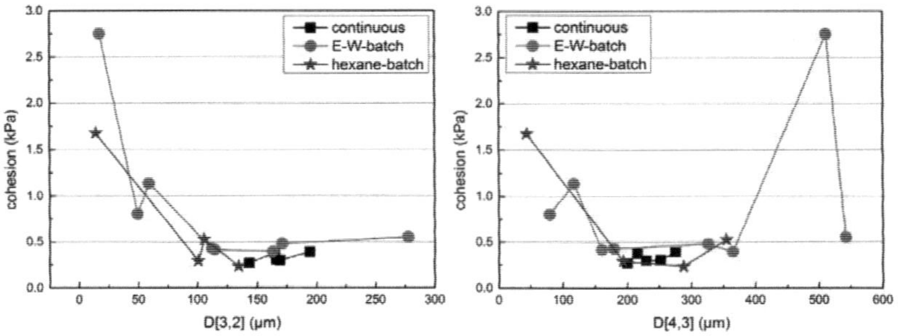

Density in shear cell vs D[3,2] and D[4,3]

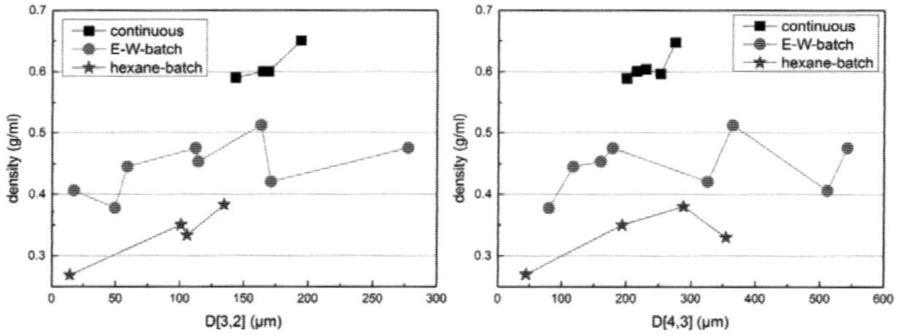

Stability & VFR test

BFE vs D[3,2] and D[4,3]

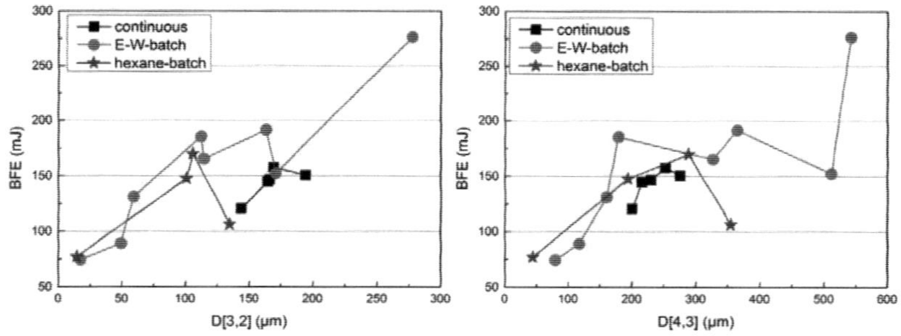

SE vs D[3,2] and D[4,3]

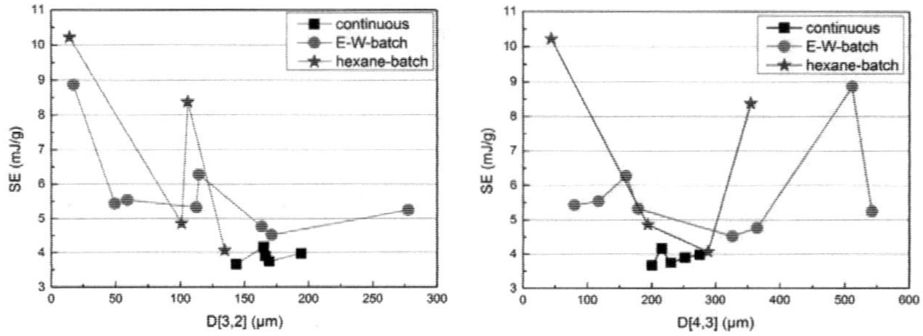

CBD vs D[3,2] and D[4,3]

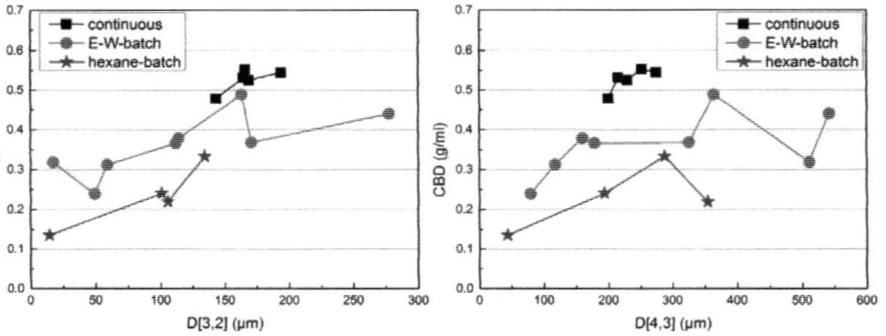

Conclusion

- Continuous products show equant shape, free-flowing, high density and small variance. Continuous crystallization is promising in preparing free-flow powders.
- Flow properties are better correlated with D[3,2] than D[4,3]
- Flow function is mainly determined by size while density is more influenced by shape than size. Rod crystals could be free-flowing by increasing size above 100 μm.
- Flowability(FF) doesn't simply increase with size. Conglomerate and shape also play critical roles in flowability.

Further study

- Investigate the direct compression and tableting behavior of tailor-made ibuprofen
- Set up a continuous manufacture process from crystallization to tableting

Thanks for your attention!

Any Questions?

497920 Formation and Dissolution Characteristics of Paracetamol Granules

Abstract:

The ability of active pharmaceutical ingredients (APIs) to affectively influence the targeted region is of great importance. Therefore, there is a need to offer lasting solutions and improve upon the current methods of formulation. This research analyzes physical and dissolution characteristics to determine the bonding tendencies of paracetamol granules formed using the drop method. Paracetamol is a common API and is considered a Category III drug. Initially, a series of experiments were conducted to form similarly sized granules. The process involved creating a paracetamol powder bed and dropping liquid binder onto the bed. Granules were separated from the powder bed using sieve trays. The cohesiveness of the granules was tested by dropping the granules onto a flat surface from various heights. The weights of each granule was measured before and after dropping to determine a percent attrition. The final investigation involved the ability of the granule to dissolve over time. This testing method involved the use of a USP Apparatus 3 to prevent coning seen in Apparatus 2. Further testing was conducted using UV Vis. Various binders were considered for this research including natural and synthetic polymers. Results show that using 8 drops of binder to form granules will deliver close to the normal dosage (325 â€“ 500 mg) of the API. Synthetic binders produced more cohesive granules during the drop testing. Granules formed using polyvinylpyrrolidone (PVP) binder had the smallest percent attrition (0.1%) when dropped from a height of 24 inches. The rate of dissolution differed for each material examined; however, polyethylene glycol had a very consistent dissolution rate and reached a percent dissolution of 85% in one hour.

492697 Size Reduction through Flash Nanoprecipitation to Improve Solubility, Dissolution, and Bioavailability of Clofazimine

Abstract:

Clofazimine is a poorly soluble riminophenazine antibiotic (logP= 7.66) approved by Food and Drug Administration (FDA) with good safety record. Thus, a reliable, efficacious, and fast-release formulation of clofazimine is highly desired. To improve the drug bioavailability and dissolution rate, clofazimine was encapsulated in nanoparticles prepared with three different surface stabilizers, hypromellose acetate succinate (HPMCAS), lecithin, and zein, through the Flash Nanoprecipitation (FNP) process with high encapsulation efficiency. To convert liquid colloidal suspensions to powders, lyophilization and spray drying, were applied to solidify nanoparticles. Dissolution tests were carried out to evaluate release kinetics of clofazimine nanoparticles prepared by different surface coatings, comparing to raw clofazimine powder and its currently market product Lamprene®. Remarkably improved dissolution rate and supersaturation levels were observed with all clofazimine nanoparticles despite different release kinetics. Additionally, calorimetry technique was explored to study the crystallinity of clofazimine in nanoparticles. The results strongly suggest that the new clofazimine nanoparticles prepared with affordable materials can be used as viable cryptosporidiosis therapeutics.

496138 Formulation Strategies for Solid Dispersions Containing Tablets

Abstract:

Formulation strategies for solid dispersions containing tablets

S. Doktorovova, E. Voney, J. Henriques

R&D Drug Product Development, Hovione Farmaciência S.A, Sete Casas, 2674-506 Loures, Portugal

In this work we present the development of formulation strategies for amorphous solid dispersions containing hypromellose acetate-succinate (HPMC-AS-M). Solid dispersions, especially spray dried, frequently have poor tableting properties due to low bulk densities and small particle size [1]. Tablet formulation then relies on excipients with capacity to improve compactability and blend flow. Three strategies are outlined: i) using a plastic filler/binder that greatly improves blend compactability and assure good mechanical strength of the tablets, ii) using a free flowing filler to improve powder flow properties and guarantee low mass and content variation iii) use a combination of a plastic and brittle excipient. The later should assure a good balance between acceptable powder flow and compactability. For the purpose of this work, a simple formulation compromising the solid dispersion (here represented by sieved HPMC-AS-M), a diluent or combination of two diluents, a disintegrant and a lubricant was considered.

The strategy based on formulating HPMC-AS-M with free flowing diluent largely improved powder flow properties of the blend without compromising compactability, but the improvement of powder flow properties did not translate into good tableting properties. Formulation of HPMC-AS-M with plastic diluent/binder resulted in excellent mechanical strength of the produced tablets but increasing ratio of the diluent compromised mass uniformity. A combination of a plastic and a well flowing diluent was therefore required to achieve a compromise between mechanical strength and mass variability. Fine tuning of the required ratio of diluents depended on the percentage of polymer present in the blend, tablet tensile strength being impacted to greater extent by increased polymer fraction than mass variability.

A model using bulk properties of excipients and polymer was set up as a tool to optimize the formulation powder flow properties of blends and quality attributes of tablets. Various mixture models were used for each property, the one that showed the best correlation between predicted and observed values were selected for the model. Prediction of blend powder flow and tablet properties using the model will enable sparing material in solid dispersion tablet development.

Figure 1: Modeling of a) powder flow properties using specific energy (SE, mJ/g) and c) tableting properties using tablet tensile strenght (TS, MPa) at 12 kN compression force and b) a sweet spot between acceptable powder flow and tableting properties.

References

[1] Henriques, J., Valente, P. & Winters, C. "Formulating Amorphous Solid Dispersions: Bridging Particle Engineering and Formulation", American Pharmaceutical Review, Nov/Dec. 2016.

493167 Modeling the Effects of Material Properties on Tablet Compaction a Case Study for Development

Abstract:

Interest in pharmaceutical continuous manufacturing (PCM) is rapidly growing. As it has been extensively reported PCM can deliver enormous advantages, including faster product development using less material, smaller equipment, superior process control, optimized performance, and more reliable quality performance [1-5]. However, to achieve these advantages in full, we need to develop accurate models of the process able of accounting for effects of material properties. Such models can be used to support faster and more economical process design as well as better process and quality control. As the pharmaceutical industry modernizes its manufacturing practices and incorporates more efficient processing approaches, it is important to reevaluate which process design elements affect product quality and the means to study these systems. The purpose of this presentation is to provide insight on a methodology to correlate and link the effect of raw material properties to equipment and process performance from using both data-driven and semi-empirical models. Such methodology aims at providing modeling tools to predict process performance *in silico*, minimizing the amount of experiments and/or narrowing the experimental scope. The presentation will include a case study where this protocol was applied to the characterization of tablet compression.

Lubricated blends of three (3) commonly used in the pharmaceutical industry: Microcrystalline Cellulose (Avicel) 101, Microcrystalline Cellulose (Avicel) 301, and Milled Lactose. Blends where made as formulations with varying levels of the lubricant, Magnesium Stereate, ranging from 0.25-1.5%. Material characterization was performed for all twelve (12) blends. The characterization included compressibility, permeability, density, particle size, and cohesion measurements. The characterization was part of the development of a reduce size Raw Material Properties Database (RMPD), which was later used to develop correlations between semi-empirical unit operation model parameters and material properties. Using Principal Component Analysis, the raw material properties were collapsed to two (2) major principle components, in order to understand the operating material space. All blends were then compressed using a two by three (2x3) experimental design, varying the fill depth and tablet thickness of the tablets produced, respectively. Tablet properties (e.g., weight, hardness, thickness, and compression force) were collected for all tablets and collected in a results matrix.

Results from the PCA analysis showed that raw material properties were slightly affected by the addition of the lubricant to the three (3) blends of Microcrystalline Cellulose (Avicel) 101, Microcrystalline Cellulose (Avicel) 301, and Milled Lactose. From the results matrix, we regressed the constant model coefficients for the Kuentz Equation, which is a model relating tablet hardness to changes in porosity. Two coefficients (i.e., the maximum tensile strength and the critical relative density) were regressed for each formulationâ€™s tablet compression profile. We evaluated several empirical models to correlate the regressed coefficient values to the original blend material properties from the RMPD. Models evaluated included Regression PCA (R-PCA), Partial Least Square (PLS), and simple linear regressions. From the analysis, we found that each of the coefficients regressed related to a material property from the RMPD. The models will be presented in this presentation. The empirical model between blend material properties and

the regressed constant values for the Kuentz equations showed less than 8% error for the formulations. The parity plots indicated a coefficient of variability (R-Squared) over 95% for both empirical model regressions along with p-values under 0.05 for all coefficients in the regression.

The results from this study indicate that correlations between material properties and semi-empirical equation regression constants are sufficiently accurate. This proof of concept of developing correlations between raw material properties and unit operation models can aid in process development, especially in the area of design space characterization and robustness analysis. Overall, we have shown with this approach that it is possible to develop these correlations and have predictive capabilities of process performance based on material property values.

References:

1. Plumb, K., *Continuous Processing in the Pharmaceutical Industry: Changing the Mind Set.* Chemical Engineering Research and Design, 2005. **83**(6): p. 730-738.

2. Ooi, S.M., et al., *Continuous processing and the applications of online tools in pharmaceutical product manufacture: developments and examples.* Therapeutic Delivery, 2013. **4**(4): p. 463-470.

3. Rantanen, J. and J. Khinast, *The Future of Pharmaceutical Manufacturing Sciences.* Journal of Pharmaceutical Sciences, 2015: p. n/a-n/a.

4. Lee, S., et al., *Modernizing Pharmaceutical Manufacturing: from Batch to Continuous Production.* Journal of Pharmaceutical Innovation, 2015. **10**(3): p. 191-199.

5. Mascia, S., et al., *End-to-End Continuous Manufacturing of Pharmaceuticals: Integrated Synthesis, Purification, and Final Dosage Formation.* Angewandte Chemie International Edition, 2013. **52**(47): p. 12359-12363.

496597 Quality By Design Approaches for Dry Powder Inhalation Products: A Regulatory Perspective

Abstract:

FDA has advocated companies to implement Quality by Design (QbD) into product lifecycles, with particular emphasis during product development. The goal of this presentation is to discuss the benefit and advantage the QbD approach brings to the design and manufacture of orally inhaled drug product, specifically the dry powder inhaler (DPI) variety. Since DPI is also a combination product, its successful development requires thorough understanding of the complex interaction between the powder formulation and the delivery device. We will illustrate how the QbD paradigm can aid in this process through case studies, including device-metered (reservoir) DPI and pre-metered (blister) DPI. We will start by identifying some key Quality Target Product Profile as well as Critical Quality Attributes (CQA). Through risk assessment and experimentation during development, Critical Material Attributes and Critical Process Parameters can be linked to the CQAs, thus allowing the design and implementation of an adequate control strategy, and reducing the need of intensive in-process testing. This is critical for DPI manufacturing since the standard assessment methods (Aerodynamic Particle Size Distribution and Delivered Dose Content Uniformity) have low testing efficiency. We will conclude with a discussion of regulatory submission expectations with respect to in-process control strategy based on commonly seen deficiencies in New Drug Applications and Abbreviated New Drug Applications.

497742 Effect of Bulk Properties of the Tracer on the Measurement of Residence Time Distributions in Continuous Powder Based Unit Operations

Abstract:

In continuous pharmaceutical manufacturing environment, the development of process mixing models is paramount to characterize the impact of upstream disturbances on the product, establish traceability of raw materials, and assess the process's performance and robustness. Upstream disturbance dissipation is key to determine whether material will remain is specification given a process deviation. The number of potentially affected products by a disturbance can be quickly quantified based on the disturbance's dissipation (i.e., back-mixing with in-specification material). Tracing raw material composition allows for product lot-to-lot delineation based on time and lots of raw material used. Process performance and robustness evaluation can allow us to determine potential failure modes and aid in the selection of appropriate process control action to correct deviations.

A commonly used tool in the characterization of mixing systems as well as mixing model development is Residence Time Distribution (RTD). Introduced in the 1950s by Danckwerts, RTD serves as a mean to describe non-ideal mixing in unit operations by characterizing the in-and-out time probability distribution function of a particle entering the unit at steady state [1]. To perform an RTD evaluation, a set of base and tracer materials need to be selected. The base material can be a single or multicomponent ingredient set, which resembles the actual process stream. The tracer is a material that can be quantified at the outlet using visual or chemical methods. Two major methods can be used for introducing a tracer into the system: pulse (or impulse) and step change. The selection of the methodology is very important for both the evaluation and characterization of the mixing behavior. Pulse experiments are effective at keeping the tracer material requirements low, while step changes often require large amounts of material to effectively produce a change in concentration at the system's outlet. Nonetheless, pulse experiments are more difficult to correctly perform due to the nature of the assumption. Failure to introduce pulse materials *evenly* and *instantaneously* may lead to misrepresentations of the system's mixing behavior [2].

RTD methods have been extensively used in liquid systems, and served the purpose of determining the mixing behavior inside of the unit. Over the last decade, RTD methods have been used to characterize and understand mixing in continuous powder (i.e., granular material) units, such as blenders and feed frames [1, 3, 4]. However, as we study in closer detail how the application of RTD methods compares between liquid and solids systems, we see a clear are of research that requires further evaluation: the selection of a tracer material. Similarly, to the selection of materials for performing unit characterization experiments, tracers are often selected based on a formulation, *ad hoc*, or based on the availability of materials with easy means of detection (e.g., easy to recognize spectral bands on an Near Infrared Spectra). Although this latter property is a key requirement for tracer materials, it is important to remember that in order to accurately measure the RTD, tracers introduced either through a pulse or step, *must* not alter the dynamic behavior of the bulk material within the unit operation [5].

Selecting the appropriate tracer requires understanding of the system material properties. Properties of a good tracer include: (1) an ability to be detectable from the bulk, (2) similar physical properties to the process stream, (3) no or minimal adherence to the equipment's walls or surfaces, (4) can be well mixed in the system, and (5) is representative of the materials involved in the system [1]. Tracers whose material properties are vastly different from the bulk can alter granular dynamics within the unit, resulting in inaccurate measure of the RTD. Similarly, a step change, which causes substantial change in the behavior of the bulk powder within the unit operation, may result in an inaccurate characterization of the RTD. Tracer materials in several published works for the characterization of mixers in pharmaceutical manufacturing include pure active pharmaceutical ingredients (API), which tend to have poor flowing characteristics and adhere to the walls and surfaces of vessels [4, 6-9]. Nevertheless, the selection of API is often the most relevant given the fact that it is the only material whose composition is measured in the product.

This work constitutes the development of a scientific framework to select appropriate tracers (i.e., meet all five criterions) for measuring RTDs. The framework is based on the concept of quantifying disparity between bulk and tracer physical material properties using multivariate analysis tools, such as Principal Component Analysis (PCA) [10]. Using the orthogonal decomposition of PCA, several material property values were collapsed into a reduced set of principal components that express material vast set of properties. Materials were placed on a multi-dimensional principal component space, where their weighted Euclidian distance from each other was measured. The goal of creating this space was to determine the level of similarity (or dissimilarity) between materials. Greater distances between two materials on the multi-dimensional space indicated a greater disparity between material property values and henceforth, the materials themselves. Our work proposes an acceptable multi-dimensional principal component region in which materials that lie within it are considered as acceptable tracers.

Both pulse and step change RTD experiments were performed to assess the effectiveness of our tracer selection based on the multivariate material property strategy. A single bulk material was run through a continuous blender (Gericke GCM-250) at a fixed mass flowrate. The impeller speed, blade arrangement and angle of the exit weir were kept constant. Nine unique materials, at variable distances from the bulk material on the PCA scores plot, and consequently exhibiting variable disparity between its own properties and those of the bulk, were selected as tracers. RTD curves were obtained from each of these nine materials. Each experiment was repeated in triplicate. Statistical methods were used to compare the RTD curves to the '*true*' RTD of the process. The true RTD was obtained by using dyed bulk material as the tracer. Materials, which resulted in curves that were statistically similar to the true RTD, were labelled as acceptable tracers, and vice versa. An acceptable Euclidian radius was proposed based on these findings. Step changes of increasing magnitude were performed for a binary mixture, and their RTD curves were obtained. The properties of the material post step were characterized and mapped on the PCA space. Step changes resulting in curves which were statistically similar to the true RTD were identified, and their Euclidean distance noted. An acceptable Euclidean radius and thus, an acceptable magnitude of the step change were proposed.

495948 Assessing Suspension Homogeneity Using CFD-DEM for Enhanced Content Uniformity of Spray Dried Intermediates

Abstract:

Ensuring a uniform concentration of API in the feed suspension/slurry of a spray drying process is a crucial step towards achieving a uniform dosage of the drug product intermediate. To accomplish this, an adequate feed tank design and process conditions are required to avoid heterogeneity of the suspension driven by, e.g., density differences between the solid and fluid phases. However, this heterogeneity can only be assessed *a posteriori* and the process optimization can only be achieved by trial and error. Here we present an alternative to this iterative and cost-ineffective development strategy by employing computational tools such as computational fluid dynamics coupled to discrete element methods (CFD-DEM) that allow offline experiments to be performed for various process conditions. These tools can also enable optimization of the agitation elements, such as number of impellers and distance to the bottom.

Here we focus on a spray drying case study where the feed suspension consists on a solid phase (the active pharmaceutical ingredient, or API) and the fluid phase is a solvent with a dissolved binder and other excipients. The process is run in batch mode while controlling key operation parameters, such as reactor wall temperature and agitator rotating speed, to obtain a homogeneous distribution of particles in the volume of the tank. Mixing in stirred tank reactors is enabled by fluid convection (at a larger scale) and turbulent exchanges (at a smaller scale) [1,2]. Therefore detailed information on the vessel, baffles and impeller (see Figure 1a), which determined the flow patterns, is required [2], [3].

A computational fluid dynamics (CFD-DEM) methodology (see Figure 1b), based on the OpenFOAM® framework, is used to aid process development and foresee potential issues related to accumulation zones within the reactor.

Upon choosing the reactor, its configuration and operating conditions, a steady-state simulation of the turbulent fluid flow, modeled via the Reynolds averaged Navier-Stokes (RANS) equations and shear stress transport (SST) turbulence model is conducted. The selected transient solver then uses this pre-calculated velocity field to update particles velocities considering the effect the forces exerted on them by fluid drag; gravity; and Coriolis acceleration. Furthermore, other capabilities such as particle-wall and particle-particle collision are available and may be incorporated into the solution (at the cost of extra computational effort). This way, the present workflow offers insight on how the interchangeable parts and operating conditions of the reactor might influence the steady-state particle concentration distribution.

[1] A. Ochieng and M. S. Onyango, "CFD simulation of solids suspension in stirred tanks: Review," *Hem. Ind.*, vol. 64, no. 5, pp. 365–374, 2010.

[2] G. R. Kasat, A. R. Khopkar, V. V. Ranade, and A. B. Pandit, "CFD simulation of liquid-phase mixing in solid-liquid stirred reactor," *Chem. Eng. Sci.*, vol. 63, no. 15, pp. 3877–3885, 2008.

[3] P. K. Biswas, S. C. Dev, K. M. Godiwalla, and C. S. Sivaramakrishnan, "Effect of some design parameters on the suspension characteristics of a mechanically agitated sand–water slurry system," *Mater. Des.*, vol. 20, no. 5, pp. 253–265, 1999.

492410 Advanced Model Predictive Control of Powder Level in Continuous Pharmaceutical Manufacturing Pilot-Plant

Abstract:

Currently, pharmaceutical companies are going under paradigm shift from batch to continuous manufacturing. In continuous pharmaceutical manufacturing (CPM), real-time effective control of Critical Process Parameters (CPPs) and Critical Quality Attributes (CQAs) is necessary to consistently produce a very high quality product as mandated by regulatory authority and needed for drug efficiency and patient safety [1]. Efficient automated coordination among different unit operations involved in continuous manufacturing is prerequisite to achieve consistent flow rate, inlet composition, line holdup, and thereby consistent product quality using QbD principles. In CPM, blender is connected with tablet press through a transfer pipe (chute). The chute also provides a suitable interface to integrate sensors (e.g. NIR, Raman) for real time monitoring and control. Maintaining consistent powder level in chute is needed to assure the presence of powder in front of sensors, to avoid over flow and under flow of the powder and to achieve the consistent product quality. However, different level of disturbances, change in raw materials and product specifications and startup, shutdown, speedup, slowdown processes could affect the powder level in chute significantly and thereby process operation, monitoring and product quality. Therefore, real time monitoring and control of powder level is necessary for continuous pharmaceutical manufacturing.

In this work, an advanced model predictive control strategy as well as a PID based control strategy for powder level control in a chute placed in between blender and tablet press unit operation of continuous tablet manufacturing process has been developed, implemented and evaluated. This work has been accomplished in three phases. In **first phase**, the *Insilco* study employing the process and sensor model has been performed to identify the best sensing and control strategy for powder level. Two types of the sensors, the one that gives discrete signal (on/off) and the one that gives continuous signal can be used for real time monitoring of powder level in chute. Discrete and continuous control strategy has been developed specifically for these two types of the sensors. The performance of two types of the sensor for real time monitoring and control of powder level has been compared *Insilco*. The performance of advanced model predictive controller (MPC) for powder level has been also compared with PID controller. We found that a continuous control strategy is more effective than a discrete control strategy, and that a MPC control strategy is more effective than a PID control strategy. In **second phase**, a novel noninvasive technique based on change in electric field concept has been developed, integrated and evaluated for real time monitoring of powder level in continuous manufacturing pilot-plant. The sensor has been integrated with a control panel (DeltaV (Emerson)) through relays and charms. The control panel then send the signal (4-20 mA) to control platform where the control strategy, a program to operate the sensor and a program to convert the sensor signal into powder level have been implemented. The performance of electric field based sensor for real time monitoring of powder level has been compared with a webcam based sensor. In **third phase**, an advanced model predictive controller as well as PID controller have been implemented into our continuous pharmaceutical manufacturing pilot-plant facility. The performance of both control strategies have been practically evaluated for set point tracking and disturbance rejection via utilizing electric filed based sensor. The MPC performs better than PID.

The objective of this presentation is two-fold; first to highlight the *Insilco* studies for identification of best sensing and control strategy for power level and then demonstrate the performance of best sensing and control strategies into our continuous pharmaceutical manufacturing pilot-plant facility

Reference

1. Singh, R., Román-Ospino, A. D., Romañach, R. J., Ierapetritou, M., Ramachandran, R. (2015). Real time monitoring of powder blend bulk density for coupled feed-forward/feed-back control of a continuous direct compaction tablet manufacturing process. International Journal of Pharmaceutics, 495, 612-625. http://dx.doi.org/10.1016/j.ijpharm.2015.09.029

AUTHOR INDEX

AUTHOR INDEX

AUTHOR INDEX

AUTHOR INDEX

AUTHOR INDEX

AUTHOR INDEX

AUTHOR INDEX